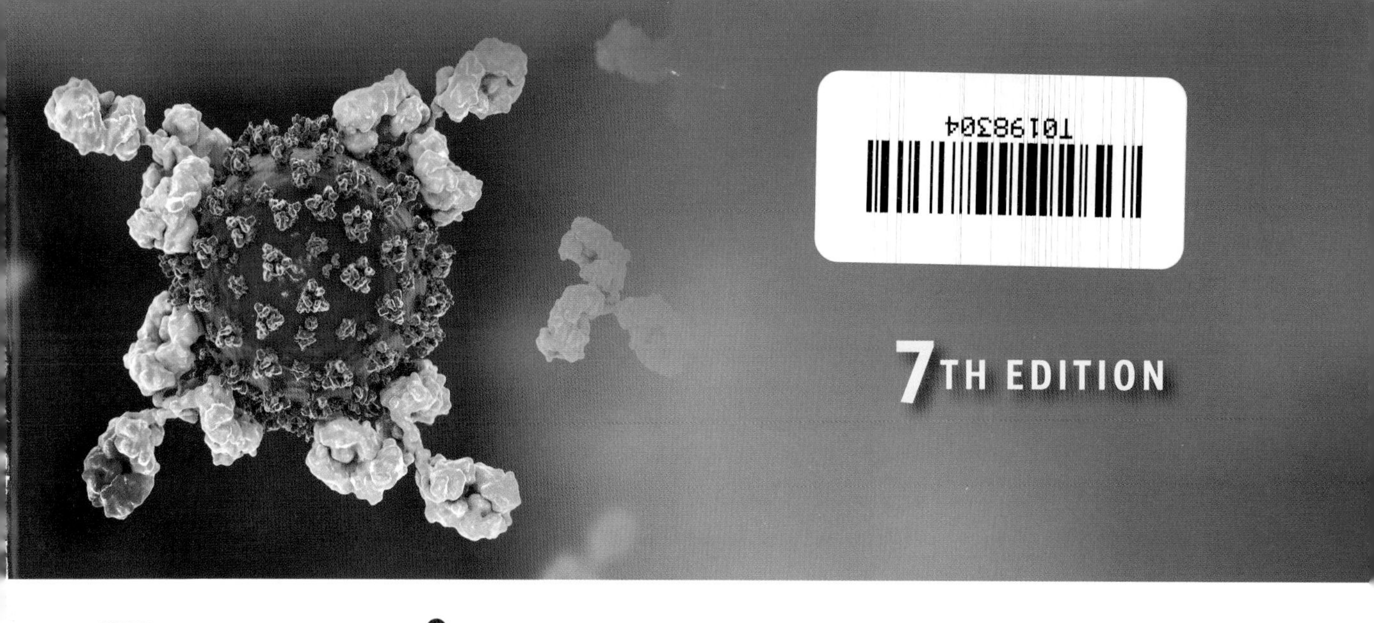

Basic IMMUNOLOGY

FUNCTIONS AND DISORDERS OF THE IMMUNE SYSTEM

Abul K. Abbas, MBBS
Emeritus Professor
Department of Pathology
University of California San Francisco
San Francisco, California

Andrew H. Lichtman, MD, PhD
Professor of Pathology
Harvard Medical School
Brigham and Women's Hospital
Boston, Massachusetts

Shiv Pillai, MBBS, PhD
Professor of Medicine and Health Sciences and Technology
Harvard Medical School
Massachusetts General Hospital
Boston, Massachusetts

Illustrations by David L. Baker, MA

ELSEVIER

1600 John F. Kennedy Blvd.
Ste. 1600
Philadelphia, PA 19103-2899

BASIC IMMUNOLOGY: FUNCTIONS AND DISORDERS OF THE IMMUNE SYSTEM, SEVENTH EDITION
ISBN: 978-0-443-10519-7

Previous editions copyrighted 2020, 2016, 2014, 2011, 2009, 2006, 2004, and 2001.

Publisher: Jeremy Bowes
Director, Content Development: Rebecca Gruliow
Publishing Services Manager: Julie Eddy
Book Production Specialist: Clay S. Broeker
Design Direction: Brian Salisbury

Printed in India

Last digit is the print number: 9 8 7 6 5 4 3 2

To our students

PREFACE

The seventh edition of *Basic Immunology* has been revised to include recent advances in our knowledge of the immune system. The original goals of this book, from the earliest edition, were to present current concepts in immunology cogently and in sufficient detail that they would be understood by students of the discipline, as well as to emphasize clinical aspects, including disease pathogenesis and the development of novel therapies based on the basic science of immunology. These are the goals that we continue to strive for. With improved understanding of the normal immune response, we believe it is possible to present the fundamental knowledge in a concise way. In addition, there has been exciting progress in applying basic principles to understanding and treating human diseases, a topic that is of paramount interest for students of medicine and allied health sciences. Foremost among these recent advances are the development of cancer immunotherapy and new information about pandemics, herd immunity, and vaccines, which illustrate how foundational science can be translated into clinical practice.

More specifically, we have focused on the following objectives. First, we have presented the most important principles governing the function of the immune system by synthesizing key concepts from the vast amount of experimental data that have emerged in the field of immunology. We have also prioritized content that is relevant to human health and disease. We have realized that in any concise discussion of complex phenomena, it is inevitable that exceptions and caveats cannot be considered in detail, so these have often been omitted. Second, we have focused on immune responses against infectious microbes, and most of our discussions of the immune system are in this context. Third, we have made liberal use of illustrations to highlight important principles, and we have reduced factual details that may be found in more comprehensive textbooks. Fourth, we have discussed immunologic diseases from the perspective of principles, emphasizing their relation to normal immune responses and avoiding details of clinical syndromes and treatments. We have included selected clinical cases in an appendix to illustrate how the principles of immunology may be applied to common human diseases. Finally, to make each chapter readable on its own, we have repeated key ideas in different places in the book. We feel such repetition will help students grasp the most important concepts.

We hope that students will find this new edition of *Basic Immunology* clear, cogent, manageable, and enjoyable to read. We hope the book will convey our sense of excitement about how the field has evolved and how it continues to grow in relevance to human health and disease. Finally, although we were spurred to tackle this project because of our associations with medical school courses, we hope the book will be valued by students of allied health and biology as well. We will have succeeded if the book can answer many of the questions these students have about the immune system and, at the same time, encourage them to delve even more deeply into immunology.

Several individuals played key roles in the writing of this book. Our talented illustrator, David Baker, continues to effectively convert our ideas into pictures that are informative and aesthetically pleasing. Our development editor, Rebecca Gruliow, and our editor, Jeremy Bowes, have kept the project organized and on track despite pressures of time and logistics. Clay Broeker has moved the book through the production process in an efficient and professional manner. To all of them we owe our many thanks. Finally, we owe an enormous debt of gratitude to our families, whose support and encouragement have been unwavering.

Abul K. Abbas
Andrew H. Lichtman
Shiv Pillai

CONTENTS

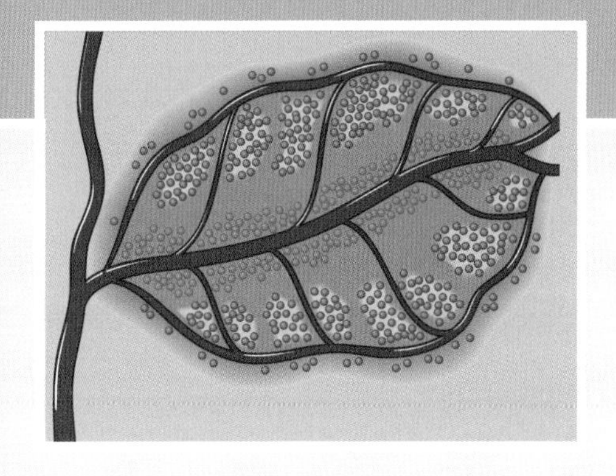

1

Introduction to the Immune System
Nomenclature, General Properties, and Components

CHAPTER OUTLINE

The term *immunity* usually refers to protection against infectious pathogens. However, reactions to some noninfectious substances, including harmless environmental molecules, tumors, and even one's own molecules are also considered forms of immunity (allergy, tumor immunity, and autoimmunity, respectively). The collection of cells, tissues, and molecules that mediate these reactions is called the **immune system**, and the coordinated response of these cells and molecules to pathogens and other substances makes up an **immune response**. Immunology is the study of the immune system and its functions. This field has captured the attention of scientists, physicians, and the lay public for many reasons (Fig. 1.1). As we will discuss later, the immune system is the primary defense against infections. The devastating consequences of pandemics such as COVID-19 have highlighted the importance of learning how to harness immune responses. The immune system prevents the growth of some tumors, and some cancers can be treated by stimulating immune responses against tumor cells. These concepts are the foundation of cancer immunotherapy, a therapeutic modality that has transformed the treatment of many cancer patients. Immune responses may become abnormal and cause inflammatory diseases with serious morbidity and mortality. Allergies and autoimmune diseases are examples of such disorders. The immune response damages transplanted tissues and is the major barrier to the success of organ transplantation. The widespread adoption of transplantation as a therapy has become possible because of the development of effective drugs to suppress these immune responses.

This chapter introduces the nomenclature of immunology, important general properties of all

Role of the immune system	Implications
Defense against infections	Deficient immunity results in increased susceptibility to infections; exemplified by AIDS Vaccination boosts immune defenses and protects against infections
Defense against tumors	Potential for immunotherapy of cancer
Control of tissue regeneration and scarring	Repair of damaged tissues
Cell injury and pathologic inflammation	Immune responses are the cause of allergic, autoimmune, and other inflammatory diseases, and for some of the harmful consequences of infections
Recognition of and injury to tissue grafts and newly introduced proteins	Immune responses are barriers to transplantation and gene therapy

Fig. 1.1 Importance of the immune system in health and disease. This table summarizes some of the physiologic functions of the immune system and its role in disease. *AIDS,* Acquired immunodeficiency syndrome.

immune responses, and the cells and tissues that are the principal components of the immune system. In particular, the following questions are addressed:

- What types of immune responses protect individuals from infections?
- What are the important characteristics of immunity, and what mechanisms are responsible for these characteristics?
- How are the cells and tissues of the immune system organized to find and respond to microbes in ways that lead to their elimination?

The basic principles introduced here set the stage for more detailed discussions of immune responses in later chapters. A Glossary of the important terms used in this book is provided near the end of the book.

INFECTIONS AND IMMUNITY

The most important physiologic function of the immune system is to prevent and eradicate infections. Individuals with defective immune responses are at increased risk for serious, often life-threatening infections. Conversely, stimulating immune responses against microbes through vaccination is the most effective method for protecting individuals against infections; this approach has led to the worldwide eradication of smallpox, the only disease that has been eliminated from civilization by human intervention (Fig. 1.2). The influenza pandemic of 1918, the appearance of acquired immunodeficiency syndrome (AIDS) in the 1980s, and COVID-19 in 2019 have tragically emphasized the importance of the immune system for defending individuals against infection. These newly emerged pathogens caused widespread infections mainly because populations had not been previously exposed to them and were hence not immune. Pandemics generally subside when a large fraction of the population develops immunity (called *herd immunity*) as a result of vaccination or by natural infection. In the case of human immunodeficiency virus (HIV)/AIDS, there is no herd immunity or effective vaccine, and control of the infection in many parts of

Disease	Maximum number of cases (year)	Number of cases in 2019
Diptheria	206,939 (1921)	2
Measles	894,134 (1941)	1,192
Mumps	152,209 (1968)	3,780
Pertussis	265,269 (1934)	18,617
Polio (paralytic)	21,269 (1952)	0
Rubella	57,686 (1969)	6
Tetanus	1,560 (1923)	26
Hemophilus influenza type B	~20,000 (1984)	18
Hepatitis B	26,611 (1985)	3,563

	Average daily deaths per 100,000 people who tested positive:	
	Unvaccinated	Vaccinated
Covid-19	1.3	0.1

Fig. 1.2 Effectiveness of vaccination for some common infectious diseases in the United States. Many infectious diseases for which effective vaccines have been developed have been virtually eradicated in the United States and other developed countries. Vaccines against SARS-CoV-2 have dramatically reduced the risks for developing a severe case of COVID-19. The COVID-19 death rate data are from a 6-month period in 2021. (Modified from Orenstein WA, Hinman AR, Bart KJ, Hadler SC. Immunization. In: Mandell GL, Bennett JE, Dolin R, editors: *Principles and Practices of Infectious Diseases*, 4th ed. New York, NY: Churchill Livingstone, 1995; and *Nationally Notifiable Infectious Diseases and Conditions, United States: 2018 Annual Tables.*)

the world relied on the development of effective antiviral drugs.

IMMUNOLOGIC DISORDERS

The immune system reacts against potentially harmful infectious pathogens and cancers, but it does not normally respond to self molecules or harmless foreign antigens. (The basis of this nonreactivity is discussed in later chapters.) In some genetically predisposed individuals, the immune system mounts damaging reactions against self structures causing autoimmune diseases, or against common environmental substances causing allergies. These disorders are characterized by reactions of host cells, called **inflammation** (see Chapter 2). The inflammation in these diseases is usually chronic (prolonged) because the inciting antigens cannot be eliminated, resulting in damage to normal tissues. Some of the most successful treatments for such chronic inflammatory diseases are specifically targeted to components of the immune response. For instance, current therapy for autoimmune diseases, such as rheumatoid arthritis and psoriasis, and for allergic diseases, such as asthma, rely on therapeutic blockade of molecules called cytokines that are responsible for many of the harmful effects of immune responses. Sometimes, protective immune responses to infections may lead to tissue damage and organ dysfunction. For instance, in COVID-19, a significant part of the morbidity is the result of inflammatory responses to the virus, not the damage caused by the virus itself.

STAGES OF HOST DEFENSE: INNATE AND ADAPTIVE IMMUNITY

Defense against infections is provided by the early reactions of innate immunity and the later, more powerful, reactions of adaptive immunity (Fig. 1.3). Innate immunity, also called natural immunity or native immunity, is always present in healthy individuals (hence the term *innate*), prepared to block the entry of microbes and to rapidly eliminate microbes that do succeed in entering host tissues. Adaptive immunity, also called specific immunity or acquired immunity, requires proliferation and differentiation of lymphocytes in response to microbes before it can provide effective defense (i.e., it adapts to the presence of microbial invaders). The potency of adaptive immune responses is because of the tremendous increase in the number of microbe-specific lymphocytes in response to infection, the highly specialized functions of different classes of lymphocytes, and the enhanced responses seen upon repeat exposures to the same microbe (the phenomenon of immunologic memory, discussed later). The adaptive immune response takes a few days to develop, and innate immunity provides defense in this critical early window after infection. Innate immunity is phylogenetically older, and the more specialized adaptive immune system evolved later.

In innate immunity, the first lines of defense are the epithelial barriers of the skin and mucosal tissues, antimicrobial substances produced by the epithelial barrier cells, and other cells located within or under the epithelium, all of which function to block the entry of microbes. If microbes do breach epithelia and enter the tissues or circulation, several other components of the innate immune system defend against them, including phagocytes and plasma proteins such as the complement system. In addition to providing early defense against infections, innate immune responses are required to initiate adaptive immune responses against the infectious agents. The cells and molecules of innate immunity recognize a limited number of molecular

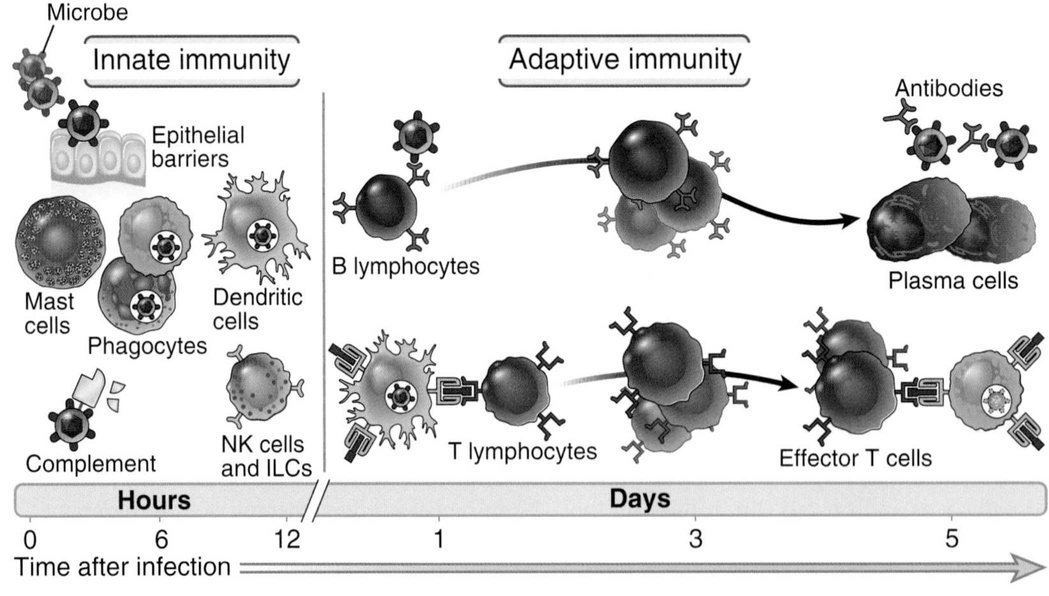

Fig. 1.3 Principal components of innate and adaptive immunity. The components of innate immunity provide the initial defense against infections. Some of these (e.g., epithelial barriers) prevent infections, and others (e.g., phagocytes, natural killer [NK] cells, innate lymphoid cells [ILCs], the complement system) eliminate microbes. Adaptive immune responses develop later and are mediated by lymphocytes and their products. Antibodies block infections and eliminate extracellular microbes, and T lymphocytes eradicate intracellular microbes. The kinetics of the innate and adaptive immune responses are approximations and may vary in different infections.

structures shared by classes of microbes. The components and mechanisms of innate immunity are discussed in detail in Chapter 2; the remainder of this chapter is an introduction to adaptive immunity.

The adaptive immune response is mediated by lymphocytes with highly diverse and variable receptors for foreign substances and the products of these cells, such as antibodies and other proteins. Adaptive immune responses are critical for defense against infectious pathogens that may have evolved to resist innate immunity. The lymphocytes of adaptive immunity express receptors that specifically recognize a wide variety of molecules produced by microbes, as well as noninfectious molecules. Any molecule that is specifically recognized by lymphocytes or antibodies is called an **antigen**.

Division of Labor: Types of Adaptive Immunity

There are two types of adaptive immunity, called humoral immunity and cell-mediated immunity, mediated by different cells and molecules, that provide defense against microbes in different locations (Fig. 1.4). Extracellular microbes (which survive outside host cells and are readily destroyed when they are ingested by phagocytes) are combated by antibodies. Microbes that have evolved to survive inside host cells, either in phagocytic vesicles or in the cytosol, are eradicated by the actions of T lymphocytes.

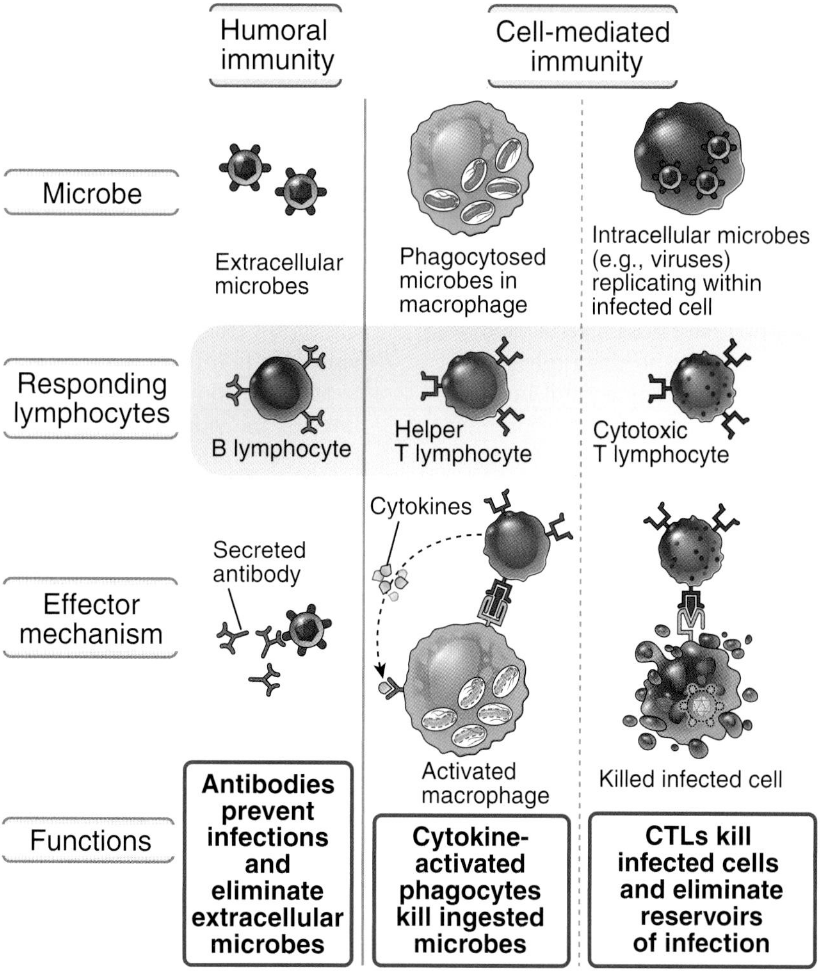

Fig. 1.4 Types of adaptive immunity. In humoral immunity, B lymphocytes secrete antibodies that eliminate extracellular microbes. In cell-mediated immunity, some T lymphocytes secrete soluble proteins called cytokines that recruit and activate phagocytes to destroy ingested microbes, and other T lymphocytes kill infected cells. *CTLs,* Cytotoxic T lymphocytes.

- **Humoral immunity** is mediated by proteins called **antibodies**, which are produced by cells called **B lymphocytes.** Secreted antibodies enter the circulation, extracellular tissue fluids, and lumens of mucosal organs such as the gastrointestinal and respiratory tracts. Antibodies defend against microbes present in these locations by preventing them from infecting tissue cells and by neutralizing toxins made by the microbes. Antibodies also enhance the uptake of extracellular microbes into phagocytes, resulting in the killing of the pathogens. In addition, antibodies are transported through the placenta into the fetal circulation and protect the fetus and newborn from infections.
- Defense against microbes that have entered host cells is called **cell-mediated immunity** because it is mediated by cells that are called **T lymphocytes.** Many intracellular microbes can live and replicate inside infected cells, including phagocytes. Although antibodies can prevent such microbes from infecting tissue cells, they are not effective after the microbes have entered the cells. Cell-mediated immunity is especially important to defend against these intracellular organisms. As we discuss later, there are two major classes of T lymphocytes. Cytokine-producing helper T lymphocytes activate phagocytes to destroy microbes that have been ingested and live within intracellular vesicles of these phagocytes. Cytotoxic T lymphocytes kill any type of host cells (including nonphagocytic cells) that harbor infectious microbes, such as viruses in the cytoplasm. Some helper T lymphocytes also promote defense against extracellular microbes by recruiting large numbers of phagocytes to sites of infection, and the phagocytes ingest and destroy the microbes.

The specificities of B and T lymphocytes differ in important respects. Most T cells recognize only peptide fragments of protein antigens presented on cell surfaces and thus sense the presence of intracellular microbes, whereas B cells and antibodies are able to recognize many different types of molecules, including proteins, carbohydrates, nucleic acids, and lipids of extracellular microbes. These and other differences are discussed in more detail later.

Immunity may be induced in an individual by infection or vaccination (active immunity) or conferred on an individual by transfer of antibodies from an actively immunized individual (passive immunity).

- In **active immunity**, an individual exposed to the antigens of a microbe mounts a response to eradicate the infection and develops resistance to later infection by that microbe. Such an individual is said to be immune to that microbe, in contrast with an individual who has not previously been exposed to that microbe's antigens and is said to be naive for that microbe.
- In **passive immunity**, a naive individual receives antibodies from another individual already immune to an infection or protective antibodies that are produced in laboratories. The recipient acquires the ability to combat the infection, but only for as long as the transferred antibodies last. Passive immunity is therefore useful for rapidly conferring immunity even before the individual is able to mount an active response, but it does not induce long-lived resistance to the infection. The only physiologic example of passive immunity is seen in newborns, whose immune systems are not mature enough to respond to many pathogens but who are protected against infections by acquiring antibodies during fetal life from their mothers through the placenta and in the neonatal period from breast milk. Clinically, passive immunity is useful for treating some immunodeficiency diseases with antibodies pooled from multiple donors and for emergency treatment of some viral infections, such as SARS-CoV-2, and snakebites using serum from immunized donors. Recently, some viral infections, including SARS-CoV-2, have been treated by administering antibodies purified from infected individuals or monoclonal antibodies produced in the laboratory. Antibodies designed to recognize tumors are now widely used for passive immunotherapy of cancers. Passive immunity by transfer of T cells between genetically nonidentical people is not possible because transferred cells will be rejected.

PROPERTIES OF ADAPTIVE IMMUNE RESPONSES

Several properties of adaptive immune responses are crucial for the effectiveness of these responses in combating infections (Fig. 1.5).

Specificity and Diversity

The adaptive immune system is capable of distinguishing millions of different antigens or portions of antigens, a feature that is referred to as specificity. It ensures that when an individual is

Feature	Functional significance
Specificity	Ensures that immune responses combat the pathogens (or tumors) that are encountered
Diversity	Enables the immune system to respond to a large variety of antigens
Memory	Leads to enhanced responses to repeated exposures to the same antigens
Nonreactivity to self	Prevents injury to the host during responses to foreign antigens

Fig. 1.5 Properties of adaptive immune responses. This table summarizes the important properties of adaptive immune responses and how each feature contributes to host defense against microbes.

infected by a microbe, the response is directed against that microbe and not, wastefully, against others that are not infecting the individual. Each lymphocyte expresses a single antigen receptor and can therefore recognize and respond to only one antigen. Because the immune system has to be able to react to a vast number of antigens from all the possible infectious pathogens, the total collection of lymphocyte specificities, sometimes called the lymphocyte repertoire, is extremely diverse. In an adult, there are about 0.5 to 1×10^{12} B and T lymphocytes, consisting of millions of clones (each clone made up of cells derived from one lymphocyte), and all the cells of one clone express identical antigen receptors, which are different from the receptors of all other clones. We now know the molecular basis for the generation of this remarkable diversity of lymphocytes (see Chapter 4). The clonal selection hypothesis, formulated in the 1950s, correctly predicted that clones of lymphocytes specific for different antigens develop before an encounter with these antigens, and each antigen elicits an immune response by selecting and activating the lymphocytes of a specific clone (Fig. 1.6).

The diversity of the lymphocyte repertoire also means that before exposure to any one antigen, very few cells, perhaps as few as 1 in 100,000 or 1 in 1,000,000 lymphocytes, are specific for that antigen. Thus, the total number of lymphocytes that can recognize and react against any one antigen ranges from approximately 1000 to 10,000 cells. To mount an effective defense against rapidly proliferating microbes, these few lymphocytes have to give rise to a large number of cells capable of eliminating the microbes. The marked proliferative expansion of the clone of lymphocytes specific for any antigen upon exposure to that antigen is called clonal expansion.

Memory

The adaptive immune system mounts faster, larger, and more effective responses to repeated exposure to the same antigen. This feature of adaptive immune responses implies that the immune system remembers every encounter with antigen, and this property of adaptive immunity is therefore called **immunologic memory**. The response to the first exposure to antigen, called the **primary immune response**, is initiated by lymphocytes called naive lymphocytes that are seeing antigen for the first time (Fig. 1.7). The term *naive* refers to these cells being immunologically inexperienced, not having previously responded to the antigen. Subsequent encounters with the same antigen lead to responses called **secondary immune responses** that usually are larger, more rapid and better able to eliminate the antigen than primary responses. Secondary responses are generated by the activation of memory lymphocytes, which are long-lived cells that were induced during the primary immune response. Immunologic memory optimizes the ability of the immune system to combat persistent and recurrent infections, because each exposure to a microbe generates more memory cells and activates previously generated memory cells. Immunologic memory is one mechanism by which vaccines confer long-lasting protection against infections.

Nonreactivity to Self

The immune system is able to react against an enormous number and variety of microbes and other foreign antigens, but it normally does not react against the host's own potentially antigenic substances, so-called self antigens. This unresponsiveness to self is called **immunologic tolerance**, referring to the ability of the immune system to coexist with (tolerate) potentially antigenic self molecules, cells, and tissues. Failure of self-tolerance is the fundamental abnormality in autoimmune diseases.

Other Features of Adaptive Immunity

Adaptive immune responses have other characteristics that are important for their functions.

- Immune responses are specialized, and different responses are designed to defend best against different

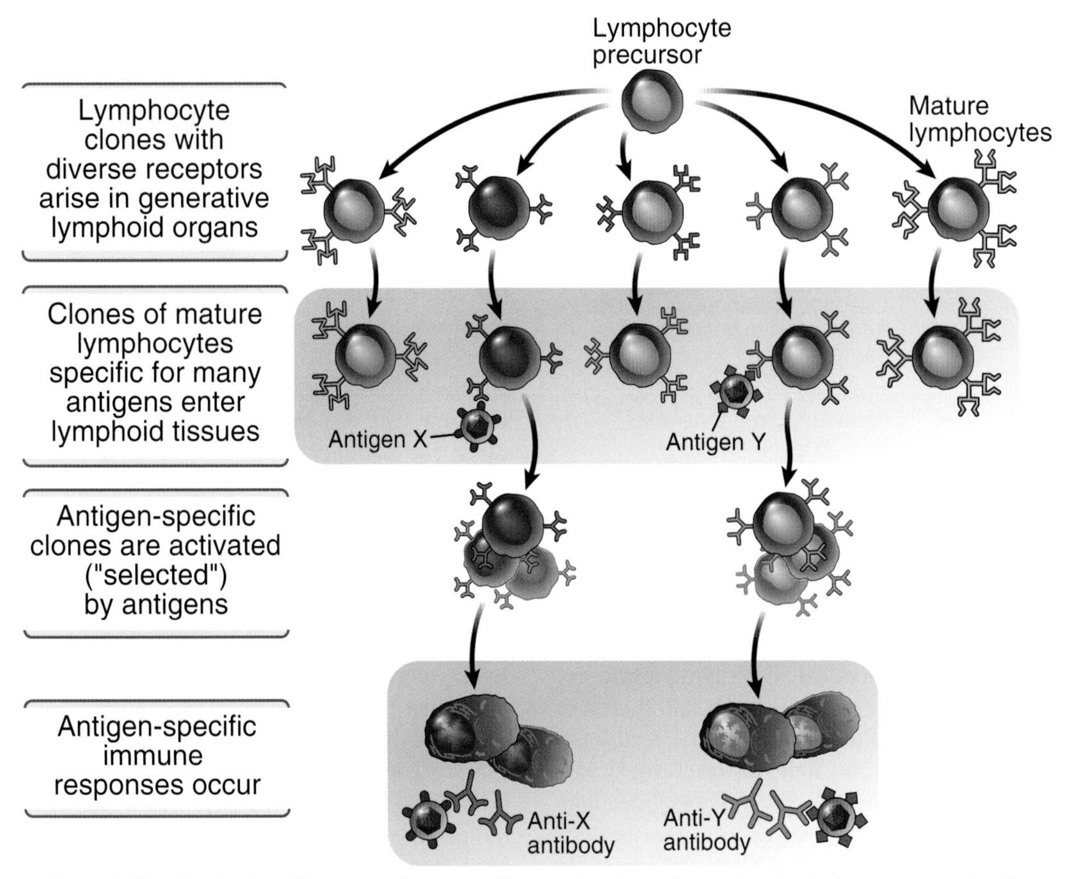

Fig. 1.6 Clonal selection. Mature lymphocytes with receptors for antigens develop before encountering the antigens. A clone refers to a population of lymphocytes with identical antigen receptors and therefore specificities; all of these cells are presumably derived from one precursor cell. Each antigen (e.g., X and Y) selects a preexisting clone of specific lymphocytes and stimulates the proliferation and differentiation of that clone. The diagram shows only B lymphocytes giving rise to antibody-secreting cells, but the same principle applies to T lymphocytes.

kinds of microbes and at different sites of infections. For example, there are several classes of secreted antibodies, each of which performs a different set of functions, and there are several subsets of T cells, each of which combats infections in different ways.

- All immune responses are self-limited and decline as the infection is eliminated, allowing the system to return to a resting state (homeostasis), prepared to respond to another infection.

CELLS OF THE ADAPTIVE IMMUNE SYSTEM

The cells of the immune system are mostly derived from progenitors in the bone marrow and are broadly classified into two groups: **myeloid cells** and **lymphoid cells (lymphocytes)** (Fig. 1.8). Myeloid cells consist mainly of phagocytes (neutrophils and macrophages), antigen-presenting cells (APCs) (e.g., dendritic cells), and mast cells. Several of these myeloid cells reside in tissues and serve as sentinels to detect the presence of microbes and to initiate immune responses. Phagocytes and mast cells are described in Chapter 2. Here, we describe lymphocytes and APCs, which serve key roles in adaptive immunity.

Lymphocytes

Lymphocytes are the only cells that produce clonally distributed receptors specific for diverse antigens and

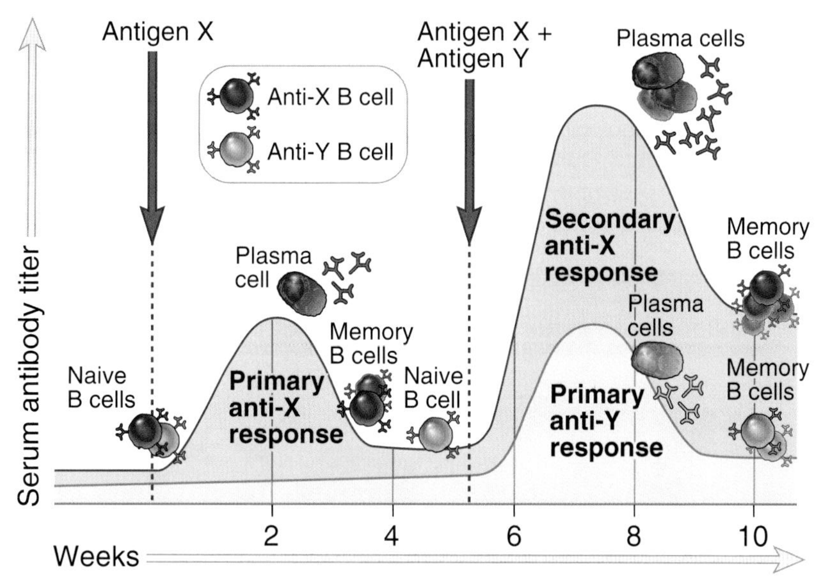

Fig. 1.7 Primary and secondary immune responses. The properties of memory and specificity can be demonstrated by repeated immunizations with defined antigens in animal experiments. Antigens X and Y induce the production of different antibodies (a reflection of specificity). The secondary response to antigen X is more rapid and larger than the primary response (illustrating memory) and is different from the primary response to antigen Y (again reflecting specificity). Antibody levels decline with time after each immunization. The level of antibody produced is shown as arbitrary values and varies with the type of antigen exposure. Only B cells are shown, but the same features are seen with T cell responses to antigens. The time after immunization may be 1 to 3 weeks for a primary response and 2 to 7 days for a secondary response, but the kinetics vary, depending on the antigen and the nature of immunization.

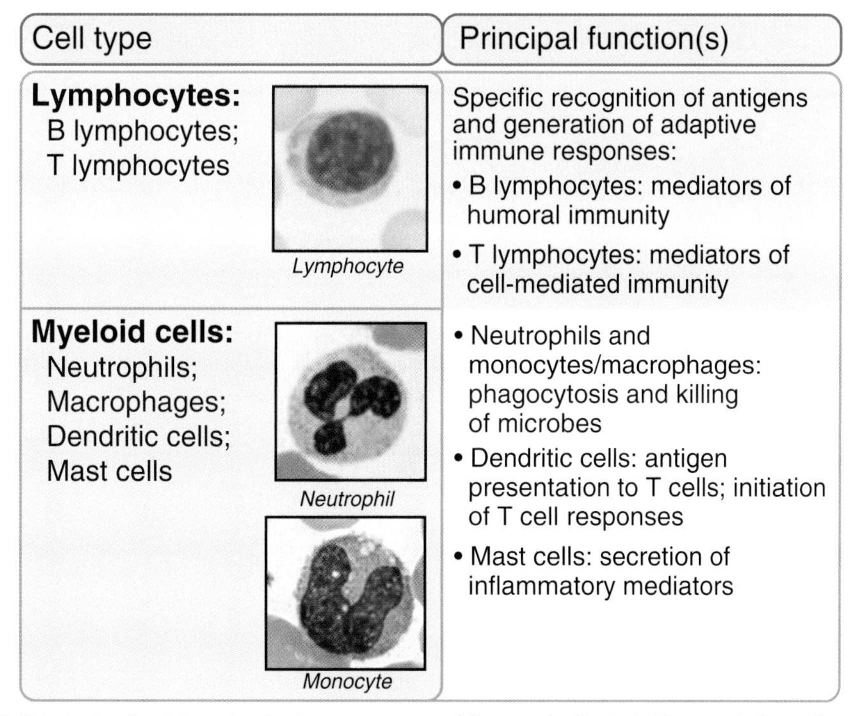

Fig. 1.8 Principal cells of the adaptive immune system. Micrographs illustrate the morphology of some cells of each type. The major functions of these cell types are listed.

are the key mediators of adaptive immunity. Although all lymphocytes are morphologically similar, they are heterogeneous in lineage, function, and phenotype (Fig. 1.9). Different types of lymphocytes (and other cells) may be distinguished by the expression of surface molecules that can be identified using monoclonal antibodies. The standard nomenclature for these proteins is the cluster of differentiation (CD) numeric designation, which is used to delineate surface proteins that define a particular cell type or stage of cell

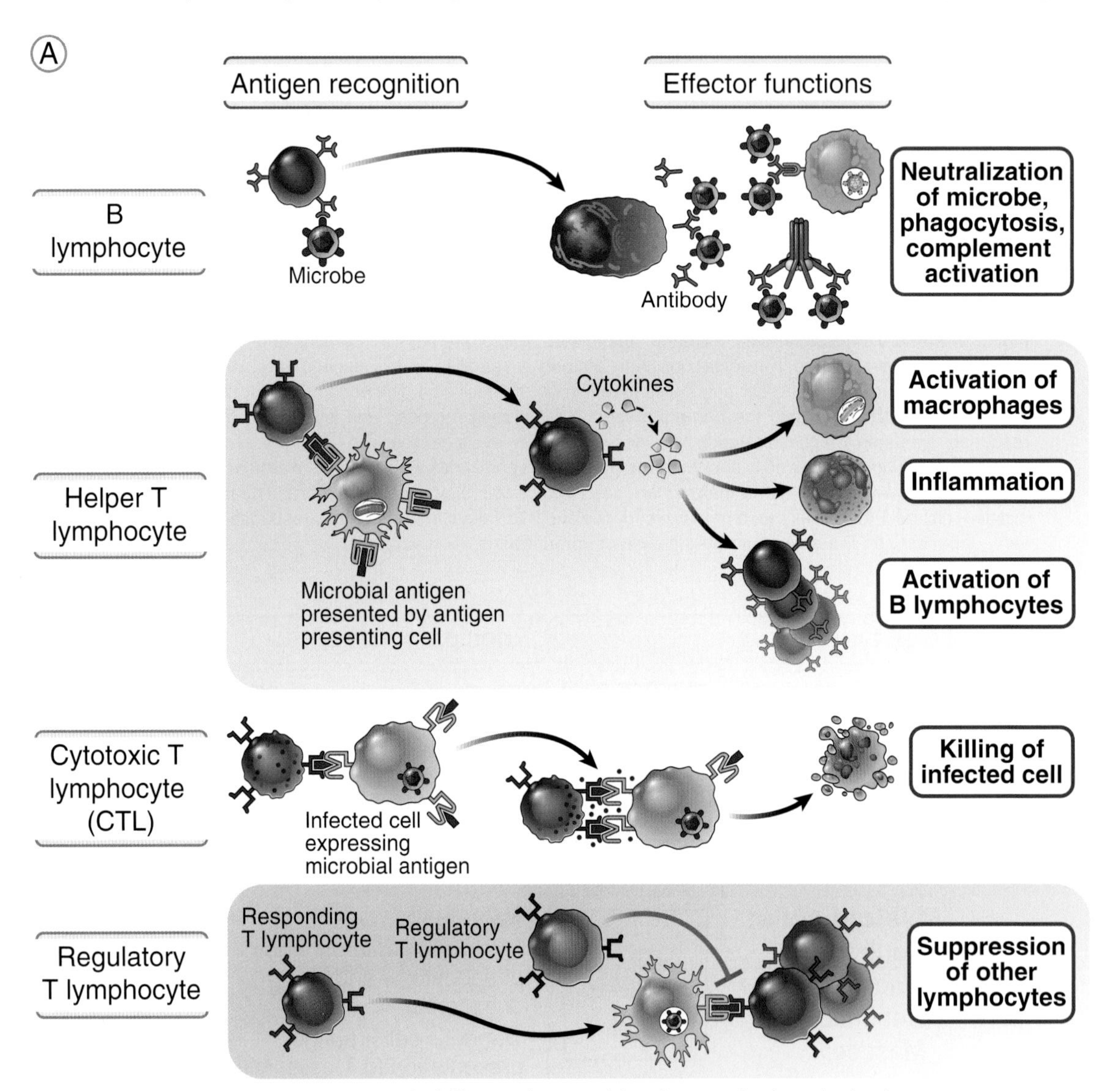

Fig. 1.9 Classes of lymphocytes. A, Different classes of lymphocytes in the adaptive immune system recognize distinct types of antigens and differentiate into effector cells whose function is to eliminate the antigens. B lymphocytes recognize soluble or microbial surface antigens and differentiate into antibody-secreting cells called plasma cells. Both helper T cells and cytotoxic T lymphocytes recognize peptides derived from intracellular microbial proteins displayed on the cell surface by major histocompatibility complex (MHC) molecules, described in Chapter 3. Helper T cells recognize these peptides displayed on the surface of macrophages or other antigen-presenting cells, and secrete cytokines that stimulate different mechanisms of immunity and inflammation. Cytotoxic T lymphocytes recognize peptides displayed by any type of infected cell type (or tumor cell) and kill these cells. Regulatory T cells limit the activation of other lymphocytes, especially of T cells, and prevent autoimmunity.

Class	Functions	Antigen receptor and specificity	Selected phenotype markers	Percentage of total lymphocytes*		
αβ T Lymphocytes				Blood	Lymph node	Spleen
$CD4^+$ helper T lymphocytes	B cell activation (humoral immunity) Macrophage activation (cell-mediated immunity) Stimulation of inflammation	αβ heterodimers Diverse specificities for peptide–class II MHC complexes	$CD3^+$ $CD4^+$ $CD8^-$	35–60	50–60	50–60
$CD8^+$ cytotoxic T lymphocytes	Killing of cells infected with intracellular microbes, tumor cells	αβ heterodimers Diverse specificities for peptide–class I MHC complexes	$CD3^+$ $CD4^-$ $CD8^+$	15–40	15–20	10–15
Regulatory T cells	Suppress function of other T cells (regulation of immune responses, maintenance of self-tolerance)	αβ heterodimers Specific for self and some foreign antigens (peptide–class II MHC complexes)	$CD3^+$ $CD4^+$ $CD25^+$ $FoxP3^+$ (most common)	0.5–2	5–10	5–10
B Lymphocytes				Blood	Lymph node	Spleen
B cells	Antibody production (humoral immunity)	Surface Ig Diverse specificities for many types of molecules	Fc receptors class II MHC CD19 CD20	5–20	20–25	40–45

Fig. 1.9 B, The table summarizes the major properties of the lymphocytes of the adaptive immune system. Not included are γδ T cells, natural killer cells and other innate lymphoid cells, which are discussed in Chapter 2. *The percentages are approximations, based on data from human peripheral blood and mouse lymphoid organs. *Ig,* Immunoglobulin; *MHC,* major histocompatibility complex.

differentiation and that are recognized by a set (cluster) of antibodies. (A list of CD molecules mentioned in the book is provided in Appendix I.)

As alluded to earlier, B lymphocytes are the only cells capable of producing antibodies; therefore, they are the cells that mediate humoral immunity. B cells express membrane-bound antibodies that serve as the receptors that recognize antigens and initiate the process of activation of the cells. Soluble antigens and antigens on the surface of microbes and other cells may bind to these B lymphocyte antigen receptors, resulting in the proliferation and differentiation of the antigen-specific B cells. This leads to the secretion of soluble forms of antibodies with the same antigen specificity as the membrane receptors.

T lymphocytes are responsible for cell-mediated immunity. The antigen receptors of most T lymphocytes recognize only peptide fragments of protein antigens that are bound to specialized peptide display molecules, called major histocompatibility complex (MHC) molecules, on the surface of other cells, called antigen-presenting cells (see Chapter 3). Among T lymphocytes, $CD4^+$ T cells are called **helper T cells** because they help B lymphocytes to produce antibodies and help phagocytes to destroy ingested microbes. $CD8^+$ T lymphocytes are called **cytotoxic T lymphocytes** (CTLs) because they kill cells harboring microbes. Some $CD4^+$ T cells belong to a special subset that functions to prevent or limit immune responses; these are called **regulatory T lymphocytes.**

All lymphocytes arise from common lymphoid progenitor cells in the bone marrow (Fig. 1.10). **B lymphocytes mature in the bone marrow, and T lymphocytes mature in an organ called the thymus.** These sites in which mature lymphocytes are produced (generated) are called the **generative** (also called central or primary) **lymphoid organs.** Mature lymphocytes leave the generative lymphoid organs and enter the circulation and **secondary** (peripheral) **lymphoid organs,** which are the major site of immune responses where lymphocytes encounter antigens and are activated.

When naive lymphocytes recognize microbial antigens and also receive additional signals induced by microbes, the antigen-specific lymphocytes proliferate and then differentiate into effector cells and memory cells (Fig. 1.11).

- **Naive lymphocytes** express receptors for antigens but do not perform the functions that are required to eliminate antigens. These cells circulate between and temporarily reside in secondary lymphoid organs, where they are positioned to respond to antigens. If they are not activated by antigen, after several months up to a few years, naive lymphocytes die by the process of apoptosis and are replaced by new cells that have developed in the generative lymphoid organs. The differentiation of naive lymphocytes into effector cells and memory cells is

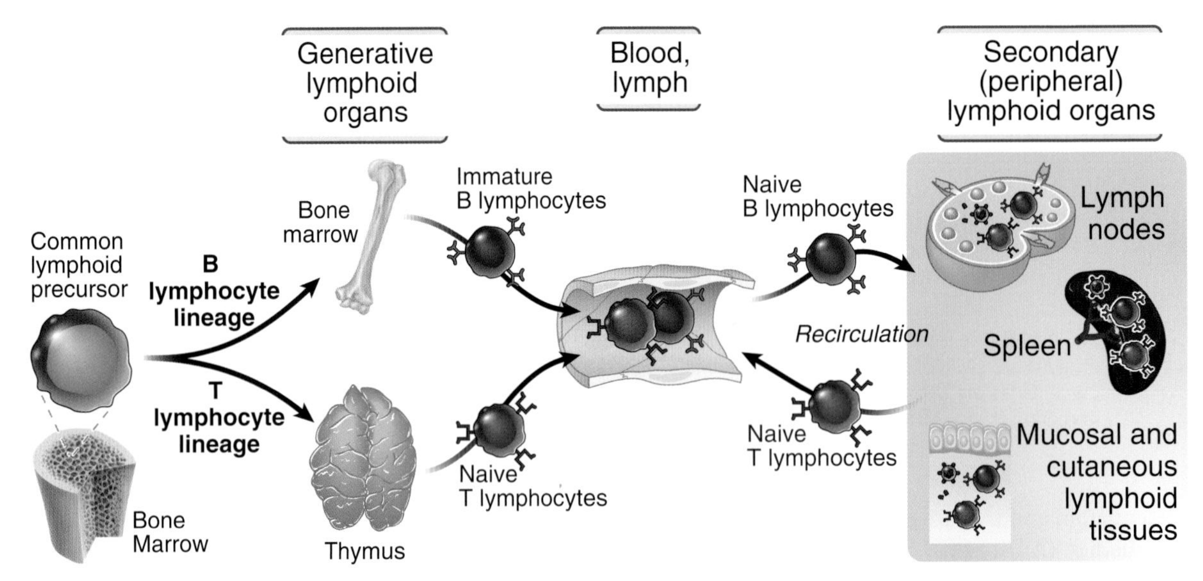

Fig. 1.10 Maturation and tissue distribution of lymphocytes. Lymphocytes develop from precursors in the generative lymphoid organs (bone marrow and thymus). Mature lymphocytes enter the secondary (peripheral) lymphoid organs, where they respond to foreign antigens, and recirculate in the blood and lymph. Some immature B cells leave the bone marrow and complete their maturation in the spleen (not shown).

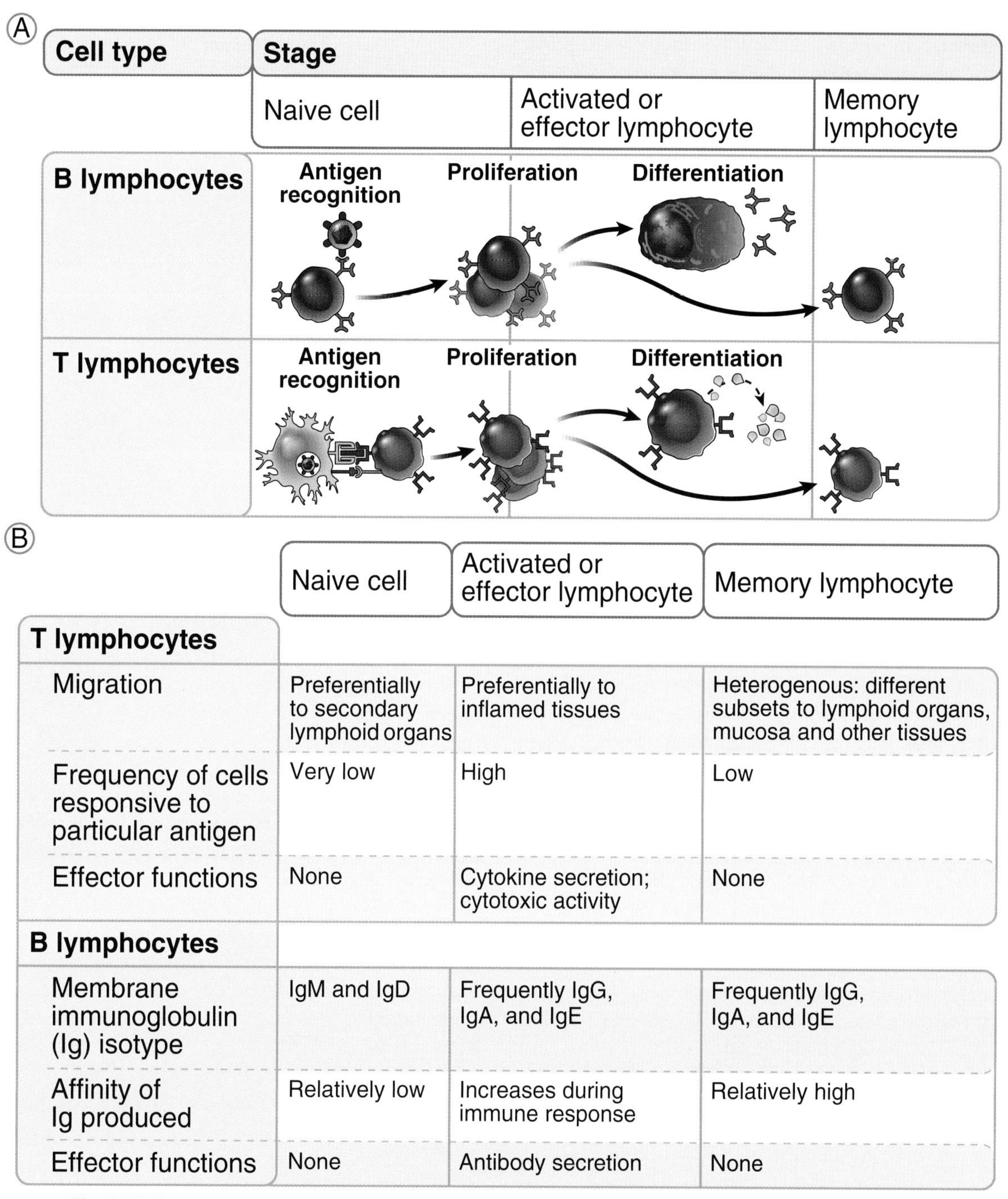

	Naive cell	Activated or effector lymphocyte	Memory lymphocyte
T lymphocytes			
Migration	Preferentially to secondary lymphoid organs	Preferentially to inflamed tissues	Heterogenous: different subsets to lymphoid organs, mucosa and other tissues
Frequency of cells responsive to particular antigen	Very low	High	Low
Effector functions	None	Cytokine secretion; cytotoxic activity	None
B lymphocytes			
Membrane immunoglobulin (Ig) isotype	IgM and IgD	Frequently IgG, IgA, and IgE	Frequently IgG, IgA, and IgE
Affinity of Ig produced	Relatively low	Increases during immune response	Relatively high
Effector functions	None	Antibody secretion	None

Fig. 1.11 Stages in the life history of lymphocytes. **A,** Naive lymphocytes recognize foreign antigens to initiate adaptive immune responses. Naive lymphocytes need signals in addition to antigens to proliferate and differentiate into effector cells; these additional signals are not shown. Effector cells, which develop from naive cells, function to eliminate antigens. The effector cells of the B lymphocyte lineage are antibody-secreting plasma cells (some of which are long lived). The effector cells of the CD4 T lymphocyte lineage produce cytokines. (The effector cells of the CD8 lineage are CTLs; these are not shown.) Other progeny of the antigen-stimulated lymphocytes differentiate into long-lived memory cells. **B,** The important characteristics of naive, effector, and memory cells in the B and T lymphocyte lineages are summarized. The generation and functions of effector cells, including changes in migration patterns and types of immunoglobulin produced, are described in later chapters.

initiated by antigen recognition, thus ensuring that the immune response that develops is specific for the antigen that is encountered.

- **Effector lymphocytes** are the differentiated progeny of naive cells that have the ability to produce molecules that function to eliminate antigens. The effector cells in the B lymphocyte lineage are antibody-secreting cells called **plasma cells.** Plasma cells develop from B cells in response to antigenic stimulation in the secondary lymphoid organs, where they may stay and produce antibodies. Small numbers of antibody-secreting cells are also found in the blood; these are called plasmablasts. These often migrate to the bone marrow, where they mature into long-lived plasma cells and continue to produce antibody for years after the infection is eradicated, providing immediate protection in case the infection recurs.

  Effector $CD4^+$ T cells (helper T cells) produce proteins called **cytokines** that activate B cells, macrophages, and other cell types, thereby mediating the helper function of this lineage. The properties of cytokines are listed in Appendix II and will be discussed in later chapters. Effector $CD8^+$ T cells (CTLs) have the machinery to kill infected host cells. The development and functions of these effector cells are also discussed in later chapters. Effector T lymphocytes are short-lived and die as the antigen is eliminated.

- **Memory cells**, also generated from the progeny of antigen-stimulated lymphocytes, can survive for long periods in the absence of antigen. Therefore, the frequency of memory cells increases with age, presumably because exposure to microbes throughout one's life has generated memory cells specific for those microbes. In fact, memory cells make up less than 5% of peripheral blood T cells in a newborn but 50% or more in an adult (Fig. 1.12). As individuals age, the gradual accumulation of memory cells compensates for the reduced output of new, naive T cells from the thymus, which involutes after puberty (see Chapter 4). Memory cells are functionally inactive; they do not perform effector functions unless stimulated by antigen. When memory cells encounter the same antigen that induced their development, they rapidly respond by becoming effector cells that initiate secondary immune responses. The signals that generate and maintain memory cells are not well understood but include cytokines.

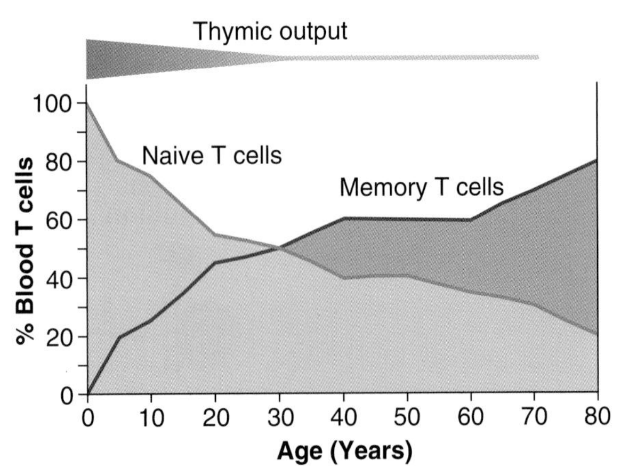

Fig. 1.12 Change in proportions of naive and memory T cells with age. The proportions of naive and memory T cells are based on data from multiple healthy individuals. The estimate of thymic output is an approximation. (Courtesy Dr. Donna L. Farber, Columbia University College of Physicians and Surgeons, New York, NY.)

Antigen-Presenting Cells

The common portals of entry for microbes—the skin and gastrointestinal, respiratory, and genitourinary tracts—contain specialized cells located at the epithelial barriers that capture antigens, transport them to secondary lymphoid organs, and display (present) them to lymphocytes. These are the first steps in the development of adaptive immune responses against antigens. This function of antigen capture and presentation is best understood for dendritic cells, the most specialized APCs in the immune system. The role of dendritic cells in presenting antigens to T lymphocytes and initiating cell-mediated immune responses is described in Chapter 3.

B lymphocytes may directly recognize the antigens of microbes (either released or on the surface of the microbes), and macrophages and dendritic cells in secondary lymphoid organs may also capture antigens and display them to B cells.

TISSUES OF THE IMMUNE SYSTEM

The tissues of the immune system consist of the generative lymphoid organs, in which T and B lymphocytes mature and become competent to respond to antigens, and the secondary lymphoid organs, in which adaptive immune responses to microbes are initiated (see Fig. 1.10). Most of the lymphocytes in a healthy human are found in lymphoid organs and other tissues (Fig. 1.13). However, as we discuss later, lymphocytes are unique among the cells of the body because of their ability to recirculate, repeatedly traveling via the blood to secondary lymphoid organs and other tissues. The generative lymphoid organs are described in Chapter 4, when we discuss the process of lymphocyte maturation. The following section highlights some of the features of secondary lymphoid organs that are important for adaptive immunity.

Secondary (Peripheral) Lymphoid Organs and Tissues

The secondary lymphoid organs and tissues, which consist of lymph nodes, the spleen, and the mucosal and cutaneous immune systems, are organized in a way that promotes the development of adaptive immune responses. T and B lymphocytes must locate microbes that enter at any site in the body, then respond to these microbes and eliminate them. The anatomic organization of secondary lymphoid organs enables APCs to concentrate antigens in these organs and lymphocytes to locate and respond to the antigens. Furthermore, different types of lymphocytes often need to communicate with each other to generate effective immune responses. For example, within secondary lymphoid organs, helper T cells specific for a protein antigen interact with and help B lymphocytes specific for the same antigen, resulting in antibody production. An important function of lymphoid organs is to bring together these rare T and B cells specific for the same antigen after stimulation by that antigen.

The major secondary lymphoid organs share many characteristics but also have some unique features.

- **Lymph nodes** are encapsulated nodular aggregates of lymphoid tissues located along lymphatic channels throughout the body (Fig. 1.14). Fluid constantly leaks out of small blood vessels in all epithelia and connective tissues and most parenchymal organs. This fluid, called **lymph**, is drained by lymphatic vessels from the tissues to the lymph nodes and eventually back into the blood circulation. Therefore, the lymph contains a mixture of substances absorbed from epithelia and tissues. As the lymph passes through lymph nodes, APCs in the nodes are able to sample the antigens of microbes that may enter through epithelia into tissues. In addition, dendritic cells pick up antigens of microbes from epithelia and other tissues and transport these antigens to the lymph nodes. The net result of these processes of antigen capture and transport is that the antigens of microbes entering through epithelia or colonizing tissues become concentrated in draining lymph nodes.
- The **spleen** is a highly vascularized abdominal organ that serves the same role in immune responses to blood-borne antigens as that of lymph nodes in responses to lymph-borne antigens (Fig. 1.15). Blood entering the spleen flows through a network of channels (sinusoids). Blood-borne antigens are captured and concentrated by dendritic cells and macrophages in the spleen. The spleen contains abundant phagocytes that line the sinusoids, which ingest and destroy microbes in the blood. These macrophages also ingest and destroy old red blood cells.
- The **cutaneous and mucosal immune systems** are specialized collections of lymphoid tissues and

Tissue	Number of lymphocytes x 10^9
Spleen	70
Lymph nodes	190
Bone marrow	50
Blood	10
Skin	20
Intestines	50
Liver	10
Lungs	30

Fig. 1.13 Distribution of lymphocytes in lymphoid organs and other tissues. Approximate numbers of lymphocytes in different organs of healthy adults are shown.

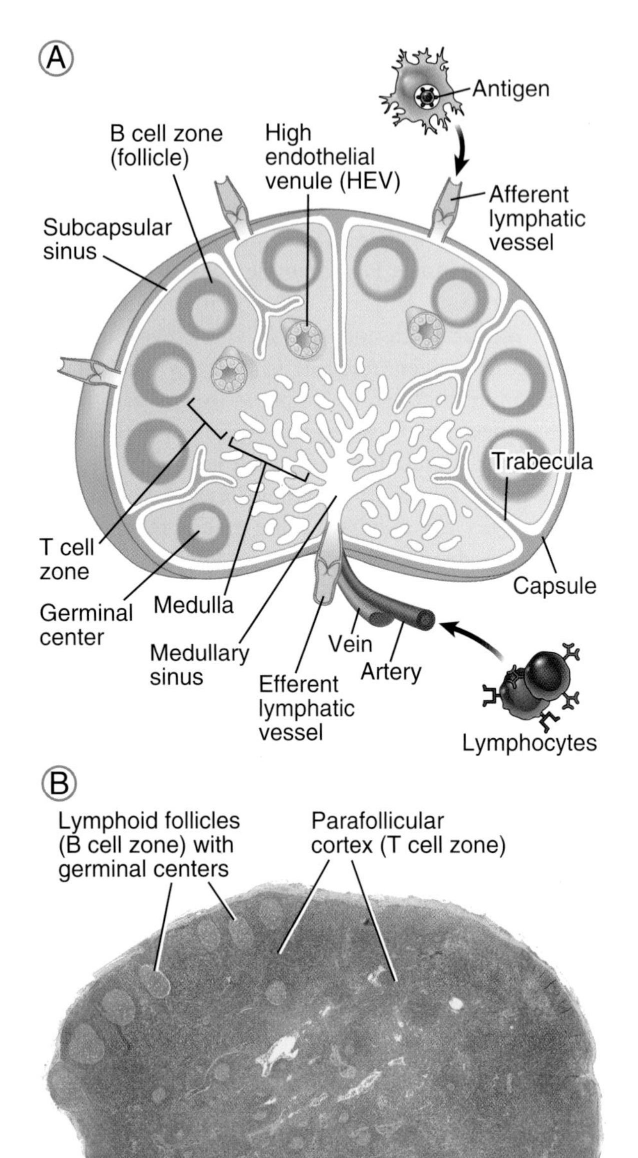

Fig. 1.14 **Morphology of lymph nodes. A,** Schematic diagram shows the structural organization of a lymph node. **B,** Light micrograph shows a cross section of a lymph node illustrating the T cell and B cell zones. The B cell zones contain numerous follicles in the cortex, some of which contain lightly stained central areas (germinal centers). (Courtesy Robert Oghami, MD, PhD, and Kaushik Sridhar, Department of Pathology, University of California San Francisco, San Francisco, CA.)

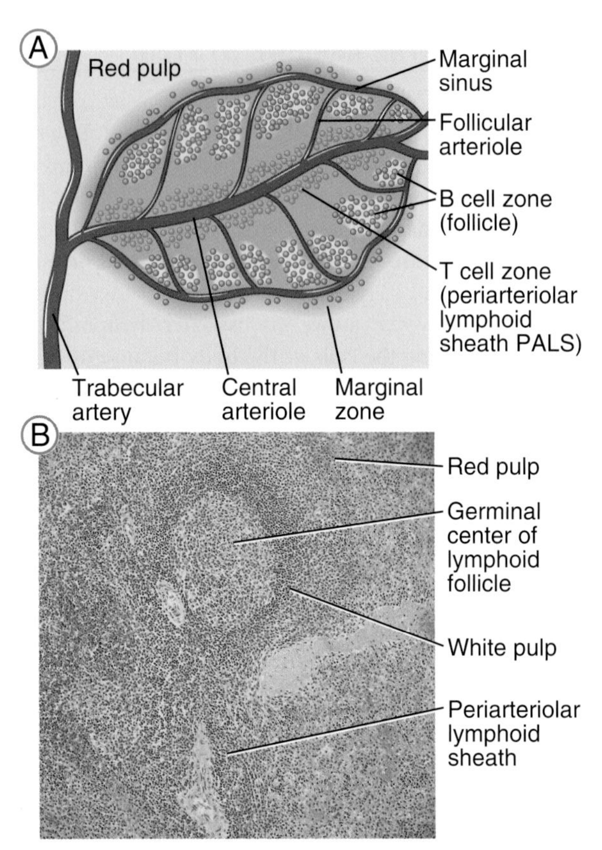

Fig. 1.15 **Morphology of the spleen. A,** Schematic diagram shows a splenic arteriole surrounded by the periarteriolar lymphoid sheath (PALS) and attached follicles. The PALS and lymphoid follicles together constitute the white pulp. The marginal zone with its sinus is the indistinct boundary between the white pulp and the red pulp. **B,** Light micrograph of a section of spleen shows an arteriole with the PALS and a follicle with a prominent germinal center. These are surrounded by the red pulp, which is rich in vascular sinusoids.

APCs located in and under the epithelia of the skin and the gastrointestinal and respiratory tracts, respectively. Although most of the immune cells in these tissues are diffusely scattered beneath the epithelial barriers, there are discrete collections of lymphocytes and APCs organized in a similar way as in lymph nodes. For example, tonsils in the pharynx and Peyer's patches in the intestine are two anatomically defined mucosal lymphoid tissues (Fig. 1.16). The immune system of the skin contains most of the cells of innate and adaptive immunity, but without any anatomically defined structures

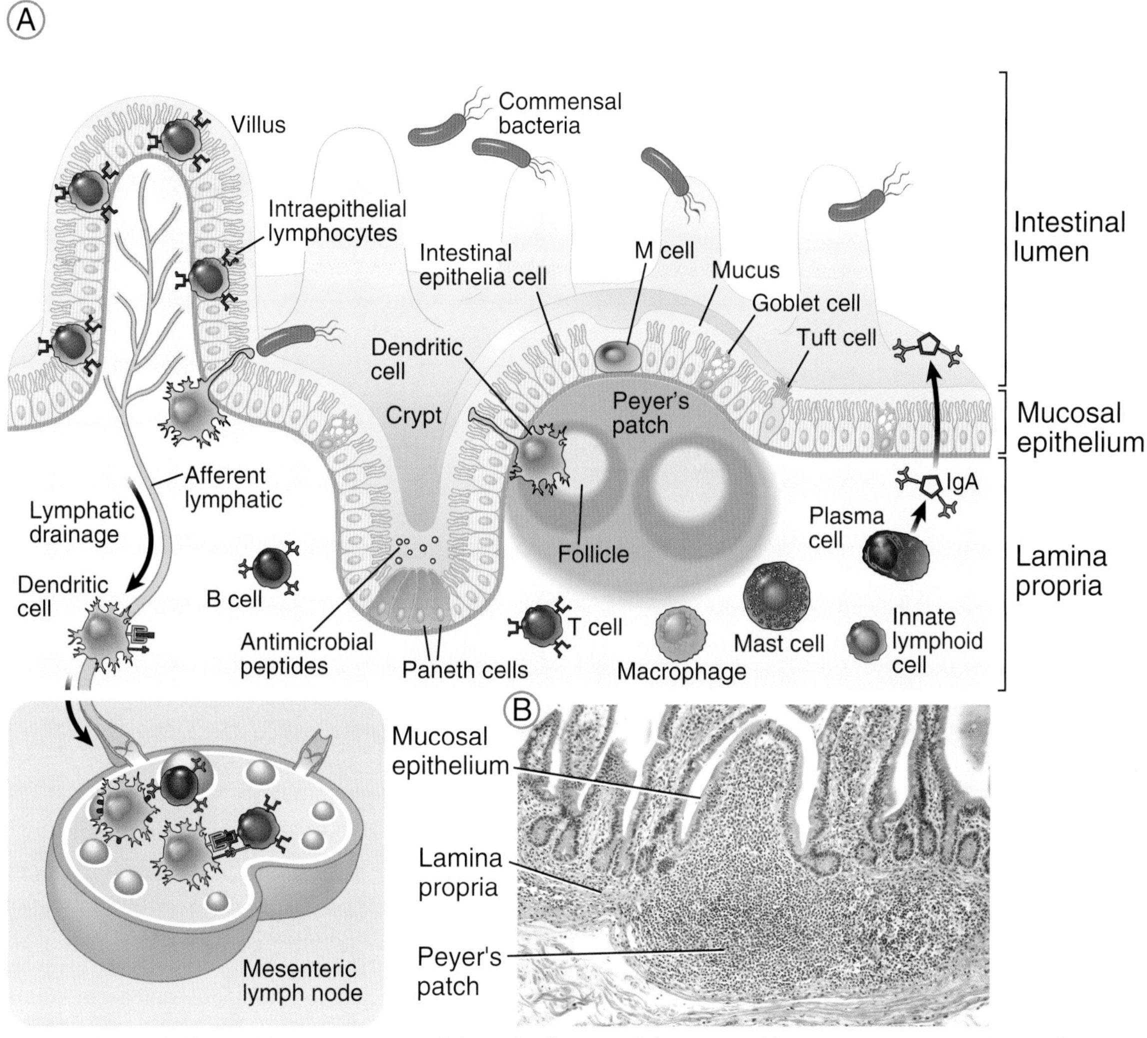

Fig. 1.16 Mucosal immune system. Schematic diagram of the mucosal immune system uses the small bowel as an example. Many commensal bacteria are present in the lumen. The mucus-secreting epithelium provides an innate barrier to microbial invasion (discussed in Chapter 2). Specialized epithelial cells, such as M cells, promote the transport of antigens from the lumen into underlying tissues. Cells in the lamina propria, including dendritic cells, T lymphocytes, and macrophages, provide innate and adaptive immune defense against invading microbes; some of these cells are organized into specialized structures, such as Peyer's patches in the small intestine. Immunoglobulin A (IgA) is a type of antibody abundantly produced in mucosal tissues that is transported into the lumen, where it binds and neutralizes microbes (see Chapter 8).

(Fig. 1.17). At any time, at least a quarter of the body's lymphocytes are in the mucosal tissues and skin (reflecting the large size of these tissues) (see Fig. 1.13), and many of these are memory cells. Cutaneous and mucosal lymphoid tissues are sites of immune responses to antigens that breach epithelia. A remarkable property of the cutaneous and mucosal immune systems is that they are able to respond to pathogens but do not react to the enormous numbers of usually harmless commensal microbes present at the epithelial barriers. This is accomplished by several mechanisms, including the action of regulatory T cells and other signals that suppress rather than activate lymphocytes.

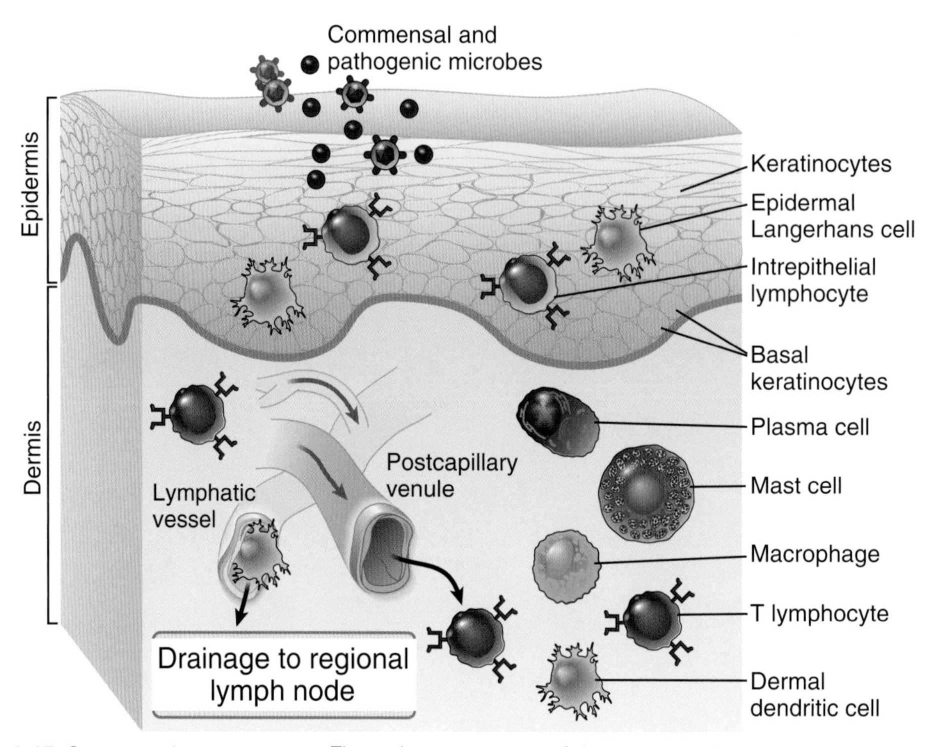

Fig. 1.17 Cutaneous immune system. The major components of the cutaneous immune system shown in this schematic diagram include keratinocytes, Langerhans cells, and intraepithelial lymphocytes, all located in the epidermis, and T lymphocytes, dendritic cells, and macrophages, located in the dermis.

Within the secondary lymphoid organs, T lymphocytes and B lymphocytes are segregated into different anatomic regions (Fig. 1.18). In lymph nodes, the B cells are concentrated in discrete structures, called **follicles**, located around the periphery, or cortex, of each node. If the B cells in a follicle have recently responded to a protein antigen and received signals from helper T cells, this follicle may contain a central lightly staining region called a **germinal center**. The germinal center has an important role in the production of highly effective antibodies and is described in Chapter 7. The T lymphocytes are concentrated outside but adjacent to the follicles, in the paracortex. APCs colocalize with the classes of lymphocytes to which they present antigens— dendritic cells with T lymphocytes in the parafollicular cortex and a type of cell called follicular dendritic cells (FDCs; see Chapter 7) with B cells in the follicles. In the spleen, T lymphocytes are concentrated in periarteriolar lymphoid sheaths surrounding small arterioles, and B cells reside in the follicles.

The anatomic organization of secondary lymphoid organs is tightly regulated to allow immune responses to develop after stimulation by antigens. B lymphocytes are attracted to and retained in the follicles because of the action of a class of cytokines called **chemokines** (chemoattractant cytokines; chemokines and other cytokines are discussed in more detail in later chapters). FDCs in the follicles secrete a particular chemokine for which naive B cells express a receptor, called CXCR5. The chemokine that binds to CXCR5 attracts B cells from the blood into the follicles of lymphoid organs. Similarly, T cells are segregated in the paracortex of lymph nodes and the periarteriolar lymphoid sheaths of the spleen because naive T lymphocytes express a receptor, called CCR7, which recognizes chemokines that are produced in these regions of the lymph nodes and spleen. When the lymphocytes are activated by antigens, they alter their expression of chemokine receptors. As a result, the antigen-activated B cells and T cells migrate toward each other and meet at the edge of follicles, where helper T cells interact with and help B

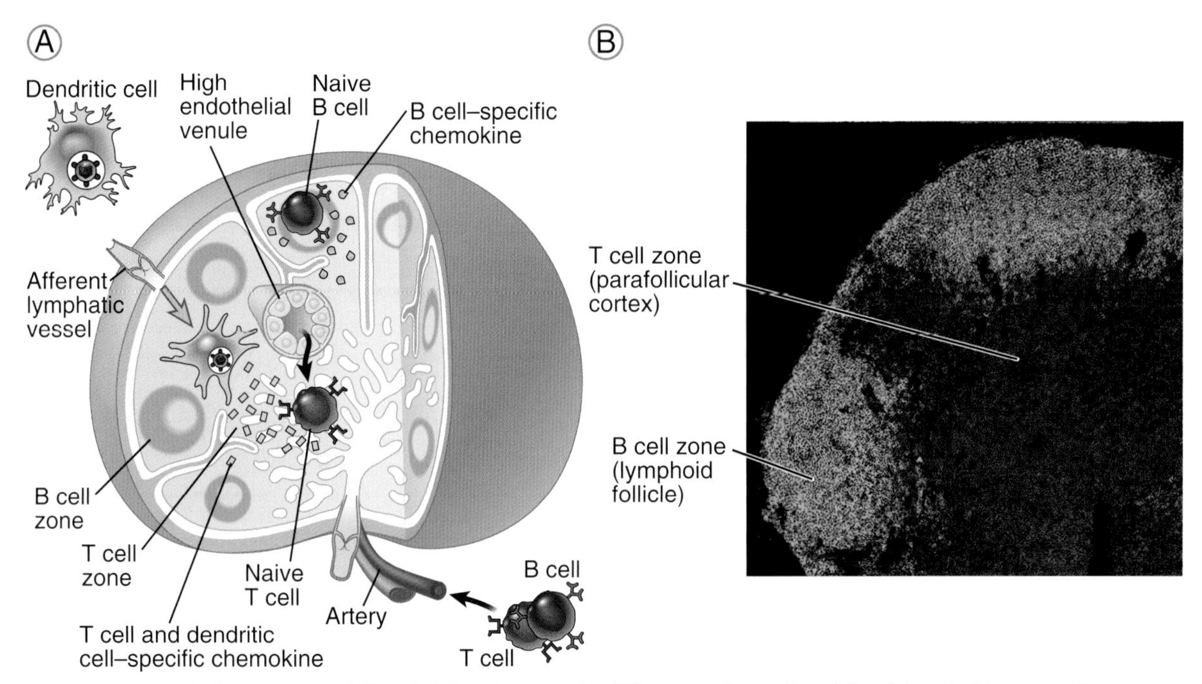

Fig. 1.18 Segregation of T and B lymphocytes in different regions of peripheral lymphoid organs. **A,** Schematic diagram illustrates the path by which naive T and B lymphocytes migrate to different areas of a lymph node. Naive B and T lymphocytes enter through a high endothelial venule (HEV), shown in cross section, and are drawn to different areas of the node by chemokines that are produced in these areas and bind selectively to only one or the other cell type. Also shown is the migration of dendritic cells, which pick up antigens from epithelia, enter through afferent lymphatic vessels, and migrate to the T cell–rich areas of the node (see Chapter 3). **B,** In this histologic section of a lymph node, the B lymphocytes, located in the follicles, are stained green, and the T cells, in the parafollicular cortex, are stained red using immunofluorescence. In this technique, a section of the tissue is incubated with antibodies specific for T or B cells coupled to fluorochromes that emit different colors when excited at the appropriate wavelengths. The anatomic segregation of T and B cells also occurs in the spleen (not shown). (B, Courtesy Drs. Kathryn Pape and Jennifer Walter, University of Minnesota Medical School, Minneapolis, MN.)

cells to differentiate into antibody-secreting plasma cells (see Chapter 7). Thus, these lymphocyte populations are kept apart from each other until it is useful for them to interact, after exposure to an antigen. This is an excellent example of how the structure of lymphoid organs ensures that the cells that have recognized and responded to an antigen interact and communicate with one another when necessary.

Many of the effector T cells exit the node through efferent lymphatic vessels and leave the spleen through veins. These activated lymphocytes end up in the circulation and can go to distant sites of infection. Some activated T cells remain in the lymphoid organ where they were generated and migrate into lymphoid follicles, where they help B cells to make high-affinity antibodies.

Lymphocyte Recirculation and Migration into Tissues

Naive lymphocytes constantly recirculate between the blood and secondary lymphoid organs, where they may be activated by antigens to become effector cells, and effector lymphocytes migrate from lymphoid tissues to sites of infection, where microbes are eliminated (Fig. 1.19). Thus, these lymphocytes with different histories of antigen exposure selectively migrate to the sites where they can perform their

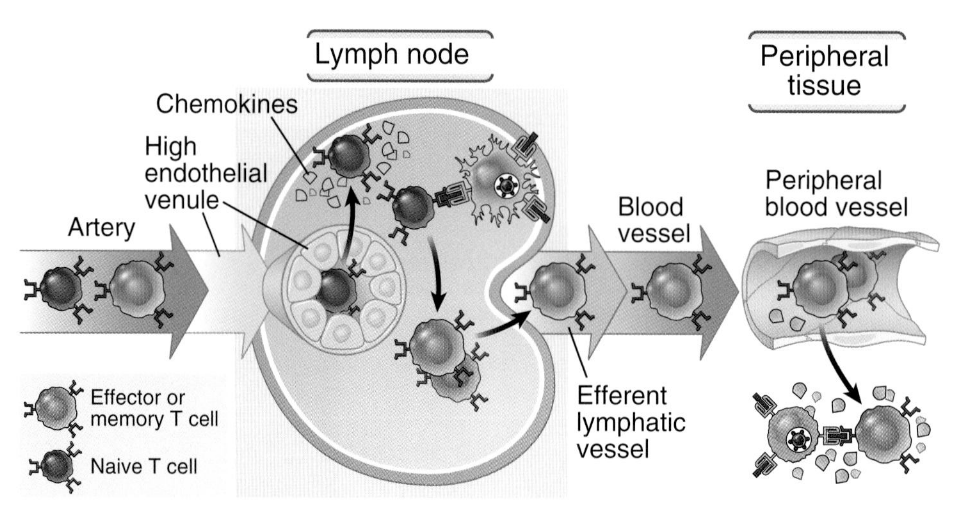

Fig. 1.19 Migration of T lymphocytes. Naive T lymphocytes migrate from the blood through high endothelial venules into the T cell zones of lymph nodes, where the cells are activated by antigens. Activated T cells exit the nodes, enter the bloodstream, and migrate preferentially to peripheral tissues at sites of infection and inflammation. The adhesion molecules involved in the attachment of T cells to endothelial cells are described in Chapter 5.

different functions. Migration of effector lymphocytes to sites of infection is most relevant for T cells because effector T cells have to locate and eliminate microbes at these sites. By contrast, B cell–derived plasma cells do not need to migrate to sites of infection; instead, they secrete antibodies, and the antibodies enter and circulate in the blood. These antibodies bind pathogens or toxins in the blood, or in tissues into which the antibodies enter. Plasma cells in mucosal organs secrete antibodies that enter the lumens of these organs, where they bind to and combat ingested and inhaled microbes.

The migration of different lymphocyte populations has distinct features and is controlled by different molecular interactions.

- Naive T lymphocytes that have matured in the thymus and entered the circulation migrate into lymph nodes by binding to adhesion molecules and chemokines on the endothelial lining of specialized postcapillary venules, called **high endothelial venules** (HEVs), located in the parafollicular cortex. The process of lymphocyte migration out of blood vessels is discussed in Chapter 5. Once outside the HEV, the T cells remain in the paracortex because they are attracted to chemokines produced there. In that location they can find antigens that are brought to the lymph nodes by dendritic cells or in free form through lymphatic vessels that drain epithelia and parenchymal organs.
- Within the lymph node paracortex, naive T cells move around rapidly along specialized connective tissue fibers, scanning the surfaces of dendritic cells, located on these fibers, for antigens. If a T cell specifically recognizes an antigen on a dendritic cell, that T cell forms stable conjugates with the dendritic cell and is activated. Such an encounter between an antigen and a specific lymphocyte is likely to be a rare and random event, but most T cells in the body circulate through some lymph nodes at least once a day. As mentioned earlier and described further in Chapter 3, the likelihood of the correct T cell finding its antigen is increased in secondary lymphoid organs, particularly lymph nodes, because microbial antigens are concentrated in the same regions of these organs through which naive T cells circulate. Thus, T cells find the antigen they can recognize, and these T cells are activated to proliferate and differentiate. Naive cells that have not encountered specific antigens leave the

lymph nodes through lymphatic vessels and reenter the circulation.

- Many of the effector cells that are generated upon T cell activation leave the lymph node by lymphatics, enter the circulation, and then preferentially migrate into infected tissues. Effector T cell migration selectively occurs at infection sites because the local innate response to the microbes induces expression of chemokines and endothelial adhesion molecules on postcapillary venules (see Chapter 5). Once in the infected tissue, T lymphocytes perform their function of eradicating the microbes.
- B lymphocytes that recognize and respond to antigen in lymph node follicles differentiate into antibody-secreting plasma cells, many of which migrate to the bone marrow or mucosal tissues (see Chapter 7).
- Memory T cells consist of different populations (see Chapter 6); some cells recirculate through lymph nodes, where they can mount secondary responses to captured antigens, and other cells migrate to sites of infection, where they can respond rapidly to eliminate the infection. Yet other memory cells permanently reside in epithelial tissues, such as mucosal tissues and the skin.

We know less about lymphocyte circulation through the spleen or other lymphoid tissues. The spleen does not contain HEVs, but the general pattern of naive lymphocyte migration through this organ is probably similar to migration through lymph nodes.

SUMMARY

- The physiologic function of the immune system is to protect individuals from infections and cancers.
- Innate immunity is the early line of defense, mediated by cells and molecules that are always present and ready to eliminate infectious microbes.
- Adaptive immunity is mediated by lymphocytes stimulated by microbial antigens, which leads to the proliferation and differentiation of the lymphocytes and the generation of effector cells, which eliminate microbes, and memory cells, which respond more effectively against each successive exposure to a microbe.
- Lymphocytes are the cells of adaptive immunity and are the only cells with clonally distributed receptors specific for different antigens.
- Adaptive immunity consists of humoral immunity, in which antibodies neutralize and eliminate extracellular microbes and toxins, and cell-mediated immunity, in which T lymphocytes eradicate intracellular microbes.
- Adaptive immune responses consist of sequential phases: antigen recognition by lymphocytes, activation of the lymphocytes to proliferate and to differentiate into effector and memory cells, elimination of the microbes, decline of the immune response, and long-lived memory.
- Different populations of lymphocytes serve distinct functions and may be distinguished by the surface expression of particular membrane molecules.
- B lymphocytes are the only cells that produce antibodies. B lymphocytes express membrane antibodies that recognize antigens, and the progeny of activated B cells, called plasma cells, secrete the antibodies that neutralize and eliminate the antigen.
- T lymphocytes recognize peptide fragments of protein antigens displayed on other cells. Helper T lymphocytes produce cytokines that activate phagocytes to destroy ingested microbes, recruit leukocytes, and activate B lymphocytes to produce antibodies. Cytotoxic T lymphocytes (CTLs) kill infected cells harboring microbes in the cytoplasm.
- Antigen-presenting cells (APCs) capture antigens of microbes that enter through epithelia, concentrate these antigens in lymphoid organs, and display the antigens for recognition by T cells.
- Lymphocytes and APCs are organized in secondary (peripheral) lymphoid organs, where immune responses are initiated and develop.
- Naive lymphocytes circulate through secondary lymphoid organs, where they may encounter foreign antigens. Effector T lymphocytes migrate to peripheral sites of infection, where they function to eliminate infectious microbes. Plasma cells remain in lymphoid organs and the bone marrow, where they secrete antibodies that enter the circulation and find and eliminate microbes.

REVIEW QUESTIONS

1. What are the major differences between innate and adaptive immunity?
2. What are the two types of adaptive immunity, and what types of microbes do these adaptive immune responses combat?
3. What are the principal classes of lymphocytes, and how do they differ in function?
4. What are the important differences among naive, effector, and memory T and B lymphocytes?
5. Where are T and B lymphocytes located in lymph nodes, and how is their anatomic separation maintained?
6. How do naive and effector T lymphocytes differ in their patterns of migration?

Answers to and discussion of the Review Questions may be found on p. 322.

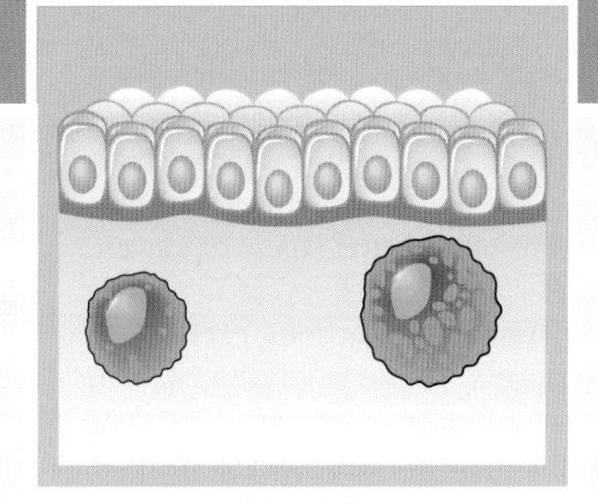

2

Innate Immunity
The Early Defense Against Infections

CHAPTER OUTLINE

The survival of multicellular organisms requires effective defense against microbial infections and the ability to eliminate damaged and necrotic cells and repair injured tissues. Many of the mechanisms that serve these roles are always present and functional within the organism, ready to recognize and eliminate microbes and dead cells. Therefore, this type of host defense is known as **innate immunity**, also called natural immunity or native immunity. The cells and molecules that are responsible for innate immunity make up the **innate immune system**. Innate immunity evolved long before adaptive immunity and is present in all multicellular organisms, including plants, invertebrates, and vertebrates.

Innate immunity is the first line of host defense against infections. It blocks microbial invasion through epithelial barriers, destroys many microbes that do enter the body, and is capable of controlling and even eradicating infections. The innate immune response is able to combat microbes immediately upon infection; by contrast, to defend against a microbe not previously encountered, the adaptive immune system requires antigen stimulation of lymphocytes, which undergo proliferation and differentiation steps, and therefore effective adaptive responses take several days to develop. Innate immunity provides essential protection against infections during this delay. The innate immune response also instructs the adaptive immune system to respond to different microbes in ways that are effective for combating those microbes. In addition, innate immunity is a key participant in the clearance of dead tissues and the initiation of repair after tissue damage.

We discuss the early defense reactions of innate immunity in this chapter, focusing on the following three questions:

- How does the innate immune system recognize microbes and damaged cells?
- How do the different components of innate immunity function to combat different types of microbes?
- How do innate immune reactions stimulate adaptive immune responses?

GENERAL FEATURES AND SPECIFICITY OF INNATE IMMUNE RESPONSES

The innate immune system performs its defensive functions with only a few types of reactions, which are more limited than the varied and specialized responses of adaptive immunity. The specificity of innate immunity is also different in several respects from the specificity of lymphocytes, the antigen-recognizing cells of adaptive immunity (Fig. 2.1).

The two principal types of reactions of the innate immune system are inflammation and antiviral defense. Inflammation consists of the accumulation and activation of leukocytes and plasma proteins at sites of infection or tissue injury. These cells and proteins act together to kill mainly extracellular microbes and to eliminate damaged tissues. Innate immune defense against intracellular viruses is mediated by natural killer (NK) cells, which kill virus-infected cells, and by cytokines called type I interferons (IFNs), which block viral replication within host cells.

The innate immune system responds in essentially the same way to repeat encounters with a microbe, whereas the adaptive immune system mounts stronger, more rapid, and thus more effective responses on successive encounters with a microbe. In other words, for the most part, the innate immune system does not remember prior encounters with microbes and resets to baseline after each such encounter, whereas memory is a cardinal feature of the adaptive immune response. Some cells of innate immunity (such as macrophages and NK cells) may be altered by encounters with microbes such that they respond better upon repeat encounters. But it is not clear if this process is seen in most or all innate immune reactions, results in improved protection against recurrent infections, or is specific for different microbes.

The innate immune system recognizes structures that are shared by various classes of microbes and are not present in or on normal host cells. The cells and molecules of innate immunity recognize and respond to a limited number of microbial structures, far fewer than the almost unlimited number of microbial and nonmicrobial antigens that can be recognized by the adaptive immune system. Each component of innate immunity may recognize many bacteria, viruses, or fungi. For example, phagocytes express receptors for bacterial endotoxin, also called lipopolysaccharide (LPS), and other receptors for peptidoglycans; these molecules are components of the outer membranes or cell walls of many bacterial species but are not produced by mammalian cells. Other receptors of phagocytes recognize terminal mannose residues, which are typical of bacterial and fungal but not mammalian glycoconjugates. Receptors in mammalian cells recognize and respond to double-stranded RNA (dsRNA), which is produced during replication of many viruses but is not produced in mammalian cells, and to unmethylated CG-rich (CpG) oligonucleotides, which are common in microbial DNA but are not abundant in mammalian DNA. The microbial molecules that stimulate innate immunity are often called **pathogen-associated molecular patterns (PAMPs)** to indicate that they are present in infectious agents (pathogens) and shared by microbes of the same type (i.e., they are molecular patterns). The receptors of innate immunity that recognize these shared structures are called **pattern recognition receptors**.

Innate immune receptors are specific for structures of microbes that are often essential for the survival and infectivity of these microbes. This characteristic of innate immunity makes it a highly effective defense mechanism because a microbe cannot evade innate immunity simply by mutating or not expressing the targets of innate immune recognition. Microbes that do not express functional forms of these structures lose their ability to infect and replicate in the host. By

Feature	Innate immunity	Adaptive immunity
Specificity	For structures shared by classes of microbes (pathogen-associated molecular patterns) 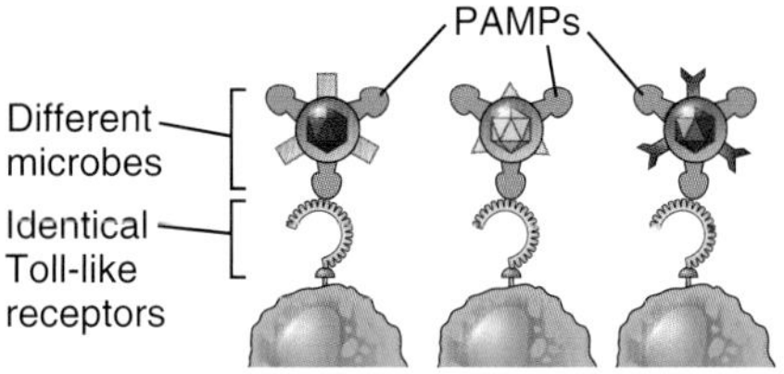 PAMPs Different microbes Identical Toll-like receptors	For structural detail of microbial molecules (antigens); may recognize nonmicrobial antigens 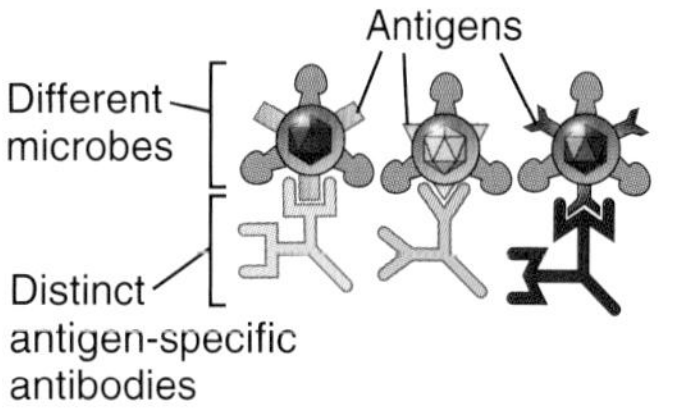 Antigens Different microbes Distinct antigen-specific antibodies
Microbial molecules recognized	About 1000 molecular patterns (estimated); essential for survival of microbes; usually cannot be mutated	>10^7 antigens; most are nonessential and can be mutated to evade immunity
Receptors	Located in plasma membrane, endosomal membrane, cytosol 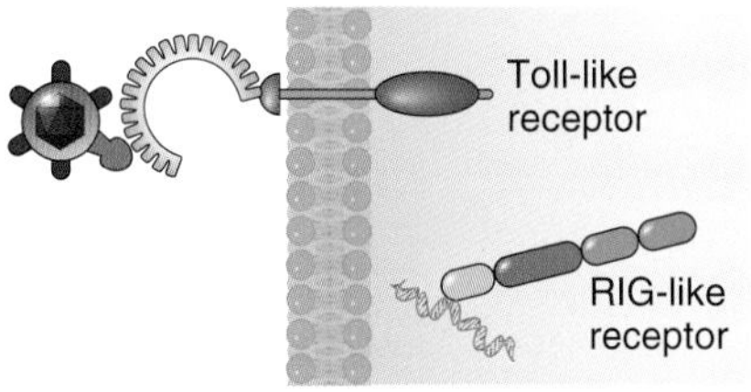 Toll-like receptor RIG-like receptor	Located only in plasma membrane 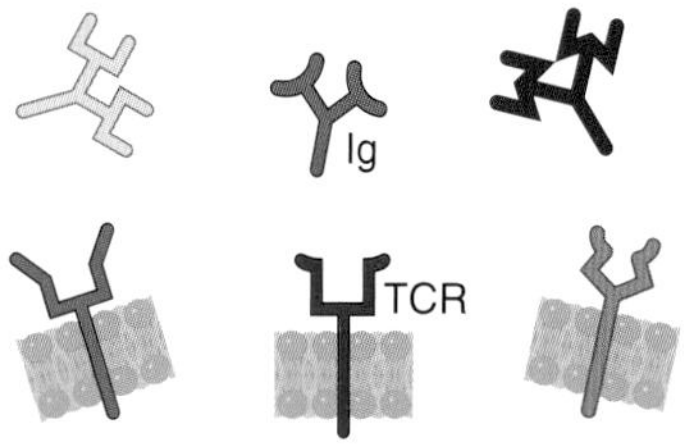 Ig TCR
Number and types of receptors	About 100 different types of invariant receptors	Only 2 types of receptors (Ig and TCR), with millions of variations of each
Distribution of receptors	Nonclonal: identical receptors on all cells that express the receptors	Clonal: clones of lymphocytes with distinct specificities express different receptors
Genes encoding receptors	Germline encoded, in all cells	Formed by somatic recombination of gene segments only in B and T cells
Discrimination of self and nonself	Yes; healthy host cells are not recognized or they may express molecules that prevent innate immune reactions	Yes; based on elimination or inactivation of self-reactive lymphocytes; may be imperfect (hence the possibility of autoimmunity)

Fig. 2.1 Specificity and receptors of innate immunity and adaptive immunity. This figure summarizes the important distinguishing features of the specificity and receptors of innate and adaptive immunity, with select examples of key immune receptors and their ligands illustrated. *Ig,* Immunoglobulin (antibody); *TCR,* T cell receptor.

contrast, microbes can evade adaptive immunity by mutating antigens because most of these are not required for the survival of the pathogens.

The innate immune system also recognizes molecules that are released from damaged or necrotic host cells. Such molecules are called **damage-associated molecular patterns (DAMPs)**. Examples include high mobility group box protein 1 (HMGB1), a histone protein that is released from cells with damaged nuclei, and extracellular adenosine triphosphate (ATP), which is released from damaged mitochondria. The subsequent responses to DAMPs serve to eliminate the damaged cells and to initiate the process of tissue repair. Thus, innate responses occur even following sterile injury, such as infarction, the death of tissue due to loss of its blood supply.

The receptors of the innate immune system are encoded by inherited genes that are identical in all cells. The pattern recognition receptors of the innate immune system are nonclonally distributed; that is, the receptors are identical on all cells that express them. Therefore, many cells of innate immunity may recognize and respond to the same microbe. This is fundamentally different from the antigen receptors of the adaptive immune system, which are encoded by genes formed by rearrangement of gene segments during lymphocyte development, resulting in many clones of B and T lymphocytes, each expressing a unique receptor. It is estimated that there are about 100 types of innate immune receptors that are capable of recognizing about 1000 PAMPs and DAMPs. In striking contrast, there are only two kinds of specific receptors in the adaptive immune system (immunoglobulin [Ig] and T cell receptors [TCRs]), but because of their diversity they are able to recognize millions of different antigens.

The innate immune system does not react against healthy cells. Several features of the innate immune system account for its inability to react against an individual's own cells and molecules. The receptors of innate immunity have evolved to be specific for microbial structures (and products of damaged cells) but not for substances in healthy cells. Healthy cells often express molecules that block innate responses; the best-defined examples of such regulatory molecules are in the complement system. The adaptive immune system also discriminates between self and nonself; in the adaptive immune system, lymphocytes capable of recognizing self antigens are produced, but they die or are inactivated upon encounter with self antigens.

Innate immunity can be considered as a set of mechanisms that provide defense at every stage of microbial infections:

- At the portals of entry for microbes: Most microbial infections are acquired through the epithelial barriers of the skin and gastrointestinal, respiratory, and genitourinary systems. The earliest defense mechanisms active at these sites are epithelia and mucus secreted at some of these sites, which provide physical barriers and antimicrobial molecules.
- In the tissues: Microbes that breach epithelia, as well as dead cells in tissues, are detected by resident macrophages, dendritic cells, and mast cells. Some of these cells react by secreting cytokines, which initiate the process of inflammation. Phagocytes residing in the tissues or recruited from the blood engulf and destroy the microbes and damaged cells.
- In the blood: Plasma proteins, including proteins of the complement system, react against microbes that enter the circulation and promote their destruction.

We will return to a more detailed discussion of these components of innate immunity and their functions later in the chapter. We start with a consideration of how microbes, damaged cells, and other foreign substances are detected and how innate immune responses are triggered.

CELLULAR RECEPTORS FOR MICROBES AND DAMAGED CELLS

The pattern recognition receptors used by the innate immune system to detect microbes and damaged cells are expressed on phagocytes, dendritic cells, epithelial barrier cells, and many other cell types and are located in different cellular compartments where microbes or their products may be found. These receptors are present on the cell surface, where they detect extracellular microbes; in vesicles (endosomes) into which microbes may be ingested; and in the cytosol, where they function as sensors of cytoplasmic microbes and products of cell damage (Fig. 2.2). The receptors for PAMPs and DAMPs belong to several protein families.

Toll-Like Receptors

Toll-like receptors (TLRs) are so named because they are homologous to a Drosophila protein called Toll, which was discovered for its role in the development of the fly and later shown to be essential for protecting flies against fungal infections. In humans, there are 10 different TLR proteins that dimerize to form 9 different functional receptors that are specific for different components of microbes (Fig. 2.3). TLR-2 complexed with TLR-1 or TLR-6 recognizes several lipopeptides and peptidoglycans made by gram-positive bacteria and some parasites; TLR-4 is specific for bacterial LPS (endotoxin), made by gram-negative bacteria; TLR-5 is

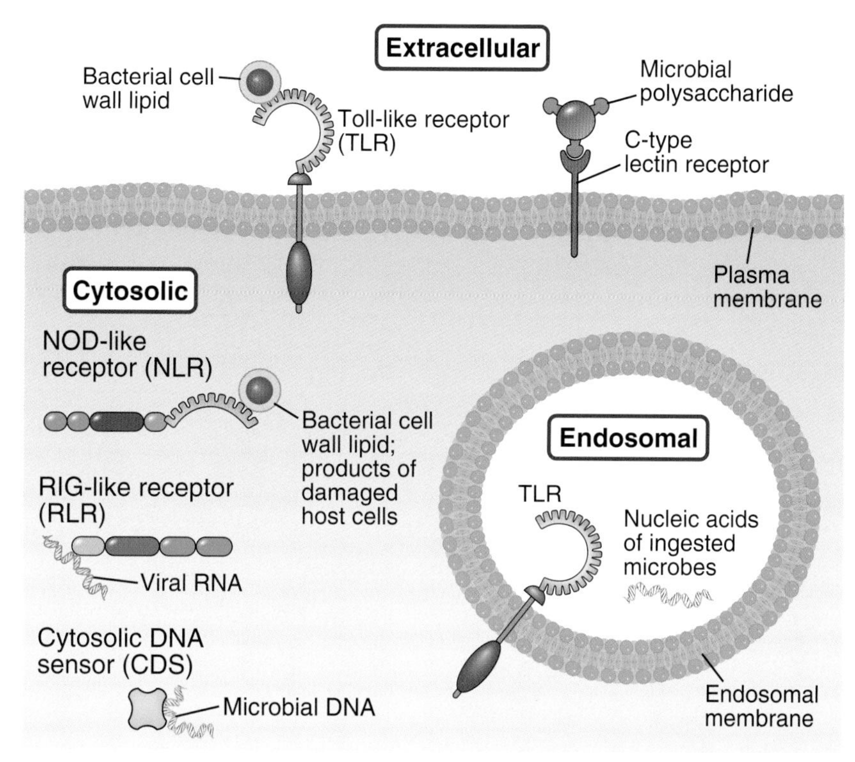

Fig. 2.2 Cellular locations of receptors of the innate immune system. Some receptors, such as certain Toll-like receptors (TLRs) and lectins, are located on cell surfaces; other TLRs are in endosomes. Some receptors for viral nucleic acids, bacterial peptides, and products of damaged cells are in the cytoplasm. NOD and RIG refer to the founding members of families of structurally homologous cytosolic receptors for bacterial and viral products, respectively. (Their full names are complex and do not reflect their functions.) There are five major families of cellular receptors in innate immunity: TLRs, C-type lectin receptors (CLRs), NOD-like receptors (NLRs), RIG-like receptors (RLRs), and cytosolic DNA sensors (CDSs).

specific for a bacterial flagellar protein called flagellin that is produced by most motile bacteria; TLR-3 is specific for dsRNA; TLR-7 and TLR-8 are specific for ssRNA; and TLR-9 recognizes unmethylated CpG DNA, which is abundant in microbial genomes. TLRs specific for microbial proteins, lipids, and polysaccharides (many of which are present in bacterial cell walls) are located on cell surfaces, where they recognize these products of extracellular microbes. TLRs that recognize nucleic acids are in endosomes, into which microbes are ingested and where they are degraded and their nucleic acids are released.

Signals generated by TLRs activate transcription factors that stimulate expression of cytokines and other proteins involved in the inflammatory or antiviral response and in the antimicrobial functions of activated phagocytes and other cells (Fig. 2.4). Among the most important transcription factors activated by TLR signals are members of the nuclear factor κB (NF-κB) family, which promote expression of various cytokines and endothelial adhesion molecules that play important roles in inflammation, and interferon regulatory factors (IRFs), which stimulate production of the antiviral cytokines type I IFNs.

Mutations affecting TLRs or their signaling molecules cause rare autosomal recessive diseases characterized by recurrent infections, highlighting the importance of these pathways in host defense against microbes. For example, individuals with mutations affecting TLR-3 are susceptible to herpes simplex virus infections, particularly encephalitis, and mutations in MyD88, a signal transduction adaptor protein downstream of several TLRs, make individuals susceptible to bacterial pneumonias.

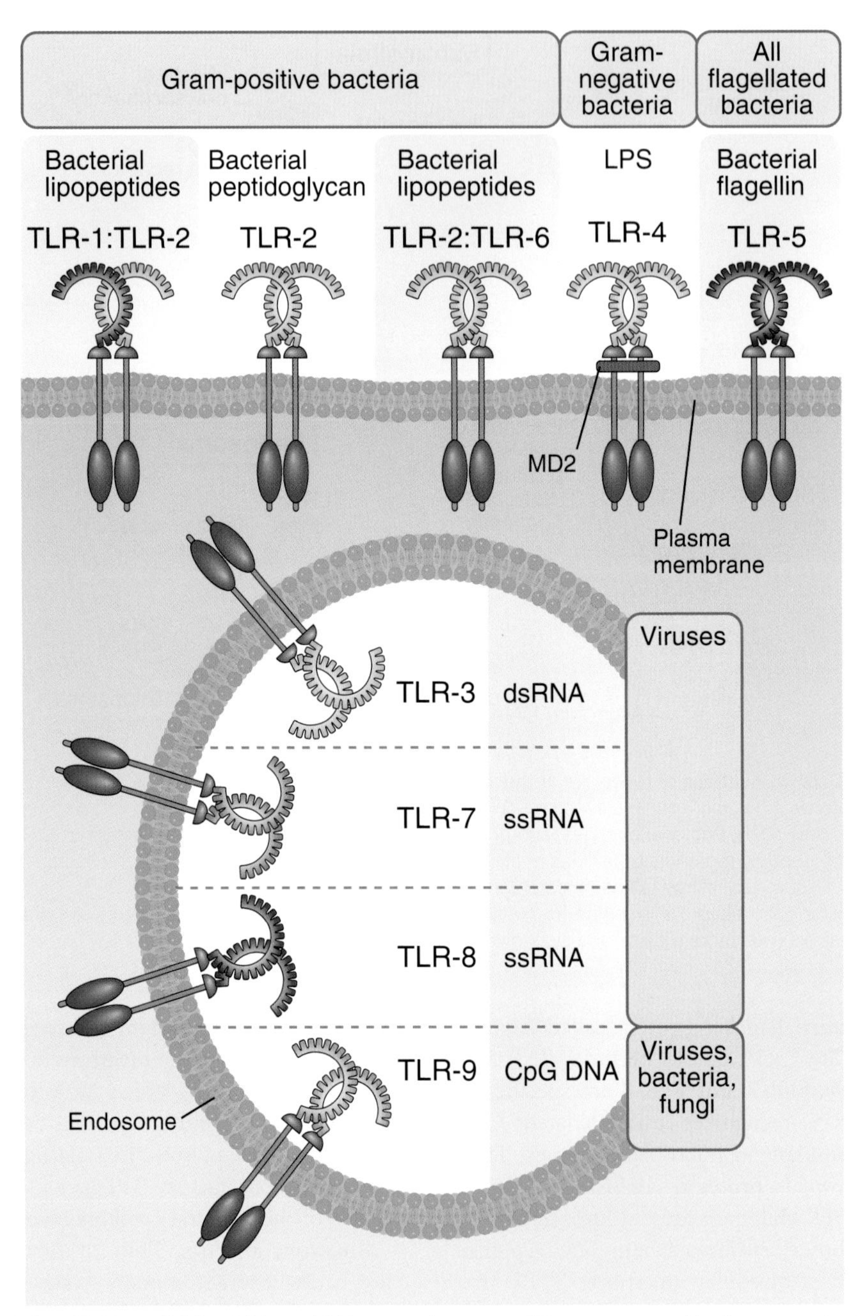

Fig. 2.3 Specificities of TLRs. Different TLRs recognize many different, structurally diverse products of microbes. Plasma membrane TLRs are specific for cell wall components of bacteria, and endosomal TLRs recognize nucleic acids. MD2 is a protein that enhances the binding of LPS to TLR-4. *ds,* Double-stranded; *LPS,* lipopolysaccharide; *ss,* single-stranded.

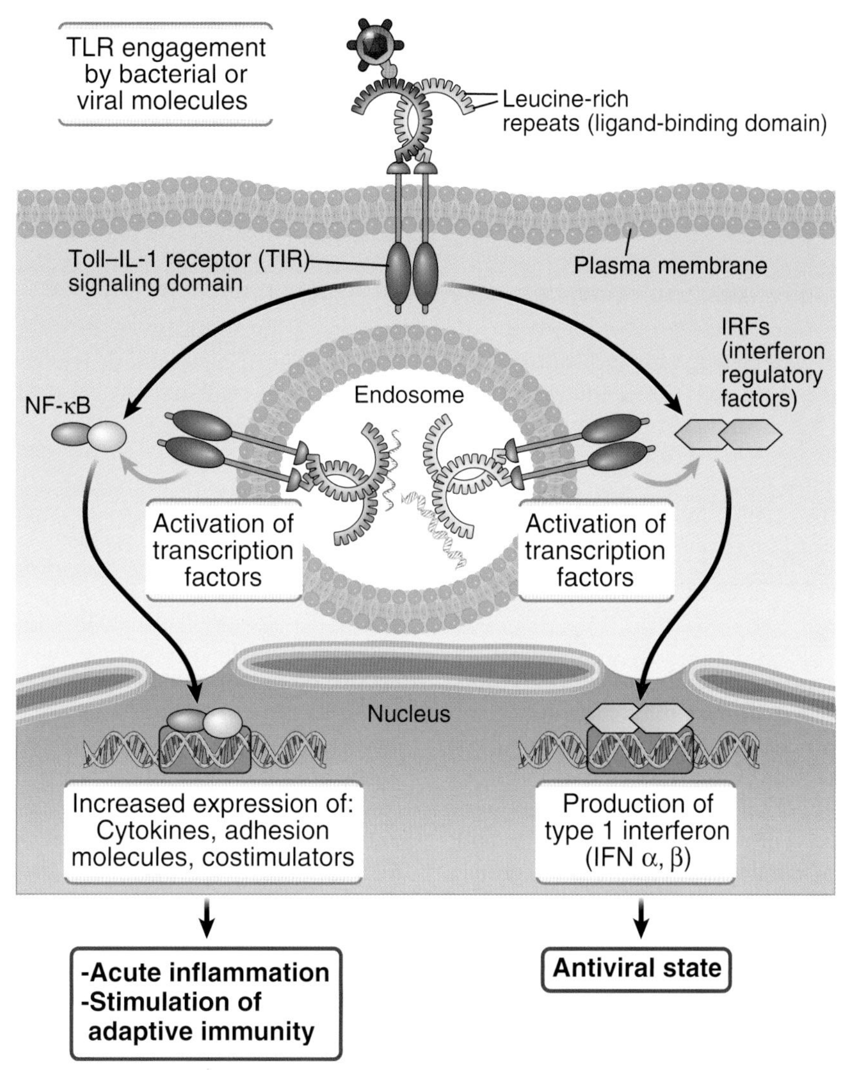

Fig. 2.4 Signaling functions of Toll-like receptors. All TLRs contain a ligand-binding domain composed of leucine-rich motifs and a cytoplasmic signaling, Toll-like interleukin-1 (IL-1) receptor (TIR) domain, so named because it is present in both types of receptors. TLRs activate signaling mechanisms that involve adaptor proteins and lead to the activation of transcription factors. These transcription factors stimulate the production of proteins that mediate inflammation and antiviral defense. *NF-κB,* Nuclear factor κB.

NOD-Like Receptors

The NOD-like receptors (NLRs) are a large family of innate receptors that sense DAMPs and PAMPs in the cytosol of cells and initiate signaling events that promote inflammation. All NLRs contain a C-terminal nucleotide oligomerization domain (NOD, named because of the activity it was originally associated with) but different NLRs have different N-terminal ligand-binding domains. Two important NLRs, NOD1 and NOD2, are expressed in several cell types, including mucosal barrier epithelial cells and phagocytes. NOD1 and NOD2 recognize different dipeptides derived from bacterial cell wall peptidoglycan, and in response, they generate signals that activate the NF-κB transcription factor, which promotes expression of genes encoding inflammatory proteins. NOD2 is highly expressed in intestinal Paneth cells in the small bowel, where it stimulates expression of antimicrobial substances called defensins in response to ingested pathogens. Some polymorphisms of the *NOD2* gene are associated with inflammatory bowel disease, perhaps because these variants have reduced function and allow luminal microbes to penetrate the intestinal wall and trigger inflammation.

Inflammasomes

Inflammasomes are multiprotein complexes that assemble in the cytosol of cells in response to microbes or changes associated with cell injury, and proteolytically generate active forms of the inflammatory cytokines interleukin-1β (IL-1β) and IL-18. IL-1β and IL-18 are synthesized as inactive precursors, which must be cleaved by the enzyme caspase-1 to become active cytokines that are released from the cell and promote inflammation. Inflammasomes are composed of a sensor, an enzyme (inactive caspase-1), and an adaptor that links the two. There are many different types of inflammasomes, some of which use one of several different NLR-family proteins as sensors. These sensors directly recognize microbial products in the cytosol or detect changes in the amount of endogenous molecules or ions in the cytosol that indirectly indicate the presence of infection or cell damage. Some inflammasomes use sensors that are not in the NLR family, such as AIM-family DNA sensors and a protein called pyrin that detects biochemical changes induced by certain bacterial toxins. After recognition of microbial or endogenous ligands, the inflammasome sensors oligomerize with an adaptor protein and the inactive (pro) form of the enzyme caspase-1, resulting in generation of the active form of caspase-1 (Fig. 2.5). Active caspase-1 cleaves pro–IL-1β, the inactive precursor form of the cytokine IL-1β, to generate biologically active IL-1β. As discussed later, IL-1 induces acute inflammation and causes fever.

One of the best characterized inflammasomes uses NLRP3 (NOD-like receptor family, pyrin domain containing 3) as a sensor. The NLRP3 inflammasome is expressed in innate immune cells, including macrophages and neutrophils, as well as keratinocytes in the skin and other cells. A wide variety of stimuli activate the NLRP3 inflammasome, including crystalline substances such as uric acid (a by-product of DNA breakdown, indicating nuclear damage) and cholesterol crystals, extracellular ATP released from nearby injured cells that binds to cell surface purinoceptors, reduced intracellular potassium ion (K^+) concentration (which indicates plasma membrane damage), and reactive oxygen species (which are produced in response to many types of cell injury). Thus, the inflammasome reacts to injury affecting various cellular components. How NLRP3 recognizes such diverse types of cellular stress or damage is not clearly understood.

Inflammasome activation also causes a form of programmed cell death of macrophages and dendritic cells called **pyroptosis**, characterized by swelling of cells, loss of plasma membrane integrity, and release of inflammatory cytokines. Activated caspase-1 cleaves a protein called gasdermin D. The N-terminal fragment of gasdermin D oligomerizes and forms a channel in the plasma membrane that allows the egress of mature IL-1β, but also permits the influx of ions, followed by cell swelling and pyroptosis.

Inflammasome activation plays important roles in several diseases. Gain-of-function mutations in NLRP3 and, less frequently, loss-of-function mutations in regulators of inflammasome activation are the cause of **autoinflammatory syndromes** characterized by uncontrolled and spontaneous inflammation. IL-1 antagonists are effective treatments for these diseases. The common joint disease **gout** is caused by deposition of urate crystals and subsequent inflammation mediated by inflammasome recognition of the crystals and IL-1β production. The inflammasome may also contribute to atherosclerosis, in which inflammation caused by cholesterol crystals may play a role.

Cytosolic RNA and DNA Sensors

The innate immune system includes several cytosolic proteins that recognize microbial RNA or DNA and respond by generating signals that lead to the production of inflammatory and antiviral cytokines.

- The RIG-like receptors (RLRs) RIG-I and MDA-5 (Fig. 2.6) are cytosolic proteins that sense viral RNA and induce the production of the antiviral cytokines, type I IFNs. RLRs recognize features of viral RNAs that are not typical of mammalian RNA, such as dsRNA that is longer than dsRNA that may be formed transiently in normal cells, or RNA with a 5′ triphosphate moiety not present in mammalian host cell cytosolic RNA. (Mammalian RNAs are modified and have a 5′ 7-methyl-guanosine cap.) RLRs are expressed in many cell types that are susceptible to infection by RNA viruses. After binding

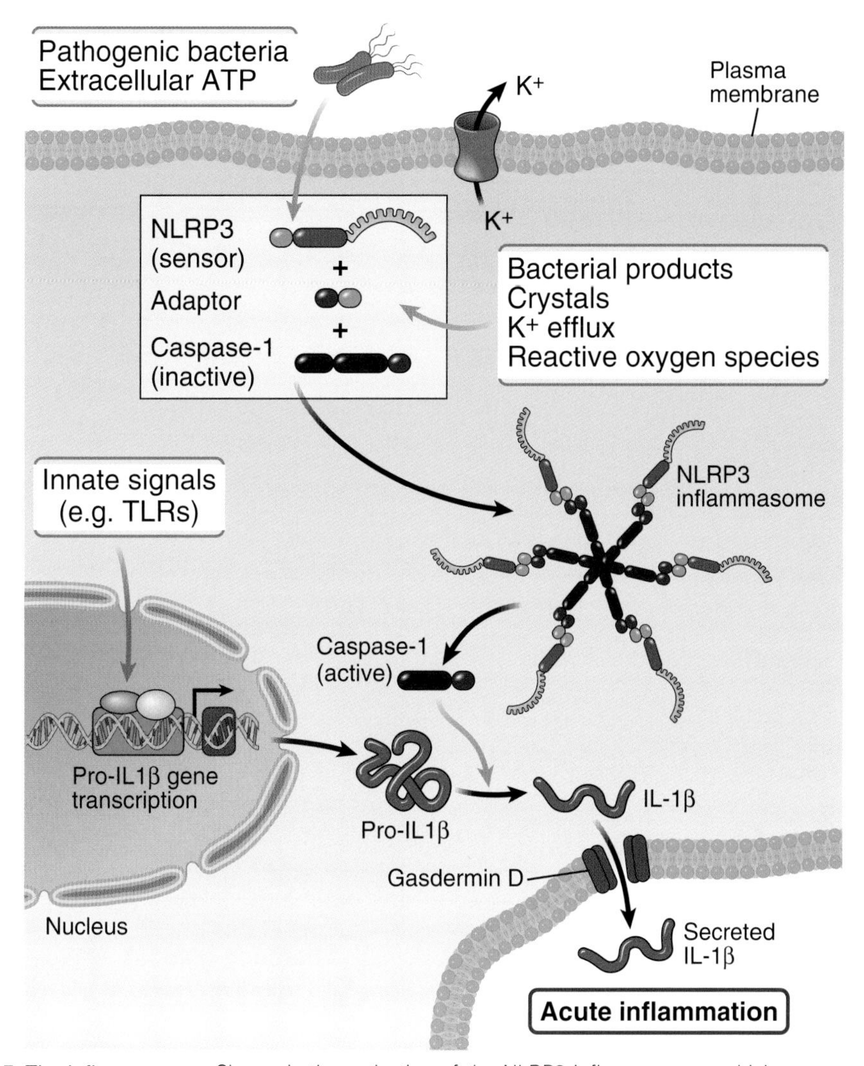

Fig. 2.5 The inflammasome. Shown is the activation of the NLRP3 inflammasome, which processes pro–interleukin-1β (pro–IL-1β) to active IL-1. The synthesis of pro–IL-1β is induced by various PAMPs or DAMPs through pattern recognition receptor signaling. Subsequent production of biologically active IL-1β is mediated by the inflammasome. The inflammasome also stimulates production of active IL-18, which is closely related to IL-1 (not shown). Other forms of the inflammasome exist that contain sensors other than NLRP3, including NLRP1, NLRC4, or AIM2. *ATP,* Adenosine triphosphate; *NLRP3,* NOD-like receptor family, pyrin domain containing 3; *TLRs,* Toll-like receptors.

viral RNAs, RLRs interact with a mitochondrial membrane protein called mitochondrial antiviral-signaling (MAVS), which is required to initiate signaling events that activate transcription factors that induce the production of type I IFNs.

- Cytosolic DNA sensors (CDSs) include several structurally related proteins that recognize microbial dsDNA in the cytosol and activate signaling pathways that initiate antimicrobial responses, including type I IFN production and autophagy. DNA may be released into the cytosol from intracellular viruses and bacteria. These sensors also recognize self DNA if it accumulates in the cytosol. Excessive accumulation (e.g., caused by mutations that reduce endonuclease function) is the basis of systemic inflammatory diseases called **interferonopathies.**

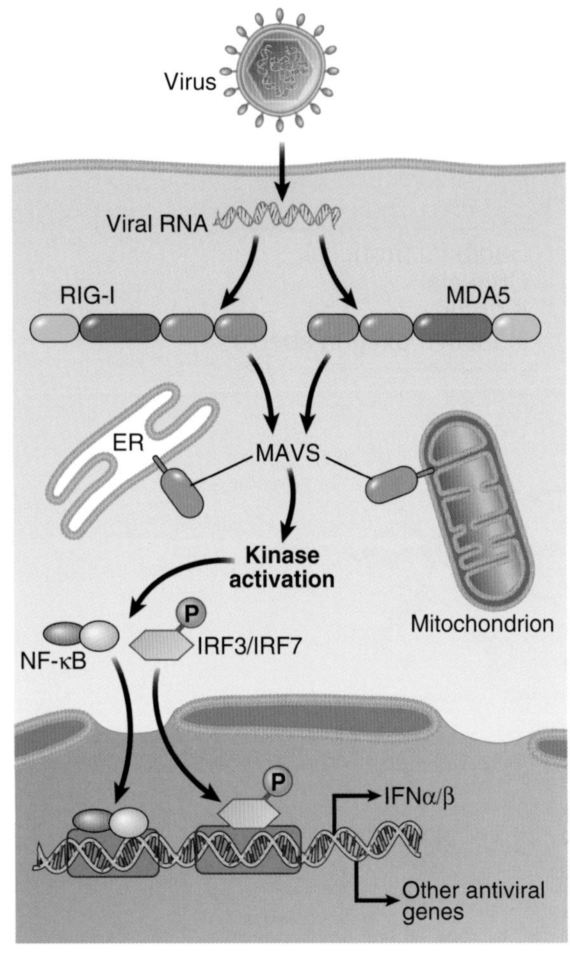

Fig. 2.6 Cytosolic RIG-like receptors. The two main receptors of this family, RIG-I and MDA5, recognize cytosolic viral RNA and trigger a signaling pathway that leads to the activation of transcription factors (IRFs) that stimulate production of the antiviral cytokine type I interferon (IFN). *ER,* Endoplasmic reticulum; *IRF,* interferon regulatory factor; *MAVS,* mitochondrial antiviral signaling.

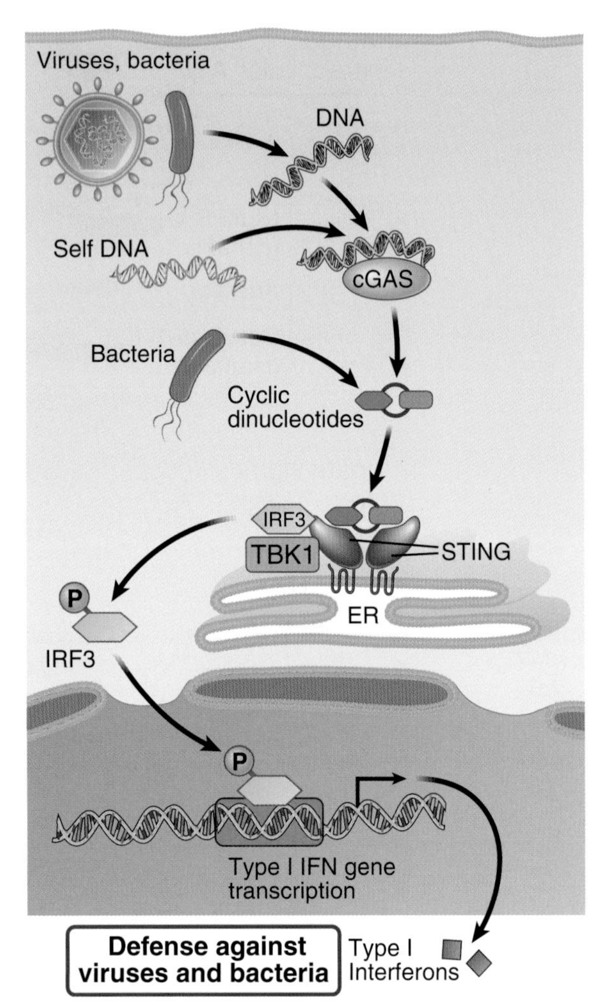

Fig. 2.7 Cytosolic DNA sensors and the STING pathway. Cytosolic microbial dsDNA activates the enzyme cGAS, which catalyzes the synthesis of cyclic GMP-AMP (cGAMP) from ATP and GTP. cGAMP binds to STING in the endoplasmic reticulum membrane, and then STING recruits and activates the kinase TBK1, which phosphorylates IRF3. Phospo-IRF3 moves to the nucleus, where it induces type I IFN gene expression. The bacterial second messenger molecules cyclic di-GMP (c-di-GMP) and cyclic di-AMP (c-di-AMP) are directly sensed by STING. STING also stimulates autophagy and lysosomal degradation of pathogens associated with cytoplasmic organelles. *cGAS,* Cyclic GMP-AMP synthase; *ER,* endoplasmic reticulum; *IFN,* interferon; *IRF3,* interferon regulatory factor 3.

Most innate cytosolic DNA sensors engage the stimulator of IFN genes (**STING**) pathway to induce type I IFN production (Fig. 2.7). For example, cytosolic dsDNA binds to the enzyme cyclic guanosine monophosphate–adenosine monophosphate (GMP-AMP) synthase (cGAS), which activates the production of a cyclic dinucleotide signaling molecule called cyclic GMP-AMP (cGAMP), which binds to an endoplasmic reticulum membrane adaptor protein called STING. In addition, some bacteria themselves produce other cyclic dinucleotides that also bind to STING. Upon binding these cyclic dinucleotides, STING initiates signaling events that lead to transcriptional activation and expression of type I IFN genes. STING also stimulates autophagy, a mechanism by which cells degrade their own organelles in lysosomes. Autophagy is used in innate immunity to deliver cytosolic microbes to the lysosome, where they are killed by proteolytic enzymes. Other cytosolic DNA sensors besides cGAS can also activate STING.

Other Cellular Receptors of Innate Immunity

Many other receptor types are involved in innate immune responses to microbes (see Fig. 2.2).

- Some lectins (carbohydrate-recognizing proteins) in the plasma membrane are receptors specific for fungal glucans (these receptors are called dectins) or for terminal mannose residues (called mannose receptors); they are involved in the phagocytosis of fungi and bacteria and in inflammatory responses to these pathogens.
- A cell surface receptor expressed mainly on phagocytes, called formyl peptide receptor 1, recognizes polypeptides with an N-terminal formylmethionine, which is a specific feature of bacterial proteins (and mammalian mitiochondrial proteins). Signaling by this receptor promotes the migration as well as the antimicrobial activities of phagocytes.

Although our emphasis thus far has been on cellular receptors, the innate immune system also contains several circulating molecules that recognize and provide defense against microbes, as discussed later.

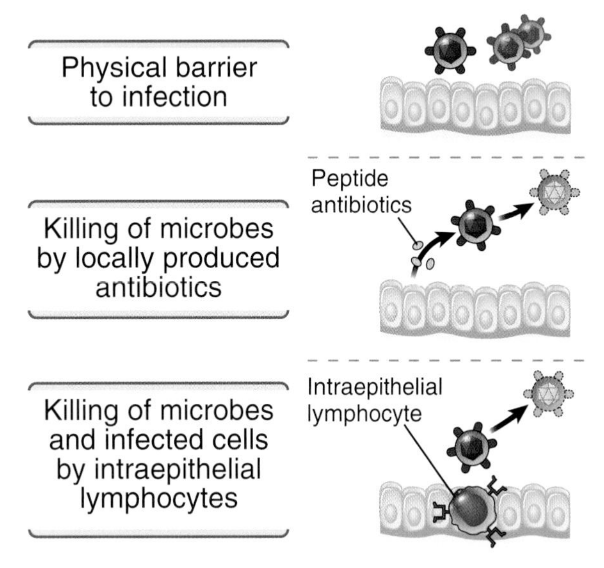

Fig. 2.8 Functions of epithelia in innate immunity. Epithelia present at the portals of entry of microbes provide physical barriers formed by keratin (in the skin) or secreted mucus (in the gastrointestinal, bronchopulmonary, and genitourinary systems) and by tight junctions between epithelial cells. Epithelia also produce antimicrobial substances (e.g., defensins and cathelicidins) and harbor lymphocytes that kill microbes and infected cells.

COMPONENTS OF INNATE IMMUNITY

The components of the innate immune system include epithelial cells; sentinel cells in tissues (resident macrophages, dendritic cells, mast cells, and others), circulating and recruited phagocytes (monocytes and neutrophils), innate lymphoid cells, NK cells, and a number of plasma proteins. We next discuss the properties of these cells and soluble proteins and their roles in innate immune responses.

Epithelial Barriers

The major interfaces between the body and the external environment—the skin, gastrointestinal tract, respiratory tract, and genitourinary tract—are protected by layers of epithelial cells that provide physical and chemical barriers against infection (Fig. 2.8). Microbes come into contact with vertebrate hosts mainly at these interfaces by external physical contact, ingestion, inhalation, and sexual activity. All these portals of entry are lined by continuous epithelia consisting of tightly adherent cells that form a mechanical barrier against microbes. Keratin on the surface of the skin and mucus secreted by mucosal epithelial cells prevent most microbes from infecting or getting through the epithelia. Epithelial cells also produce antimicrobial peptides, called defensins and cathelicidins, that kill bacteria and some viruses by disrupting their outer membranes. Thus, antimicrobial peptides provide a chemical barrier against infection. In addition, epithelia contain lymphocytes called intraepithelial lymphocytes that belong to the T cell lineage but express antigen receptors of limited diversity. Some of these T cells express receptors composed of two chains, γ and δ, that are similar but not identical to the $\alpha\beta$ TCRs expressed on the majority of T lymphocytes (see Chapters 4 and 5). Intraepithelial T lymphocytes presumably react against infectious agents that attempt to breach the epithelia, but the specificity and functions of these cells are poorly understood.

Feature	Neutrophils	Macrophages
Origin	HSCs in bone marrow	HSCs in bone marrow (in inflammatory reactions) Stem cells in yolk sac or fetal liver (early in development): Many tissue-resident macrophages
Life span in tissues	1-2 days	Inflammatory macrophages: days or weeks Tissue-resident macrophages: years
Responses to activating stimuli	Rapid, short lived, enzymatic activity	More prolonged, slower, often dependent on new gene transcription
Reactive oxygen species	Rapidly induced by assembly of phagocyte oxidase (respiratory burst)	Less prominent
Nitric oxide	Low levels or none	Induced following transcriptional activation of iNOS
Degranulation	Major response; induced by cytoskeletal rearrangement	Not prominent
Cytokine production	Low levels per cell	Major functional activity, large amounts per cell, requires transcriptional activation of cytokine genes
Extracellular traps	Rapidly induced, by extrusion of nuclear contents	Little

Fig. 2.9 Distinguishing properties of neutrophils and monocytes. This table lists the major differences between neutrophils and macrophages. These two cell types share many features, such as phagocytosis, chemotaxis, and ability to migrate through blood vessels into tissues. *HSC,* Hematopoietic stem cell; *iNOS,* inducible nitric oxide synthase.

Phagocytes: Neutrophils and Monocytes/Macrophages

The two types of circulating phagocytes, neutrophils and monocytes, are blood cells that are recruited to sites of infection, where they recognize and ingest microbes for intracellular killing (Fig. 2.9).

- **Neutrophils**, also called polymorphonuclear leukocytes (PMNs), are the most abundant leukocytes in the blood, numbering 4000 to 10,000 per μL (Fig. 2.10A). In response to certain bacterial and fungal infections, the production of neutrophils from the bone marrow increases rapidly and their numbers in the blood may rise up to 10 times the normal. The production of neutrophils is stimulated by cytokines, known as colony-stimulating factors (CSFs), which are secreted by many cell types in response to infections and act on hematopoietic cells to stimulate proliferation and maturation of neutrophil precursors. Neutrophils are the first and most numerous cell type to respond to most infections,

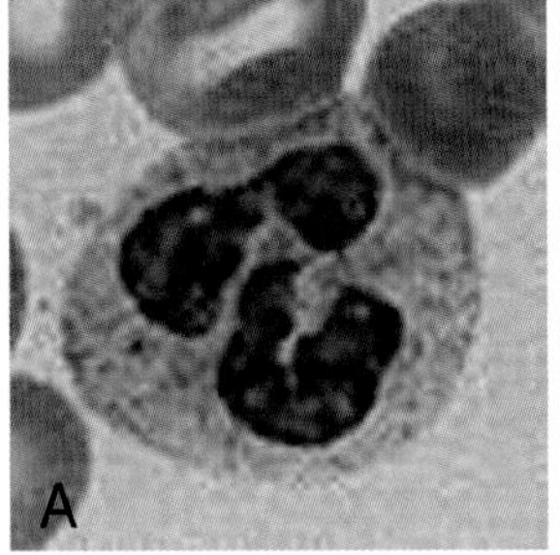

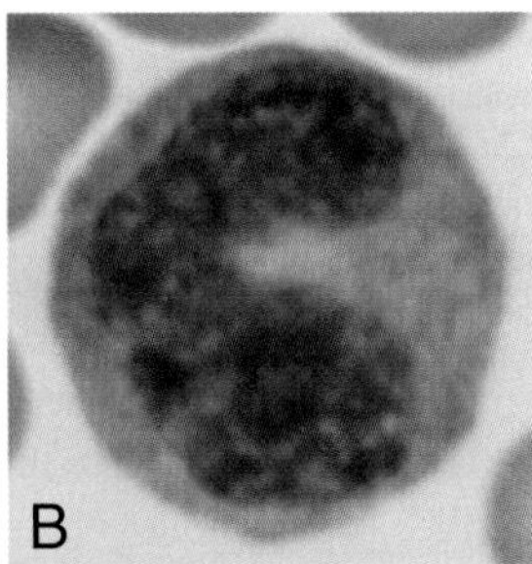

Fig. 2.10 Morphology of neutrophils and monocytes. **A,** Light micrograph of blood neutrophil shows the multilobed nucleus, which is why these cells are also called polymorphonuclear leukocytes, and the faint cytoplasmic granules, most of which are lysosomes. **B,** Light micrograph of blood monocyte shows the typical horseshoe-shaped nucleus.

particularly bacterial and fungal infections, and thus are the dominant cells of acute inflammation, as discussed later. Neutrophils ingest microbes in the circulation, and they rapidly enter extravascular tissues at sites of infection, where they also phagocytose (ingest) and destroy microbes. Neutrophils express receptors for products of complement activation and for antibodies that coat microbes. These receptors enhance phagocytosis of antibody- and complement-coated microbes and also transduce activating signals that stimulate the ability of the neutrophils to kill ingested microbes. The process of phagocytosis and intracellular destruction of microbes is described later. Neutrophils are also recruited to sites of tissue damage in the absence of infection, where they initiate the clearance of cell debris. Neutrophils live for only several hours in tissues, so they are the early responders but they do not provide prolonged defense.

- **Monocytes** are less abundant in the blood than neutrophils, numbering 500 to 1000 per μL (see Fig. 2.10B). They also ingest microbes in the blood and tissues. During inflammatory reactions, monocytes enter extravascular tissues and differentiate into cells called **macrophages**, which, unlike neutrophils, survive in these sites for long periods. Thus, blood monocytes and tissue macrophages are two stages of the same cell lineage, which often is called the mononuclear phagocyte system (Fig. 2.11). Some macrophages are always present in most healthy organs and tissues. These cells, called tissue-resident macrophages, are derived from progenitors in the yolk sac or fetal liver early during fetal development; in some tissues they may be replenished over time by bone marrow–derived monocytes.

Macrophages serve several important roles in host defense: they ingest and destroy microbes, they clear dead tissues and initiate the process of tissue repair, and they produce cytokines that induce and regulate inflammation (Fig. 2.12). A number of receptor families are expressed in macrophages and involved in the activation and functions of these cells. Pattern recognition receptors discussed earlier, including TLRs and NLRs, recognize products of microbes and damaged cells and activate the macrophages. Phagocytosis is mediated by cell surface receptors, such as mannose receptors and scavenger receptors, which directly bind microbes (and other particles), and receptors for antibodies or products of complement activation that are bound to microbes. Some of these phagocytic receptors activate the microbial killing functions of macrophages as well. In addition, macrophages can be activated by various cytokines.

There are two different pathways of macrophage activation, called classical and alternative, that serve distinct functions (see Fig. 6.9). **Classical macrophage activation** is induced by innate immune signals, such as from TLRs, and by the cytokine IFN-γ, which is produced in both innate and adaptive immune responses. Classically activated macrophages, also called M1 or pro-inflammatory, are involved in destroying microbes and in triggering inflammation. **Alternative macrophage activation** occurs in the absence of strong TLR signals and is induced by the cytokines IL-4 and IL-13; these macrophages, called M2 or pro-healing, appear to be more important for tissue repair and to terminate inflammation. The relative abundance of these two forms of activated macrophages may influence the outcome of host reactions and contribute to various disorders. (These populations are often termed M1-like and M2-like because the phenotypic markers for distinguishing them are not definitive. The M1/M2 nomenclature is used for the sake of simplicity.) We will return to the functions of these macrophage populations in Chapter 6, when we discuss cell-mediated immunity.

Although our discussion has been limited to the role of phagocytes in innate immunity, macrophages are also important effector cells in both the cell-mediated arm and the humoral arm of adaptive immunity, as discussed in Chapters 6 and 8, respectively.

Dendritic Cells

Dendritic cells function as sentinels in tissues that respond to microbes by producing numerous cytokines and serve two main functions: they initiate inflammation and they stimulate adaptive immune responses. They also capture protein antigens and display fragments of these antigens to T cells. By sensing microbes and interacting with lymphocytes, especially T cells, dendritic cells constitute an important bridge between innate and adaptive immunity. We discuss the properties and functions of dendritic cells in Chapter 3 in the context of antigen presentation.

Mast Cells

Mast cells are bone marrow–derived cells with abundant cytoplasmic granules that are present in the skin, mucosal tissues, and most connective tissues. Mast cells can be activated by microbial products binding to TLRs and by components of the complement system as part

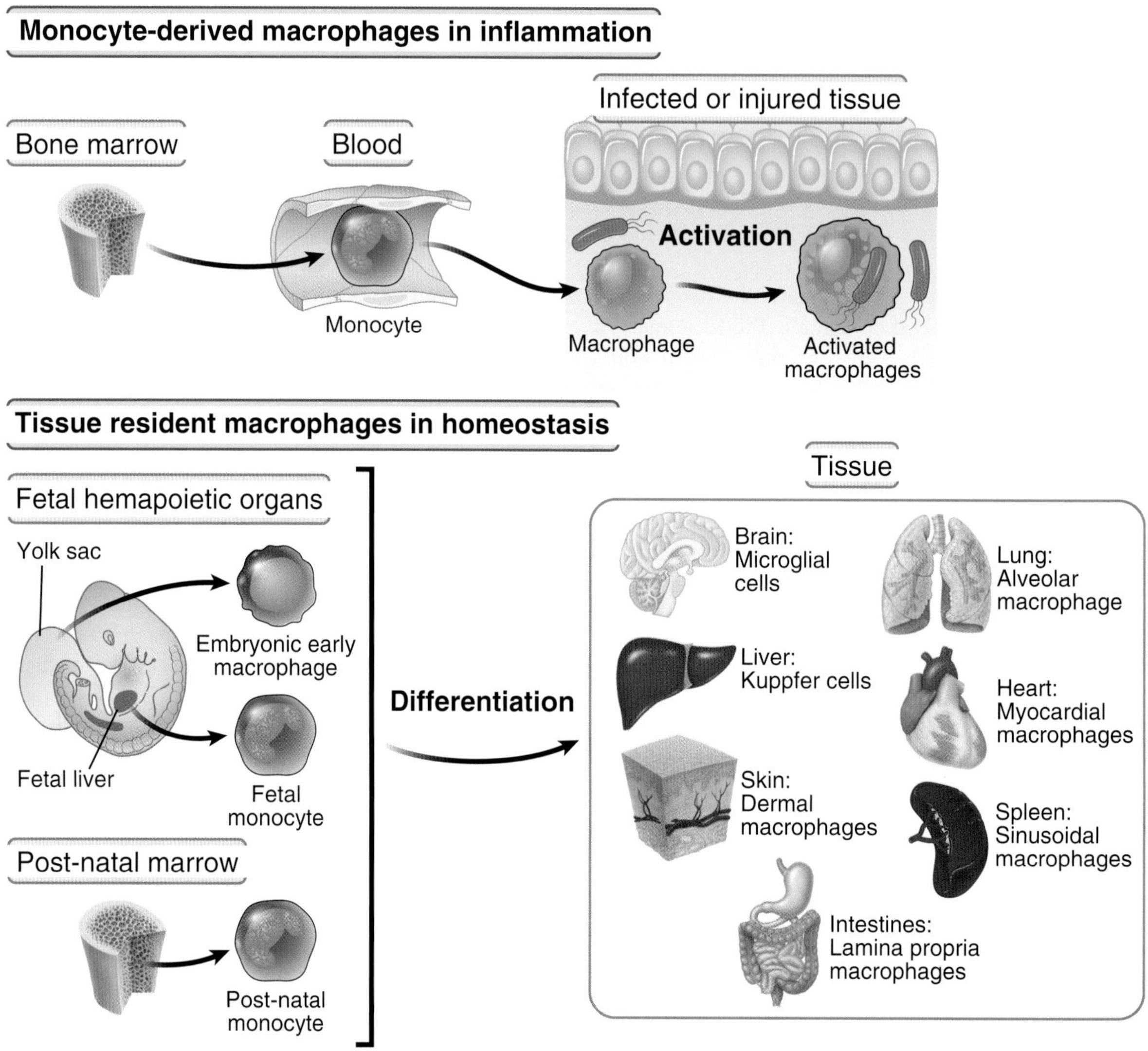

Fig. 2.11 Maturation of mononuclear phagocytes. In the steady state in adults, and during inflammatory reactions, precursors in the bone marrow give rise to circulating monocytes, which enter peripheral tissues, mature to form macrophages, and are activated locally. In fetal life, precursors in the yolk sac and liver give rise to cells that seed tissues to generate specialized tissue-resident macrophages.

of innate immunity or by an antibody-dependent mechanism in adaptive immunity. Mast cell granules contain vasoactive amines such as histamine that cause vasodilation and increased capillary permeability, as well as proteolytic enzymes that can kill bacteria or inactivate microbial toxins. Mast cells also synthesize and secrete lipid mediators (e.g., prostaglandins and leukotrienes) and cytokines (e.g., tumor necrosis factor [TNF]), which stimulate inflammation. Mast cell products provide defense against helminths and other pathogens, as well as protection against snake and insect venoms, and they are responsible for symptoms of allergic diseases (see Chapter 11).

Innate Lymphoid Cells

Innate lymphoid cells (ILCs) are tissue-resident cells that produce cytokines similar to those secreted by helper T lymphocytes but do not express T cell

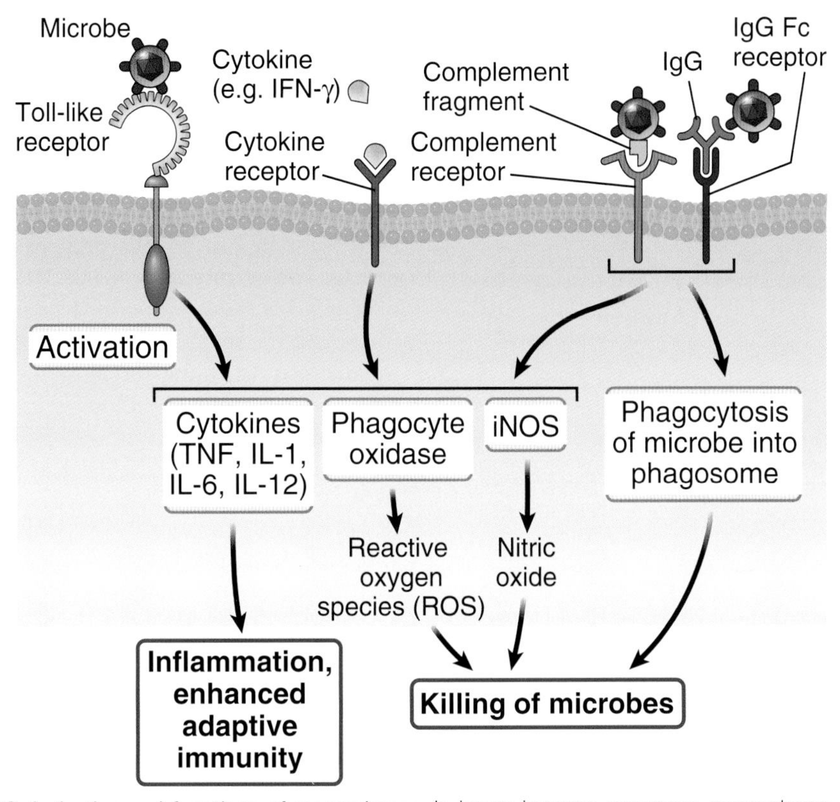

Fig. 2.12 Activation and functions of macrophages. In innate immune responses, macrophages are activated by microbial products binding to TLRs and by cytokines, such as NK cell–derived interferon-γ (IFN-γ), which lead to the production of proteins that mediate inflammatory and microbicidal functions of these cells. Cell surface complement receptors promote the phagocytosis of complement-coated microbes, as well as activation of the macrophages. Macrophage Fc receptors for IgG bind antibody-coated microbes and perform functions similar to those of the complement receptors. *IL,* Interleukin; *iNOS,* inducible nitric oxide synthase; *TNF,* tumor necrosis factor.

antigen receptors (TCRs). ILCs have been divided into three major groups based on their secreted cytokines; these groups correspond to the Th1, Th2, and Th17 subsets of $CD4^+$ T cells that we describe in Chapter 6. The responses of ILCs are often stimulated by cytokines produced by damaged epithelial and other cells at sites of infection. ILCs are always present in tissues, unlike T cells, which are activated in secondary lymphoid organs and migrate into tissues, a process that may take several days. Therefore, ILCs may provide early defense against infections in tissues. However, their contribution to host defense or immunologic diseases, especially in humans, is not clear.

Natural Killer Cells

NK cells recognize cells infected with viruses and some other microbes and stressed cells, and they respond by killing these cells and by secreting the macrophage-activating cytokine IFN-γ (Fig. 2.13). NK cells are developmentally related to group 1 ILCs and make up approximately 5% to 20% of the cells with lymphocyte morphology in the blood and secondary lymphoid organs. NK cells contain cytoplasmic granules and express some unique surface proteins but do not express immunoglobulins or TCRs, the antigen receptors of B and T lymphocytes, respectively.

On activation by infected cells, NK cells deliver the contents of their cytoplasmic granules into the infected cells, where they activate enzymes that induce apoptosis. The cytotoxic mechanisms of NK cells, which are the same as the mechanisms used by cytotoxic T lymphocytes (CTLs; see Chapter 6), result in the death of infected cells. Thus, as with CTLs, NK cells function to eliminate cellular reservoirs of infection and

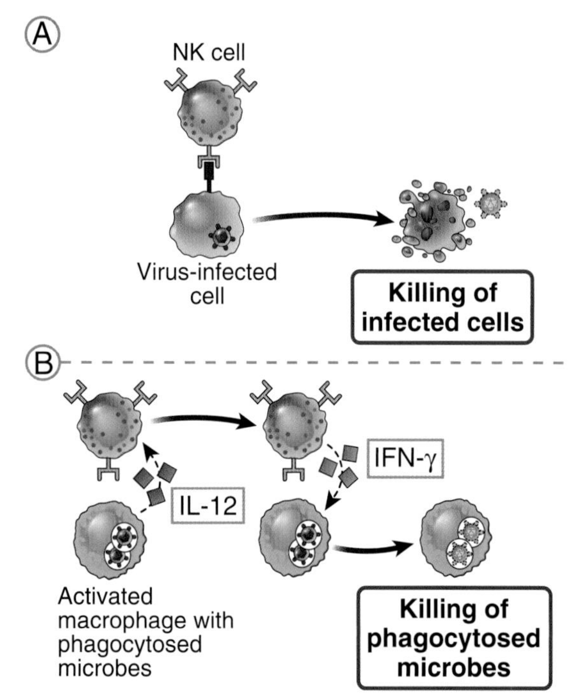

Fig. 2.13 Functions of NK cells. A, NK cells kill host cells infected by intracellular microbes, thus eliminating reservoirs of infection. **B,** NK cells respond to interleukin-12 (IL-12) produced by macrophages and secrete interferon-γ (IFN-γ), which activates the macrophages to kill phagocytosed microbes.

eradicate infections by obligate intracellular microbes, such as viruses. In addition, NK cells may contribute to the destruction of tumors.

Activated NK cells also synthesize and secrete the cytokine IFN-γ, which activates macrophages to become more effective at killing phagocytosed microbes. Cytokines secreted by macrophages and dendritic cells that have encountered microbes enhance the ability of NK cells to protect against infections. Three of these NK cell–activating cytokines are IL-15, type I IFNs, and IL-12. IL-15 is important for the development and maturation of NK cells, and type I IFNs and IL-12 enhance the killing functions of NK cells. Thus, NK cells and macrophages are examples of two cell types that function cooperatively to eliminate intracellular microbes: macrophages ingest microbes and produce IL-12, IL-12 activates NK cells to secrete IFN-γ, and IFN-γ in turn activates the macrophages to kill the ingested microbes. As discussed in Chapter 6, essentially the same sequence of reactions involving macrophages and T lymphocytes is central to the cell-mediated arm of adaptive immunity.

The responses of NK cells are determined by a balance between engagement of activating and inhibitory receptors (Fig. 2.14). The activating receptors recognize cell surface molecules typically expressed on cells infected with viruses and intracellular bacteria, some cancer cells, and cells stressed by DNA damage. These receptors enable NK cells to eliminate cells infected with intracellular microbes, as well as irreparably injured cells and tumor cells. One of the well-defined activating receptors of NK cells is called NKG2D; it recognizes molecules that resemble class I major histocompatibility complex (MHC) proteins and are expressed in response to many types of cellular stress. Another activating receptor, called CD16, is specific for the Fc region of immunoglobulin G (IgG) antibodies bound to cells. The recognition of antibody-coated cells results in killing of these cells, a phenomenon called **antibody-dependent cellular cytotoxicity** (ADCC). NK cells are the principal mediators of ADCC. The role of this reaction in antibody-mediated immunity is described in Chapter 8. Activating receptors on NK cells have signaling subunits that contain immunoreceptor tyrosine-based activation motifs (ITAMs) in their cytoplasmic tails. ITAMs, which are also present in subunits of lymphocyte antigen receptor–associated signaling molecules, become phosphorylated on tyrosine residues when the receptors recognize their activating ligands. The phosphorylated ITAMs bind and promote the activation of cytosolic protein tyrosine kinases, and these enzymes phosphorylate, and thereby activate, other substrates in several different downstream signal transduction pathways, eventually leading to cytotoxic granule exocytosis and production of IFN-γ.

The inhibitory receptors of NK cells block signaling by activating receptors and are specific for self class I major histocompatibility complex (MHC) molecules, which are expressed on all healthy nucleated cells. Therefore, class I MHC expression protects healthy cells from destruction by NK cells. (In Chapter 3, we describe the important function of MHC molecules in displaying peptide antigens to T lymphocytes.) Two major families of NK cell inhibitory receptors in humans are the killer cell immunoglobulin-like receptors (KIRs), so called because they share structural homology with Ig molecules (see Chapter 4), and

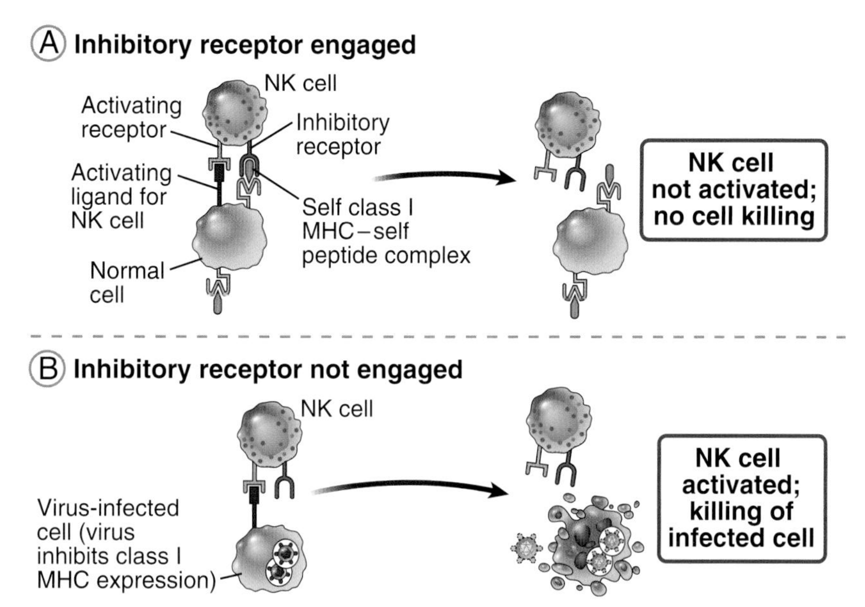

Fig. 2.14 Activating and inhibitory receptors of NK cells. **A,** Healthy host cells express self class I major histocompatibility complex (MHC) molecules, which are recognized by inhibitory receptors, thus ensuring that NK cells do not attack normal host cells. Note that healthy cells may express ligands for activating receptors (as shown) or may not express such ligands, but they are not attacked by NK cells because they engage the inhibitory receptors. **B,** NK cells are activated by infected cells in which ligands for activating receptors are expressed (often at high levels) and class I MHC expression is reduced so that the inhibitory receptors are not engaged. The result is that the infected cells are killed.

receptors consisting of a protein called CD94 and a lectin subunit called NKG2. Both families of inhibitory receptors contain structural motifs in their cytoplasmic tails called immunoreceptor tyrosine-based inhibitory motifs (ITIMs), which become phosphorylated on tyrosine residues when the receptors bind class I MHC molecules. The phosphorylated ITIMs bind and promote activation of cytosolic protein tyrosine phosphatases. These enzymes remove phosphate groups from the tyrosine residues of various signaling molecules, thereby counteracting the function of the ITAMs and blocking the activation of NK cells through activating receptors. Therefore, when the inhibitory receptors of NK cells encounter self MHC molecules on normal host cells, the NK cells are shut off (see Fig. 2.14). Many viruses have developed mechanisms to block expression of class I molecules in infected cells, which allows them to evade killing by virus-specific $CD8^+$ CTLs. When this happens, the NK cell inhibitory receptors are not engaged, and if the virus induces expression of activating ligands at the same time, the NK cells become activated and eliminate the virus-infected cells.

The role of NK cells and CTLs in defense illustrates how hosts and microbes are engaged in a constant struggle for survival. The host uses CTLs to recognize MHC-displayed viral antigens, viruses inhibit MHC expression to evade killing of the infected cells by CTLs, and NK cells can compensate for the defective CTL response because the NK cells are more effective in the absence of MHC molecules. The winner of this struggle, the host or the microbe, determines the outcome of the infection.

Lymphocytes with Limited Diversity

Several types of lymphocytes that have some features of T and B lymphocytes may also function in the early defense against microbes and are thus part of the innate immune system. A unifying characteristic of these lymphocytes is that they express somatically rearranged antigen receptors (as do T and B cells), but the receptors have limited diversity. These include cells of the T lymphocyte lineage ($\gamma\delta$ T cells, NK-T cells, and mucosal-associated invariant T cells), which are described in Chapter 6, and others of the B cell lineage (B1 cells and marginal zone B cells), described in

Chapter 7. All these cell types respond to infections in ways that are characteristic of adaptive immunity (e.g., cytokine secretion or antibody production) but have features of innate immunity (rapid responses, limited diversity of antigen recognition).

Complement System

The complement system is a collection of circulating and membrane-associated proteins that are important in defense against microbes. Many complement proteins are proteolytic enzymes, and complement activation involves the sequential activation of these enzymes. The complement cascade may be initiated by any of three pathways (Fig. 2.15):

- The **alternative pathway** is triggered when some complement proteins are activated on microbial surfaces and cannot be controlled, because complement regulatory proteins are not present on microbes (but are present on host cells). The alternative pathway is a component of innate immunity.
- The **classical pathway** is most often triggered by antibodies that bind to microbes or other antigens and is thus a component of the humoral arm of adaptive immunity.
- The **lectin pathway** is activated when carbohydrate-binding plasma proteins, mannose-binding lectin (MBL) or ficolins, bind to their carbohydrate ligands on microbes. These lectins activate proteins of the classical pathway, but because it is initiated by a microbial product in the absence of antibody, it is a component of innate immunity.

Activated complement proteins function as proteolytic enzymes to cleave other complement proteins. Such an enzymatic cascade can be rapidly amplified because each proteolytic step generates many products that are themselves enzymes in the cascade. The central component of all three complement pathways is a plasma protein called C3, which is cleaved by enzymes generated in the early steps. The major proteolytic fragment of C3, called C3b, becomes covalently attached to microbes and is able to recruit and activate downstream complement proteins on the microbial surface. The three pathways of complement activation differ in how they are initiated, but they share the late steps and perform the same effector functions.

The complement system serves three main functions in host defense:

- **Opsonization and phagocytosis.** C3b coats microbes and promotes the binding of these microbes to phagocytes by virtue of receptors for C3b that are expressed on the phagocytes. Thus, microbes that are coated with complement proteins are rapidly ingested and destroyed by phagocytes. This process of coating a microbe with molecules that are recognized by receptors on phagocytes is called **opsonization.**
- **Inflammation.** Some proteolytic fragments of complement proteins, especially C5a and C3a, are chemoattractants for leukocytes (mainly neutrophils and monocytes), and they are also activators of endothelial cells and mast cells. Thus, they promote movement of leukocytes and plasma proteins into tissues (inflammation) at the site of complement activation.
- **Cell lysis.** Complement activation culminates in the formation of a polymeric protein complex that inserts into the microbial cell membrane, disturbing the permeability barrier and causing osmotic lysis.

A more detailed discussion of the activation and functions of complement is presented in Chapter 8, where we consider the effector mechanisms of humoral immunity.

Other Plasma Proteins of Innate Immunity

Several circulating proteins in addition to complement proteins are involved in innate immune defense against infections. Plasma MBL recognizes microbial carbohydrates and can coat microbes for phagocytosis or activate the complement cascade by the lectin pathway, as discussed earlier. MBL belongs to a family of proteins called collectins, because they are structurally similar to collagen and contain a carbohydrate-binding (lectin) domain. Surfactant proteins in the lung also belong to the collectin family and protect the airways from infection. C-reactive protein (CRP) is a pentraxin (five-headed molecule) that binds to phosphorylcholine on microbes and opsonizes the microbes for phagocytosis by macrophages, which express a receptor for CRP. CRP can also activate proteins of the classical complement pathway.

The circulating levels of many of these plasma proteins increase rapidly after infection. This protective response is called the **acute-phase response** to infection.

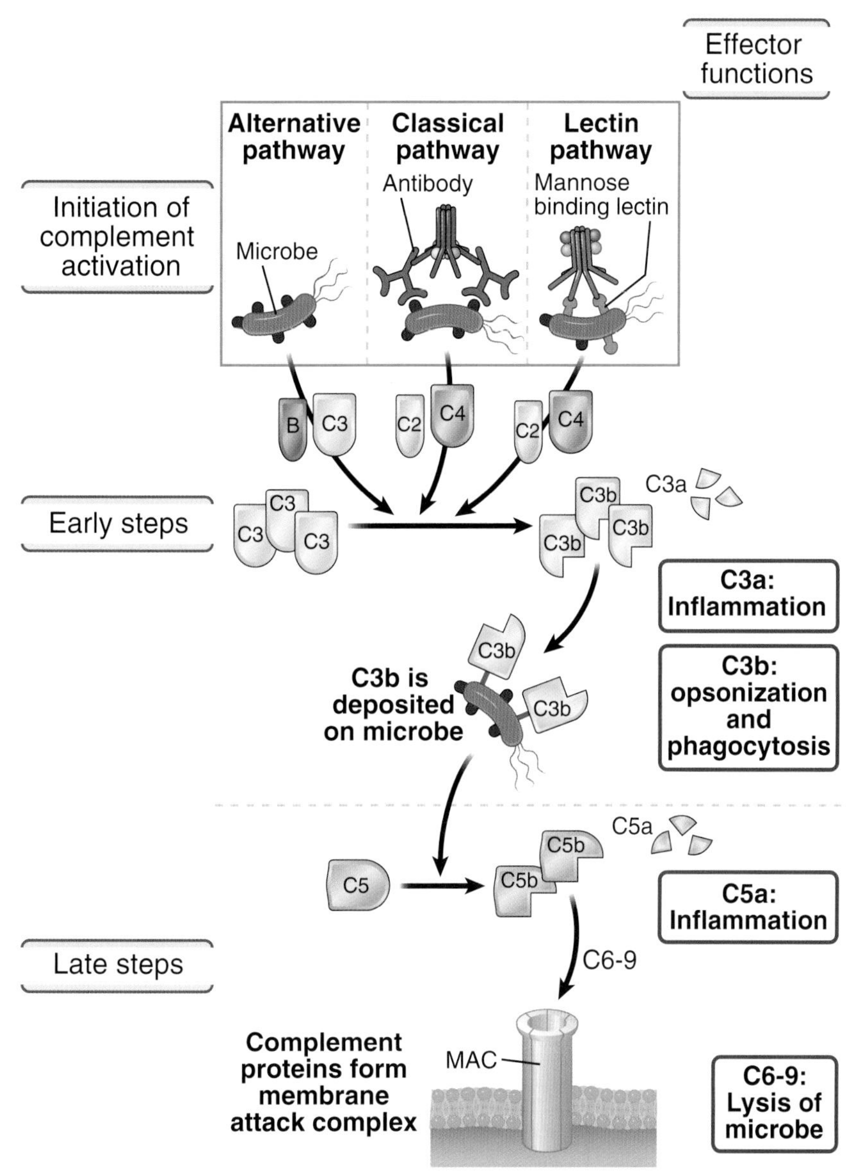

Fig. 2.15 Pathways of complement activation. The activation of the complement system (the early steps) may be initiated by three distinct pathways, all of which lead to the production of C3b. C3b initiates the late steps of complement activation, culminating in the formation of a multiprotein complex called the membrane attack complex (MAC), which is a transmembrane channel composed of polymerized C9 molecules that causes lysis of thin-walled microbes. Peptide by-products released during complement activation are the inflammation-inducing C3a and C5a. The principal functions of proteins produced at different steps are shown. The activation, functions, and regulation of the complement system are discussed in more detail in Chapter 8.

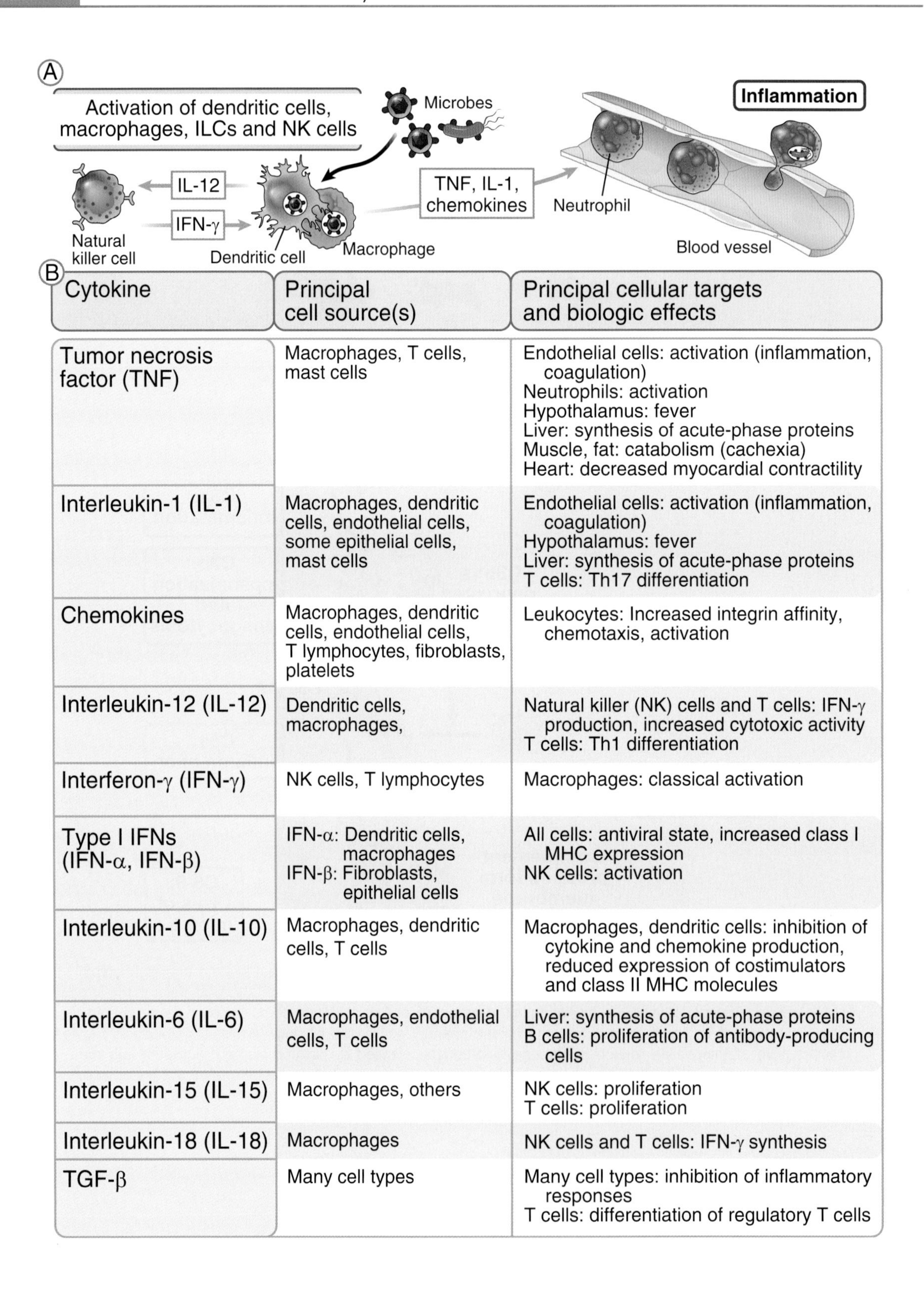

Cytokine	Principal cell source(s)	Principal cellular targets and biologic effects
Tumor necrosis factor (TNF)	Macrophages, T cells, mast cells	Endothelial cells: activation (inflammation, coagulation) Neutrophils: activation Hypothalamus: fever Liver: synthesis of acute-phase proteins Muscle, fat: catabolism (cachexia) Heart: decreased myocardial contractility
Interleukin-1 (IL-1)	Macrophages, dendritic cells, endothelial cells, some epithelial cells, mast cells	Endothelial cells: activation (inflammation, coagulation) Hypothalamus: fever Liver: synthesis of acute-phase proteins T cells: Th17 differentiation
Chemokines	Macrophages, dendritic cells, endothelial cells, T lymphocytes, fibroblasts, platelets	Leukocytes: Increased integrin affinity, chemotaxis, activation
Interleukin-12 (IL-12)	Dendritic cells, macrophages,	Natural killer (NK) cells and T cells: IFN-γ production, increased cytotoxic activity T cells: Th1 differentiation
Interferon-γ (IFN-γ)	NK cells, T lymphocytes	Macrophages: classical activation
Type I IFNs (IFN-α, IFN-β)	IFN-α: Dendritic cells, macrophages IFN-β: Fibroblasts, epithelial cells	All cells: antiviral state, increased class I MHC expression NK cells: activation
Interleukin-10 (IL-10)	Macrophages, dendritic cells, T cells	Macrophages, dendritic cells: inhibition of cytokine and chemokine production, reduced expression of costimulators and class II MHC molecules
Interleukin-6 (IL-6)	Macrophages, endothelial cells, T cells	Liver: synthesis of acute-phase proteins B cells: proliferation of antibody-producing cells
Interleukin-15 (IL-15)	Macrophages, others	NK cells: proliferation T cells: proliferation
Interleukin-18 (IL-18)	Macrophages	NK cells and T cells: IFN-γ synthesis
TGF-β	Many cell types	Many cell types: inhibition of inflammatory responses T cells: differentiation of regulatory T cells

Cytokines of Innate Immunity

Dendritic cells, macrophages, mast cells, ILCs and other cells secrete cytokines that initiate many of the cellular reactions of innate immunity (Fig. 2.16). As mentioned earlier, cytokines are soluble proteins that mediate immune and inflammatory reactions and are responsible for communications between leukocytes and between leukocytes and other cells. Most of the molecularly defined cytokines are called interleukins with a number, for example interleukin-1, but several have other names, for example, tumor necrosis factor, for historical reasons related to how they were discovered. In innate immunity, the principal sources of cytokines are dendritic cells, macrophages, ILCs, and mast cells that are activated by recognition of microbes or by other cytokines; epithelial cells and other cell types also secrete cytokines. Recognition of bacterial cell wall components such as LPSs and peptidoglycans by TLRs and recognition of microbial nucleic acids by TLRs, RLRs, and CDSs are powerful stimuli for cytokine secretion by macrophages, dendritic cells, and many tissue cells. In adaptive immunity, helper T lymphocytes are a major source of cytokines (see Chapters 5 and 6).

Cytokines are secreted in small amounts in response to an external stimulus and bind to high-affinity receptors on target cells. Most cytokines act on nearby cells (paracrine actions), and some act on the cells that produce them (autocrine actions). In innate immune reactions against infections, enough dendritic cells and macrophages may be activated that large amounts of cytokines are produced, and they may be active distant from their site of secretion (endocrine actions).

The cytokines of innate immunity serve various functions in host defense. TNF, IL-1, and chemokines (chemoattractant cytokines) are the principal cytokines involved in recruiting blood neutrophils and monocytes to sites of infection (described later). TNF and IL-1 also have systemic effects, including inducing fever by acting on the hypothalamus, and these cytokines as well as IL-6 stimulate liver cells to produce various proteins of the acute phase response, such as CRP and fibrinogen, which contribute to microbial killing and walling off infectious sites. At high concentrations, TNF promotes thrombus formation on the endothelium and reduces blood pressure by a combination of reduced myocardial contractility and vascular dilation and leakiness. Severe bacterial and fungal infections sometimes lead to a potentially lethal clinical syndrome called **septic shock**, which is characterized by low blood pressure (the defining feature of shock), disseminated intravascular coagulation, and metabolic disturbances. The early clinical and pathologic manifestations of septic shock may be caused by high levels of TNF, which is produced in response to bacterial PAMPs. Dendritic cells and macrophages also produce IL-12 in response to LPS, peptidoglycans, and other microbial molecules. The role of IL-12 in activating NK cells, leading to increased killing activity and macrophage activation, was mentioned previously. NK cells produce IFN-γ, whose function as a macrophage-activating cytokine was also described earlier. In viral infections, a subset of dendritic cells, and to a lesser extent other infected cells, produce type I IFNs, which inhibit viral replication and prevent spread of the infection to uninfected cells.

INNATE IMMUNE REACTIONS

The innate immune system eliminates microbes mainly by inducing the acute inflammatory response and by antiviral defense mechanisms. Different types of innate immune responses are particularly effective against different types of microbes:

- Extracellular bacteria and fungi are defended against mainly by the acute inflammatory response, in

Fig. 2.16 Cytokines of innate immunity. **A,** Dendritic cells, macrophages, and other cells (such as mast cells and ILCs, not shown) respond to microbes by producing cytokines that stimulate inflammation (leukocyte recruitment) and activate natural killer (NK) cells to produce the macrophage-activating cytokine interferon-γ (IFN-γ). **B,** Some important characteristics of the major cytokines of innate immunity are listed. Note that IFN-γ and transforming growth factor beta (TGF-β) are cytokines of both innate and adaptive immunity, and ILCs produce several cytokines that are better known as products of $CD4^+$ T cells (see Chapters 5 and 6). More information about these cytokines and their receptors is provided in Appendix II. *MHC,* Major histocompatibility complex.

which neutrophils and monocytes are recruited to the site of infection, aided by the complement system.

- Intracellular bacteria, which can survive inside phagocytes, are eliminated by phagocytes that are activated by TLRs and other innate sensors as well as by cytokines.
- Protection against viruses is provided by type I IFNs and NK cells.

Inflammation

Inflammation is a tissue reaction that delivers mediators of host defense—circulating cells and proteins—to sites of infection and tissue damage (Fig. 2.17). The process of inflammation consists of recruitment of cells and leakage of plasma proteins through blood vessels and activation of these cells and proteins in the extravascular tissues. The initial release of histamine, TNF, prostaglandins, and other mediators

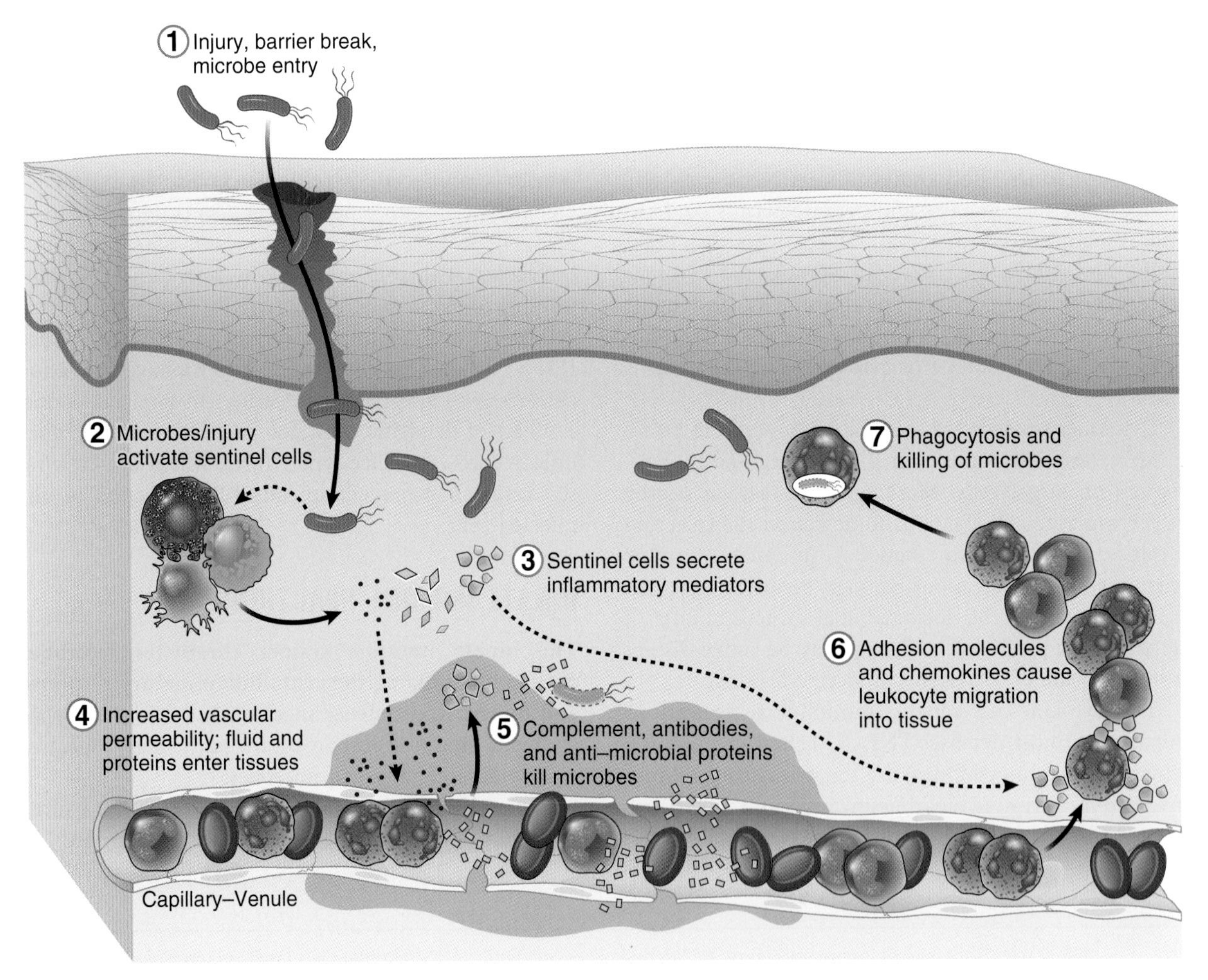

Fig. 2.17 Acute inflammatory response. Cytokines and other mediators are produced by macrophages, dendritic cells, mast cells, and other cells in tissues in response to microbial products and damaged host cells. Some of these mediators (e.g., histamine, prostaglandins) increase the permeability of blood vessels, leading to the entry of plasma proteins (e.g., complement proteins) into the tissues, and others (IL-1, TNF) increase expression of endothelial adhesion molecules and chemokines that promote the movement of leukocytes from the blood into the tissues, where the leukocytes destroy microbes, clear damaged cells, and promote more inflammation and repair.

by mast cells and macrophages causes an increase in local blood flow and exudation of plasma proteins. These contribute to redness, warmth, and swelling, which are characteristic features of acute inflammation. This is often followed by local accumulation in the tissue of phagocytes, mainly neutrophils and blood monocyte–derived macrophages, in response to cytokines, discussed later. Activated phagocytes engulf microbes and necrotic material and destroy these potentially harmful substances. We next describe the cellular events in a typical inflammatory response.

Recruitment of Phagocytes to Sites of Infection and Tissue Damage

Neutrophils and monocytes migrate to extravascular sites of infection or tissue damage by binding to venular endothelial adhesion molecules and in response to chemoattractants produced by tissue cells reacting to infection or injury. Leukocyte migration from the blood into tissues is a multistep process in which initial weak adhesive interactions of the leukocytes with endothelial cells are followed by firm adhesion and then transmigration through the endothelium (Fig. 2.18).

If an infectious microbe breaches an epithelium and enters the subepithelial tissue, resident dendritic cells, macrophages, and other cells recognize the microbe and respond by producing cytokines. Two of these cytokines, TNF and IL-1, act on the endothelium of venules near the site of infection and initiate the sequence of events in leukocyte migration into tissues.

- **Rolling of leukocytes.** In response to TNF and IL-1, venular endothelial cells express an adhesion

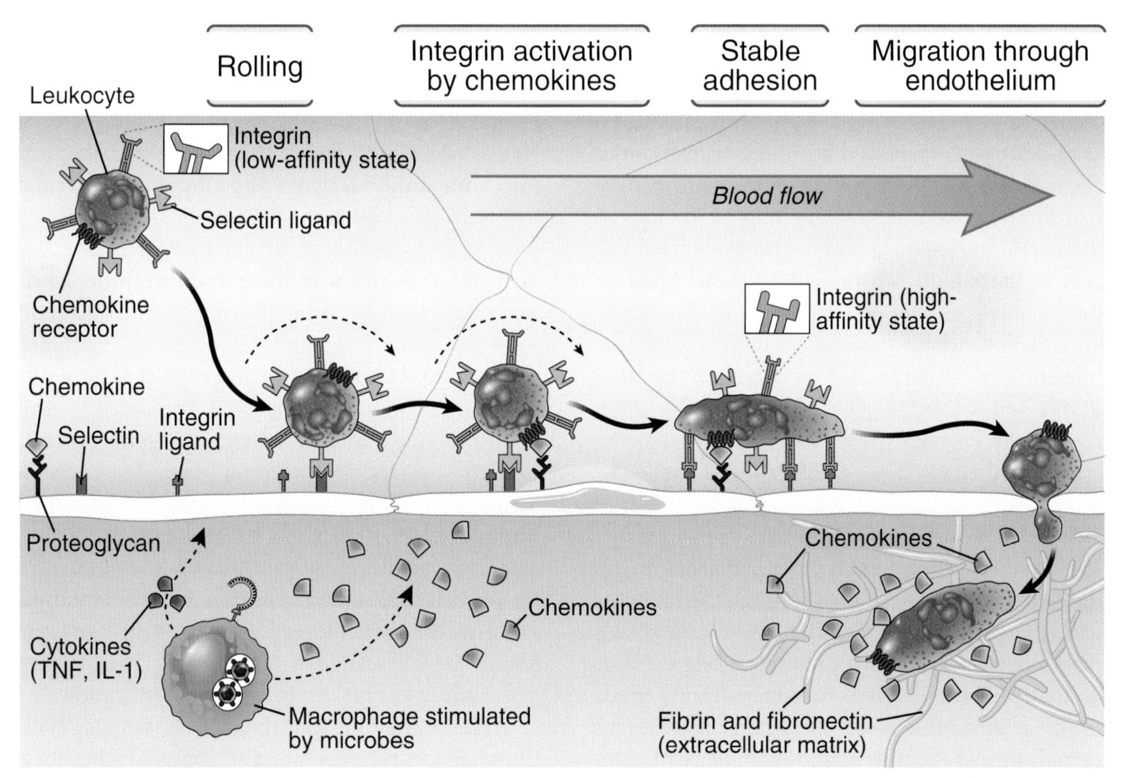

Fig. 2.18 Sequence of events in migration of blood leukocytes to sites of infection. At sites of infection, macrophages, dendritic cells, and other cells that have encountered microbes produce cytokines such as tumor necrosis factor (TNF) and interleukin-1 (IL-1) that activate the endothelial cells of nearby venules to express selectins and ligands for integrins and to secrete chemokines. Selectins mediate weak tethering and rolling of blood neutrophils on the endothelium, integrins mediate firm adhesion of neutrophils, and chemokines activate the neutrophils and stimulate their migration through the endothelium to the site of infection. Blood monocytes and activated T lymphocytes use the same mechanisms to migrate to sites of infection.

molecule of the **selectin** family called E-selectin. Other stimuli, including thrombin, cause rapid translocation of P-selectin to the endothelial surface. (The term *selectin* refers to the carbohydrate-binding, or lectin, property of these molecules.) Circulating neutrophils and monocytes express surface carbohydrates that bind specifically to the selectins. The neutrophils become tethered to the endothelium, flowing blood disrupts this binding, the bonds reform downstream, and this repetitive process results in the rolling of the leukocytes along the endothelial surface. Rolling slows down the leukocytes enough that they can interact with chemokines and other adhesion molecules in the next steps.

- **Firm adhesion.** Leukocytes express another set of adhesion molecules that are called **integrins** because they integrate extrinsic signals into cytoskeletal alterations. Leukocyte integrins, such as LFA-1 and VLA-4, are present in a low-affinity state on unactivated cells. Within a site of infection, tissue macrophages and other cells produce **chemokines**, which bind to proteoglycans on the luminal surface of endothelial cells and are thus displayed at a high concentration to the leukocytes that are rolling on the endothelium. These immobilized chemokines bind to chemokine receptors on the leukocytes and stimulate a rapid increase in the affinity of the leukocyte integrins for their ligands on the endothelium. Concurrently, TNF and IL-1 act on the endothelium to stimulate expression of ligands for integrins, including ICAM-1 and VCAM-1. The firm binding of integrins to their ligands arrests the rolling leukocytes on the endothelium. The cytoskeleton of the leukocytes is reorganized, and the cells spread out on the endothelial surface.
- **Leukocyte migration.** Leukocytes adherent to the endothelium crawl to and then through the junctions between endothelial cells, exiting the blood vessels. Within the tissue, leukocytes migrate along extracellular matrix fibers, directed by concentration gradients of chemoattractants, including chemokines, bacterial formyl peptides, and complement fragments C5a and C3a. The concentrations of these chemoattractants are highest where the microbes are located, and leukocytes have receptors for these molecules that stimulate migration toward their source.

The sequence of selectin-mediated rolling, integrin-mediated firm adhesion, and chemokine-mediated motility leads to the migration of blood leukocytes to an extravascular site of infection within minutes after the infection. (As discussed in Chapters 5 and 6, the same sequence of events is responsible for the migration of activated T lymphocytes into infected tissues.) Inherited deficiencies in integrins and selectin ligands lead to defective leukocyte recruitment to sites of infection and increased susceptibility to infections. These disorders are called **leukocyte adhesion deficiencies** (LADs).

The phagocytes work together with plasma proteins that have entered the site of inflammation, such as complement proteins, to destroy the offending agents. In some infections, such as those with helminthic parasites, eosinophils may be recruited to sites of infection and provide defense against the pathogens.

Phagocytosis and Destruction of Microbes

Neutrophils and macrophages ingest (phagocytose) microbes and destroy the ingested microbes in intracellular vesicles (Fig. 2.19). Phagocytosis is a process of ingestion of particles larger than 0.5 μm in diameter. It begins with membrane receptors binding to the microbe. The principal phagocytic receptors are some pattern recognition receptors, such as mannose receptors and other lectins, and receptors for antibodies and complement. Microbes that are coated (opsonized) with antibodies and complement fragments can bind avidly to specific receptors on phagocytes, resulting in greatly enhanced internalization (see Chapter 8). Binding of the microbe to the cell is followed by extension of the phagocyte plasma membrane around the particle. The membrane then closes up and pinches off, and the microbe is internalized in a membrane-bound vesicle, called a phagosome. The phagosomes fuse with lysosomes to form phagolysosomes.

At the same time that the microbe is being bound by the phagocyte's receptors and ingested, the phagocyte receives signals from various receptors that activate several enzymes. One of these enzymes, called phagocyte oxidase, is abundant in neutrophils, rapidly

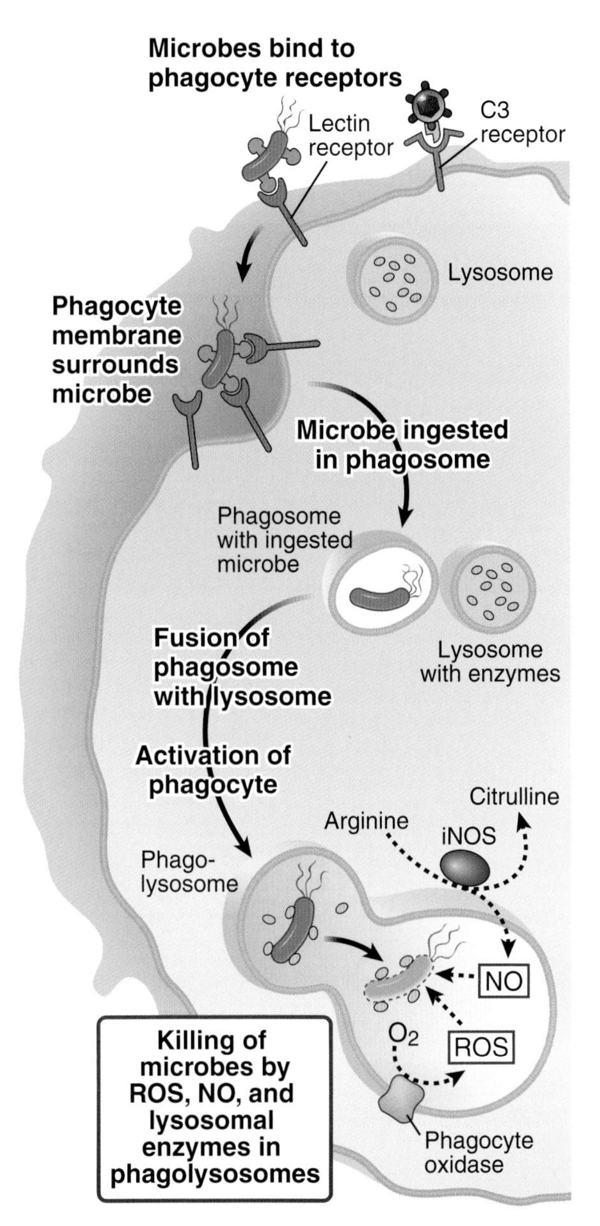

Fig. 2.19 Phagocytosis and intracellular killing of microbes. Macrophages and neutrophils express many surface receptors that may bind microbes for subsequent phagocytosis; select examples of such receptors are shown. Microbes are ingested into phagosomes, which fuse with lysosomes, and the microbes are killed by enzymes and several toxic substances produced in the phagolysosomes. The same substances may be released from the phagocytes and may kill extracellular microbes (not shown). *iNOS,* Inducible nitric oxide synthase; *NO,* nitric oxide; *ROS,* reactive oxygen species.

assembles in the phagolysomal membrane, and converts molecular oxygen into superoxide anion and free radicals, a process called the oxidative burst (or respiratory burst). These free radicals are called **reactive oxygen species** (ROS) and are toxic to the ingested microbes. A second enzyme, inducible nitric oxide synthase (iNOS), is produced mainly in macrophages and catalyzes the conversion of arginine to **nitric oxide** (NO), also a microbicidal substance. A third set of enzymes, the lysosomal proteases, break down microbial proteins. All these microbicidal substances are produced mainly within lysosomes and phagolysosomes, where they act on the ingested microbes but do not damage the phagocytes.

In addition to intracellular killing, neutrophils use additional mechanisms to destroy microbes. They can release microbicidal granule contents into the extracellular environment. In response to pathogens and inflammatory mediators, neutrophils die, and during this process they extrude their nuclear contents to form networks of chromatin called neutrophil extracellular traps (NETs), which contain antimicrobial substances that are normally sequestered in neutrophil granules. These NETs trap bacteria and fungi and kill the organisms. In some cases, the enzymes and ROS that are liberated into the extracellular space may injure host tissues. This is the reason why inflammation, normally a protective host response to infections, may cause tissue injury as well.

Inherited deficiency of the phagocyte oxidase enzyme is the cause of an immunodeficiency disorder called **chronic granulomatous disease** (CGD). In CGD, neutrophils are unable to eradicate intracellular microbes, and the host tries to contain the infection by calling in more macrophages, resulting in collections of activated macrophages around the microbes called granulomas.

Tissue Repair

In addition to eliminating pathogenic microbes and damaged cells, cells of the immune system initiate the process of tissue repair. Macrophages, especially of the alternatively activated type, produce growth factors that stimulate the proliferation of residual tissue cells and fibroblasts, resulting in regeneration of the tissue and scarring of what cannot be regenerated. Other immune cells, such as helper T cells and ILCs, may serve similar roles.

Antiviral Defense

Defense against viruses is a special type of host response that involves IFNs, NK cells, and other mechanisms, which may occur concomitantly with, but are distinct from, inflammation.

Type I IFNs inhibit viral replication and induce an antiviral state in which cells become resistant to productive infection. Type I IFNs, which include several forms of IFN-α and one of IFN-β, are secreted by many cell types infected by viruses. A major source of these cytokines is a type of dendritic cell called the plasmacytoid dendritic cell (so named because these cells morphologically resemble plasma cells), which secretes type I IFNs in response to recognition of viral nucleic acids by TLRs, RLRs, and other pattern recognition receptors. When type I IFNs secreted from dendritic cells or other infected cells bind to the type I IFN receptor on the infected or adjacent uninfected cells, signaling pathways are activated that inhibit viral replication and destroy viral genomes (Fig. 2.20). This action is the basis for the use of IFN-α to treat some forms of chronic viral hepatitis. Inherited or acquired deficiency of type I IFN production or signaling is associated with severe cases of COVID-19, highlighting the importance of this cytokine in combating viral infections.

Virus-infected cells may be destroyed by NK cells, as described earlier. Type I IFNs enhance the ability of NK cells to kill infected cells. Recognition of viral DNA by CDSs also induces autophagy, by which cellular organelles containing viruses are engulfed by lysosomes and proteolytically destroyed (see Fig. 2.7). In addition, part of the innate response to viral infections includes increased apoptosis of infected cells, which also helps to eliminate the reservoir of infection.

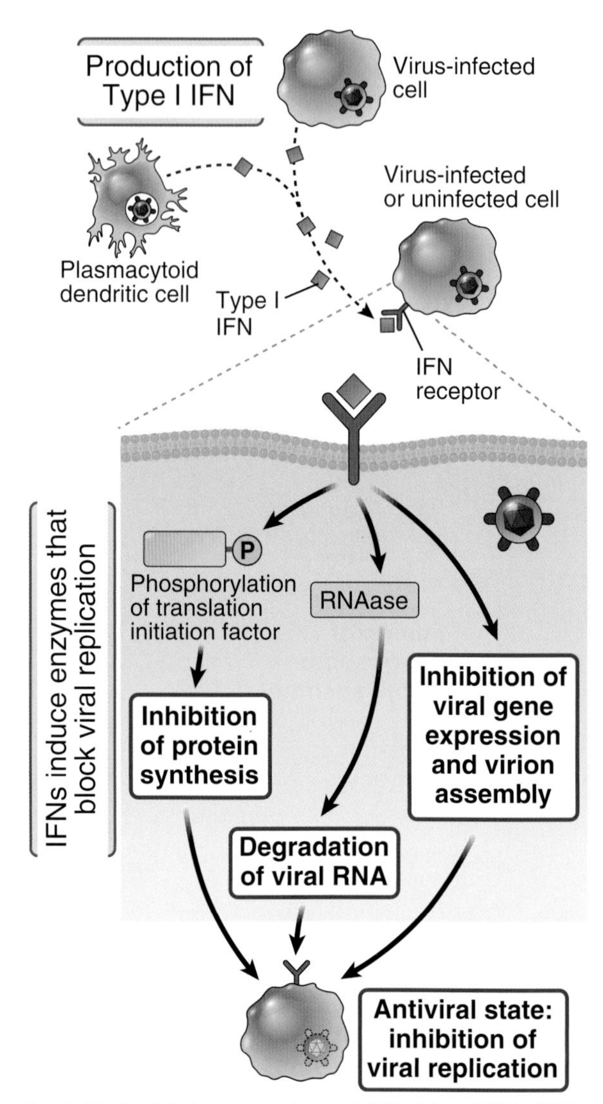

Fig. 2.20 Antiviral actions of type I IFNs. Type I IFNs (IFN-α, IFN-β) are produced by plasmacytoid dendritic cells and virus-infected cells in response to intracellular TLR signaling and other sensors of viral nucleic acids. Type I interferons bind to receptors on the infected and uninfected cells and activate signaling pathways that induce expression of enzymes that interfere with viral replication at different steps, including inhibition of viral protein translation, increasing viral RNA degradation, and inhibition of viral gene expression and virion assembly. Type I IFNs also increase the infected cell's susceptibility to CTL-mediated killing (not shown).

Regulation of Innate Immune Responses

Innate immune responses are regulated by a variety of mechanisms that have evolved to prevent excessive damage to tissues. These regulatory mechanisms include the production of anti-inflammatory cytokines by macrophages and dendritic cells, including IL-10, which inhibits the microbicidal and proinflammatory functions of macrophages (the classical pathway of macrophage activation), and IL-1 receptor antagonist, which blocks the actions of IL-1. There are also many feedback mechanisms in which signals that induce proinflammatory cytokine production also induce

Mechanism of immune evasion	Organism (example)	Mechanism
Resistance to phagocytosis	Pneumococci	Capsular polysaccharide inhibits phagocytosis
Resistance to reactive oxygen intermediates in phagocytes	Staphylococci	Production of catalase, which breaks down reactive oxygen intermediates
Resistance to complement activation (alternative pathway)	*Neisseria meningitidis*	Sialic acid expression inhibits C3 and C5 convertases
	Streptococci	M protein blocks C3 binding to organism, and C3b binding to complement receptors
Resistance to antimicrobial peptide antibiotics	*Pseudomonas*	Synthesis of modified LPS that resists action of peptide antibiotics
Evasion of recognition by viral RNA sensors	Coronaviruses	Chemical modifications of viral RNA

Fig. 2.21 Evasion of innate immunity by microbes. Selected examples of the mechanisms by which microbes may evade or resist innate immunity. *LPS,* Lipopolysaccharide.

expression of inhibitors of cytokine signaling. For example, TLR signaling stimulates expression of proteins called suppressors of cytokine signaling (SOCS), which block the responses of cells to various cytokines, including IFNs. Inflammasome activation is tightly controlled by posttranslational modifications such as ubiquitination and phosphorylation, which block inflammasome assembly or activation, and some microRNAs, which inhibit NLRP3 messenger RNA.

Microbial Evasion of Innate Immunity

Pathogenic microbes have evolved to resist the mechanisms of innate immunity and are thus able to enter and colonize their hosts (Fig. 2.21). Some intracellular bacteria resist destruction inside phagocytes. *Listeria monocytogenes* produces a protein that enables it to escape from phagocytic vesicles and enter the cytoplasm of infected cells, where it is no longer susceptible to ROS or NO (which are produced mainly in phagolysosomes). The cell walls of mycobacteria contain a lipid that inhibits fusion of phagosomes containing ingested bacteria with lysosomes. Other microbes have cell walls that are resistant to the actions of complement proteins. Several viruses encode proteins that block the induction of type I IFNs by infected cells or the signals induced by the type I IFN receptor, and thus these viruses evade the host antiviral state.

ROLE OF INNATE IMMUNITY IN STIMULATING ADAPTIVE IMMUNE RESPONSES

Thus far, we have focused on how the innate immune system recognizes microbes and combats infections. We mentioned at the beginning of this chapter that, in addition to its roles in host defense, the innate immune response to microbes serves an important warning function by alerting the adaptive immune system that an effective immune response is needed. In this final section, we summarize some of the mechanisms by which innate immune responses stimulate adaptive immune responses.

Innate immune responses generate molecules that provide signals, in addition to antigens, that are required to activate naive T and B lymphocytes. In

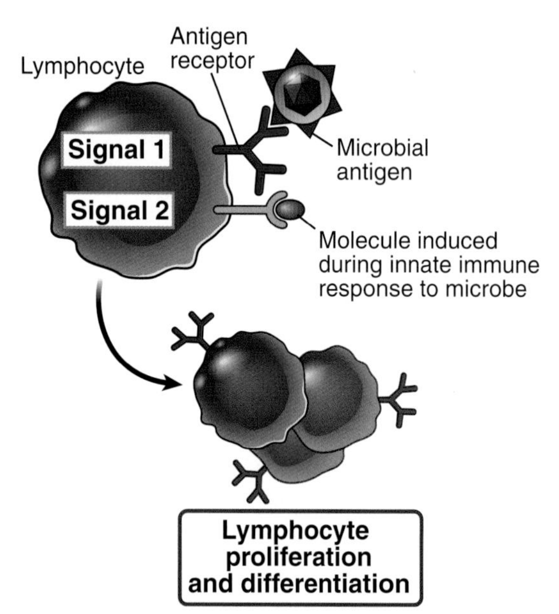

Fig. 2.22 Two-signal requirement for lymphocyte activation. Antigen recognition by lymphocytes provides signal 1 for activation of the lymphocytes, and substances produced during innate immune responses to microbes (or components of microbes) provide signal 2. In this illustration, the lymphocytes could be T cells or B cells. By convention, the major second signals for T cells are called costimulators because they function together with antigens to stimulate the cells. The nature of second signals for T and B lymphocytes is described further in later chapters.

Chapter 1, we introduced the concept that full activation of antigen-specific lymphocytes requires two signals. Antigen may be referred to as signal 1, and innate immune responses to microbes and to host cells damaged by microbes may provide signal 2 (Fig. 2.22). The stimuli that warn the adaptive immune system that it needs to respond have also been called danger signals. This requirement for microbe-dependent second signals ensures that lymphocytes respond to infectious pathogens and not to harmless, noninfectious substances. In experimental situations or for vaccination, adaptive immune responses may be induced by antigens without microbes. In all such instances, the antigens need to be administered with substances called adjuvants that elicit the same innate immune reactions as microbes do. In fact, many potent adjuvants are the products of microbes. The nature and mechanisms of action of second signals are described in the discussion of the activation of T and B lymphocytes in Chapters 5 and 7, respectively. Here we describe two illustrative examples of second signals that are generated during innate immune reactions.

In infected tissues, microbes activate dendritic cells and macrophages to increase their expression of surface molecules called **costimulators**, which bind to receptors on naive T cells and function together with antigen recognition to activate the T cells. The best-defined costimulators for T cells are called B7-1 (CD80) and B7-2 (CD86); these are discussed in Chapter 5.

Blood-borne microbes activate the complement system by the alternative pathway. One of the proteins produced during complement activation by proteolysis of C3b, called C3d, becomes covalently attached to the microbe. At the same time that B lymphocytes recognize microbial antigens by their antigen receptors, the B cells recognize the C3d bound to the microbe by a receptor for C3d. The combination of antigen recognition and C3d recognition initiates the process of B cell differentiation into antibody-secreting cells. Thus, a complement product serves as the second signal for humoral immune responses.

These examples illustrate an important feature of second signals: these signals not only stimulate adaptive immunity but also guide the nature of the adaptive immune response. Intracellular and phagocytosed microbes need to be eliminated by cell-mediated immunity, the adaptive response mediated by T lymphocytes. Microbes that are encountered and ingested by dendritic cells or macrophages induce the second signals—that is, costimulators—that stimulate T cell responses. By contrast, blood-borne microbes need to be combated by antibodies, which are produced by B lymphocytes during humoral immune responses. Blood-borne microbes activate the plasma complement system, which in turn stimulates B cell activation and antibody production. Thus, different types of microbes induce innate immune responses that stimulate the types of adaptive immunity that are best able to combat different infectious pathogens.

SUMMARY

- All multicellular organisms have intrinsic mechanisms of defense against infections, which constitute innate immunity.
- The innate immune system uses germline-encoded membrane and cytosolic receptors and extracellular circulating molecules to sense and respond to structures that are characteristic of various classes of microbes and also molecules released by damaged host cells. Innate immune reactions usually are not enhanced by repeat exposures to microbes.
- The principal responses of innate immunity are inflammation and the antiviral state.
- Toll-like receptors (TLRs), expressed on plasma membranes and endosomal membranes of many cell types, are a major class of innate immune system receptors that recognize different microbial products, including bacterial cell wall constituents and microbial nucleic acids, and generate signals that activate inflammatory and/or antiviral responses.
- Several types of innate immune sensors respond to microbial molecules in the cytosol and stimulate inflammatory and antiviral responses, including members of the NOD-like receptor (NLR) family that recognize microbial cell wall lipoproteins, RIG-like receptors that recognize viral RNAs, and cytosolic DNA sensors that detect dsDNA.
- Inflammasomes are multiprotein complexes in the cytosol that assemble in response to products of damaged cells and cytosolic changes typical of infection or cell injury, and generate the active form the proinflammatory cytokine interleukin-1β (IL-1β).
- The principal components of innate immunity are epithelial barrier cells in the skin, gastrointestinal tract, and respiratory tract; phagocytes; dendritic cells; mast cells; natural killer cells; cytokines; and plasma proteins, including the proteins of the complement system.
- Epithelia provide physical barriers against microbes; produce antimicrobial peptides, including defensins and cathelicidins; and contain lymphocytes that may prevent infections.
- The principal phagocytes—neutrophils and monocytes/macrophages—are blood cells that are recruited to sites of infection; macrophages also may be tissue-resident. Phagocytes are activated by engagement of different receptors and destroy microbes and dead cells. Some macrophages limit inflammation and initiate tissue repair.
- Innate lymphoid cells (ILCs) secrete various cytokines that induce inflammation. Natural killer (NK) cells kill host cells infected by intracellular microbes and produce the cytokine interferon-γ (IFN- γ), which activates macrophages to kill phagocytosed microbes.
- The complement system is a family of proteins that are activated on encounter with some microbes (in innate immunity) and by antibodies (in the humoral arm of adaptive immunity). Complement proteins coat (opsonize) microbes for phagocytosis, stimulate inflammation, and lyse microbes.
- Cytokines of innate immunity function to stimulate inflammation (tumor necrosis factor [TNF], interleukin-1 [IL-1], IL-6, chemokines), activate NK cells (IL-12), activate macrophages (IFN-γ), and prevent viral infections (type I IFNs).
- In inflammation, phagocytes are recruited from the circulation to sites of infection and tissue damage. The cells bind to endothelial adhesion molecules that are induced by the cytokines TNF and IL-1 and migrate in response to soluble chemoattractants, including chemokines, complement fragments, and bacterial peptides. The leukocytes are activated, and they ingest and destroy microbes and damaged cells.
- Antiviral defense is mediated by type I IFNs, which inhibit viral replication, and by NK cells, which kill infected cells.
- In addition to providing early defense against infections, innate immune responses provide signals that work together with antigens to activate B and T lymphocytes. The requirement for these second signals ensures that adaptive immunity is elicited by microbes (the most potent inducers of innate immune reactions) and not by nonmicrobial substances.

REVIEW QUESTIONS

1. How does the specificity of innate immunity differ from that of adaptive immunity?
2. What are examples of microbial substances recognized by the innate immune system, and what are the receptors for these substances?
3. What is the inflammasome, and how is it stimulated?
4. What are the mechanisms by which the epithelia of the skin and gastrointestinal tract prevent the entry of microbes?
5. How do phagocytes ingest and kill microbes?
6. What is the role of MHC molecules in the recognition of infected cells by NK cells, and what is the physiologic significance of this recognition?
7. What are the roles of the cytokines TNF, IL-1, IL-12, and type I IFNs in defense against infections?
8. How do innate immune responses enhance adaptive immunity?

Answers to and discussion of the Review Questions may be found on p. 322.

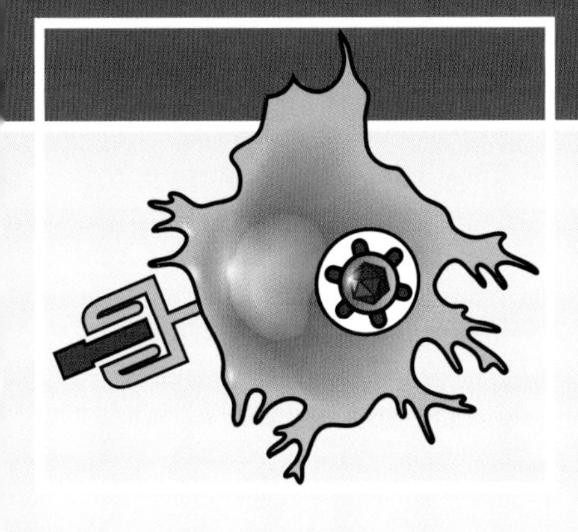

3

Antigen Presentation to T Lymphocytes and the Functions of Major Histocompatibility Complex Molecules

What T Lymphocytes See

CHAPTER OUTLINE

Adaptive immune responses are initiated by the recognition of antigens by antigen receptors on lymphocytes. B and T lymphocytes differ in the types of antigens they recognize. The antigen receptors of B lymphocytes, which are plasma membrane-bound antibodies, can recognize structural features of a variety of macromolecules (proteins, polysaccharides, lipids, nucleic acids), in soluble form or on cell surfaces, as well as small molecules. Therefore, B cell–mediated humoral immune responses may be generated against many types of microbial surface structures, released internal molecules, and other soluble antigens. The antigen receptors of most T lymphocytes, on the other hand, can see only peptide fragments of protein antigens, and only when these peptides are displayed on host cell surfaces bound to specialized proteins called major histocompatibility complex (MHC) molecules. Because the association of antigenic peptides and MHC molecules occurs inside cells, T cell–mediated immune responses may be generated only against protein antigens that are either produced in or taken up by host cells. Thus, T cells detect the presence of intracellular foreign antigens, the first step in cell-mediated immunity. The peptide display function of the MHC allows for surface display of parts of microbes that get inside host cells, and this enables T cells to detect the infected cells amidst a fog of cell-free microbial antigens. This chapter focuses on the nature of the antigens that are recognized by T lymphocytes; antigen recognition by B cells is discussed in Chapter 7. Chapter 4 describes the receptors used by lymphocytes to recognize and respond to these antigens.

The induction of immune responses by antigens is a highly efficient process with a number of remarkable features. The number of naive lymphocytes specific for any one antigen is very low, as few as 1 in 10^5 or 10^6 circulating lymphocytes, and this small fraction of the body's lymphocytes needs to locate and react rapidly to the antigen, wherever it is introduced. Moreover, different types of T lymphocytes are required to defend against different types of microbes. In fact, the immune system has to react in different ways even to the same microbe at different stages of the microbe's life cycle. For example, defense against a microbe (e.g., a virus) that has entered the bloodstream depends on antibodies that bind the microbe, prevent it from infecting host cells, and help to eliminate it. The production of potent antibodies requires the activation of $CD4^+$ helper T cells. Many microbes infect host cells, however, where they are safe from antibodies, which cannot enter cells. As a result, activation of $CD8^+$ cytotoxic T lymphocytes (CTLs) may be necessary to kill the infected cells and eliminate the reservoir of infection. Thus, we are faced with two important questions:

- How do the rare naive lymphocytes specific for any microbial antigen find that microbe, especially considering that microbes may enter anywhere in the body?
- How do different types of T cells recognize microbes in different cellular compartments?

The answer to both questions is that the immune system has developed a specialized system for capturing and displaying antigens to lymphocytes. Research by immunologists, cell biologists, and biochemists has led to a sophisticated understanding of how protein antigens are captured, broken down, and displayed for recognition by T lymphocytes, and the role of MHC molecules in this process.

ANTIGENS RECOGNIZED BY T LYMPHOCYTES

The majority of T lymphocytes recognize peptide antigens that are bound to and displayed by the MHC molecules of antigen-presenting cells (APCs). The MHC is a genetic locus whose principal protein products function as the peptide display molecules of the immune system. $CD4^+$ and $CD8^+$ T cells can see peptides only when these peptides are displayed by that individual's MHC molecules. This property of T cells is called MHC restriction. The T cell receptor (TCR) recognizes some amino acid residues of the peptide antigen and simultaneously also recognizes residues of the MHC molecule that is displaying that peptide (Fig. 3.1). Each TCR, and hence each clone of $CD4^+$ or $CD8^+$ T cells, recognizes one peptide displayed by one of the many different MHC molecules in every individual. The properties of MHC molecules and the significance of MHC restriction are described later in this chapter. How we generate T cells that recognize peptides presented only by self MHC molecules is described in Chapter 4. Also, some small populations of T cells recognize lipid and other nonpeptide antigens either presented by class I MHC–like molecules or without a requirement for a specialized antigen display system.

The cells that capture microbial antigens and display them for recognition by T lymphocytes are called **antigen-presenting cells** (APCs). Naive T lymphocytes see protein antigens presented by dendritic cells to initiate clonal expansion and differentiation of the T cells into effector and memory cells. Differentiated effector T cells again need to see antigens, which may be presented by various kinds of APCs in addition to dendritic cells, to activate the effector functions of the T cells in both humoral and cell-mediated immune

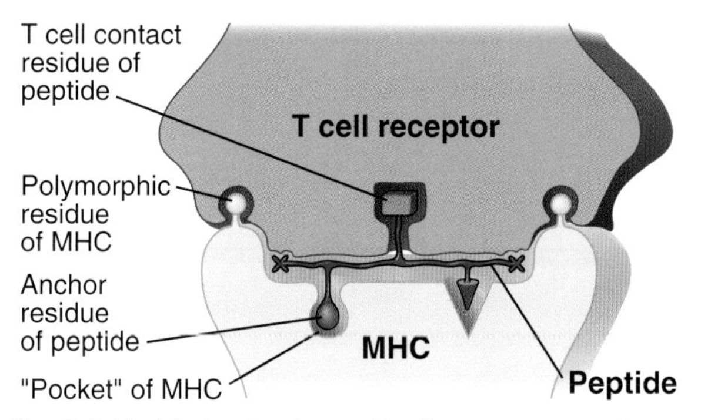

Fig. 3.1 Model showing how a T cell receptor recognizes a complex of peptide antigen displayed by an MHC molecule. MHC molecules are expressed on antigen-presenting cells and function to display peptides derived from protein antigens. Peptides bind to the MHC molecules by anchor residues, which attach the peptides to pockets in the MHC molecules. The antigen receptor of every T cell recognizes some amino acid residues of the peptide and some (polymorphic) residues of the MHC molecule.

responses. We first describe how APCs capture and present antigens to trigger immune responses and then examine the role of MHC molecules in antigen presentation to T cells.

CAPTURE OF PROTEIN ANTIGENS BY ANTIGEN-PRESENTING CELLS

Protein antigens of microbes that enter the body are captured mainly by dendritic cells and concentrated in the secondary (peripheral) lymphoid organs, where immune responses are initiated (Fig. 3.2). Microbes usually enter the body through the skin (by contact), the gastrointestinal tract (by ingestion), the respiratory tract (by inhalation), and the genitourinary tract (by sexual contact). Some microbes may enter the bloodstream. Microbial antigens can also be produced in any infected tissue. Because of the vast surface area of the epithelial barriers and the large volume of blood, connective tissues, and internal organs, it would be impossible for lymphocytes of all possible specificities to efficiently patrol all these sites searching for foreign invaders; instead, antigens are taken to the lymphoid organs through which lymphocytes recirculate.

Antigens are taken to secondary lymphoid organs in two ways.

- Dendritic cells in epithelia, connective tissues, and organs capture microbial antigens and transport them to lymph nodes draining these tissues. This process involves a series of events following the encounter of dendritic cells with microbes—ingestion of antigens, activation of the dendritic cells, migration of the antigen-carrying cells to lymph nodes, and display of the antigen to T cells. These steps are described below.
- Microbes or their antigens also may be carried to lymph nodes in the lymph, or to the spleen in the blood, and are then captured by resident dendritic cells in these secondary lymphoid organs and presented to T cells.

All the interfaces between the body and the external environment are lined by continuous epithelia, which provide barriers to infection. The epithelia and subepithelial tissues contain a network of cells with long processes, called **dendritic cells**; these cells are also present in the T cell–rich areas of secondary lymphoid organs and, in smaller numbers, in most other nonlymphoid organs (Fig. 3.3). There are two major populations of dendritic cells, called conventional (or classical) and plasmacytoid, which differ in their locations and responses. The majority of dendritic cells in tissues and lymphoid organs belong to the conventional subset; these are the cells that capture and present most protein antigens to T lymphocytes. In the skin, the epidermal dendritic cells are called Langerhans cells. Plasmacytoid dendritic cells are named because of their morphologic resemblance to plasma cells; they are present in the blood and tissues. Plasmacytoid dendritic cells are also major source of type I interferons in innate immune responses to viral infections (see Chapter 2).

Dendritic cells use various membrane receptors to bind microbes. These microbes or their antigens are taken up by dendritic cells by phagocytosis or receptor-mediated endocytosis. At the same time that the dendritic cells are capturing antigens, products of the microbes stimulate innate immune reactions by binding to Toll-like receptors (TLRs) and to other innate pattern-recognition receptors in the dendritic cells, tissue epithelial cells, and tissue-resident macrophages (see Chapter 2). This results in the production of inflammatory cytokines such as tumor necrosis factor (TNF) and interleukin-1 (IL-1). The combination of innate receptor signaling and cytokines activates the dendritic cells, resulting in several changes in their phenotype, migration, and function.

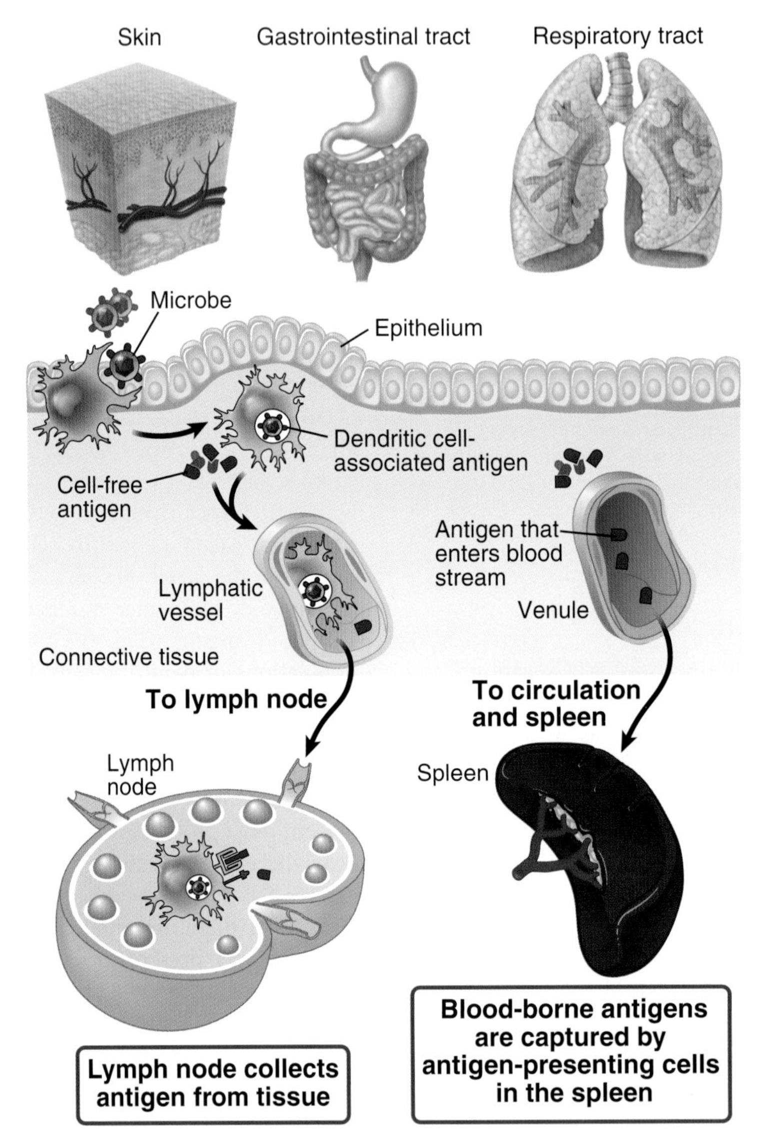

Fig. 3.2 Capture and display of microbial antigens. Microbes enter through an epithelial barrier and are captured by dendritic cells resident in the tissue, or microbes enter lymphatic vessels or blood vessels. The microbes and their antigens are transported to secondary lymphoid organs, the lymph nodes and the spleen, where peptide fragments of protein antigens are displayed by dendritic cell (MHC) molecules for recognition by naive T lymphocytes.

Upon activation, conventional dendritic cells lose their adhesiveness for epithelia and peripheral tissues and begin to express the chemokine receptor CCR7, which is specific for chemoattracting cytokines (chemokines) produced by lymphatic endothelium and by stromal cells in the T cell zones of lymph nodes. These chemokines direct the dendritic cells to exit the epithelium and migrate through lymphatic vessels to the lymph nodes draining that epithelium (Fig. 3.4). During the process of migration, the dendritic cells mature from cells designed to capture antigens into APCs capable of stimulating naive T lymphocytes. This maturation is reflected by increased synthesis and stable expression of MHC molecules, which display antigens to T cells, and also of costimulators,

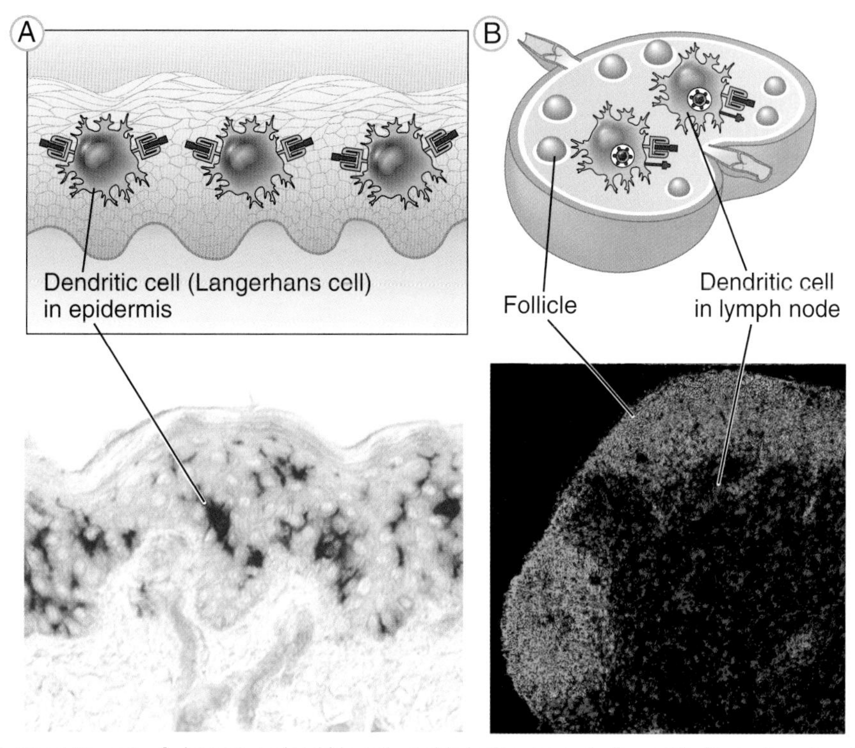

Fig. 3.3 Dendritic cells. A, Immature dendritic cells reside in tissues, including epithelia, such as the skin, and form a network of cells with interdigitating processes, seen as blue cells on the section of skin stained with an antibody that recognizes dendritic cells. **B,** Mature dendritic cells reside in the T cell–rich areas of lymph nodes (and spleen, not shown) and are seen in the section of a lymph node stained with fluorochrome-conjugated antibodies against dendritic cells *(red)* and B cells in follicles *(green)*. Note that the dendritic cells are in the same regions of the lymph node as T cells (see Fig. 1.18B). (A, Micrograph of skin courtesy Dr. Y.-J. Liu, MD Anderson Cancer Center, Houston, TX. B, Courtesy Drs. Kathryn Pape and Jennifer Walter, University of Minnesota Medical School, Minneapolis, MN.)

which were introduced in Chapter 2 as molecules required for the effective induction of T cell responses.

The net result of this sequence of events is that the protein antigens of microbes that enter the body are transported to and concentrated in the regions of the lymph nodes (and spleen) where the antigens are most likely to encounter T lymphocytes. Recall that naive T lymphocytes continuously recirculate through lymph nodes and also express CCR7, which promotes their entry into the T cell zones of lymph nodes (see Chapter 1). Therefore, dendritic cells bearing captured antigen and naive T cells poised to recognize antigens come together in lymph nodes. This process is remarkably efficient; it is estimated that if a microbial antigen is introduced at any site in the body, a T cell response to the antigen begins in the lymph nodes draining that site within 12 to 18 hours.

Different types of APCs serve distinct functions in T cell–dependent immune responses (Fig. 3.5).

- Dendritic cells are the principal inducers of T-dependent responses because these cells are located at sites of microbe entry, are capable of migrating to lymph nodes through which naive T cells circulate, and are the most potent APCs for activating naive T lymphocytes because they express high levels of MHC molecules and costimulators.
- One important type of APC for effector T cells, especially of the helper T cell lineage, is the macrophage, including macrophages resident in all tissues and monocyte-derived macrophages, which accumulate at sites of infection in innate immune responses. In cell-mediated immune reactions, macrophages phagocytose microbes and display the antigens of

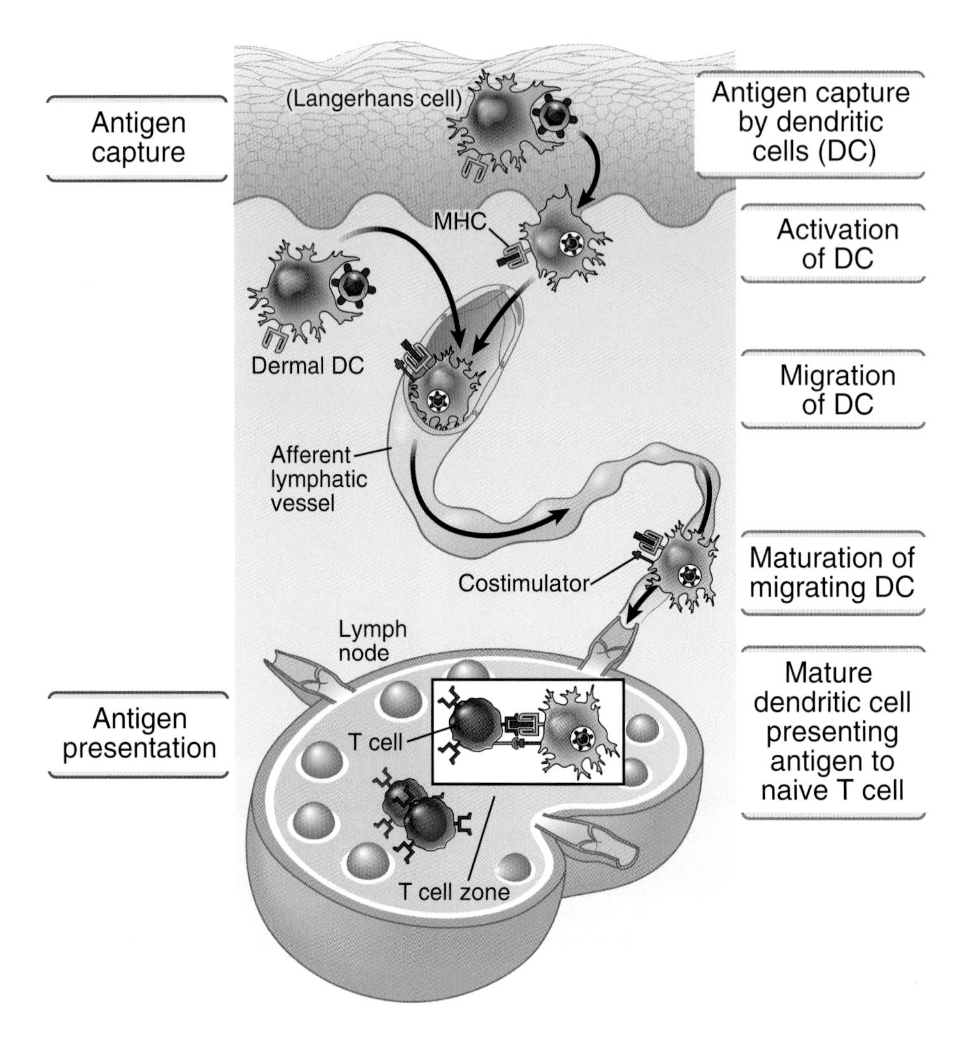

Fig. 3.4 Capture, transport, and presentation of protein antigens by dendritic cells. Immature dendritic cells in epithelial barrier tissues, such as the epithelium or dermis of the skin, shown here, capture microbial antigens, are activated, express CCR7 (not shown), and leave the epithelium. The dendritic cells migrate to draining lymph nodes, being attracted there by chemokines produced in the lymphatic vessels and nodes. In response to signals induced by the microbe, such as Toll-like receptor (TLR) signals and cytokines, the dendritic cells mature and acquire the ability to present antigens to naive T lymphocytes in the lymph nodes. Dendritic cells at different stages of their maturation may express different membrane proteins. Immature dendritic cells express surface receptors that capture microbial antigens, whereas mature dendritic cells express high levels of MHC molecules and costimulators, which function to stimulate T cells.

these microbes to effector T cells, which are then reactivated and activate the macrophages to kill the ingested microbes (see Chapter 6).

- B lymphocytes endocytose protein antigens and display them to helper T cells within lymphoid tissues; this process is important for the development of humoral immune responses to protein antigens (see Chapter 7).
- As discussed later in this chapter, any nucleated cell containing foreign (microbial or tumor) protein antigens in the cytosol can present peptides derived from these antigens to $CD8^+$ effector T cells.

Cell type	Expression of		Principal function
	Class II MHC	Costimulators	
Dendritic cells	Constitutive; increases with maturation; increased by IFN-γ	Constitutive; increases with maturation; increased by TLR ligands, IFN-γ, and T cells (CD40-CD40L interactions)	Antigen presentation to naive T cells in the initiation of T cell responses to protein antigens (priming)
Macrophages	Low or negative; inducible by IFN-γ	Low, inducible by TLR ligands, IFN-γ, and T cells (CD40-CD40L interactions)	Antigen presentation to CD4+ effector T cells in the effector phase of cell-mediated immune responses
B lymphocytes	Constitutive; increased by cytokines (e.g., IL-4)	Induced by T cells (CD40-CD40L interactions), antigen receptor cross-linking	Antigen presentation to CD4+ helper T cells in humoral immune responses (T cell-B cell interactions)

Fig. 3.5 Principal antigen-presenting cells (APCs). The properties of the most important class II major histocompatibility complex (MHC)–expressing APCs, which present antigens to CD4+ helper T cells, are summarized. Other cell types, such as vascular endothelial cells, also express class II MHC, but their roles in initiating immune responses to microbes are not established. In the thymus, epithelial cells express class II MHC molecules and play a role in the maturation and selection of T cells. All nucleated cells can present class I MHC–associated peptides to CD8+ T cells. *IFN-γ,* Interferon-γ; *IL-4,* interleukin-4; *TLR,* Toll-like receptor.

Now that we have described how protein antigens are captured, transported to, and concentrated in secondary lymphoid organs, we next ask how these antigens are displayed to T lymphocytes. To answer this question, we first need to describe the structure of MHC molecules and examine how they function in immune responses.

STRUCTURE AND FUNCTION OF MAJOR HISTOCOMPATIBILITY COMPLEX MOLECULES

MHC genes were discovered from observations of immunologic reactions to cells and tissues transferred between genetically nonidentical animals, before the peptide display functions of the proteins they encoded were known. The MHC was defined as the genetic locus that is the principal determinant of acceptance or rejection of tissue grafts exchanged between individuals (tissue, or histo, compatibility). In other words, individuals who are identical at their MHC locus (inbred animals and identical twins) will accept grafts from one another, and individuals who differ at their MHC loci will reject such grafts. Because transplantation is not a natural phenomenon, MHC genes and the molecules they encode must have evolved to perform other functions. We now know that the physiologic role of MHC molecules is to display peptides derived from microbial protein antigens to antigen-specific T lymphocytes as a first step in protective T cell–mediated immune responses to microbes. This function of MHC molecules explains the phenomenon of MHC restriction of T cells, mentioned earlier.

All vertebrates possess maternally and paternally inherited MHC loci, which include genes encoding the MHC proteins (and other proteins involved in immune responses) (Fig. 3.6). MHC molecules were first identified as proteins encoded by the murine MHC locus involved in graft rejection. They were rediscovered in humans when it was found that women who had multiple pregnancies, or individuals who received multiple blood transfusions, made antibodies

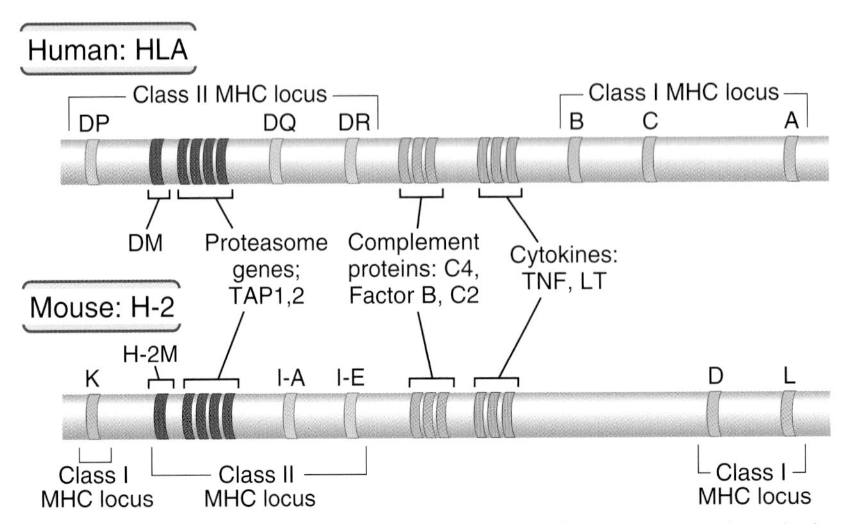

Fig. 3.6 Genes of the major histocompatibility complex (MHC). Schematic maps show the human MHC, called the human leukocyte antigen (HLA) complex, and the mouse MHC, called the H-2 complex, illustrating the major genes that code for molecules involved in immune responses. Sizes of genes and intervening DNA segments are not drawn to scale. Class II genes are shown as single blocks, but each consists of two genes encoding the α and β chains, respectively. The products of some of the genes (DM [H-2M in mice], proteasome components, TAP) are involved in antigen processing. The MHC also contains genes that encode molecules other than peptide display molecules, including some complement proteins and cytokines. *LT*, Lymphotoxin; *TAP*, transporter associated with antigen processing; *TNF*, tumor necrosis factor.

that recognized proteins on the white blood cells (leukocytes) of paternal or donor origin, respectively. These proteins were called **human leukocyte antigens** (HLAs) and were soon shown to be analogous to the MHC molecules identified in mice. (Pregnancy and transfusions expose individuals to cellular antigens of other individuals, so antibodies produced against these cells reflect histoincompatibility, as in the mouse transplantation experiments.) In all vertebrates, the MHC contains two sets of highly polymorphic genes, called the class I and class II MHC genes. (As discussed later, polymorphism refers to the presence of many variants of these genes in the population.) These genes encode the class I and class II MHC molecules that display peptides to T cells. In addition to the polymorphic genes, the MHC contains many nonpolymorphic genes, some of which code for proteins involved in antigen presentation.

Structure of MHC Molecules

Class I and class II MHC molecules are membrane proteins that each contains an extracellular peptide-binding cleft. Although the two classes of molecules differ in subunit composition, they are very similar in overall structure (Fig. 3.7).

Class I MHC Molecules

Each **class I MHC molecule** consists of an α chain noncovalently associated with a protein called β_2-microglobulin that is encoded by a gene outside the MHC locus. The α chain consists of three extracellular domains followed by transmembrane and cytoplasmic domains.

- The amino-terminal α1 and α2 domains of the α chain form two walls and a floor that together make up a peptide-binding cleft, or groove, that can accommodate peptides typically 8 to 11 amino acids long. The floor of the peptide-binding cleft contains amino acid residues that bind peptides for display to T lymphocytes, and the tops of the cleft walls make contact with the T cell receptor (which also contacts part of the displayed peptide; see Fig. 3.1). The polymorphic residues of class I molecules—that is, the amino acids that differ among different individuals' MHC molecules—are located in the α1 and α2 domains of the α chain. Most of these polymorphic residues contribute to

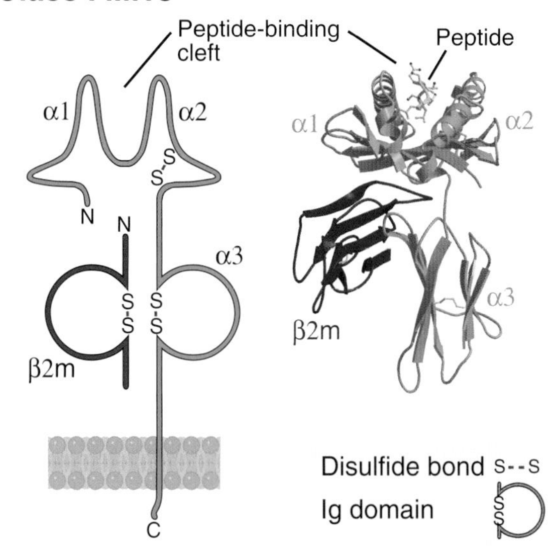

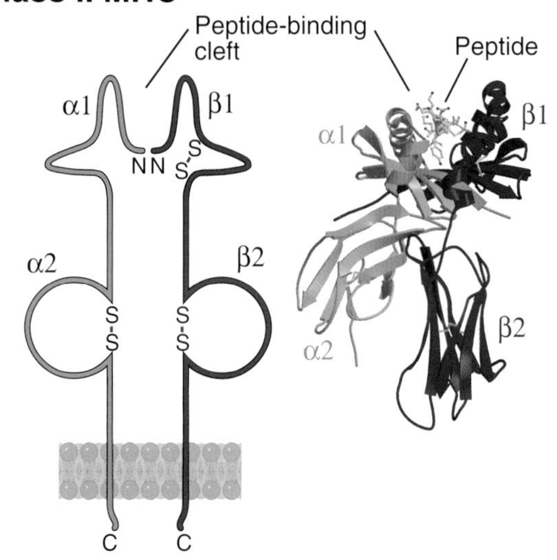

Fig. 3.7 Structure of class I and class II major histocompatibility complex (MHC) molecules. Schematic diagrams *(at left)* and models of the crystal structures *(at right)* of class I MHC and class II MHC molecules illustrate the domains of the molecules and the fundamental similarities between them. Both types of MHC molecules contain peptide-binding clefts and invariant portions that bind CD8 (the α3 domain of class I) or CD4 (the α2 and β2 domains of class II). *β2m*, β_2-microglobulin; *Ig*, Immunoglobulin. Crystal structures courtesy Dr. P. Bjorkman, California Institute of Technology, Pasadena, CA.

variations in the floor of the peptide-binding cleft and thus influence the ability of different MHC molecules to bind distinct sets of peptides.

- The α3 domain is invariant, associates with β2-microglobulin, and contains a site that binds the CD8 T cell coreceptor but not CD4. As discussed in Chapter 5, T cell activation requires recognition of MHC-associated peptide antigen by the TCR and simultaneous recognition of the MHC molecule by the coreceptor. Therefore, $CD8^+$ T cells can respond only to peptides displayed by class I MHC molecules, the MHC molecules to which the CD8 coreceptor binds.

Class II MHC Molecules

Each **class II MHC molecule** consists of two transmembrane chains, called α and β. Each chain has two extracellular domains, followed by the transmembrane and cytoplasmic regions.

- The amino-terminal regions of both chains, called the α1 and β1 domains, contain polymorphic residues and together form a cleft (made up of walls and a floor as in class I MHC molecules) that is large enough to accommodate peptides of 10 to 30 residues.
- The nonpolymorphic α2 and β2 domains contain the binding site for the CD4 T cell coreceptor. Because CD4 binds to class II MHC molecules but not to class I, $CD4^+$ T cells can respond only to peptides presented by class II MHC molecules.

Properties of MHC Genes and Proteins

Several features of MHC genes and proteins are important for the normal function of these molecules (Fig. 3.8):

- **MHC genes are highly polymorphic**, meaning that many different alleles (variants) are present among the different individuals in the population. The total number of different HLA proteins in the population is estimated to be more than 18,000, with about 13,000 class I and 5400 class II polypeptides, making MHC molecules the most polymorphic of all proteins in mammals. The polymorphism of MHC proteins is so great that any two individuals in an outbred population are extremely unlikely to have exactly the same MHC molecules. These

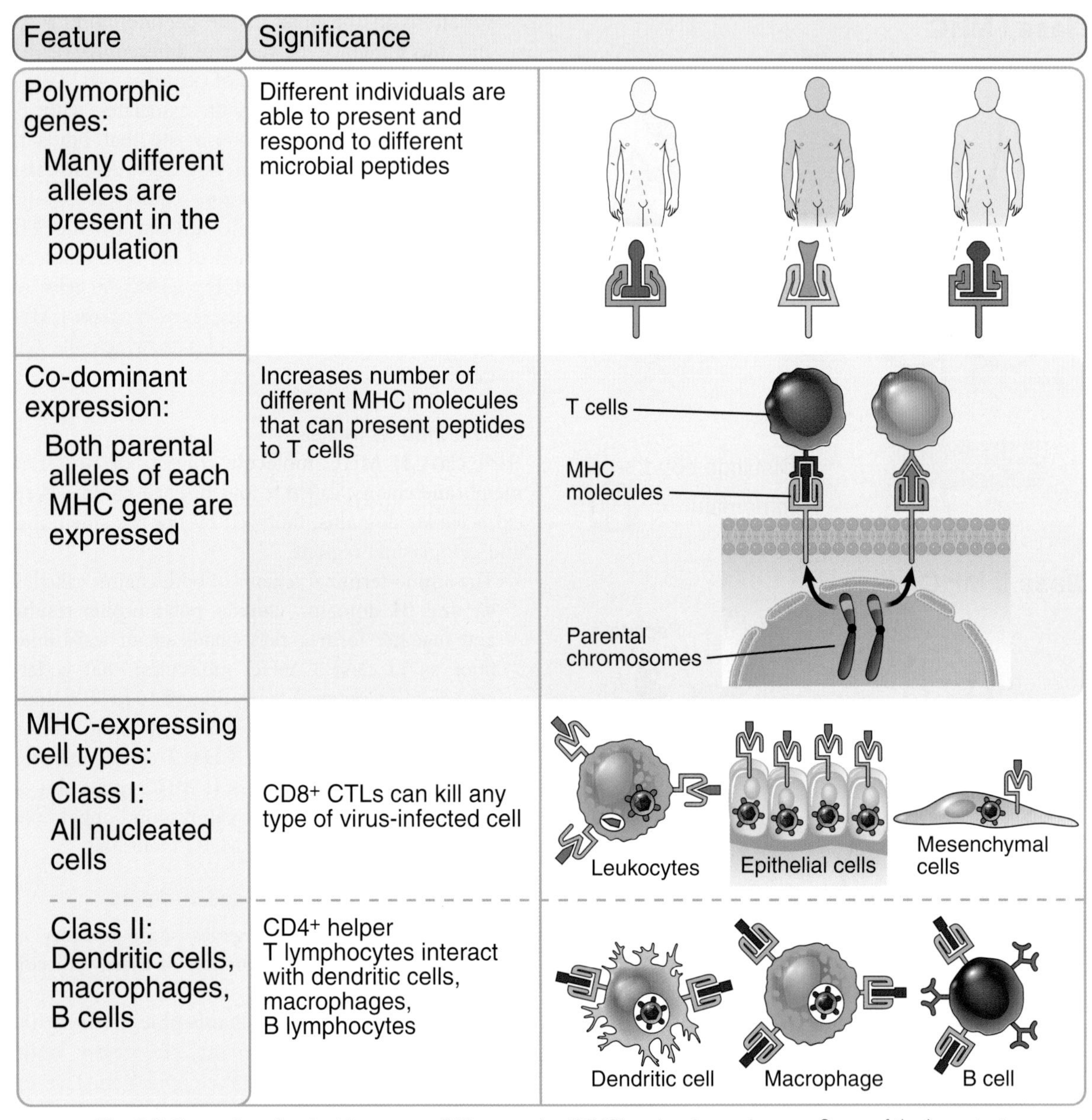

Fig. 3.8 **Properties of major histocompatibility complex (MHC) molecules and genes.** Some of the important features of MHC molecules and their significance for immune responses. *CTLs,* Cytotoxic T lymphocytes.

different polymorphic variants are inherited and not generated de novo in individuals by somatic gene recombination, as are antigen receptors (see Chapter 4). Any one individual inherits and expresses only two alleles of each MHC gene (one from each parent), which represent very few of the many variants in the population. Because the polymorphic residues determine which peptides are presented by specific MHC molecules, the existence of multiple alleles likely ensures that there are always some members of the population who will be able to present some peptide from any microbial protein antigen. Therefore, MHC polymorphism ensures that a population will be able to deal with the diversity of microbes, and at least some individuals will be able to mount effective immune responses to the peptide antigens of these microbes. Furthermore, because of the existence of many different MHC molecules in the population, everyone will not succumb to a newly encountered or mutated microbe due to an inability to present the microbe's proteins to T cells.

- **MHC genes are codominantly expressed, meaning that the alleles inherited from both parents are expressed equally.** Codominant expression maximizes the number of HLA proteins expressed by each individual and thus enables each individual to display a large number of peptides.
- **Class I molecules are expressed on all nucleated cells, but class II molecules are expressed mainly on dendritic cells, macrophages, and B lymphocytes.** The physiologic significance of this strikingly different expression pattern is described later. Class II molecules are also expressed on thymic epithelial cells and endothelial cells and can be induced on other cell types by the cytokine interferon-γ.

Inheritance Patterns and Nomenclature of HLA Genes

Humans have three polymorphic class I genes, called human leukocyte antigen-A *(HLA-A), HLA-B,* and *HLA-C,* and each person inherits one of these genes from each parent, so any cell can express six different class I molecules. In the class II locus, individuals inherit, from each chromosome, separate genes encoding the α and β chains of DP and DQ, the gene for DRα, and variable numbers of genes encoding DRβ (typically 1 to 3). Because of this variation, and because the α chain from one chromosome may pair with the β chain derived from the other chromosome, the number of class II molecules expressed is typically more than six.

The set of MHC genes present on each chromosome (and their encoded MHC proteins) is called an **MHC haplotype**. The genes in an MHC haplotype are tightly linked and inherited together in a Mendelian fashion. Therefore, the chance that two siblings will inherit identical sets of HLA alleles is 25%. This is why siblings are often tested before unrelated individuals for their suitability as donors for transplantation—the chance of finding an HLA match with the recipient is much greater for siblings. In humans, each HLA allele is given a numeric designation. For example, an HLA haplotype of an individual could be HLA-A2, B5, DR3, and so on. In the modern terminology, based on molecular typing, individual alleles may be called HLA-A*0201, referring to the 01 subtype of HLA-A2, or HLA-DRB1*0401, referring to the 01 subtype of the *DR4B1* gene, and so on.

Peptide Binding to MHC Molecules

The peptide-binding clefts of MHC molecules bind peptides derived from protein antigens and display these peptides for recognition by T cells (Fig. 3.9). There are pockets in the floors of the peptide-binding clefts of most MHC molecules. Side chains of some of the amino acid residues in the peptide antigens fit into these MHC pockets and thereby anchor peptides in the cleft; these amino acid residues are called anchor residues. Other residues of the bound peptide project upward and are recognized by the antigen receptors of T cells.

Several features of the interaction of peptide antigens with MHC molecules are important for understanding the peptide display function of MHC molecules (Fig. 3.10):

- Each MHC molecule can present only one peptide at a time, because there is only one peptide-binding cleft, but each MHC molecule is capable of presenting many different peptides. As long as the pockets of the MHC molecule can accommodate the anchor residues of the peptide, that peptide can be displayed by the MHC molecule. Therefore, only one or two residues in a peptide determine if that peptide will bind to the cleft of a particular MHC molecule. Thus, MHC molecules are said to have a broad specificity for peptide binding; each MHC molecule can bind many peptides as long as they have the optimal length and amino acid sequence. This broad specificity is essential for the antigen display function of MHC molecules, because each individual has only a few different MHC molecules that must be able to present peptides derived from a vast number and variety of protein antigens.
- MHC molecules bind mainly peptides. Among various classes of antigens, only peptides have the

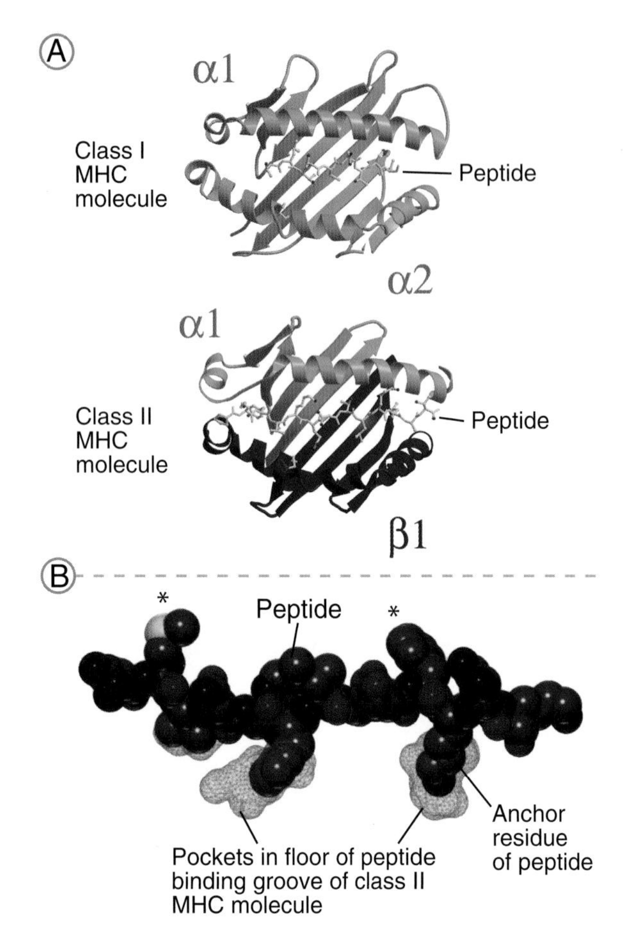

Fig. 3.9 Binding of peptides to major histocompatibility complex (MHC) molecules. A, The *top views* of the crystal structures of MHC molecules show how peptides (in *yellow*) lie on the floors of the peptide-binding clefts and are available for recognition by T cells. **B,** The side view of a cutout of a peptide bound to a class II MHC molecule shows how anchor residues of the peptide hold it in the pockets in the cleft of the MHC molecule. Other residues *(asterisks)* project up from the cleft and are recognized by T cells. (A, Courtesy Dr. P. Bjorkman, California Institute of Technology, Pasadena, CA. B, From Scott CA, Peterson PA, Teyton L, Wilson IA: Crystal structures of two I-A^d-peptide complexes reveal that high affinity can be achieved without large anchor residues, *Immunity* 8:319–329, 1998. Copyright Cell Press; with permission.)

structural and charge characteristics that permit binding to the clefts of MHC molecules. This is why MHC-restricted CD4$^+$ T cells and CD8$^+$ T cells can recognize and respond to protein antigens, the natural source of peptides. The MHC is also involved in the responses of T cells to some nonpeptide antigens, such as small molecules and metal ions. The recognition of such antigens is discussed briefly later in the chapter.

- MHC molecules acquire their peptide cargo during their biosynthesis, assembly, and transport inside cells. Therefore, MHC molecules display peptides derived from protein antigens that are inside host cells (produced inside cells or ingested from the extracellular environment). This explains why MHC-restricted T cells recognize cell-associated antigens and not cell-free antigens in the circulation, tissue fluids, or mucosal lumens.
- Only peptide-loaded MHC molecules are stably expressed on cell surfaces, and empty molecules are degraded inside cells. The reason for this is that MHC molecules must assemble both their chains and bound peptides to achieve a stable structure that is resistant to intracellular proteolysis. This requirement for peptide binding ensures that only useful MHC molecules—that is, those displaying peptides—are expressed on cell surfaces for recognition by T cells. Once peptides bind to MHC molecules, they stay bound for a long time, up to days for some peptides. The slow off-rate ensures that after an MHC molecule has acquired a peptide, it will display the peptide long enough to allow a particular T cell that can recognize the peptide-MHC complex to find the bound peptide and initiate a response.
- In each individual, the MHC molecules can display peptides derived from the individual's own proteins, as well as peptides from foreign (i.e., microbial) proteins. This inability of MHC molecules to discriminate between self antigens and foreign antigens raises two questions. First, at any time the quantity of self proteins in an APC is likely to be much greater than that of any microbial proteins. Why, then, are the available MHC molecules not constantly occupied by self peptides and unable to present foreign antigens? The likely answer is that new MHC molecules are constantly being synthesized, ready to accept peptides, and they are adept at capturing any peptides that are present in cells. Also, a single T cell may need to see a peptide displayed by only as few as 0.1% to 1% of the approximately 10^5 MHC molecules on the surface of an APC, so that even rare MHC molecules displaying a peptide are enough to initiate an immune response. In addition, during infections, especially by some viruses, host protein synthesis is suppressed and viral

Feature	Significance	
Broad specificity	Many different peptides can bind to the same MHC molecule	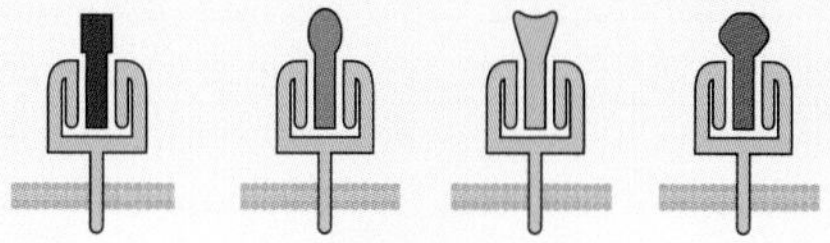
Each MHC molecule displays one peptide at a time	Each T cell responds to a single peptide bound to an MHC molecule	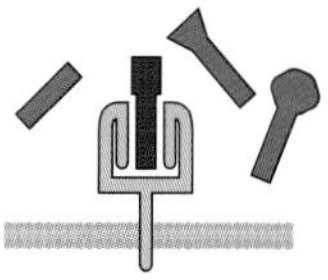
MHC molecules bind only peptides	MHC-restricted T cells respond mainly to protein antigens	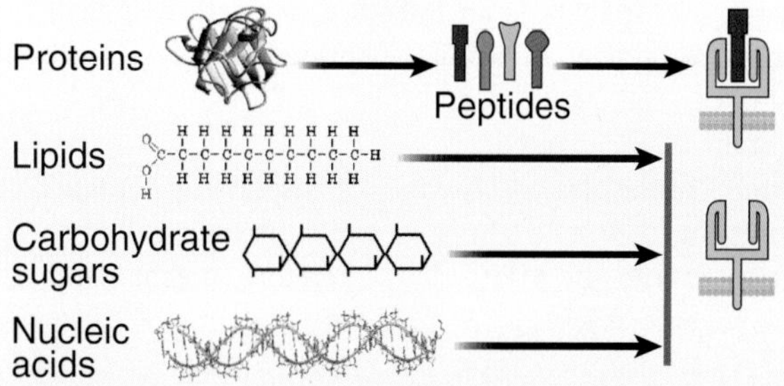 Proteins → Peptides →; Lipids; Carbohydrate sugars; Nucleic acids
Class I and class II MHC molecules display peptides from different cellular compartments	Class I and class II MHC molecules facilitate immune surveillance for microbes in different locations	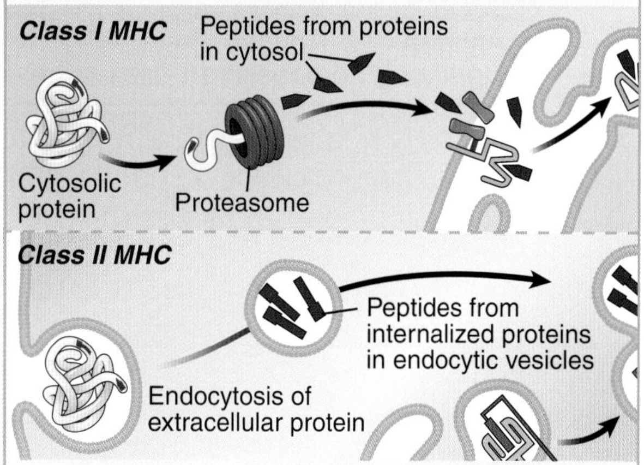 *Class I MHC*: Cytosolic protein; Proteasome; Peptides from proteins in cytosol. *Class II MHC*: Endocytosis of extracellular protein; Peptides from internalized proteins in endocytic vesicles
Stable surface expression of MHC molecule requires bound peptide	Only peptide-loaded MHC molecules are expressed on the cell surface for recognition by T cells	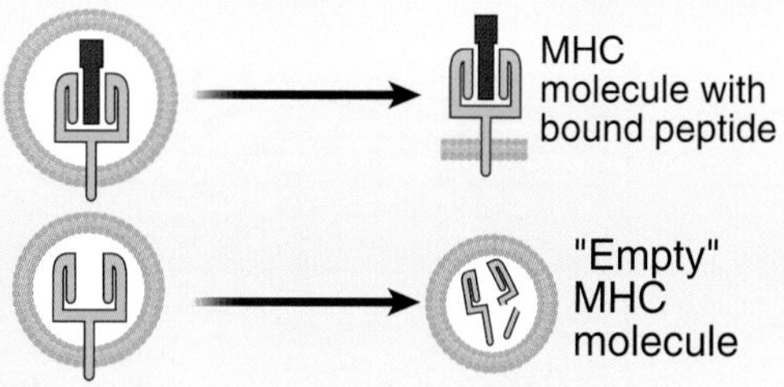 MHC molecule with bound peptide; "Empty" MHC molecule
Very slow off-rate	MHC molecule displays bound peptide for long enough to be located by T cell	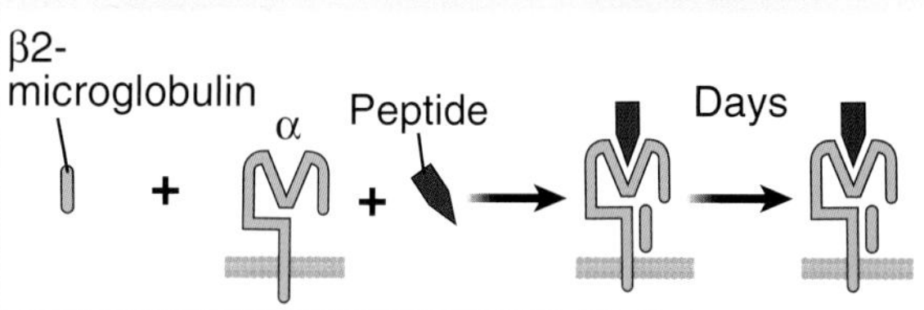 β2-microglobulin + α + Peptide → → Days

Fig. 3.10 Features of peptide binding to MHC molecules. Some of the important features of peptide binding to MHC molecules, with their significance for immune responses. *ER,* Endoplasmic reticulum; I_i, invariant chain.

proteins dominate and are therefore preferentially presented by MHC molecules. The second problem is that if MHC molecules are constantly displaying self peptides, why do we not develop immune responses to self antigens, so-called autoimmune responses? The answer is that most T cells specific for self antigens have previously been killed or inactivated (see Chapter 9). Thus, T cells are constantly patrolling the body, looking at MHC-associated peptides, and if there is an infection, only those T cells that recognize microbial peptides will respond, whereas self peptide–specific T cells will either be absent or inactivated.

MHC molecules are capable of displaying peptides but not intact protein antigens, which are generally too large to fit into the MHC cleft. Therefore, mechanisms must exist for converting naturally occurring proteins into peptides able to bind to MHC molecules. This conversion, called **antigen processing**, is described next.

PROCESSING AND PRESENTATION OF PROTEIN ANTIGENS

In any nucleated cell, peptide fragments of proteins generated by proteolytic complexes in the cytosol called proteasomes are displayed by class I MHC molecules, whereas in specialized APCs (dendritic cells, macrophages, B cells), peptides proteolytically generated in late endosomes and lysosomes are displayed by class II MHC molecules (Figs. 3.11 and 3.12). Together, these two pathways can sample all the proteins present in the extracellular and intracellular

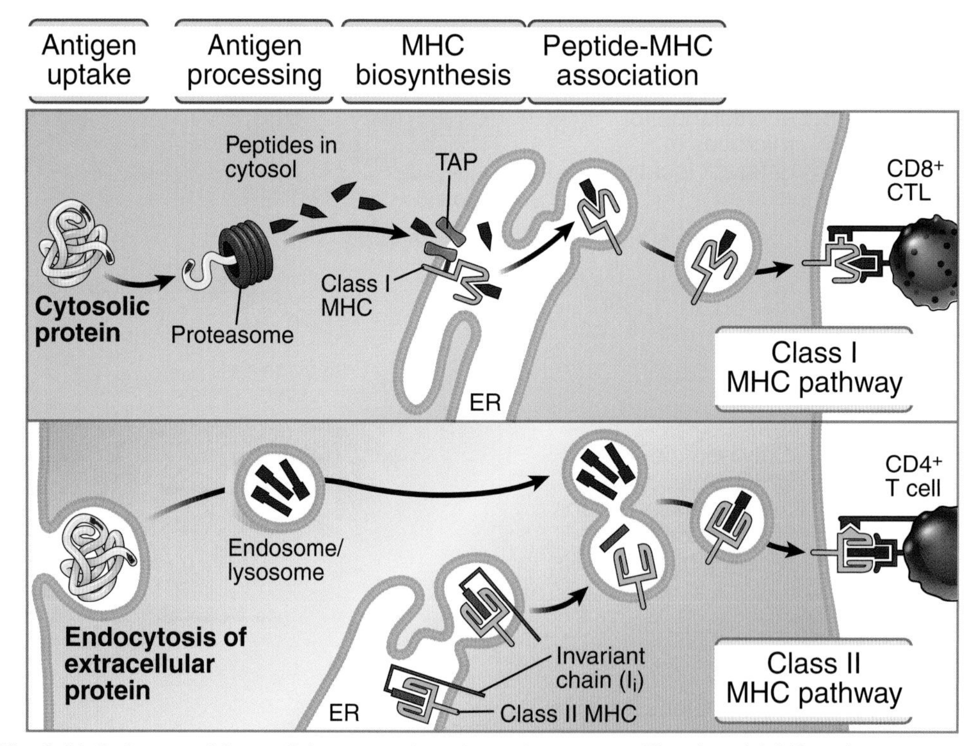

Fig. 3.11 Pathways of intracellular processing of protein antigens. The class I MHC pathway converts proteins in the cytosol into peptides that bind to class I MHC molecules for recognition by $CD8^+$ T cells. The class II MHC pathway converts protein antigens that are endocytosed into vesicles of antigen-presenting cells into peptides that bind to class II MHC molecules for recognition by $CD4^+$ T cells. *CTL,* Cytotoxic T lymphocyte; *ER,* endoplasmic reticulum; *TAP,* transporter associated with antigen processing.

Feature	Class I MHC pathway	Class II MHC Pathway
Composition of stable peptide-MHC complex	Polymorphic α chain of MHC, β2-microglobulin, peptide 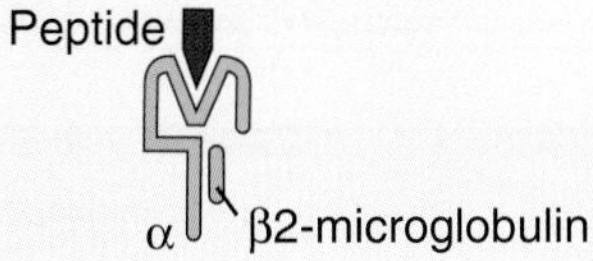 Peptide α β2-microglobulin	Polymorphic α and β chains of MHC, peptide 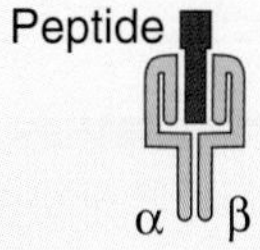 Peptide α β
Cells that express that MHC	All nucleated cells	Dendritic cells, mononuclear phagocytes, B lymphocytes; endothelial cells, thymic epithelium
Responsive T cells	CD8+ T cells 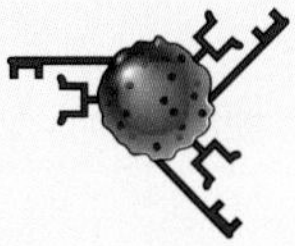	CD4+ T cells 
Source of protein antigens	Cytosolic proteins (mostly synthesized in the cell; may enter cytosol from phagosomes)	Endosomal/lysosomal proteins (mostly internalized from extracellular environment)
Enzymes responsible for peptide generation	Protease components of cytosolic proteasome	Endosomal and lysosomal proteases (e.g. cathepsins)
Site of peptide loading of MHC	Endoplasmic reticulum	Late endosomes and lysosomes
Molecules involved in transport of peptides and loading of MHC molecules	TAP	Invariant chain, DM

Fig. 3.12 Features of the pathways of antigen processing. Some of the comparative features of the two major antigen processing pathways. *MHC,* Major histocompatibility complex; *TAP,* transporter associated with antigen processing.

environments. Because of the segregation of class I and class II MHC pathways with proteasomal versus endosomal/lysosomal processing respectively, the functional type of T cell activated by each pathway differs. Proteasomal processing of proteins in the cytosol ensures that $CD8^+$ T lymphocytes will be able to see antigens from microbes living in almost any host cell type. The proteins found in late endosomes and lysosomes are largely derived from the extracellular environment, and thus the endosomal/lysosomal processing pathway ensures $CD4^+$ T lymphocytes will be able to recognize antigens from microbes originally present outside the APC. Next, we discuss the mechanisms of antigen processing, beginning with the class I MHC pathway.

Processing of Cytosolic Antigens for Display by Class I MHC Molecules

The main steps in antigen presentation by class I MHC molecules are the tagging of antigens in the cytosol or nucleus for proteolysis, proteolytic generation of peptide fragments of the antigen by a specialized cytosolic enzyme complex called the proteasome, transport of peptides into the endoplasmic reticulum (ER), binding of peptides to newly

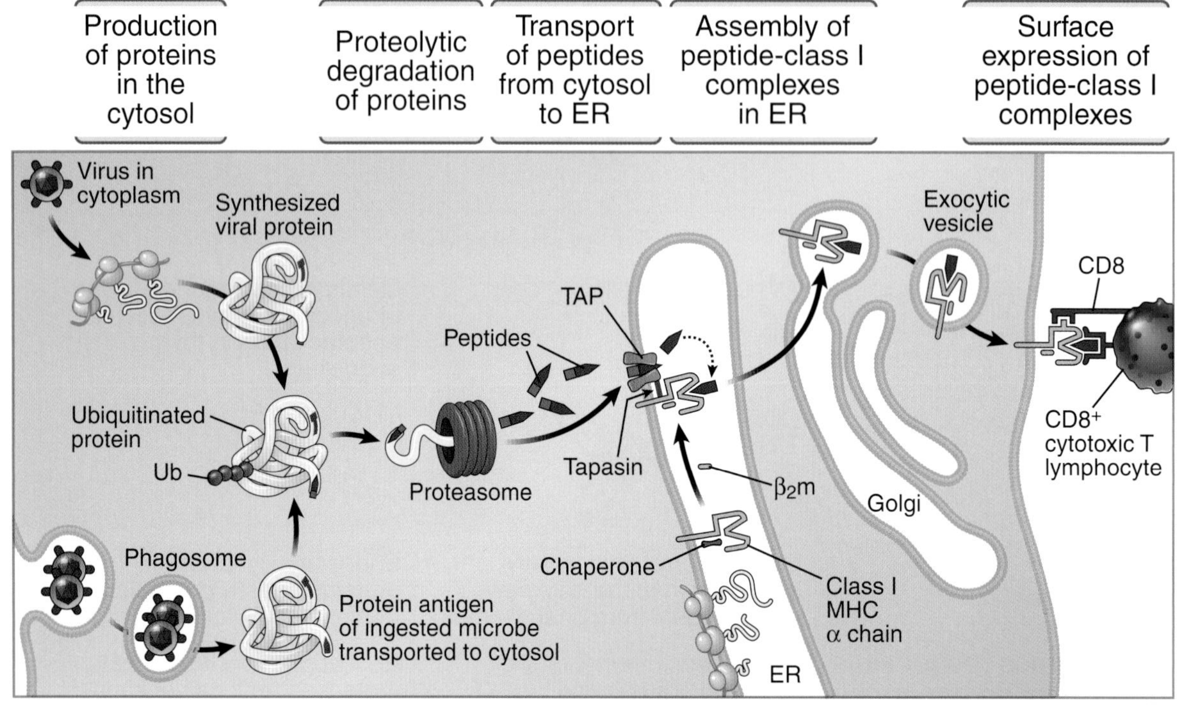

Fig. 3.13 Class I MHC pathway of processing of cytosolic antigens. Proteins enter the cytoplasm of cells either from endogenous synthesis by microbes, such as viruses, that reside in the cytosol (or nucleus, not shown) of infected cells or from microbes that are ingested but whose antigens are transported into the cytosol (the process of cross-presentation, described later). Cytosolic proteins are unfolded, ubiquitinated, and degraded in proteasomes. The peptides that are produced are transported by the transporter associated with antigen processing (TAP) into the endoplasmic reticulum (ER), where the peptides may be further trimmed. Newly synthesized class I MHC molecules are initially stabilized by chaperones and attached to TAP by a linker protein called tapasin, so the MHC molecules are strategically located to receive peptides that are transported into the ER by TAP. The peptide–class I MHC complexes are transported to the cell surface and are recognized by $CD8^+$ T cells. *β_2m,* β_2-microglobulin; *Ub,* Ubiquitin.

synthesized class I molecules, and transport of peptide-MHC complexes to the cell surface (Fig. 3.13).

Proteolysis of Cytosolic Proteins

The peptides that bind to class I MHC molecules are derived from proteins following digestion by the ubiquitin-proteasome pathway. The **proteasome** is a complex composed of rings cotaining proteolytic enzymes that in most cells constitutively perform the housekeeping function of degrading misfolded proteins into peptides. In order for proteins to be processed by proteasomes, they are first unfolded, covalently tagged with multiple copies of a peptide called ubiquitin, and then threaded through the rings of the proteasome. In cells that have been exposed to inflammatory cytokines (as in an infection), the enzymatic composition of the proteasomes changes. As a result, these cells become very efficient at cleaving proteins into peptides with the size and sequence properties that enable them to bind well to class I MHC molecules.

There are several sources of microbial proteins that may be processed by cytosolic proteasomes. These include viral proteins synthesized in the cytosol, but also microbial proteins taken into cells via endocytosis or phagocytosis, many of which then leak out or are transported to the cytosol in a process called cross-presentation (discussed later). All host cell proteins (i.e., nuclear, plasma membrane, secreted) are at some time present in the cytosol, regardless of their site of function. Peptides are derived from proteins from all these locations by proteasomal processing, and then these peptides bind to and are displayed by class I MHC molecules. The host cell proteins that are processed by the ubiquitin-proteasome pathway include microbial proteins, misfolded proteins, and antigenic proteins in cancer cells.

Binding of Peptides to Class I MHC Molecules

In order to form peptide-MHC complexes, the peptides must be transported into the ER. The peptides produced by proteasomal digestion are in the cytosol, and the MHC molecules are being synthesized in the ER; the two need to come together. This transport function is provided by a molecule, called the **transporter associated with antigen processing** (TAP), located in the ER membrane. TAP is an ATP-dependent molecular pump that binds proteasome-generated peptides on the cytosolic side of the ER membrane, then actively moves them into the interior of the ER. Newly synthesized class I MHC molecules, which do not contain bound peptides, associate with a bridging protein called tapasin that links them to TAP molecules in the ER membrane, and plays a key role in preferentially loading peptides that bind with high affinity to the clefts of class I MHC molecules. (As we discuss later, in the ER newly synthesized class II MHC molecules are not able to bind peptides because of the associated invariant chain.)

Transport of Peptide-MHC Complexes to the Cell Surface

High-affinity peptide loading stabilizes class I MHC molecules, which are then exported to the cell surface. Once the class I MHC molecule binds tightly to one of the peptides generated from proteasomal digestion and delivered into the ER by TAP, this peptide-MHC complex becomes stable, is released from binding to tapasin, and is delivered to the cell surface. Peptides that bind weakly to class I MHC molecules cannot mediate the release of the molecules from tapasin. If the MHC molecule does not find a peptide it can bind strongly, the empty molecule is unstable and is eventually degraded. One protein antigen may give rise to many peptides, only a few of which (perhaps only one or two from each antigen) can bind strongly to the MHC molecules present in the individual and have the potential to stimulate immune responses in that individual. The class I MHC–peptide complexes are recognized by $CD8^+$ T cells.

The evolutionary struggle between microbes and their hosts is well illustrated by the numerous strategies that viruses have developed to block the class I MHC pathway of antigen presentation. These strategies include removing newly synthesized MHC molecules from the ER, inhibiting the transcription of MHC genes, and blocking peptide transport by the TAP. By inhibiting the class I MHC pathway, viruses reduce presentation of their own antigens to $CD8^+$ T cells and are thus able to evade the adaptive immune system. These mechanisms of immune evasion are discussed in Chapter 6.

Cross-Presentation of Internalized Antigens to $CD8^+$ T Cells

Some dendritic cells can present ingested antigens on class I MHC molecules to $CD8^+$ T lymphocytes. The initial response of naive $CD8^+$ T cells, similar to $CD4^+$ cells, requires that the antigens be presented by mature dendritic cells in lymph nodes through which the naive T cells circulate. However, some viruses may infect only particular cell types in various tissues and not dendritic cells, and these infected tissue cells may not be capable of traveling to lymph nodes or producing all the signals needed to initiate T cell activation. How, then, are naive $CD8^+$ T lymphocytes in lymph nodes able to respond to the intracellular antigens of infected cells? Similarly, tumors arise from many different types of cells, so how can diverse tumor antigens be presented to naive $CD8^+$ T cells in lymph nodes by dendritic cells?

Many conventional dendritic cells have the ability to ingest infected host cells, dead tumor cells, microbes, and microbial and tumor antigens and transport the ingested antigens into the cytosol, where they are processed by the proteasome. The antigenic peptides that are generated then enter the ER and bind to class I molecules, which display the antigens for recognition by $CD8^+$ T lymphocytes (Fig. 3.14). This process is called **cross-presentation** (or cross-priming) to indicate that one type of cell, dendritic cells, can present the antigens of other infected or dying cells or cell fragments and prime (or activate) naive $CD8^+$ T lymphocytes specific for these antigens. Once the $CD8^+$ T cells have differentiated into CTLs, they kill infected host cells or tumor cells without the need for dendritic cells or signals other than recognition of antigen (see Chapter 6). The same pathway of cross-presentation is involved in initiating

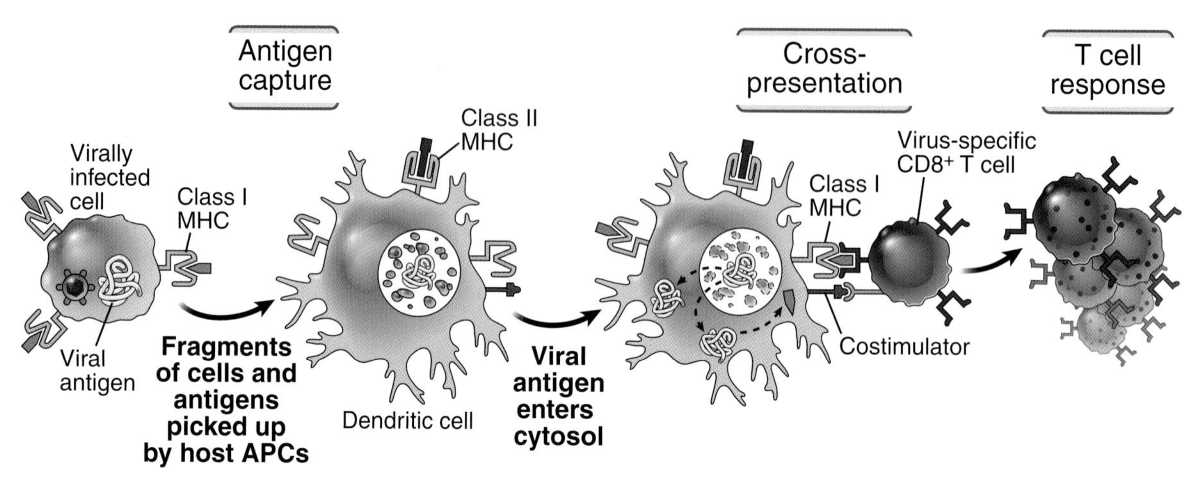

Fig. 3.14 Class I MHC-restricted cross-presentation of microbial antigens from infected cells by dendritic cells. Fragments of cells infected with intracellular microbes (e.g., viruses) or antigens produced in these cells are ingested by dendritic cells, and the antigens of the infectious microbes are transported to the cytosol, processed into peptides by the protaeasomal pathway *(not shown)*, and presented in association with class I MHC molecules of the antigen-presenting cells (APCs). T cells recognize the microbial antigens expressed on the APCs, and the T cells are activated. By convention, the term *cross-presentation* (or *cross-priming*) is applied to $CD8^+$ T cells (cytotoxic T lymphocytes) recognizing class I MHC–associated antigens (as shown); the same cross-presenting APC may display class II MHC–associated antigens from the microbe for recognition by $CD4^+$ helper T cells.

$CD8^+$ T cell responses to some antigens in organ transplants (see Chapter 10).

Processing of Internalized Antigens for Display by Class II MHC Molecules

The main steps in the presentation of peptides by class II MHC molecules are internalization of the antigen, proteolysis in endocytic vesicles, association of peptides with class II molecules, and transport of peptide-MHC complexes to the cell surface (Fig. 3.15).

Internalization and Proteolysis of Antigens

Antigens destined for the class II MHC pathway are usually internalized from the extracellular environment. Dendritic cells and macrophages may ingest extracellular microbes or microbial proteins by several mechanisms. Microbes may bind to surface receptors specific for microbial products or to receptors that recognize antibodies or products of complement activation (opsonins) attached to the microbes. B lymphocytes efficiently internalize proteins that specifically bind to the cells' antigen receptors (see Chapter 7). Certain APCs, especially dendritic cells, may also pinocytose proteins without any specific recognition event. After internalization into APCs by any of these pathways, the microbial proteins enter acidic intracellular vesicles, called endosomes or phagosomes, which fuse with lysosomes. In these vesicles, the proteins are broken down by proteolytic enzymes, generating many peptides of varying lengths and sequences.

Binding of Peptides to Class II MHC Molecules

Peptides bind to newly synthesized class II MHC molecules in specialized vesicles. Class II MHC–expressing APCs constantly synthesize these MHC molecules in the ER. Each newly synthesized class II molecule carries with it an attached protein called the **invariant chain** (I_i), which contains a sequence called the class II invariant chain peptide (CLIP) that binds to the peptide-binding cleft of the class II molecule. Thus, the cleft of the newly synthesized class II molecule is occupied and prevented from accepting peptides in the ER that are destined to bind to class I MHC

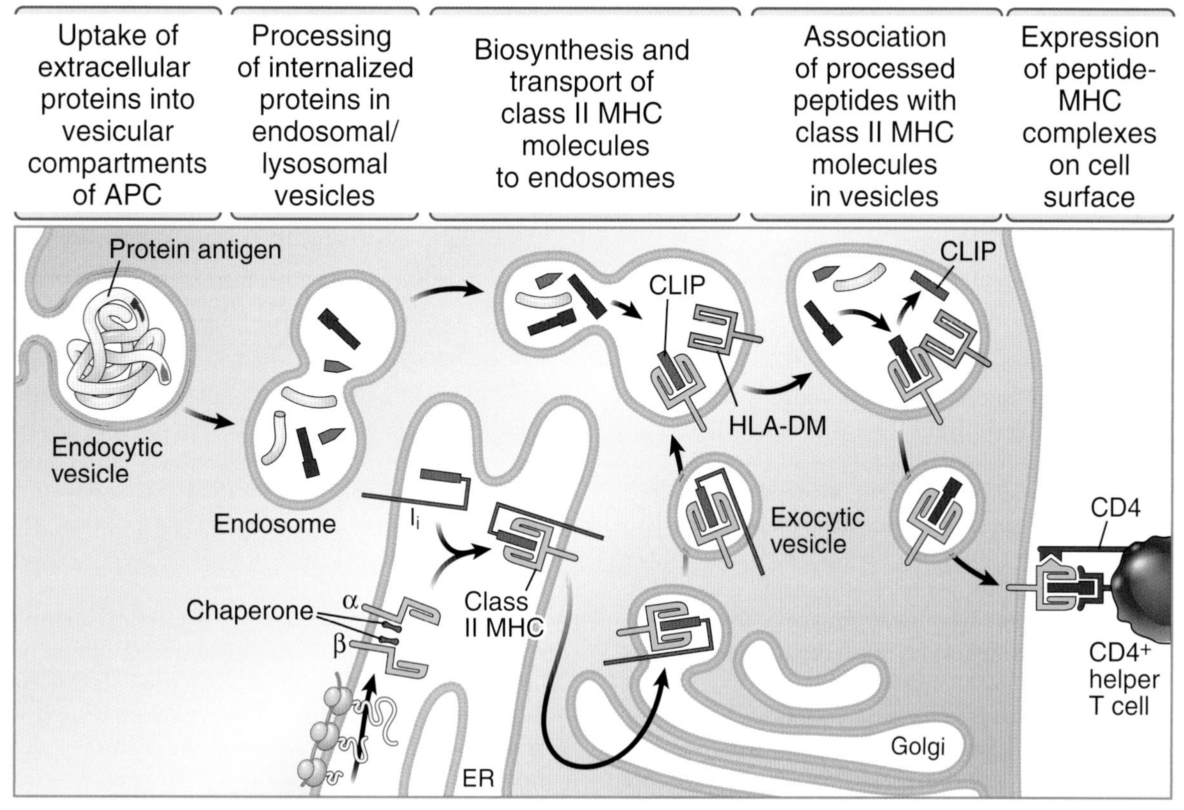

Fig. 3.15 Class II major histocompatibility complex (MHC) pathway of processing of internalized vesicular antigens. Protein antigens are ingested by antigen-presenting cells (APCs) into vesicles, where they are degraded into peptides. Class II MHC molecules enter the same vesicles, where the class II invariant chain peptide (CLIP) that occupies the cleft of newly synthesized class II molecules is removed. These class II molecules are then able to bind peptides derived from the endocytosed protein. The DM molecule facilitates the removal of CLIP and subsequent binding of the antigenic peptide. The peptide–class II MHC complexes are transported to the cell surface and are recognized by $CD4^+$ T cells. *ER*, Endoplasmic reticulum; *HLA-DM*, human leukocyte antigen; I_i, invariant chain.

molecules. This class II molecule with its associated I_i migrates from the ER through the Golgi stacks and then, instead of traveling directly to the plasma membrane, is directed by the cytosolic tail of the invariant chain to move to the membranes of acidic vesicles (endosomes and lysosomes). In this compartment, the invariant chain is degraded, leaving only CLIP in the peptide-binding cleft. Ingested proteins are digested into peptides in the same compartment. The vesicles also contain a class II MHC–like protein called DM, whose function is to exchange CLIP in the class II MHC molecule with other peptides that may be available in this compartment that can bind to the MHC molecule with higher affinity.

Transport of Peptide-MHC Complexes to the Cell Surface

Peptide loading stabilizes class II MHC molecules, which are exported to the cell surface. If a class II molecule binds a peptide with high affinity, the complex is stabilized and transported to the cell surface, where it can be recognized by a $CD4^+$ T cell. Class II molecules that do not find peptides they can bind are eventually degraded by lysosomal proteases. As for the class I pathway, only a few of the peptides produced from any protein antigen can bind to class II MHC molecules present in the individual and stimulate immune responses in each individual.

Physiologic Significance of MHC-Associated Antigen Presentation

Many fundamental features of T cell–mediated immunity are closely linked to the peptide display function of MHC molecules:

- The restriction of T cell recognition to MHC-associated peptides ensures that T cells see and respond only to cell-associated antigens and not to cell-free soluble antigens. This is because MHC molecules are cell membrane proteins and because peptide loading and subsequent expression of MHC molecules depend on intracellular biosynthetic and assembly steps. In other words, MHC molecules can be loaded with peptides only inside cells, where intracellular and ingested antigens are present. Therefore, T lymphocytes can recognize the antigens of intracellular microbes, which require T cell–mediated effector mechanisms, as well as antigens ingested from the extracellular environment, such as those against which antibody responses are generated.
- By segregating the class I and class II pathways of antigen processing, the immune system is able to respond to extracellular and intracellular microbes in different ways that are specialized to defend against these microbes (Fig. 3.16). Protein antigens present in the cytosol are processed and displayed by class I MHC molecules, which are expressed on all nucleated cells—as expected, because all nucleated cells can be infected with one or more species of viruses. Class I–associated peptides are recognized by $CD8^+$ T lymphocytes, which differentiate into CTLs. The CTLs kill the infected cells and eradicate the infection, this being the most effective mechanism for eliminating cytosolic microbes. CTLs also kill tumor cells, which produce cytosolic proteins from mutated genes or oncogenic viruses. Many bacteria, fungi, and even extracellular viruses are typically captured and ingested by macrophages, and their antigens are presented by class II molecules. Because of the specificity of CD4 for class II, class II–associated peptides are recognized by $CD4^+$ T lymphocytes, which function as helper cells. These T cells help the macrophages to destroy ingested microbes, thereby activating an effector mechanism that can eliminate microbes that are internalized from the extracellular environment. B lymphocytes use their antigen receptors to recognize and ingest protein antigens of microbes and also present processed peptides for recognition by $CD4^+$ helper T cells. These helper cells activate the antigen-specific B cells and stimulate the production of antibodies, which serve to eliminate extracellular microbes. Neither phagocytes nor antibodies are effective against intracellular viruses and other pathogens that can survive and replicate in the cytoplasm of host cells; cells harboring these cytosolic microbes are eliminated by $CD8^+$ CTLs.

Thus, the nature of the protective immune response to different microbes is optimized by linking several features of antigen presentation and T cell recognition: the pathways of processing of vesicular and cytosolic antigens, the cellular expression of class I and class II MHC molecules, the specificity of CD8 and CD4 coreceptors for class I and class II molecules, and the functions of $CD8^+$ cells as CTLs and of $CD4^+$ cells as helper cells. The function of linking the type of microbe to one of the two antigen-processing pathways is important because the antigen receptors of T cells cannot distinguish between intracellular and extracellular microbes. In fact, as previously mentioned, the same virus can be extracellular early after infection and becomes intracellular once the infection is established. During its extracellular life, the virus is combatted by antibodies and phagocytes, whose production or functions are stimulated by helper T cells, but once the virus has found a haven in the cytoplasm of cells, it can be eradicated only by CTL-mediated killing of the infected cells. The segregation of class I and class II antigen presentation pathways ensures the correct, specialized immune response against microbes in different locations.

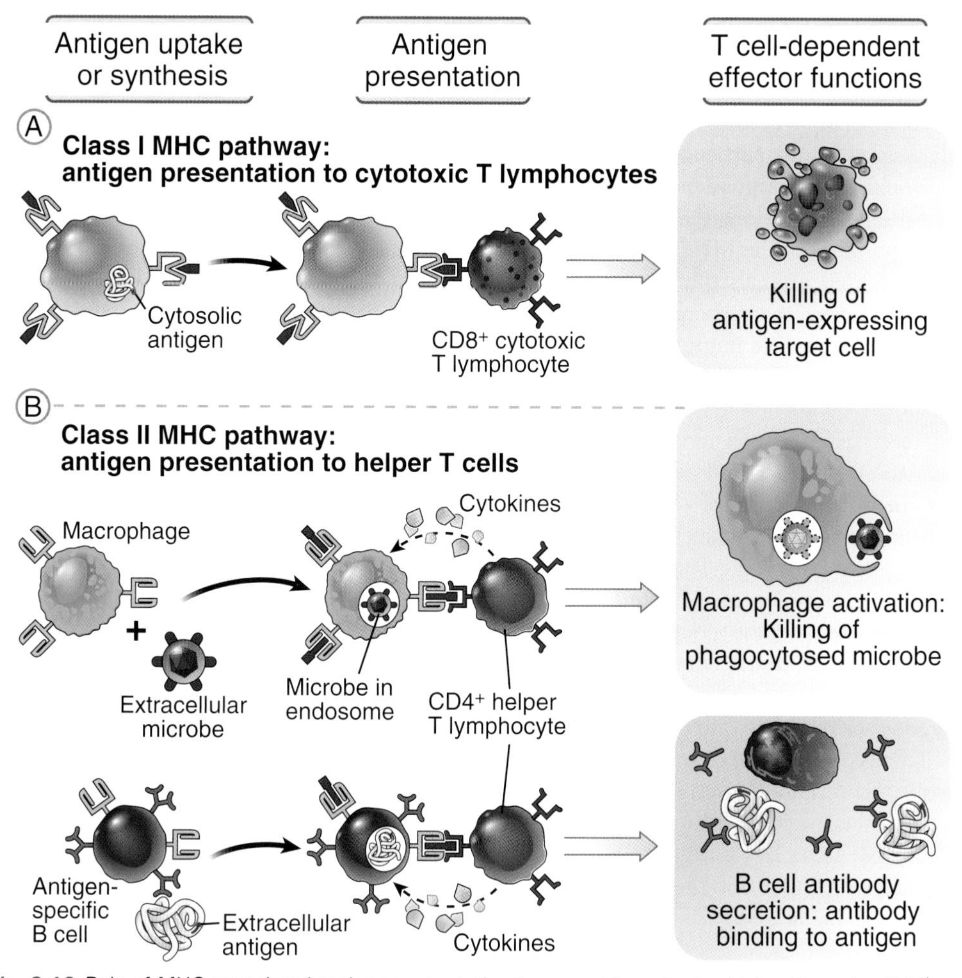

Fig. 3.16 Role of MHC-associated antigen presentation in recognition of microbial antigens by CD8$^+$ and CD4$^+$ effector T cells. **A,** Protein antigens of microbes that live in the cytoplasm of infected cells enter the class I MHC pathway of antigen processing. As a result, these proteins are recognized by CD8$^+$ cytotoxic T lymphocytes, whose function is to kill infected cells. **B,** Protein antigens of microbes that are endocytosed from the extracellular environment by macrophages and B lymphocytes enter the class II MHC pathway of antigen processing. As a result, these proteins are recognized by CD4$^+$ helper T lymphocytes, whose functions are to activate macrophages to destroy phagocytosed microbes and activate B cells to produce antibodies against extracellular microbes and toxins.

- The structural constraints on peptide binding to different MHC molecules, including length and anchor residues, account for the immunodominance of some peptides derived from complex protein antigens and for the inability of some individuals to respond to certain protein antigens. When any protein is proteolytically degraded in APCs, many peptides may be generated, but only those peptides able to bind strongly to the MHC molecules in that individual can be presented for recognition by T cells. These MHC-binding peptides are the immunodominant peptides of the antigen. Even microbes with complex protein antigens express a limited number of immunodominant peptides. Many attempts have been made to identify these peptides in order to develop vaccines, but it is difficult to select a small number of peptides from any microbe that would be immunogenic in a large number of people, because of the enormous polymorphism of MHC molecules in the population. The polymorphism of the MHC also means that some individuals may not express MHC molecules capable of binding any peptide derived from a particular antigen. These individuals would be nonresponders to that antigen. One of the earliest observations that established the physiologic importance of the MHC was the discovery that some inbred animals did not respond to simple protein antigens and responsiveness (or lack of) mapped to genes called immune response *(Ir)* genes, later shown to be class II MHC genes.

Finally, it should be mentioned that T cells also recognize and react against small molecules and even metal ions in an MHC-restricted manner. In fact, exposure to some small molecules that are used as therapeutic drugs and to metals such as nickel and beryllium often leads to pathologic T cell reactions (so-called hypersensitivity reactions; see Chapter 11). There are several ways in which these nonpeptide antigens may be recognized by MHC-restricted CD4$^+$ and CD8$^+$ T cells. Some of the chemicals are thought to covalently modify self peptides or the MHC molecules themselves, creating altered molecules that are recognized as foreign. Other chemicals may bind noncovalently to MHC molecules and alter the structure of the peptide-binding cleft such that the MHC molecule can display peptides that are not normally presented and these peptide-MHC complexes are seen as being foreign.

This chapter began with two questions: how do rare antigen-specific lymphocytes find antigens, and how are the appropriate immune responses generated against extracellular and intracellular microbes? Understanding the biology of APCs and the role of MHC molecules in displaying the peptides of protein antigens has provided satisfying answers to both questions, specifically for T cell–mediated immune responses.

FUNCTIONS OF ANTIGEN-PRESENTING CELLS IN ADDITION TO ANTIGEN DISPLAY

APCs not only display peptides for recognition by T cells but, in response to microbes, also express additional signals for T cell activation. The two-signal hypothesis of lymphocyte activation was introduced in Chapters 1 and 2 (see Fig. 2.19), and we will return to this concept when we discuss the responses of T and B cells in Chapters 5 and 7. Recall that antigen is the necessary signal 1, and for T cells, signal 2 is provided by APCs reacting to microbes. The expression of molecules in APCs that serve as second signals for lymphocyte activation is part of the innate immune response to different microbial products. For example, many bacteria produce a substance called lipopolysaccharide (LPS, endotoxin). When the bacteria are captured by APCs for presentation of their protein antigens, LPS acts on the same APCs, through a TLR, and stimulates the expression of costimulators and the secretion of cytokines. The costimulators and cytokines act in concert with antigen recognition by the T cell to stimulate the proliferation of the T cells and their differentiation into effector and memory cells.

ANTIGEN RECOGNITION BY OTHER T LYMPHOCYTES

This chapter has focused on peptide recognition by MHC-restricted CD4$^+$ and CD8$^+$ T cells, but there are other, smaller populations of T cells that recognize different types of antigens. Natural killer T cells (called NK-T cells), which are distinct from the natural killer (NK) cells described in Chapter 2, are specific for lipids displayed by class I–like CD1 molecules. Mucosal associated invariant T cells (MAIT cells) are specific for bacteria-derived vitamin B metabolites displayed by class I–like MR1 molecules. $\gamma\delta$ T cells recognize a wide variety of molecules, some displayed by class I–like molecules and others apparently requiring no specific processing or display. The functions of these cells and the significance of their unusual specificities are poorly understood.

SUMMARY

- The induction of immune responses to the protein antigens of microbes depends on a specialized system for capturing and displaying these antigens for recognition by the rare naive T cells specific for any antigen. Microbes and microbial antigens that enter the body through epithelia are captured by dendritic cells located in or under the epithelia and transported to regional lymph nodes or captured by dendritic cells in lymph nodes and spleen. The protein antigens of the microbes are displayed by the antigen-presenting cells (APCs) to naive T lymphocytes that recirculate through the lymphoid organs.

- Molecules encoded in the major histocompatibility complex (MHC) perform the function of displaying peptides derived from protein antigens.
- MHC genes are highly polymorphic. Their major products are class I and class II MHC molecules, which contain peptide-binding clefts, where the polymorphic residues are concentrated, and invariant regions, which bind the coreceptors CD8 and CD4, respectively.
- Proteins that are produced in the cytosol of infected and tumor cells, or that enter the cytosol from phagosomes, are degraded by proteasomes, transported into the endoplasmic reticulum by TAP, and bind to the clefts of newly synthesized class I MHC molecules, and the peptide–class I MHC complexes are displayed on the cell surface where T cells can recognize them. CD8 on T cells binds the invariant part of class I MHC molecules, so naive $CD8^+$ T cells and $CD8^+$ cytotoxic T lymphocytes can be activated only by class I MHC–associated peptides derived from proteasomal degradation of cytosolic proteins.
- Proteins that are ingested by APCs from the extracellular environment are proteolytically degraded within the vesicles of the APCs, the peptides generated bind to the clefts of newly synthesized class II MHC molecules, and the peptide–class I IMHC complexes are displayed on the cell surface, where T cells can recognize them. CD4 on T cells binds to invariant parts of class II MHC, because of which naive $CD4^+$ T cells and differentiated $CD4^+$ helper T cells can be activated only by class II MHC–associated peptides derived mainly from proteins degraded in vesicles, which are typically ingested extracellular proteins.
- The role of MHC molecules in antigen display ensures that T cells recognize only cell-associated protein antigens and that the correct type of T cell (helper or cytotoxic) responds to the type of microbe the T cell is best able to combat.
- Microbes activate APCs to express membrane proteins (costimulators) and to secrete cytokines that provide signals that function in concert with antigens to stimulate specific T cells. The requirement for these signals ensures that T cells respond to microbial antigens and not to harmless, nonmicrobial substances.

REVIEW QUESTIONS

1. When antigens enter through epithelial barriers, such as the skin or intestinal mucosa, in what organs are they concentrated? What cell type(s) plays an important role in this process of antigen capture?
2. What are MHC molecules? What are human MHC molecules called? How were MHC molecules discovered, and what is their function?
3. What are the differences between the antigens that are displayed by class I and class II MHC molecules?
4. Describe the sequence of events by which class I and class II MHC molecules acquire antigens for display.
5. Which subsets of T cells recognize antigens presented by class I and class II MHC molecules? What molecules on T cells contribute to their specificity for either class I or class II MHC–associated peptide antigens?

Answers to and discussion of the Review Questions may be found on p. 322.

4

Antigen Recognition in the Adaptive Immune System

Structure of Lymphocyte Antigen Receptors and Development of Immune Repertoires

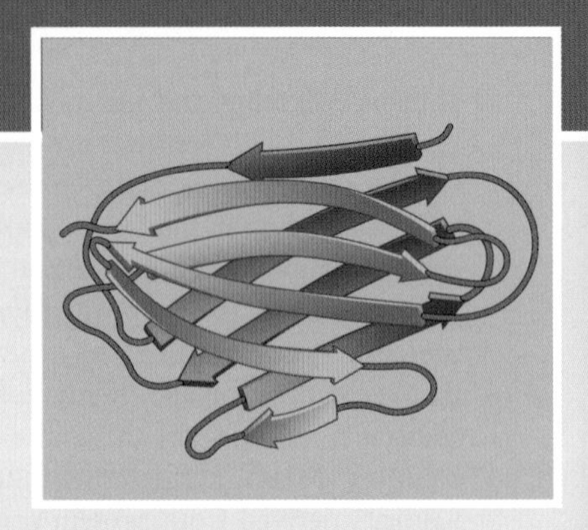

CHAPTER OUTLINE

Antigen receptors serve critical roles in the development and functions of the adaptive immune system. During lymphocyte maturation, antigen receptors guide the selection of B and T cells to preserve the useful specificities and eliminate potentially harmful self reactivity. In adaptive immune responses, naive lymphocytes recognize antigens to initiate the responses and effector T cells and antibodies recognize antigens to eliminate microbes and tumors.

B and T lymphocytes express different receptors that recognize antigens: membrane-bound antibodies on B cells and T cell receptors (TCRs) on T lymphocytes. The principal function of cellular receptors in the immune system, as in other biologic systems, is to detect external stimuli and trigger responses of the cells on which the receptors are expressed. The B cell antigen receptor also binds and internalizes protein antigens for processing and presentation to helper T cells (see Chapter 7). In Chapter 1, we introduced the concept that to recognize a large variety of different antigens, the antigen receptors of lymphocytes must be able to bind to and distinguish between many, often closely related, chemical structures. Antigen receptors are clonally distributed, meaning that each lymphocyte clone is specific for a distinct antigen and has a unique receptor, different from the receptors of all other clones.

(Recall that a clone consists of a parent cell and its progeny.) The total number of distinct lymphocyte clones is very large, and this entire collection makes up the immune repertoire. Although each clone of B lymphocytes or T lymphocytes recognizes a different antigen, the antigen receptors transmit biochemical signals that are fundamentally the same in all lymphocytes and are unrelated to specificity. These features of lymphocyte recognition and antigen receptors raise the following questions:

- How do the antigen receptors of lymphocytes recognize extremely diverse antigens and transmit activating signals to the cells?
- What are the differences in the recognition properties of antigen receptors on B cells and T cells?
- How is the vast diversity of receptors in the lymphocyte repertoire generated? The diversity of antigen recognition implies the existence of many structurally different antigen receptor proteins, more than can be encoded in the inherited genome (germline). Therefore, special mechanisms must exist for generating this diversity.

In this chapter, we describe the structures of the antigen receptors of B and T lymphocytes and how these receptors recognize antigens. We also discuss how diverse antigen receptors are generated during the process of lymphocyte development, giving rise to the repertoire of mature lymphocytes. The process of antigen-induced lymphocyte activation is described in later chapters.

ANTIGEN RECEPTORS OF LYMPHOCYTES

The antigen receptors of B and T lymphocytes have several features that are important for their functions in adaptive immunity (Fig. 4.1). Although these two types of receptors have many similarities in terms of structure and mechanisms of signaling, there are fundamental differences related to the types of antigenic structures they recognize.

- Membrane-bound antibodies, which serve as the antigen receptors of B lymphocytes, can recognize many types of chemical structures, while T cell antigen receptors recognize only peptides bound to major histocompatibility complex (MHC) molecules. B lymphocyte antigen receptors and the antibodies that B cells secrete can recognize the shapes (or conformations) of macromolecules, including proteins, lipids, carbohydrates, and nucleic acids, as well as simpler, smaller chemical moieties. When B cells are activated, they secrete antibodies with the same specificity as the receptors. This ability of antibodies to recognize structurally different types of molecules in their native form enables the humoral immune system to bind to and eliminate microbes and toxins. In striking contrast, the antigen receptors on T cells see only peptides that are derived from intracellular protein antigens and are displayed on antigen-presenting cells (APCs) bound to MHC molecules. This specificity ensures that T cells do not interact with free or soluble antigens but recognize only infected cells or tumor cells, which contain intracellular antigens seen as foreign, or cells that have taken up extracellular protein antigens.
- Antigen receptor molecules consist of regions (domains) involved in antigen recognition—therefore varying between clones of lymphocytes—and other regions required for structural integrity and binding to signaling molecules—and thus relatively conserved among all clones. These domains were first identified in secreted antibodies, discussed later. The antigen-recognizing domains of the receptors are called variable (V) regions, and the conserved portions are the constant (C) regions. Even within each V region, most of the sequence variation is concentrated within short stretches, which are called hypervariable regions, or complementarity-determining regions (CDRs), because they vary from one receptor to the next and are complementary to the shapes of antigens. By concentrating sequence variation in small regions of the receptor, it is possible to maximize the variability of the antigen-binding part while retaining the basic structure of the receptors. As discussed later, special mechanisms exist in developing lymphocytes to create genes that encode different variable regions of antigen receptor proteins in individual clones.
- Antigen receptor polypeptides are associated with invariant membrane proteins whose function is to deliver intracellular signals following antigen recognition (see Fig. 4.1). These signals, which are transmitted to the cytosol and the nucleus, may cause a lymphocyte to divide, to differentiate, to perform effector functions, or in certain circumstances to die. Thus, lymphocytes use different proteins to recognize antigens and to transduce signals in response to antigens. This again allows variability

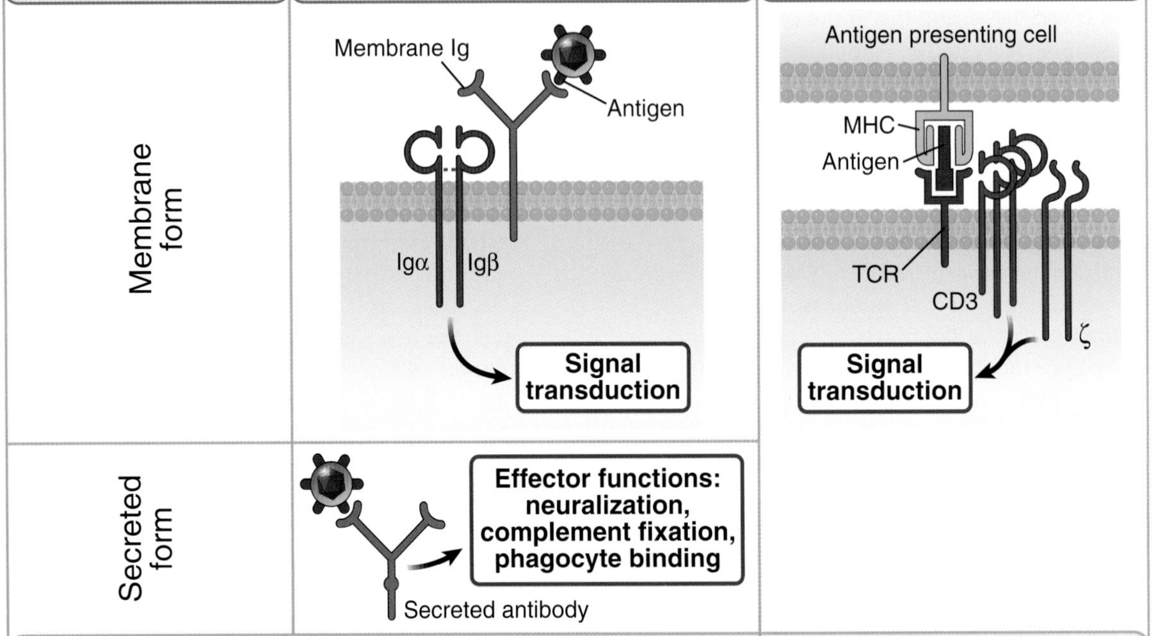

Feature or function	Antibody (Immunoglobulin)	T cell receptor (TCR)
Membrane form	Membrane Ig; Antigen; Igα; Igβ; Signal transduction	Antigen presenting cell; MHC; Antigen; TCR; CD3; ζ; Signal transduction
Secreted form	Effector functions: neuralization, complement fixation, phagocyte binding; Secreted antibody	
Types of antigens recognized	Macromolecules (proteins, polysaccharides, lipids, nucleic acids), small chemicals Conformational and linear epitopes	Mainly peptides displayed by MHC molecules on APCs Linear epitopes
Diversity	Each clone has a unique specificity; potential for ~10^{11} distinct specificities	Each clone has a unique specificity; potential for ~10^{16} distinct specificities
Antigen recognition is mediated by:	Variable (V) regions of heavy and light chains of membrane Ig	Variable (V) regions of α and β chains of the TCR
Antigen binding site	Made up of three CDRs in V_H and three CDRs in V_L	Made up of three CDRs in Vα and three CDRs in Vβ
Affinity of antigen binding	K_d 10^{-7}-10^{-11} M; average affinity of Igs increases during immune responses to protein antigens	K_d 10^{-5}-10^{-7} M; No change during immune responses
Signaling functions are mediated by:	Proteins (Igα and Igβ) associated with membrane Ig	Proteins (CD3 and ζ) associated with the TCR
Effector functions are mediated by:	Constant (C) regions of secreted Ig	TCR does not perform effector functions
Changes in constant regions	Heavy chain class switching and change from membrane to secretory Ig	None

Fig. 4.1 Properties of antibodies and T cell antigen receptors (TCRs). Antibodies (also called immunoglobulins) may be expressed as membrane receptors or secreted proteins; TCRs only function as membrane receptors. When immunoglobulin (Ig) or TCR molecules recognize antigens, signals are delivered to the lymphocytes by proteins associated with the antigen receptors. The antigen receptors and associated signaling proteins form the B cell receptor (BCR) and TCR complexes. Note that single antigen receptors are shown recognizing antigens, but signaling by BCRs typically requires the binding of two or more receptors to adjacent antigen molecules. The important characteristics of these antigen-recognizing molecules are summarized; some of these are discussed later in the chapter. *The total number of possible receptors with unique binding sites is very large, but only $\sim 10^7$–10^9 clones with distinct specificities are present in adults. *APCs,* Antigen-presenting cells; *CDRs,* complementarity-determining regions; *MHC,* major histocompatibility complex.

to be segregated in one set of molecules—the antigen receptors themselves—while leaving the conserved function of signal transduction to other invariant proteins. The set of plasma membrane antigen receptor and associated signaling molecules in B lymphocytes is called the B cell receptor (BCR) complex, and in T lymphocytes the analogous set of proteins is called the TCR complex. When antigens bind to the extracellular portions of the antigen receptors of lymphocytes, intracellular portions of the associated signaling proteins are phosphorylated on conserved tyrosine residues by enzymes called protein tyrosine kinases. Phosphorylation triggers signaling cascades that culminate in the transcriptional activation of many genes and the production of numerous proteins that mediate the responses of the lymphocytes. We return to the processes of T and B lymphocyte activation in Chapters 5 and 7, respectively.

- Antibodies exist in two forms (as membrane-bound antigen receptors on B cells and as secreted proteins) but TCRs exist only as membrane receptors on T cells. Secreted antibodies are present in the blood and mucosal secretions, where they provide protection against microbes (i.e., they are the effector molecules of humoral immunity). Antibodies are also called immunoglobulins (Igs), referring to immunity-conferring proteins with the physical characteristics of globulins. Secreted antibodies recognize microbial antigens and toxins by their variable domains, the same as the membrane-bound antigen receptors of B lymphocytes. The constant regions of some secreted antibodies have the ability to bind to other molecules that participate in the elimination of antigens; these molecules include proteins of the complement system and receptors on other cells (phagocytes, mast cells, NK cells). Thus, antibodies serve different functions at different stages of humoral immune responses: membrane-bound antibodies on B cells recognize antigens to initiate B cell activation, and secreted antibodies neutralize and eliminate microbes and their toxins in the effector phase of humoral immunity. In cell-mediated immunity, the effector function of microbe elimination is performed by T lymphocytes themselves and by other leukocytes responding to the T cells. The antigen receptors of T cells are involved only in antigen recognition and T cell activation, and these proteins are not secreted and do not mediate effector functions.

With this introduction, we describe next the antigen-recognizing molecules of lymphocytes, first antibodies and then TCRs.

Antibodies

An antibody molecule is composed of four polypeptide chains—two identical heavy (H) chains and two identical light (L) chains—with each chain containing a variable region and a constant region (Fig. 4.2). The four chains are assembled to form a Y-shaped molecule. Each light chain is attached to one heavy chain, and the two heavy chains are attached to each other, all by disulfide bonds. A light chain is made up of one V and one C domain, and a heavy chain has one V and three or four C domains. Each domain folds into a characteristic three-dimensional shape, called the immunoglobulin (Ig) domain (see Fig. 4.2D). An Ig domain consists of two layers of a β-pleated sheet held together by a disulfide bridge. The adjacent strands of each β-sheet are connected by short, protruding loops; in the V regions of Ig molecules, three of these loops make up the three CDRs responsible for antigen recognition. Ig domains without hypervariable regions are present in many other proteins in the immune system, as well as outside the immune system, and most of these proteins are involved in responding to stimuli from the

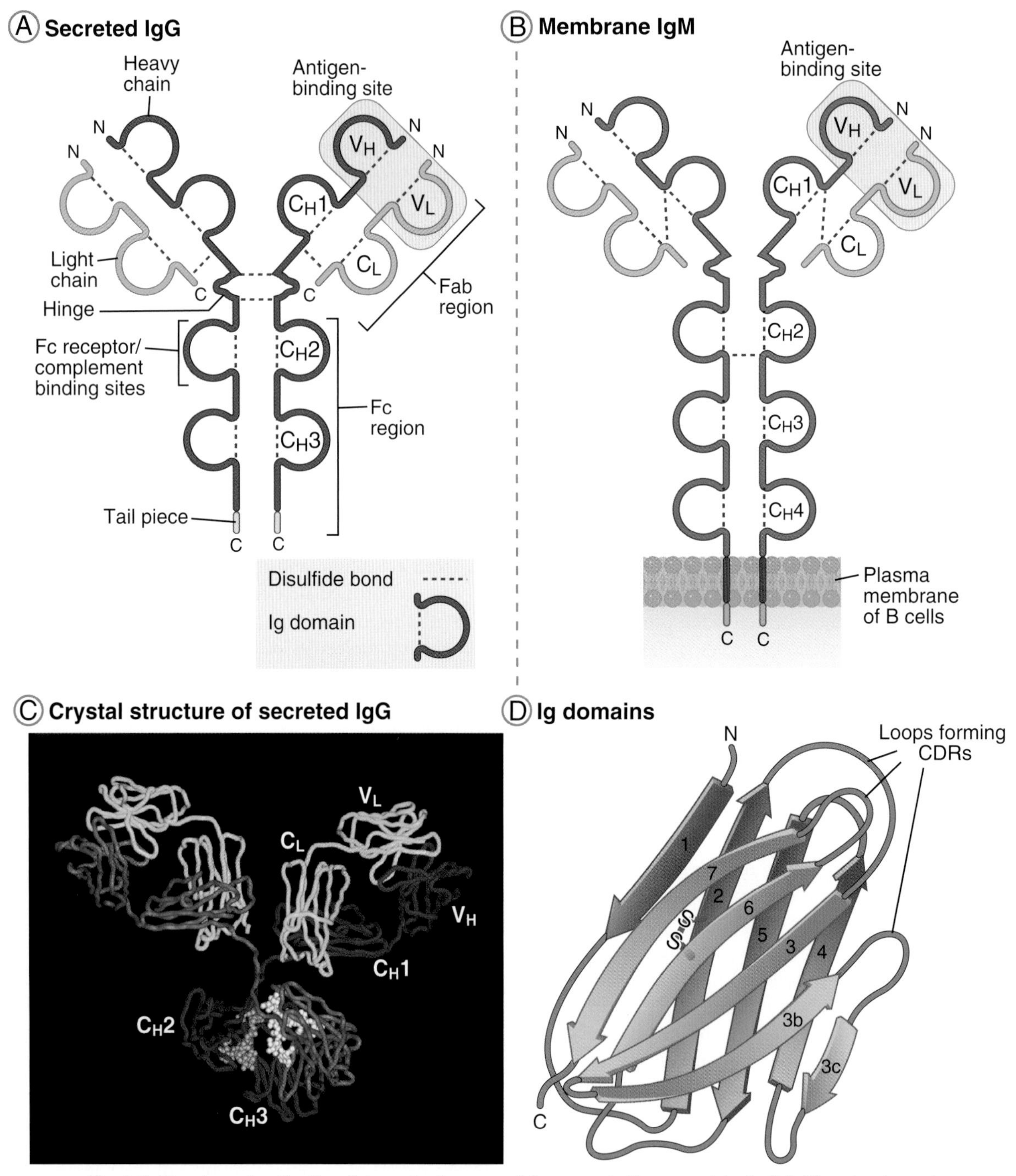

Fig. 4.2 Structure of antibodies. Schematic diagrams of **A,** a secreted immunoglobulin G (IgG) molecule, and **B,** a molecule of a membrane-bound form of IgM, illustrating the domains of the heavy and light chains and the regions of the proteins that participate in antigen recognition and effector functions. N and C refer to the amino-terminal and carboxy-terminal ends of the polypeptide chains, respectively. **C,** The crystal structure of a secreted IgG molecule illustrates the domains and their spatial orientation; the heavy chains are colored *blue* and *red*, the light chains are *green*, and carbohydrates are *gray*. **D,** The ribbon diagram of the Ig V domain shows the basic β-pleated sheet structure and the projecting loops that form the three complementarity-determining regions (CDRs.) **C,** Courtesy Dr. Alex McPherson, University of California, Irvine, CA.

environment and from other cells or in adhesive interactions between cells. All of these proteins are said to be members of the immunoglobulin superfamily.

The antigen-binding site of an antibody is composed of the V regions of both the heavy chain and the light chain, and the core antibody structure contains two identical antigen binding sites (see Fig. 4.2). Of the three hypervariable regions, or CDRs, that are present in the antigen binding sites, the greatest variability is in CDR3, which is located at the junction of the variable (V) and constant (C) regions. As may be predicted from this variability, CDR3 is also the portion of the Ig molecule that contributes most to antigen binding.

Functionally distinct portions of antibody molecules were first identified based on proteolysis of secreted antibodies, which generated fragments that were composed of different parts of antibody proteins. The fragment of an antibody that contains a whole light chain (with its single V and C domains) attached to the V and first C domains of a heavy chain is capable of antigen recognition and was therefore called **Fab** (fragment, antigen-binding); the corresponding part of an intact Ig molecule is the Fab region. The proteolytic fragment containing the remaining heavy-chain C domains is identical in all antibody molecules of a particular type and tends to crystallize in solution and was therefore called the **Fc** (fragment, crystalline); the corresponding part of an intact Ig molecule is the Fc region. In each Ig molecule, there are two identical Fab regions that bind antigen attached to one Fc region that is responsible for most of the biologic activity; antigen binding by the Fab part can neutralize microbes and toxins, and the Fc region activates other effector functions. (As discussed later, some types of antibodies exist as multimers of two or five Ig molecules attached to one another.) Linking the Fab and Fc regions of most antibody molecules is a flexible portion called the hinge region. The hinge allows the two antigen-binding Fab regions of each antibody molecule to move independent of each other, enabling them to simultaneously bind antigen epitopes that are separated from one another by varying distances.

The C-terminal end of the heavy chain may be anchored in the plasma membrane, as seen in BCRs, or it may terminate in a tail piece that lacks the membrane anchor so that the antibody is produced as a secreted protein. Light chains in Ig molecules are not attached to cell membranes.

There are five types of Ig heavy chain proteins, called μ, δ, γ, ε, and α, which differ in their C regions; in humans, there are four subtypes of γ chain, called γ1, γ2, γ3, γ4, and two of the α chain, called α1 and α2. Antibodies that contain different heavy chains belong to different **classes**, or **isotypes**, and are named according to their heavy chains (IgM, IgD, IgG, IgE, and IgA). Each isotype has distinct physical and biologic properties and effector functions (Fig. 4.3; also see Chapter 8). The IgG subtypes also differ from one another in functional properties, but the IgA subtypes do not. The antigen receptors of naive B lymphocytes (mature B cells that have not encountered antigen) are membrane-bound IgM and IgD. After stimulation by antigen and helper T lymphocytes, the antigen-specific B lymphocyte clone may expand and differentiate into progeny that secrete antibodies. Some of the progeny of IgM and IgD expressing B cells may secrete IgM, and other progeny of the same B cells may produce antibodies of other heavy-chain classes. This change in Ig isotype production is called **heavy-chain class** (or **isotype**) **switching**; its mechanism and importance are discussed in Chapter 7.

There are significant differences in the half-lives of antibodies of different isotypes. Some IgG antibodies have half-lives in the plasma of 3 to 4 weeks, whereas the other isotypes have half-lives of 3 to 6 days. The mechanism that controls the half-life of antibodies is discussed in Chapter 8.

The two types of light chains, called κ and λ, differ in their C regions. Each antibody has either κ or λ light chains but not both, and the antibodies made by a single B cell clone have the same type of light chain. Each type of light chain may complex with any type of heavy chain to form an antibody molecule. The light-chain class (κ or λ) also remains fixed throughout the life of each B cell clone, regardless of whether or not heavy-chain class switching has occurred. The function of light chains is to form the antigen-binding surface of antibodies, along with the heavy chains; light chains contribute to binding and neutralizing microbes and toxins but do not participate in other effector functions. There is no difference in the antigen-binding functions of κ or λ light chain–containing antibodies.

Binding of Antigens to Antibodies

Antibodies are capable of binding a wide variety of antigens, including macromolecules and small chemicals. The reason for this is that the antigen-binding hypervariable loops of antibody molecules can either

Isotype of antibody	Subtypes (H chain)	Serum concentration (mg/ml)	Serum half-life (days)	Secreted form	Functions
IgA	IgA1,2 (α1 or α2)	3.5	6	Mainly dimer, also monomer, trimer 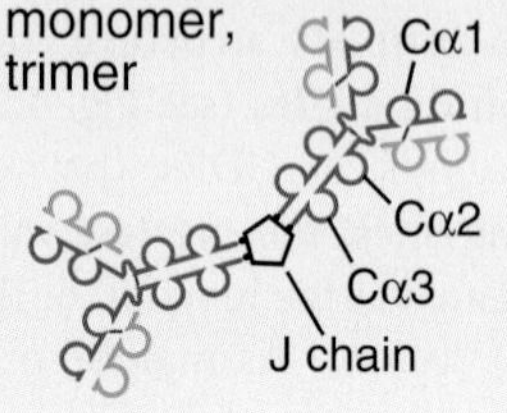	Mucosal immunity (neutralization)
IgD	None (δ)	Trace	3	Monomer	Naive B cell antigen receptor
IgE	None (ε)	0.05	2	Monomer 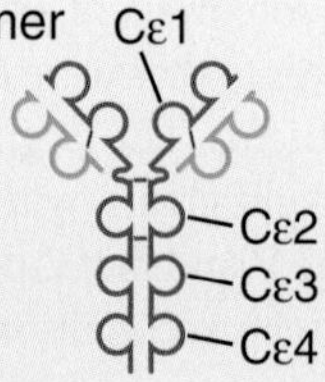	Immediate hypersensitivity (mast cell activation)
IgG	IgG1-4 (γ1, γ2, γ3 or γ4)	13.5	23	Monomer 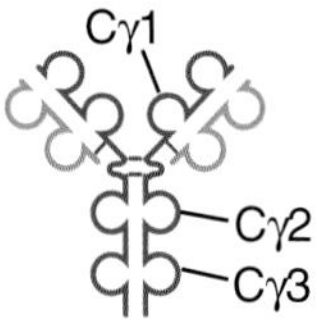	Defense against microbes in all tissues (neutralization, opsonization and phagocytosis, complement activation, antibody-dependent cellular cytotoxicity); neonatal immunity
IgM	None (μ)	1.5	5	Pentamer 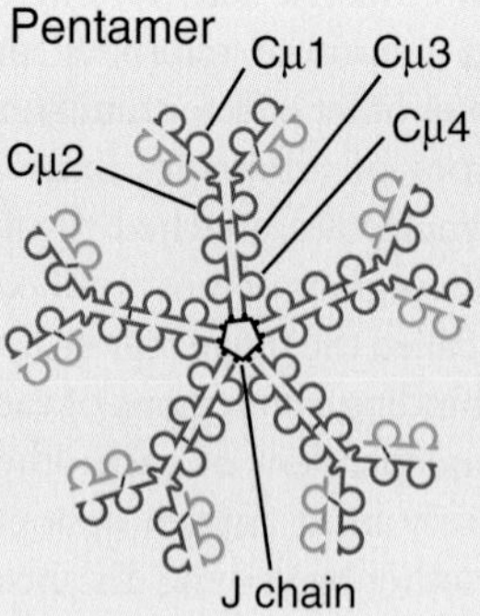	Naive B cell antigen receptor (monomeric form); defense against microbes (complement activation, neutralization)

Fig. 4.3 Features of the major isotypes (classes) of antibodies. This figure summarizes some important features of the major antibody isotypes of humans. Isotypes are classified on the basis of their heavy *(H)* chains; each isotype may contain either κ or λ light chain. The schematic diagrams illustrate the distinct shapes of the secreted forms of these antibodies. Note that IgA consists of two subclasses, called IgA1 and IgA2, and IgG consists of four subclasses, called IgG1, IgG2, IgG3, and IgG4. Most of the opsonizing and complement fixation functions of IgG are attributable to IgG1 and IgG3. The domains of the heavy chains in each isotype are labeled. The plasma concentrations and half-lives are average values in normal individuals. *Ig,* Immunoglobulin.

come together to form clefts capable of accommodating small molecules or form more extended surfaces capable of accommodating larger molecules (Fig. 4.4). Antibodies bind to antigens by reversible, noncovalent interactions, including hydrogen bonds, hydrophobic interactions, and charge-based interactions. The parts of antigens that are recognized by antibodies are called **epitopes**, or determinants. Some epitopes of protein antigens may be a contiguous stretch of amino acids in the primary structure of the protein; these are called linear epitopes. In other cases, regions of a folded protein may form distinct shapes that are recognized by antibodies; these are called conformational epitopes.

The strength with which one antigen-binding site of an antibody binds to one epitope of an antigen is called the **affinity** of the interaction. Affinity often is expressed as the dissociation constant (K_d), which is the molar concentration of an antigen required to occupy half the available antibody molecules in a solution; the lower the K_d, the higher the affinity. Most antibodies produced in a primary immune response have a K_d in the range of 10^{-6} to 10^{-9} M, but with repeated stimulation (e.g., in a secondary immune response), the affinity increases to a K_d of 10^{-8} to 10^{-11} M. This increase in antigen-binding strength is called **affinity maturation** (see Chapter 7).

Each IgG, IgD, and IgE antibody molecule has two antigen-binding sites. Secreted IgA is a dimer of two linked IgA molecules and therefore has 4 antigen-binding sites, and secreted IgM is a pentamer with 10 antigen-binding sites. Therefore, each antibody molecule can bind 2 to 10 epitopes of an antigen, or epitopes on 2 or more neighboring antigens. The total strength of binding is much greater than the affinity of a single antigen-antibody bond and is called the **avidity** of the interaction. Antibodies produced against one antigen may bind other, structurally similar antigens. Such binding to similar epitopes is called a **cross-reaction**.

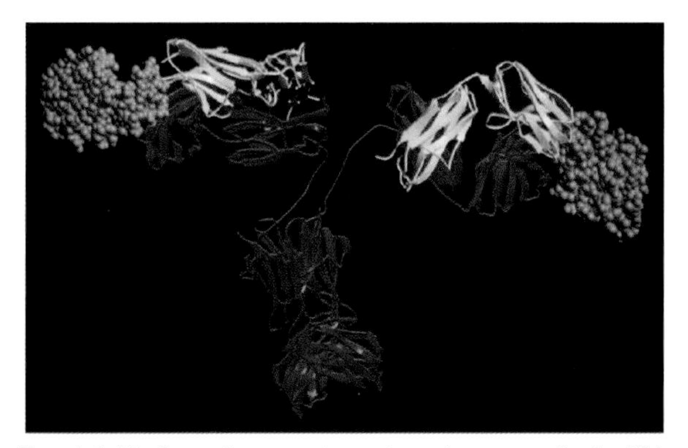

Fig. 4.4 Binding of a protein antigen by an antibody. This model of a protein antigen bound to an antibody molecule shows how the antigen-binding site can accommodate large macromolecules in their native (folded) conformation. The heavy chains of the antibody are *red*, the light chains are *yellow*, and the antigens are *blue*. Courtesy Dr. Dan Vaughn, Cold Spring Harbor Laboratory, Cold Spring Harbor, NY.

In B lymphocytes, membrane-bound Ig molecules are noncovalently associated with two other proteins, called Igα and Igβ, that combine with the membrane Ig to make up the BCR complex. When the BCR recognizes antigen, Igα and Igβ transmit signals to the interior of the B cell that initiate the process of B cell activation. These and other signals in humoral immune responses are discussed in Chapter 7.

Monoclonal Antibodies

The realization that one clone of B cells makes an antibody of only one specificity has been exploited to produce **monoclonal antibodies**, one of the most important technical advances in immunology, with far-reaching applications in clinical medicine and research. To produce monoclonal antibodies, B cells, which have a short life span in vitro, are obtained from an animal immunized with an antigen and fused in vitro with myeloma cells (tumors of plasma cells), which can be propagated indefinitely in tissue culture (Fig. 4.5). The myeloma cell line is constructed to lack a specific enzyme, as a result of which these cells cannot grow in the presence of a certain toxic drug; fused cells, containing both myeloma and normal B cell nuclei, however, do grow in the presence of this drug because the normal B cells provide the missing enzyme. Thus, by fusing the two cell populations and culturing them with the drug, it is possible to grow out fused cells that are hybrids of the B cells and the myeloma and are called **hybridomas**. These hybridoma cells produce antibodies, like normal B cells, and grow continuously, having acquired the immortal property of the myeloma tumor. From a population of hybridomas, one can select and expand individual cells that secrete the antibody of desired specificity; such antibodies, derived from a single B cell clone, are homogeneous monoclonal antibodies. Monoclonal antibodies against virtually any epitope on any antigen can be produced using this technology.

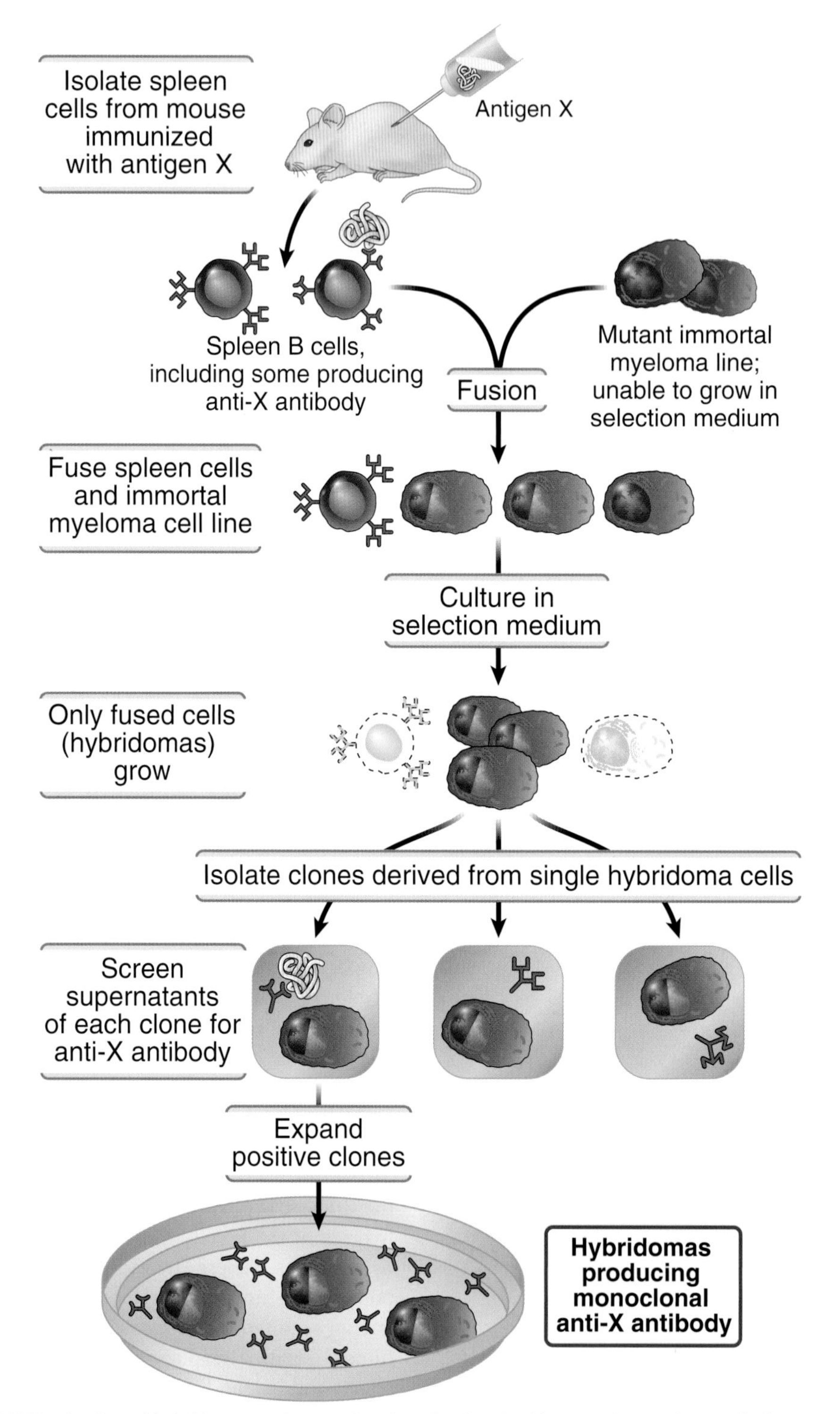

Fig. 4.5 Production of hybridomas and monoclonal antibodies. In this procedure, spleen cells from a mouse that has been immunized with a known antigen are fused with an enzyme-deficient myeloma cell line that does not secrete its own immunoglobulins. The fused cells are then placed in a selection medium that permits the survival of only immortalized hybrids; the normal B cells provide the enzyme that the myeloma lacks, and unfused B cells cannot survive indefinitely. These hybrid cells are then grown as single-cell clones and tested for the secretion of antibody of the desired specificity. The clone producing this antibody is expanded and becomes a source of the monoclonal antibody.

Most monoclonal antibodies specific for molecules of interest are made by fusing cells from mice immunized with that antigen with a mouse myeloma cell line. Such mouse monoclonal antibodies cannot be injected repeatedly into human subjects, because the human immune system sees the mouse Ig as foreign and mounts an immune response against the injected antibodies. This problem has been partially overcome by genetic engineering approaches that retain the antigen-binding regions of the mouse monoclonal antibody and replace the rest of the antibody with human Ig; such humanized antibodies are less immunogenic and more suitable for administration to people. More recently, monoclonal antibodies have been generated by using recombinant DNA technology to clone the DNA encoding human antibodies of desired specificity. Another approach is to replace the Ig genes of mice with human antibody genes and then immunize these mice with an antigen to produce specific human antibodies. Monoclonal antibodies are now in widespread use as therapeutic agents and diagnostic reagents for many diseases in humans (Fig. 4.6).

T Cell Receptor for Antigens

The TCR, which recognizes peptide antigens displayed by MHC molecules, is a membrane-bound heterodimeric protein composed of an α chain and a β chain, each chain containing one V region and one C region (Fig. 4.7A, B). The V and C regions are homologous to immunoglobulin V and C regions. In the V region of each TCR chain, there are three hypervariable, or complementarity-determining, regions, each corresponding to a loop in the V domain. As in antibodies, CDR3 is the most variable among different TCRs. The C regions of both α and β chains continue into short hinge regions, which contain cysteine residues where a disulfide bond links the two chains, followed by hydrophobic transmembrane portions and short carboxy-terminal cytoplasmic tails.

Antigen Recognition by the T Cell Receptor

Both the α chain and β chain of the TCR participate in specific recognition of MHC molecules and bound peptides (Fig. 4.7C). Recognition of the peptide is responsible for the antigen specificity of the TCR, and recognition of the MHC for MHC restriction (Chapter 3).

The TCR recognizes antigen, but as with membrane Ig on B cells, it is incapable of transmitting signals to the T cell on its own. Associated with the TCR is a group of proteins, called the CD3 and ζ proteins, which together with the TCR make up the TCR complex (see Fig. 4.1). The CD3 and ζ chains are crucial for the initiation of signaling when the TCR recognizes antigen. In addition, T cell activation requires engagement of the coreceptor molecule CD4 or CD8, which recognizes nonpolymorphic portions of MHC molecules. The functions of these TCR-associated proteins and coreceptors are discussed in Chapter 5.

Antigen recognition by B and T lymphocyte receptors differs in important ways (see Fig. 4.1). Antibodies can bind many different types of chemical structures, often with high affinities, which allows antibodies to bind to and neutralize many different microbes and toxins even if they are present at low concentrations in the circulation or in the lumens of mucosal organs. TCRs only recognize peptide-MHC complexes and bind these with relatively low affinity, which may be why the binding of T cells to APCs has to be strengthened by additional cell surface adhesion molecules (see Chapter 5). The three-dimensional structure of the TCR is similar to that of the Fab region of an Ig molecule. In contrast to membrane antibodies, in which only the heavy chain is membrane-anchored, both TCR chains are anchored in the plasma membrane. Unlike BCRs, TCRs are not produced in a secreted form, do not undergo affinity maturation during the life of a T cell, and do not have isotypes to which switching could occur.

About 5% to 10% of T cells in the body express receptors composed of γ and δ chains. These receptors are structurally similar to the αβ TCR but have very different specificities. The γδ TCR may recognize a variety of protein and nonprotein antigens that are not displayed by MHC molecules. T cells expressing γδ TCRs are abundant in epithelia. This observation suggests that γδ T cells recognize microbes usually encountered at epithelial surfaces. Another subpopulation of T cells, making up less than 5% of all T cells, express αβ TCRs as well as surface molecules found on

Inflammatory (immunological) diseases		
Target	**Effect**	**Diseases**
CD20	Depletion of B cells	B cell lymphomas, rheumatoid arthritis, multiple sclerosis, other autoimmune diseases
IgE	Blocking mast cell sensitization	Asthma, peanut allergy
IL-4/IL-13 receptor	Blocking inflammation	Asthma, atopic dermatitis
IL-5	Blocking eosinophil activation	Asthma
IL-6 receptor	Blocking inflammation	Rheumatoid arthritis, cytokine storm
IL-17	Blocking inflammation	Psoriasis
TNF	Blocking inflammation	Rheumatoid arthritis, Crohn's disease, psoriasis
α4 integrin	Blocking leukocyte migration	Inflammatory bowel disease; multiple sclerosis
C5	Blocking complement-mediated cell lysis	Paroxysmal nocturnal hemoglobinuria, hemolytic uremic syndrome
Cancer		
Target	**Effect**	**Diseases**
CD52	Depletion of lymphocytes	Chronic lymphocytic leukemia
CTLA-4	Activation of T cells	Melanoma, lung cancer
EGFR	Growth inhibition of epithelial tumors	Colorectal, lung, and head and neck cancers
HER2/NEU	Inhibition of EGF signaling; depletion of tumor cells	Breast cancer
PD-1, PD-L1	Activation of effector T cells	Many tumors
VEGF	Blocking tumor angiogenesis	Breast cancer, colon cancer, age-related macular degeneration
Other diseases		
Target	**Effect**	**Diseases**
Glycoprotein IIb/IIIa	Inhibition of platelet aggregation	Cardiovascular disease
SARS-CoV2 spike protein	Blocking SARS-CoV2 infection of cells	COVID-19

Fig. 4.6 Selected therapeutic monoclonal antibodies in clinical use. The figure lists some of the monoclonal antibodies that are approved for the treatment of various types of diseases. Note that anti-CD20 is listed under inflammatory diseases but is also used to treat cancers of B cells. Anti-VEGF is listed under cancer but is also used for the treatment of age-related macular degeneration. *EGFR,* Epidermal growth factor receptor; *Ig,* immunoglobulin; *IL,* interleukin; *TNF,* tumor necrosis factor; *VEGF,* vascular endothelial growth factor.

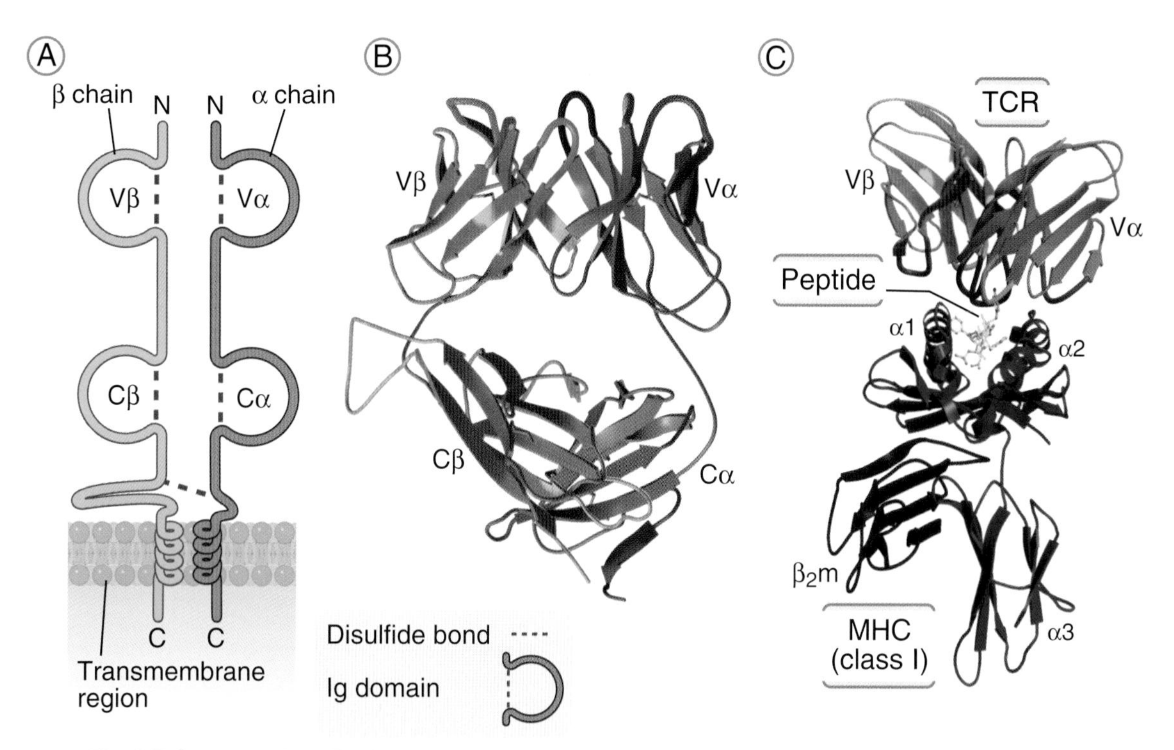

Fig. 4.7 Structure of the T cell antigen receptor (TCR) and TCR recognition of a peptide–major histocompatibility complex (MHC) complex. **A,** The schematic diagram of the αβ TCR *(left)* shows the domains of a TCR specific for a peptide-MHC complex. The antigen-binding portion of the TCR is formed by the V domains of the α and β chains. *N* and *C* refer to the amino-terminal and carboxy-terminal ends of the polypeptides. **B,** The ribbon diagram *(right)* shows the structure of the extracellular portion of a TCR as revealed by x-ray crystallography. **C,** The crystal structure of the extracellular portion of a peptide-MHC complex bound to a TCR that is specific for the peptide displayed by the MHC molecule. The peptide can be seen attached to the cleft at the *top* of the MHC molecule, and one residue of the peptide contacts the V region of a TCR. The structure of MHC molecules and their function as peptide display proteins are described in Chapter 3. *β_2m,* β_2-microglobulin; *Ig,* immunoglobulin. From Bjorkman PJ: MHC restriction in three dimensions: a view of T cell receptor/ligand interactions, *Cell* 89:167–170, 1997. Copyright Cell Press; with permission.

natural killer cells, and are therefore called natural killer T cells (NK-T cells). NK-T cells express αβ TCRs with limited diversity, and they recognize lipid antigens displayed by nonpolymorphic class I MHC–like molecules called CD1. A third subset of T cells called mucosal associated invariant T (MAIT) cells also express αβ TCRs with limited diversity, some of which are specific for bacterially derived vitamin B metabolites bound to an MHC-like protein called MR1. MAIT cells account for only about 5% of blood T cells in humans but up to 20% to 40% of liver T cells. The physiologic functions of γδ cells, NK-T cells, and MAIT cells are not well understood.

DEVELOPMENT OF B AND T LYMPHOCYTES

Now that we have discussed the structure of antigen receptors of B and T lymphocytes and how these receptors recognize antigens, the next question is how the enormous diversity of these receptors is generated. As the clonal selection hypothesis predicted, there are many clones of B and T lymphocytes, perhaps as many as 10^7

to 10^9, each with a distinct specificity, and these arise before encounter with antigen. There are not enough genes in the human genome for every possible receptor to be encoded by a different gene. In fact, the immune system has developed mechanisms for generating extremely diverse antigen receptors from a limited number of inherited genes, and the generation of diverse receptors is intimately linked to B and T lymphocyte maturation.

The process of lymphocyte maturation first generates a large number of cells, each with a different antigen receptor, and then preserves the cells with useful receptors. The generation of millions of receptors is a molecular process that cannot be influenced by what the receptors recognize because recognition can occur only after receptor generation and expression. Once these antigen receptors are expressed on developing lymphocytes, selection processes come into play that promote the survival of cells with receptors that can recognize antigens, including microbial antigens, and eliminate cells that can recognize self antigens strongly enough to pose danger of autoimmunity. We discuss each of these events next.

Lymphocyte Development

The development of lymphocytes from bone marrow stem cells involves commitment of hematopoietic progenitors to the B or T cell lineage, the proliferation of these progenitors, the rearrangement (recombination) and expression of antigen receptor genes, and selection events to preserve and expand cells that express potentially useful antigen receptors (Fig. 4.8). These steps are common to B and T lymphocytes, even though B lymphocytes mature in the bone marrow and T lymphocytes mature in the thymus. Each of the processes that occurs during lymphocyte maturation plays a special role in the generation of the lymphocyte repertoire.

- The maturation of common lymphoid progenitors in the bone marrow results in commitment to the B cell lineage. Similar progenitors migrate to the thymus and commit to the T cell lineage there. Commitment is associated with the activation of several lineage-specific transcription factors and increased accessibility of Ig and TCR genes to the gene recombination machinery, described later.
- Developing lymphocytes undergo proliferation at several stages during their maturation. Proliferation of developing lymphocytes is necessary to ensure that an adequate number of cells will be available to express different antigen receptors and mature into functionally competent lymphocytes. Survival and proliferation of the earliest lymphocyte precursors are stimulated mainly by growth factors that are produced by stromal cells in the bone marrow and the thymus. In humans, IL-7 maintains and expands the number of T lymphocyte progenitors before they express antigen receptors. The growth factors required for expansion of human B cell progenitors are not defined. This proliferative expansion generates a large pool of cells in which diverse antigen receptors may be produced. Even greater proliferation of the B and T cell lineages occurs after the developing lymphocytes have completed their first antigen receptor gene rearrangement and assembled a so-called preantigen receptor (described later). Proliferation of the cells at this stage is driven by signals from the preantigen receptor. Expression of the receptor is a quality control checkpoint in lymphocyte development that ensures preservation of cells with functional receptors.
- Lymphocytes are selected at multiple steps during their maturation to preserve useful specificities. As discussed later, many attempts to generate antigen receptors fail because of errors during the gene recombination process. Therefore, checkpoints are needed at which only cells that can express functional components of antigen receptors are selected to survive and proliferate. Prelymphocytes and immature lymphocytes that fail to express antigen receptor proteins die by apoptosis (see Fig. 4.8). The gene rearrangements in the developing lymphocytes randomly generate antigen receptors with highly diverse specificities. In order to ensure that the specificities of lymphocytes that emerge from the generative organs are useful, selection processes work to allow only a subset of the developing lymphocytes to fully mature. These selection processes are based on the expression of functional receptors and, in the case of T cells, the recognition of self molecules present in the thymus. Positive selection allows only cells that express functional receptors to complete the maturation steps. Negative selection kills off cells with high affinity for self antigens, thus eliminating

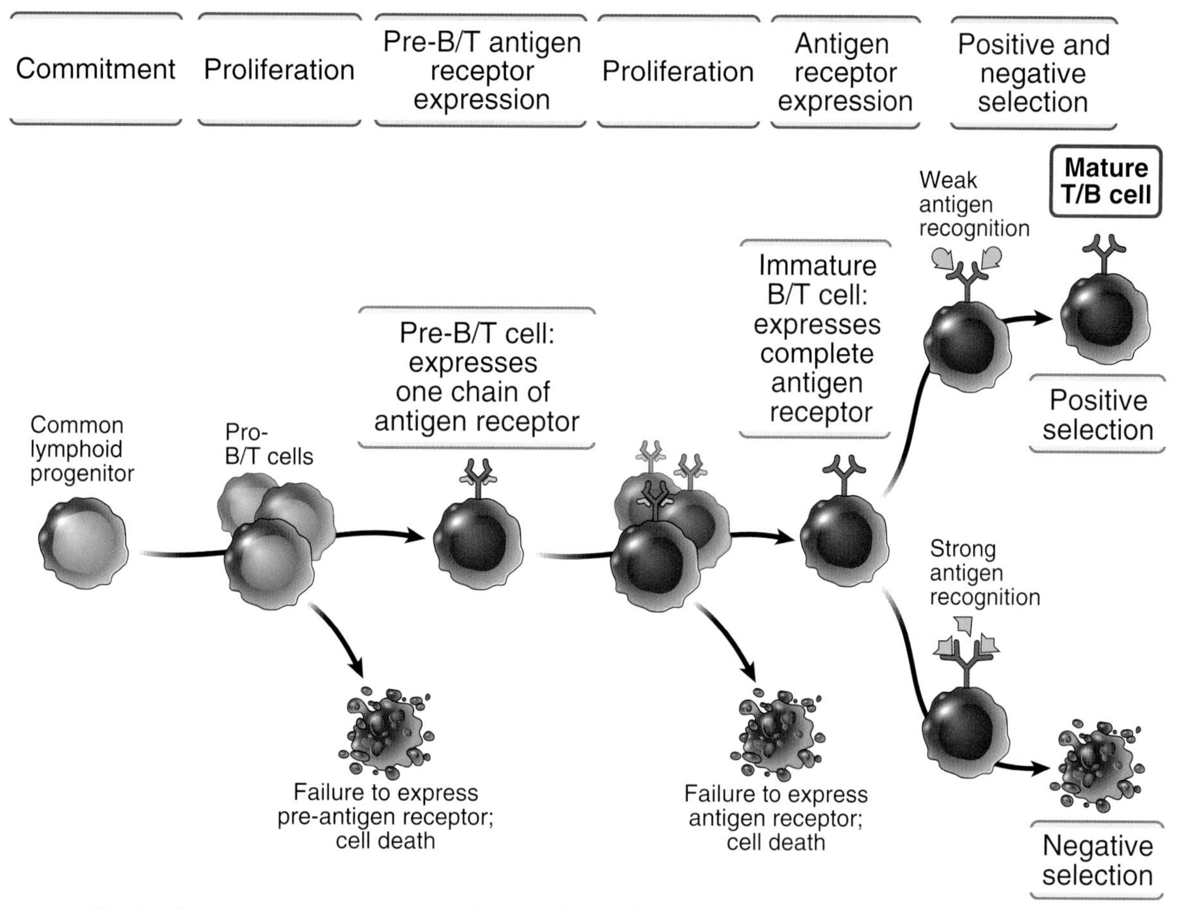

Fig. 4.8 **Steps in the maturation of lymphocytes.** During their maturation, B and T lymphocytes go through cycles of proliferation and expression of antigen receptor proteins by gene recombination. Cells that fail to express intact, functional receptors die by apoptosis, because they do not receive the necessary survival signals. At the end of the process, the cells undergo positive and negative selection. The lymphocytes shown may be B or T cells; T cells recognize peptide-MHC complexes but these are not shown for simplicity.

potentially dangerous lymphocytes that could cause autoimmune disease.

The processes of B and T lymphocyte maturation and selection share some important features but also differ in many respects. We start with the central event that is common to both lineages: the recombination and expression of antigen receptor genes.

Production of Diverse Antigen Receptors

The formation of functional genes that encode B and T lymphocyte antigen receptors is initiated by somatic recombination of gene segments that code for the variable regions of the receptors, and diversity is generated during this process.

Inherited Antigen Receptor Genes

Hematopoietic stem cells in the bone marrow and early lymphoid progenitors contain Ig and TCR gene segments in their inherited, or germline, configuration. In this configuration, Ig heavy-chain and light-chain loci and the TCR α chain and β chain loci each contain multiple V region gene segments, numbering about 30 to 45, and one or a few C region genes (Fig. 4.9). Between the V and C gene segments are groups of several short coding sequences called diversity (D) and joining (J) gene segments. (All antigen receptor gene loci in B cells and αβ T cells contain V, J, and C gene segments, but only the Ig heavy chain and TCR β chain loci also contain D gene segments.) These separated gene

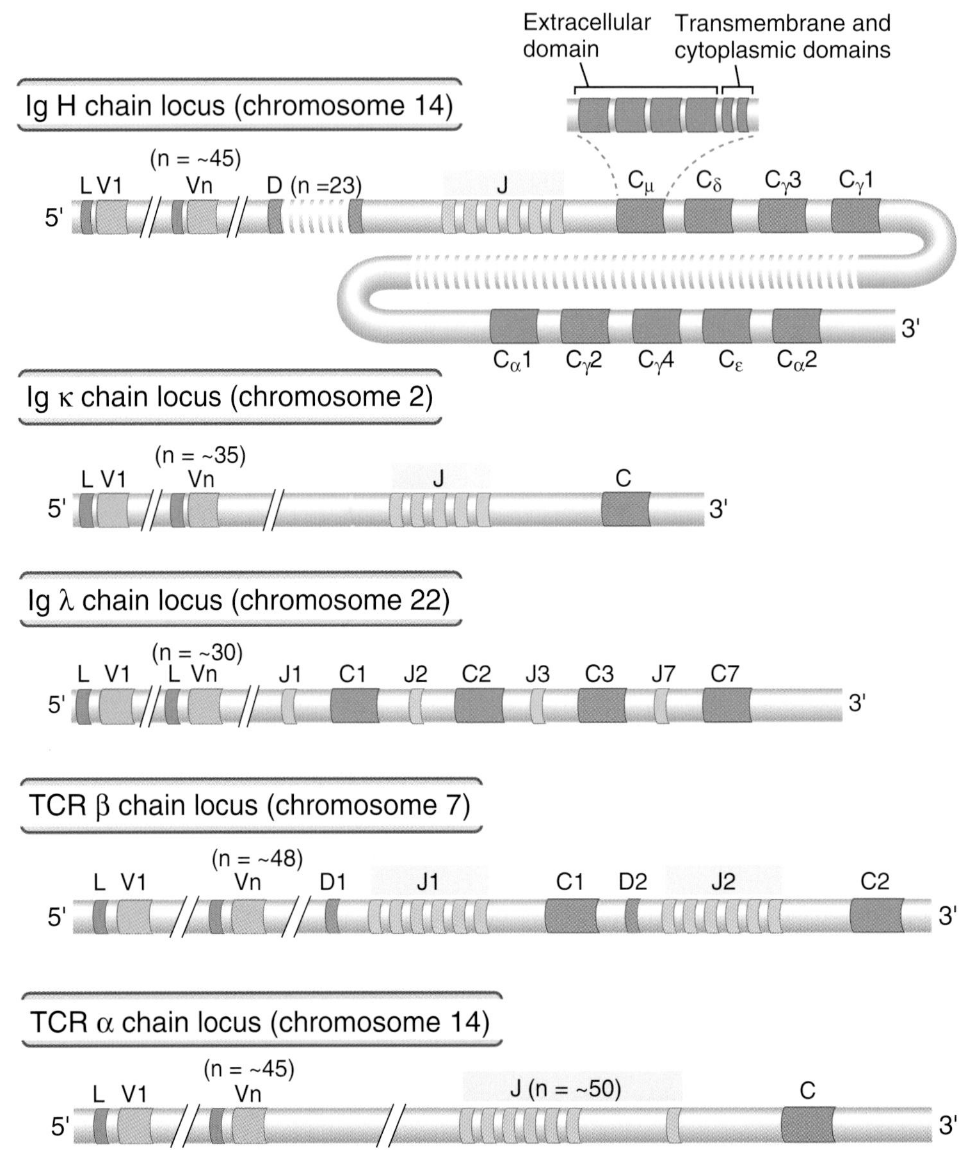

Fig. 4.9 Germline organization of antigen receptor gene loci. In the germline, inherited antigen receptor gene loci contain coding segments (exons, shown as *colored blocks* of various sizes) that are separated by segments that are not expressed (introns, shown as *gray* sections). Each immunoglobulin (Ig) heavy-chain C region and T cell receptor (TCR) C region consists of multiple exons, which are not shown, that encode the domains of the C regions; the organization of the Cμ exons in the Ig heavy-chain locus is shown as an example. The diagrams illustrate the antigen receptor gene loci in humans; the basic organization is the same in all species, although the precise order and number of gene segments may vary. The numbers of V, D, and J gene segments are estimates of functional gene segments (those that can code for proteins). The sizes of the segments and the distances between them are not drawn to scale. *C*, Constant; *D*, diversity; *J*, joining; *L*, leader sequence (a small stretch of nucleotides that encodes a peptide that guides proteins through the endoplasmic reticulum and is cleaved from the mature proteins); *V*, variable.

segments cannot code for functional antigen receptor proteins, so they have to be brought together as lymphocytes mature.

Somatic Recombination and Expression of Antigen Receptor Genes

Following the commitment of a lymphocyte progenitor to the B lymphocyte lineage, a randomly selected Ig heavy-chain D gene segment recombines with a J segment to form a DJ complex. Subsequently, a V segment is joined to the fused DJ complex (Fig. 4.10). Thus, the committed but still-developing B cell now has a recombined VDJ exon in the heavy-chain locus. This gene is transcribed, and in the primary RNA transcript, the VDJ exon is spliced to the C-region exons of the μ chain, the most 5′ C region in the Ig heavy chain gene locus, to form a complete μ messenger RNA (mRNA). The μ mRNA is translated to produce the μ heavy chain, which is the first Ig protein synthesized during B cell maturation.

Essentially the same sequence of DNA recombination, transcription and RNA splicing leads to production of a light chain in B cells, except that the light-chain loci lack D segments, so a V region gene segment recombines directly with a J segment. The rearrangement of TCR α chain and β chain genes in T lymphocytes is similar to that of Ig L and H chains, respectively.

Mechanisms of V(D)J Recombination

The somatic recombination of V and J, or of V, D and J, gene segments is mediated by a lymphoid-specific enzyme, the VDJ recombinase, and additional enzymes, most of which are not lymphocyte specific and are involved in repair of double-stranded DNA breaks introduced by the recombinase. The VDJ recombinase is composed of the recombination-activating gene 1 and 2 (RAG-1 and RAG-2) proteins. These proteins form a dimer that recognizes seven and nine nucleotide DNA sequences (heptamer and nonamer, respectively) that flank all antigen receptor V, D, and J gene segments. As a result of this recognition, the recombinase brings two Ig or TCR gene segments close together and cleaves the DNA at specific sites. The recognition signal sequences are located such that in the Ig light chain and TCRα loci, a V segment recombines with a J segment, and in the Ig heavy chain and TCRβ loci, a D segment joins with a J and then a V segment is joined to the previously recombined DJ. The VDJ recombinase–mediated formation of double-stranded DNA (dsDNA) breaks first involves the formation of hairpin loops that are opened asymmetrically by an enzyme called ARTEMIS. The two dsDNA breaks downstream and upstream, respectively, of the two involved gene segments are then enzymatically ligated, producing a full-length recombined VJ or VDJ exon without the intervening DNA segments (see Fig. 4.10).

The VDJ recombinase is expressed only in immature B and T lymphocytes. Although the same enzyme can mediate recombination of all Ig and TCR genes, intact Ig heavy-chain and light-chain genes are rearranged and expressed only in B cells, and TCR α and β genes are rearranged and expressed only in T cells. The lineage specificity of receptor gene rearrangement appears to be linked to the expression of lineage-specific transcription factors. In B cells, B lineage–specific transcription factors make Ig gene loci accessible to the VDJ recombinase but not the TCR loci, whereas in developing T cells, transcriptional regulators help open the TCR loci but not the Ig loci. The "open" loci are the ones that are accessible to the recombinase.

Generation of Ig and TCR Diversity

Diversity of antigen receptors is produced by the use of different combinations of V, D, and J gene segments in different clones of lymphocytes (called combinatorial diversity) and even more by changes in nucleotide sequences introduced at the junctions of the recombining V, D, and J gene segments (called junctional diversity) (Fig. 4.11). Combinatorial diversity is limited by the number of available V, D, and J gene segments, but junctional diversity is almost unlimited. Junctional diversity is produced by three mechanisms, which generate more sequences than are present in the germline genes:

- Exonucleases may remove nucleotides from V, D, and J gene segments at the sites of recombination.
- A lymphocyte-specific enzyme called terminal deoxyribonucleotidyl transferase (TdT) catalyzes the random addition of nucleotides that are not part of germline genes to the junctions between V and D segments and D and J segments, forming so-called N regions.
- During the DNA repair process, ARTEMIS cuts the hairpin loops asymmetrically, forming overhanging DNA sequences. These overhangs are filled in with new nucleotides, which are called P-nucleotides, creating new coding sequences not in the germline

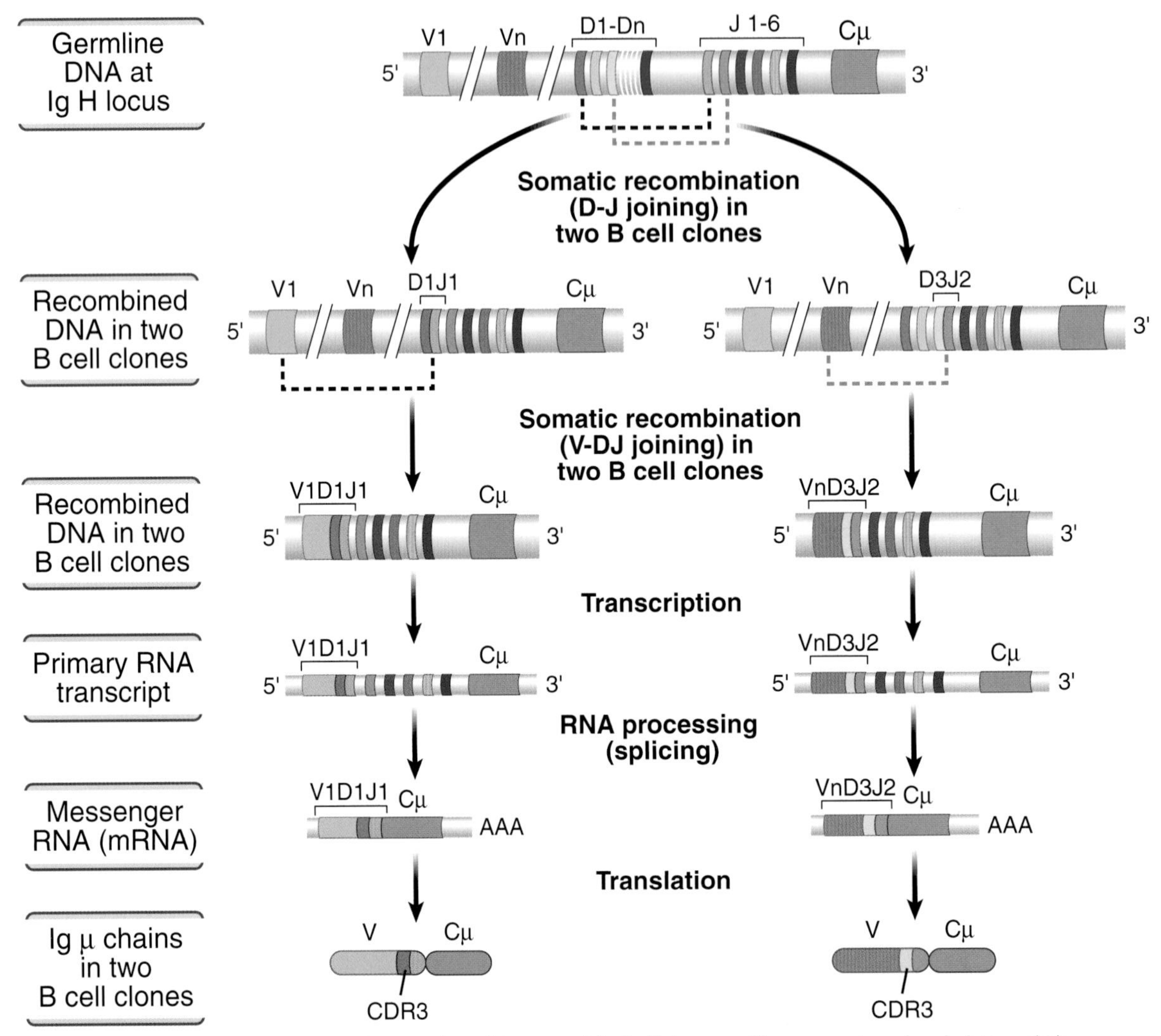

Fig. 4.10 **Recombination and expression of immunoglobulin (Ig) genes.** The expression of an Ig heavy chain involves two gene recombination events (D-J joining, followed by joining of a V region to the DJ complex, with deletion of intervening gene segments). The recombined gene is transcribed, and the VDJ complex is spliced onto the C region exons of the first heavy-chain RNA (which is μ), to give rise to the μ messenger RNA (mRNA). The mRNA is translated to produce the μ heavy-chain protein. The recombination of other antigen receptor genes—that is, the Ig light chain and the T cell receptor (TCR) α and β chains—follows essentially the same sequence, except that in loci lacking D segments (Ig light chains and TCR α), a V gene recombines directly with a J gene segment.

DNA, thereby introducing additional variability at the sites of recombination.

As a result of these mechanisms, the nucleotide sequence of the V(D)J exons in antibody or TCR genes in one clone of lymphocytes differs from the sequence of the V(D)J exons of antibody or TCR molecules in every other clone. The junctional sequences and the D and J segments encode the amino acids of the CDR3 loop, mentioned earlier as the most variable of the CDRs and the one that contributes most to the fine specificity of antigen recognition. Thus, junctional diversity maximizes the variability in the antigen-recognizing portions of antibodies and TCRs. In the process of creating junctional diversity, many genes

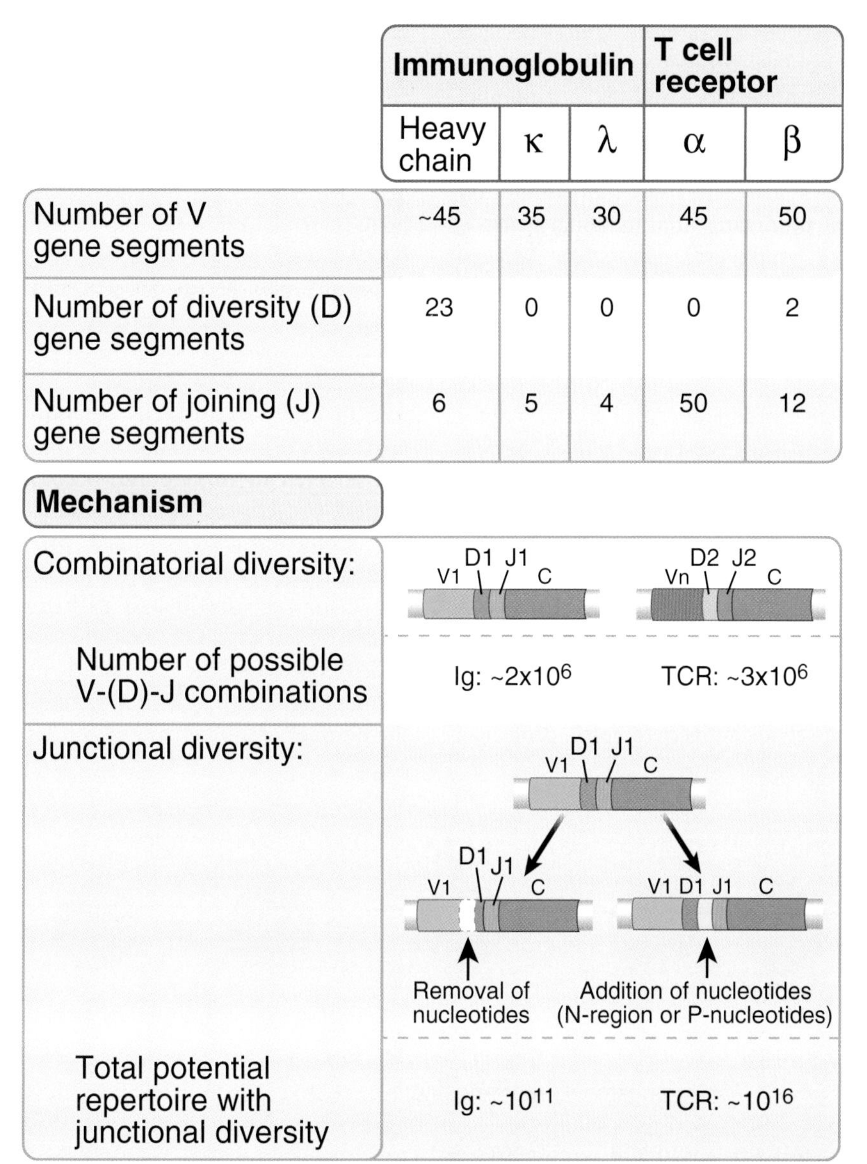

Fig. 4.11 Mechanisms of diversity in antigen receptors. Diversity in immunoglobulins (Igs) and T cell receptors (TCRs) is produced by random combinations of V, D, and J gene segments, which is limited by the numbers of these segments and by removal and addition of nucleotides at the V-J or V-D-J junctions, which is almost unlimited. The numbers of gene segments refer to the average numbers of functional genes (which are known to be expressed as RNA or protein) in humans. Junctional diversity maximizes the variations in the CDR3 regions of the antigen receptor proteins because CDR3 includes the junctions at the site of V-J and V-D-J recombination. The diversity is further enhanced by the juxtaposition of the V regions of the two types of chains in Ig or TCRs to form the complete antigen binding sites, and thus the total diversity is theoretically the product of the total diversity of each of the juxtaposed V regions. The estimated contributions of these mechanisms to the total possible numbers of distinct B and T cell antigen receptors are shown. Although the upper limit on the number of immunoglobulin (Ig) and TCR proteins that may be expressed is extremely large, each individual contains on the order of only 10^7 to 10^9 clones of B cells and T cells with distinct specificities and receptors; in other words, only a fraction of the potential repertoire may actually be expressed. Modified from Davis MM, Bjorkman PJ: T-cell antigen receptor genes and T-cell recognition, *Nature* 334:395–402, 1988.

may be produced with out-of-frame sequences because precisely three nucleotides encode one amino acid, and sometimes junctional nucleotides add up to a number not divisible by 3; these sequences cannot code for proteins and are therefore useless. This is the price the immune system pays for generating tremendous diversity. The risk of producing nonfunctional genes is why the process of lymphocyte maturation contains checkpoints at which only cells with useful receptors are selected to survive.

The uniqueness of CDR3 sequences in every lymphocyte clone can be exploited to distinguish neoplastic and reactive proliferations of B and T lymphocytes. In tumors arising from these cells, all the cells of the tumor will have the same CDR3 (because they all arose from a single B or T cell clone), but in proliferations that are reactions to external stimuli, many lymphocyte clones may proliferate, so many CDR3 sequences will be present. The same principle can be used to define the magnitude of an immune response—when a specific antigen triggers lymphocyte proliferation, many of the cells will express the same CDR3 sequence. If DNA sequencing of an antigen receptor gene in a mixture of lymphocytes reveals relative abundance of a specific CDR3 sequence, this serves as a tool to identify clonal expansion.

Maturation and Selection of B Lymphocytes

The maturation of B lymphocytes occurs mainly in the bone marrow (Fig. 4.12). Progenitors committed to the B cell lineage proliferate, giving rise to a large number of precursors of B cells. Subsequent maturation involves antigen receptor gene expression and selection.

Early Steps in B Cell Maturation

The Ig heavy-chain locus rearranges first, and only cells that are able to make an Ig μ heavy-chain protein are selected to survive and become pre-B cells. The earliest progenitors committed to the B cell lineage are proliferating cells called **pro-B cells**. Ig gene recombination occurs after pro-B cells stop dividing, producing a DJ and then a VDJ complex, as described earlier. Given that junctional nucleotides are randomly added both when the D-J joint is made and when a V segment fuses with a DJ unit, in roughly half of the cells, the number of junctional nucleotides will not add up to

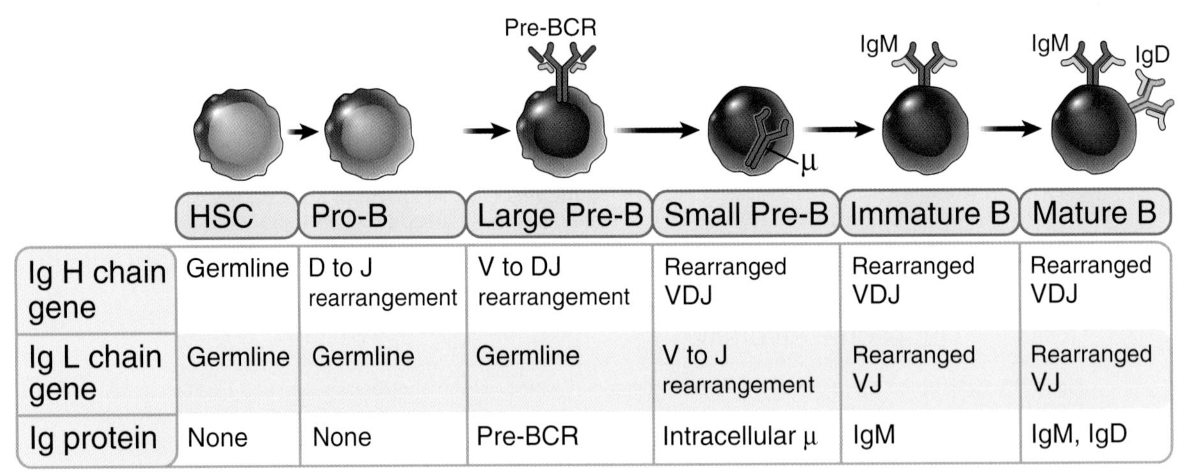

	HSC	Pro-B	Large Pre-B	Small Pre-B	Immature B	Mature B
Ig H chain gene	Germline	D to J rearrangement	V to DJ rearrangement	Rearranged VDJ	Rearranged VDJ	Rearranged VDJ
Ig L chain gene	Germline	Germline	Germline	V to J rearrangement	Rearranged VJ	Rearranged VJ
Ig protein	None	None	Pre-BCR	Intracellular μ	IgM	IgM, IgD

Fig. 4.12 Steps in the maturation and selection of B lymphocytes. The maturation of B lymphocytes proceeds through sequential steps, each of which is characterized by particular changes in immunoglobulin *(Ig)* gene expression and in the patterns of Ig protein expression. Pro-B cells begin to rearrange Ig heavy-chain genes, and large pre-B cells are selected to survive and proliferate if they successfully rearrange an Ig heavy-chain gene and assemble a pre-BCR. The pre-BCR consists of a membrane-associated Ig μ protein attached to two other proteins called surrogate light chains because they take the place of the light chain in a complete Ig molecule. Small pre-B cells initiate Ig light-chain gene rearrangement, immature B cells assemble a complete membrane IgM receptor, and mature B cells coexpress IgD, with the same V regions and specificity as in the first Ig produced. *BCR,* B cell receptor; *HSC,* hematopoietic stem cell; *mRNA,* messenger RNA.

a multiple of three. If recombination in one of the two inherited IgH loci is not successful because it produces out-of-frame sequences, recombination will take place in the other locus. The cells that successfully make functional heavy-chain gene rearrangements and synthesize the Ig heavy-chain μ protein are called **pre-B cells.** Pre-B cells are therefore defined by the presence of the Ig μ heavy-chain protein. They also express the μ protein on the cell surface in association with two other invariant proteins called surrogate light chains because they resemble light chains and associate with the μ heavy chain. The complex of μ chain and surrogate light chains associates with the Igα and Igβ signaling molecules to form the pre-B cell receptor (pre-BCR) complex.

Role of the Pre-BCR Complex in B Cell Maturation

The assembled pre-BCR complex serves essential functions in the maturation of B cells:

- Signals from the pre-BCR complex promote the survival and proliferation of B lineage cells that have made a productive rearrangement at the Ig H chain locus. This is the first checkpoint in B cell development, and it selects and expands the pre-B cells that express a functional μ heavy chain (which is an essential component of the pre-BCR and BCR). Pre-B cells that make out-of-frame (nonproductive) rearrangements at the heavy-chain locus fail to make the μ protein, cannot express a pre-BCR or receive pre-BCR signals, and die by programmed cell death (apoptosis). The pre-BCR signaling pathway includes a downstream tyrosine kinase called BTK, which is encoded on the X chromosome. Mutations in *BTK* in boys results in the failure of pre-B cells to survive and the subsequent absence of B cells. This disease is called **X-linked agammaglobulinemia.**
- The pre-BCR complex signals to shut off recombination of Ig heavy-chain genes on the second chromosome, so each B cell can express an Ig heavy chain from only one of the two inherited parental alleles. This process is called allelic exclusion, and it ensures that each cell can only express a receptor of a single specificity.
- The pre-BCR complex helps induce V to J rearrangement of the κ light-chain gene, leading to production of the κ protein and the assembly of cell surface IgM. The cells at this next stage of differentiation are called **immature B cells.** The λ light chain is produced only if both rearranged κ chain loci fail to express a functional protein or if the κ chain generates a potentially harmful self-reactive receptor and has to be eliminated by a process called receptor editing, described later.

In immature B cells, the BCR complex delivers signals that promote survival, thus preserving cells that express complete antigen receptors; this is the second checkpoint during B cell maturation. Signals from the antigen receptor also shut off production of the recombinase enzyme and further recombination at light-chain loci. As a result, each B cell produces either one κ or one λ light chain from one of the inherited parental alleles. The presence of two sets of light-chain genes in the genome simply increases the chance of completing successful gene recombination and receptor expression.

Late Steps of B Cell Maturation

Further maturation occurs after the immature B cells leave the bone marrow and enter the spleen. The final maturation step involves coexpression of IgD with IgM; this occurs because in any given B cell, the recombined heavy-chain VDJ unit may be spliced either to Cμ or Cδ exons in the primary RNA transcript, giving rise to μ or δ mRNA, respectively. We know that the ability of B cells to respond to antigens develops together with the coexpression of IgM and IgD, but why both classes of receptor are needed is not known. The $IgM^{+}IgD^{+}$ cell is the **mature B cell,** able to respond to antigen in peripheral lymphoid tissues.

Selection of Mature B Cells

Developing B cells are positively selected based mainly on expression of complete antigen receptors. The repertoire of developing B cells is further shaped by negative selection. In this process, if an immature B cell binds an antigen in the bone marrow with high affinity, it may re-express the VDJ recombinase enzyme, undergo additional light-chain V-J recombination, generate a different light chain, and thus change the specificity of the antigen receptor, a process called **receptor editing** (see Chapter 9). Immature B cells called transitional B cells that encounter antigens in the periphery may die by apoptosis, also known as **deletion**. The antigens that developing B cells may recognize in the bone marrow and early in development in the periphery are mostly self antigens that are abundantly expressed throughout the body (i.e., are ubiquitous), such as blood proteins, and membrane molecules common to all cells. Negative selection therefore eliminates potentially dangerous cells that can recognize and react against ubiquitous self antigens.

The process of Ig gene recombination is random and cannot be inherently biased toward recognition of microbes. However, the receptors produced are able to recognize the antigens of many, varied microbes that the immune system must defend against. The repertoire of B lymphocytes is selected positively for expression of functional receptors and selected negatively against strong recognition of self antigens. What is left after these selection processes is a large collection of mature B cells, which include cells that are able to recognize almost any microbial antigen that may be encountered.

Subsets of Mature B Cells

Most mature B cells are called follicular B cells because they are found within lymph node and spleen follicles. Marginal-zone B cells, which are found at the margins of splenic follicles, develop from bone marrow–derived hematopoietic stem cells, as do follicular B cells. B-1 lymphocytes, a distinct population found at mucosal sites and the peritoneal cavity, develop earlier from fetal liver–derived hematopoietic stem cells. The diversity of marginal zone and B1 B cells is much more limited than that of follicular B cells, but the mechanisms constraining diversity of these subsets are not well understood. The role of these B cell subsets in humoral immunity is described in Chapter 7.

Maturation and Selection of T Lymphocytes

T cell progenitors migrate from the bone marrow to the thymus, where the entire process of maturation occurs (Fig. 4.13). The process of T lymphocyte maturation has some unique features, primarily related

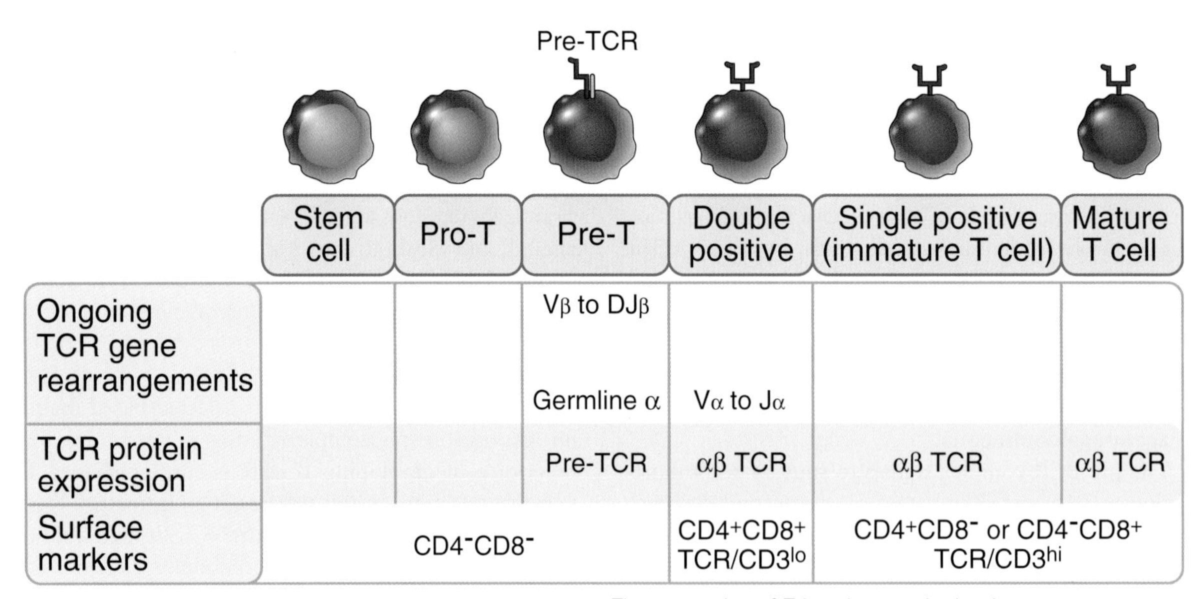

Fig. 4.13 Steps in the maturation of T lymphocytes. The maturation of T lymphocytes in the thymus proceeds through sequential steps often defined by the expression of the CD4 and CD8 coreceptors. The T cell receptor (TCR) β chain is first expressed at the double-negative pre-T cell stage, and the complete T cell receptor is expressed in double-positive cells. The pre-TCR consists of the TCR β chain associated with a protein called pre-Tα. Maturation culminates in the development of $CD4^+$ and $CD8^+$ single-positive T cells. As with B cells, failure to express antigen receptors at any stage leads to death of the cells by apoptosis.

to the development of different classes of T cells that have receptors that recognize different classes of MHC molecules.

Early Steps in T Cell Maturation

The least developed progenitors in the thymus are called **pro-T cells** or **double-negative T cells** because they do not express CD4 or CD8. These cells expand in number mainly under the influence of IL-7 produced in the thymus. TCR β gene recombination, mediated by the VDJ recombinase, occurs in some of these double-negative cells. (The γδ T cells undergo similar recombination involving TCR γ and δ loci, but they belong to a distinct lineage and are not discussed further.) If VDJ recombination is successful in one of the two inherited TCR β chain loci and a TCR β chain protein is synthesized, it is expressed on the cell surface in association with an invariant protein called pre-Tα, to form the pre-TCR complex of **pre-T cells**. If the recombination in one of the two inherited loci is not successful, recombination will take place on the other locus. If that also fails and a complete TCR β chain is not produced in a pro-T cell, the cell dies.

The pre-TCR complex delivers intracellular signals once it is assembled, similar to the signals from the pre-BCR complex in developing B cells. These signals promote survival, proliferation, and TCR α gene recombination and inhibit VDJ recombination in the second β chain locus (allelic exclusion). Failure to express the α chain and the complete TCR again results in death of the cell. The surviving cells express the complete αβ TCR and both the CD4 and CD8 coreceptors; these cells are called **double-positive T cells**.

Selection of Mature T Cells

The different αβ TCRs produced in double-positive T cells are capable of recognizing peptides displayed by any MHC allele in the population, but most of these are not present in each individual. Therefore, there have to be mechanisms that preserve the T cells that can recognize peptide-MHC complexes in each individual and eliminate cells incapable of recognizing that individual's MHC molecules. If the TCR of a T cell recognizes an MHC molecule in the thymus, which must be a self MHC molecule displaying a self peptide, and if the interaction is of low or moderate affinity, this T cell is selected to survive (Fig 4.14). T cells that do not recognize an MHC molecule in the thymus die by apoptosis; these T cells would not be functional because they would be incapable of seeing MHC-displayed antigens in that individual. This preservation of self MHC—restricted (i.e., useful) T cells is the process of **positive selection**. During this process, T cells whose TCRs recognize class I MHC—peptide complexes preserve the expression of CD8, the coreceptor that binds to class I MHC, and lose expression of CD4, the coreceptor specific for class II MHC molecules. Conversely, if a T cell recognizes class II MHC—peptide complexes, this cell maintains expression of CD4 and loses expression of CD8. Thus, what emerges are **single-positive T cells** (or single-positive thymocytes), which are either $CD8^+$ class I MHC restricted or $CD4^+$ class II MHC restricted. During positive selection, the T cells also become committed to different functional fates: the $CD8^+$ T cells will differentiate into CTLs on activation, and the $CD4^+$ cells will differentiate into cytokine-producing helper T cells.

Immature, double-positive T cells whose receptors strongly recognize MHC-peptide complexes in the thymic cortex undergo apoptosis. This is the process of **negative selection**, and it serves to eliminate potentially dangerous T lymphocytes. Negative selection continues in immature $CD4^+$ and $CD8^+$ single-positive T cells in the thymic medulla. If a T cell that recognizes a self peptide-MHC complex with high avidity in the thymus were allowed to mature, recognition of the same self antigen in the periphery could lead to harmful immune responses against self tissues, so such a T cell must be eliminated. Some immature $CD4^+$ T cells that recognize self antigens in the thymus with intermediate avidity do not die but develop into regulatory T cells (Treg) (see Chapter 9), although the basis of which cells die and which become Tregs is not known. Most of the proteins present in the thymus are self proteins, because foreign (microbial and tumor) antigens are typically captured and taken to secondary lymphoid organs. Some of these self proteins are present throughout the body, and others are proteins that are restricted to particular tissues but are expressed in thymic epithelial cells by special mechanisms, as discussed in Chapter 9 in the context of self-tolerance.

It may seem surprising that both positive selection and negative selection are mediated by recognition of the same set of self MHC—self peptide complexes in the

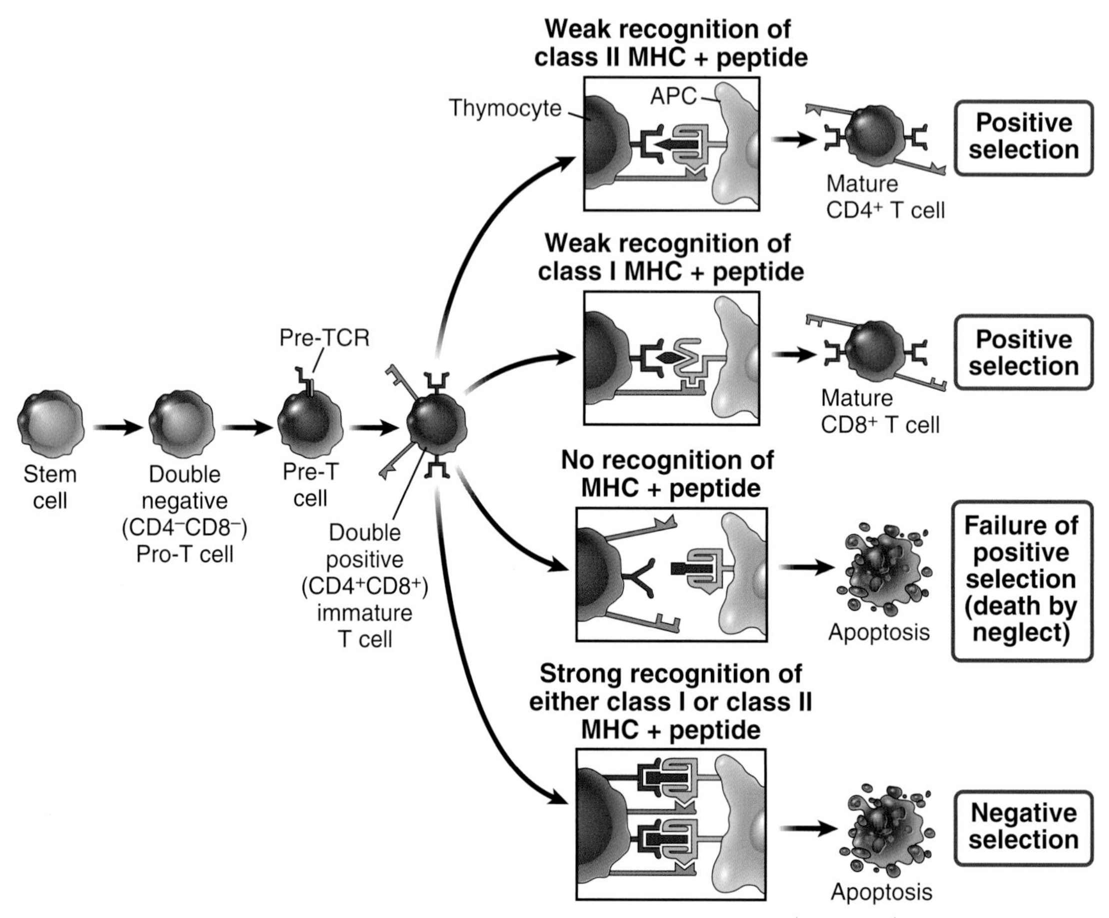

Fig. 4.14 Selection of major histocompatibility complex (MHC)–restricted CD4$^+$ and CD8$^+$ T lymphocytes. The strength of recognition of peptide-MHC complexes by immature double-positive T cells in the thymus determines the development of these cells into single-positive CD4$^+$ and CD8$^+$ cells (positive selection) and the elimination of self-reactive cells (negative selection). Only class II MHC is shown for negative selection, but the same process eliminates self-reactive class I MHC–restricted CD8$^+$ T cells. *APC,* Antigen-presenting cell.

thymus. The two factors that determine the choice between positive and negative selection are the affinity of the TCR and the concentration of the self antigen in the thymus. If a TCR strongly recognizes an abundant self antigen in the thymus, that T cell will be negatively selected, which makes sense because strong recognition of an abundant self antigen has the potential for causing autoimmunity. However, if a TCR recognizes a self peptide–self MHC complex weakly, that T cell will be positively selected because there is a reasonable chance the T cell will recognize a foreign peptide presented by self MHC strongly. This is the process that gives rise to the repertoire of functional T cells.

SUMMARY

- In the adaptive immune system, the molecules responsible for specific recognition of antigens are antibodies and T cell antigen receptors.
- Antibodies (also called immunoglobulins) may be produced as membrane receptors of B lymphocytes

and as proteins secreted by antigen-stimulated B cells that have differentiated into antibody-secreting plasma cells. Secreted antibodies are the effector molecules of humoral immunity, capable of neutralizing microbes and microbial toxins and eliminating them by activating various effector mechanisms.

- T cell receptors (TCRs) are membrane receptors and are not secreted.
- The core structure of antibodies consists of two identical heavy chains and two identical light chains, forming a disulfide-linked complex. Each chain consists of a variable (V) region, which is the portion that recognizes antigen, and a constant (C) region, which provides structural stability and, in heavy chains, performs the effector functions of antibodies. The V regions of one heavy chain and one light chain together form the antigen-binding site, and thus the core structure has two identical antigen-binding sites.
- T cell receptors consist of an α chain and a β chain. Each chain contains one V region and one C region, and both chains participate in the recognition of antigens, which for most T cells are peptides displayed by MHC molecules.
- The V regions of immunoglobulin (Ig) and TCR molecules contain hypervariable segments, also called complementarity-determining regions (CDRs), which are the regions of contact with antigens.
- The inherited genes that encode antigen receptors consist of multiple segments separated in the germline and brought together during maturation of lymphocytes. In B cells, the Ig gene segments undergo recombination as the cells mature in the bone marrow, and in T cells, the TCR gene segments undergo recombination during maturation in the thymus.
- Receptors of different specificities are generated in part by different combinations of V, D, and J gene segments. The process of recombination introduces additional variability in the nucleotide sequences at the sites of recombination by adding or removing nucleotides from the junctions. The result of this introduced variability is the development of a diverse repertoire of lymphocytes, in which clones of cells with different antigen specificities express receptors that differ in sequence and recognition, and most of the differences are concentrated at the regions of gene recombination.
- During their maturation, lymphocytes are selected to survive at several checkpoints; only cells with complete functional antigen receptors are preserved and expanded. In addition, T lymphocytes are positively selected to recognize peptide antigens displayed by self MHC molecules and to ensure that the recognition of the appropriate type of MHC molecule (class I or class II) matches the coreceptor (CD8 or CD4, respectively) that is preserved.
- Immature lymphocytes that strongly recognize self antigens are negatively selected and prevented from completing their maturation, or in B cells undergo further light chain gene rearrangements that change thier specificity, thus eliminating cells with the potential of reacting in harmful ways against self tissues.

REVIEW QUESTIONS

1. What are the functionally distinct domains (regions) of antibody and TCR molecules? What features of the amino acid sequences in these regions are important for their functions?
2. What are the differences in the types of antigens recognized by antibodies and TCRs?
3. What mechanisms contribute to the diversity of antibody and TCR molecules? Which of these mechanisms contributes the most to the diversity?
4. What are some of the checkpoints during lymphocyte maturation that ensure survival of the useful cells?
5. What is the phenomenon of negative selection, and what is its importance?

Answers to and discussion of the Review Questions may be found on p. 322.

5

T Cell–Mediated Immunity
Activation of T Lymphocytes

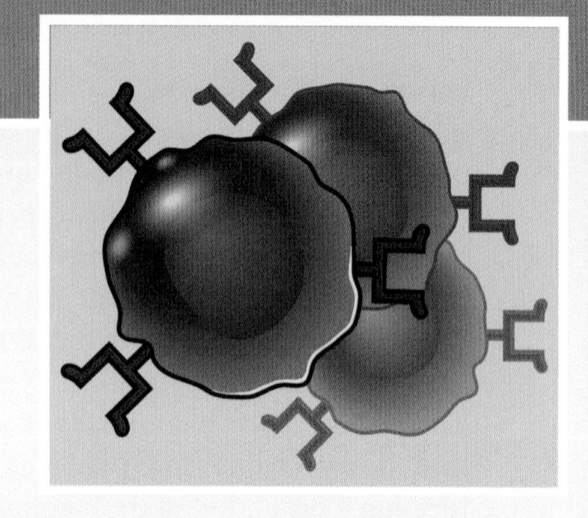

CHAPTER OUTLINE

T lymphocytes perform multiple functions in defend-illkng against infections by various kinds of microbes. A major role for T lymphocytes is in **cell-mediated immunity**, which provides defense against infections by microbes that live and replicate inside host cells. In all viral and some bacterial, fungal, and protozoan infections, microbes may find a haven inside cells, from where they must be eliminated by cell-mediated immune responses (Fig. 5.1).

- Many microbes are ingested by phagocytes as part of the early defense mechanisms of innate immunity and are killed by microbicidal mechanisms that are largely limited to phagocytic vesicles (to protect the phagocytes themselves from damage by these mechanisms). However, some of these microbes have evolved to resist the microbicidal activities of phagocytes and are able to survive, and even replicate, in the vesicles of macrophages. In such infections, $CD4^+$ helper T cells enhance the ability of macrophages to kill the ingested microbes.
- Some extracellular microbes, such as bacteria and fungi, are readily destroyed if they are phagocytosed, especially by neutrophils. Other extracellular pathogens, such as helminthic parasites, are destroyed by eosinophils. In these infections, $CD4^+$ T cells produce cytokines that recruit and activate the leukocytes that destroy the microbes.
- Some microbes, notably viruses, are able to infect and replicate inside a wide variety of cells, and parts of the life cycles of the viruses take place in the cytosol and nucleus. These infected cells often do not possess intrinsic mechanisms for destroying the

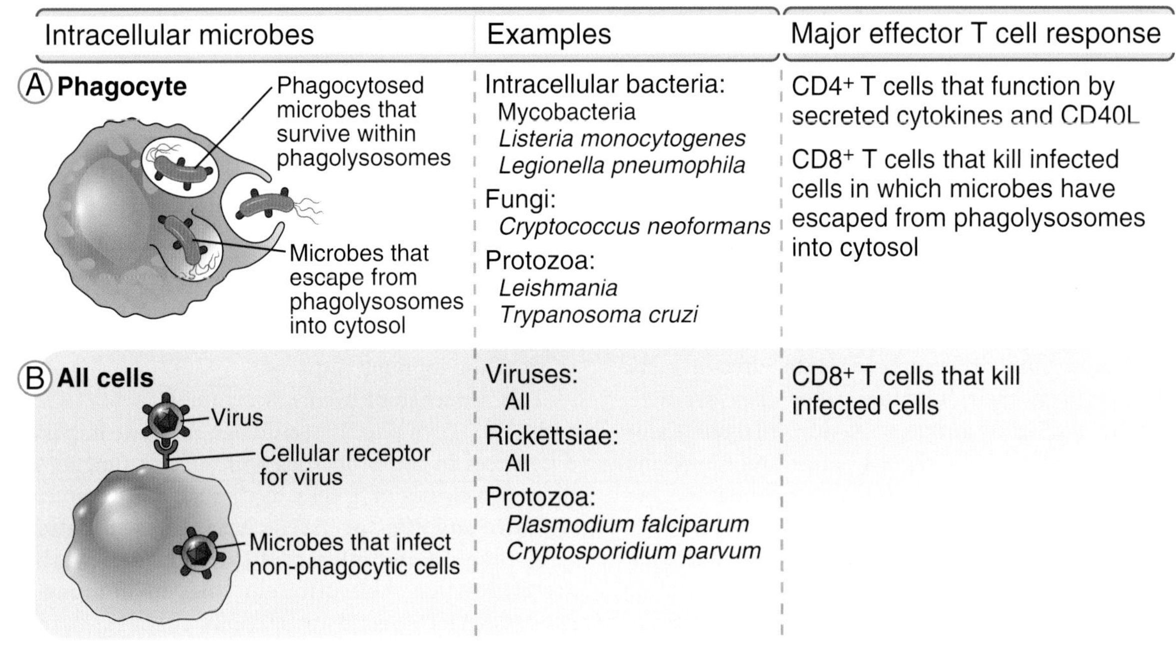

Fig. 5.1 Types of intracellular microbes combated by T cell–mediated immunity. **A,** Microbes may be ingested by phagocytes and may survive within vesicles (phagolysosomes) or escape into the cytosol, where they are not susceptible to the microbicidal mechanisms of the phagocytes. Helper T cells combat microbes by enhancing the killing functions of the phagocytes in which the microbes reside; the roles of cytokines and CD40-ligand (CD40L) are discussed later. **B,** Viruses may infect many cell types, including nonphagocytic cells, and replicate in the nucleus and cytosol of the infected cells. Rickettsiae and some protozoa are obligate intracellular parasites that reside in nonphagocytic cells. Cytotoxic T lymphocytes (CTLs) combat these microbes by killing the cells in which they reside. CTLs also destroy cells in which ingested microbes have escaped into the cytosol **(A).**

microbes. Even some phagocytosed microbes within macrophages can escape into the cytosol and evade microbicidal mechanisms, which are restricted to vesicles. $CD8^+$ cytotoxic T lymphocytes (CTLs) kill the infected cells, thus eliminating the reservoir of infection.

Other populations of $CD4^+$ T cells help B cells produce antibodies as part of humoral immune responses (see Chapter 7). Although our emphasis in this chapter is on defense against infections, the principal physiologic function of the immune system, some T cells, especially $CD8^+$ CTLs, also kill cancer cells. This role of T cells is discussed in Chapter 10.

Most of the functions of T lymphocytes—activation of phagocytes, killing of infected and tumor cells, and help for B cells—require that the T lymphocytes interact with other cells, which may be phagocytes, infected host cells, or B lymphocytes. Furthermore, the initiation of T cell responses requires that naive T cells recognize antigens displayed by dendritic cells, which capture antigens from epithelial barriers and organs and concentrate the antigens in secondary lymphoid organs. Recall that the specificity of T cells for peptides displayed by major histocompatibility complex (MHC) molecules ensures that the T cells can see and respond only to antigens inside other host cells (see Chapters 3 and 4). This chapter discusses the way in which T lymphocytes are activated by recognition of cell-associated antigens and other stimuli. We address the following questions:

- What stimuli are needed to activate naive T lymphocytes and initiate cell-mediated immune responses?
- How are the few naive T cells specific for any microbe converted into the large number of effector T cells that have specialized functions and the ability to eliminate diverse microbes?
- What biochemical signals are required for the activation of T lymphocytes?

After describing here how T cells recognize and respond to the antigens of cell-associated microbes, the development of effector T lymphocytes and their functions in cell-mediated immunity are discussed in Chapter 6 and the roles of helper T cells in antibody responses are discussed in Chapter 7.

STEPS IN T CELL RESPONSES

Naive T lymphocytes that recognize antigens in the secondary (peripheral) lymphoid organs respond by proliferation and differentiation into effector cells, which perform their functions when they are activated by the same antigens in any infected tissue (Fig. 5.2). Naive T cells express antigen receptors and coreceptors that recognize cells harboring microbes, but naive cells are incapable of performing the effector functions required for eliminating the microbes. Differentiated effector cells are capable of performing these functions, which they do at any site of infection. In this chapter, we focus on the initial responses of naive T cells to antigens in secondary lymphoid organs.

The responses of naive T lymphocytes to host cell-associated microbial antigens consist of a series of sequential steps that result in an increase in the number of antigen-specific T cells and the differentiation of naive T cells into effector and memory cells (Fig. 5.3).

- One of the earliest responses is the secretion of cytokines required for proliferation and differentiation and increased expression of receptors for various cytokines. The cytokine interleukin-2 (IL-2), which is produced by antigen-activated T cells, stimulates proliferation of these cells, resulting in a rapid increase in the number of antigen-specific lymphocytes, a process called clonal expansion.
- The activated lymphocytes differentiate, resulting in the conversion of naive T cells into a population of effector T cells that function to eliminate microbes.
- Many of the effector T cells leave the lymphoid organs, enter the circulation, and migrate to sites of infection, where they can eradicate the microbes. Some activated T cells may remain in secondary lymphoid organs, where they provide signals to B cells that promote antibody responses against the microbes.
- Some of the progeny of the T cells that have proliferated in response to antigen develop into memory T cells, which are long-lived, circulate in the blood or reside in tissues for years, and are ready to respond rapidly to subsequent exposure to the same microbe.
- As effector T cells eliminate the infectious agent, the stimuli that triggered T cell expansion and differentiation are also eliminated. As a result, most of the cells in the greatly expanded clones of antigen-specific lymphocytes die, returning the system to a resting state, with only memory cells remaining from the immune response.

This sequence of events is common to both $CD4^+$ and $CD8^+$ T lymphocytes, although there are important differences in the properties and effector functions of these two classes of T cells, as discussed in Chapter 6.

Naive and effector T cells have different patterns of circulation and migration through tissues, which are critical for their different roles in immune responses. As discussed in previous chapters, naive T lymphocytes constantly recirculate through secondary lymphoid organs searching for foreign protein antigens. Microbial antigens are transported from the portals of entry of the microbes to the same regions of the lymphoid organs where the recirculating naive T cells are located. Dendritic cells are the most efficient antigen-presenting cells (APCs) for transporting antigens to lymph nodes and for stimulating naive T cells (see Chapter 3). In the secondary lymphoid organs, dendritic cells process the antigens and display peptides bound to MHC molecules at the cell surface. When a T cell recognizes antigen, it is transiently arrested on the dendritic cell and receives signals from the antigen receptor and other receptors that activate the T cell. Activation results in proliferation and differentiation, and then the cells may leave the lymphoid organ and migrate preferentially to the inflamed tissue, the original source of the antigen. The regulation of this directed migration is discussed later in this chapter.

With this overview, we proceed to a description of the stimuli required for T cell activation and regulation. We then describe the biochemical signals that are generated by antigen recognition, the biologic responses of the lymphocytes, and how T cell responses are regulated.

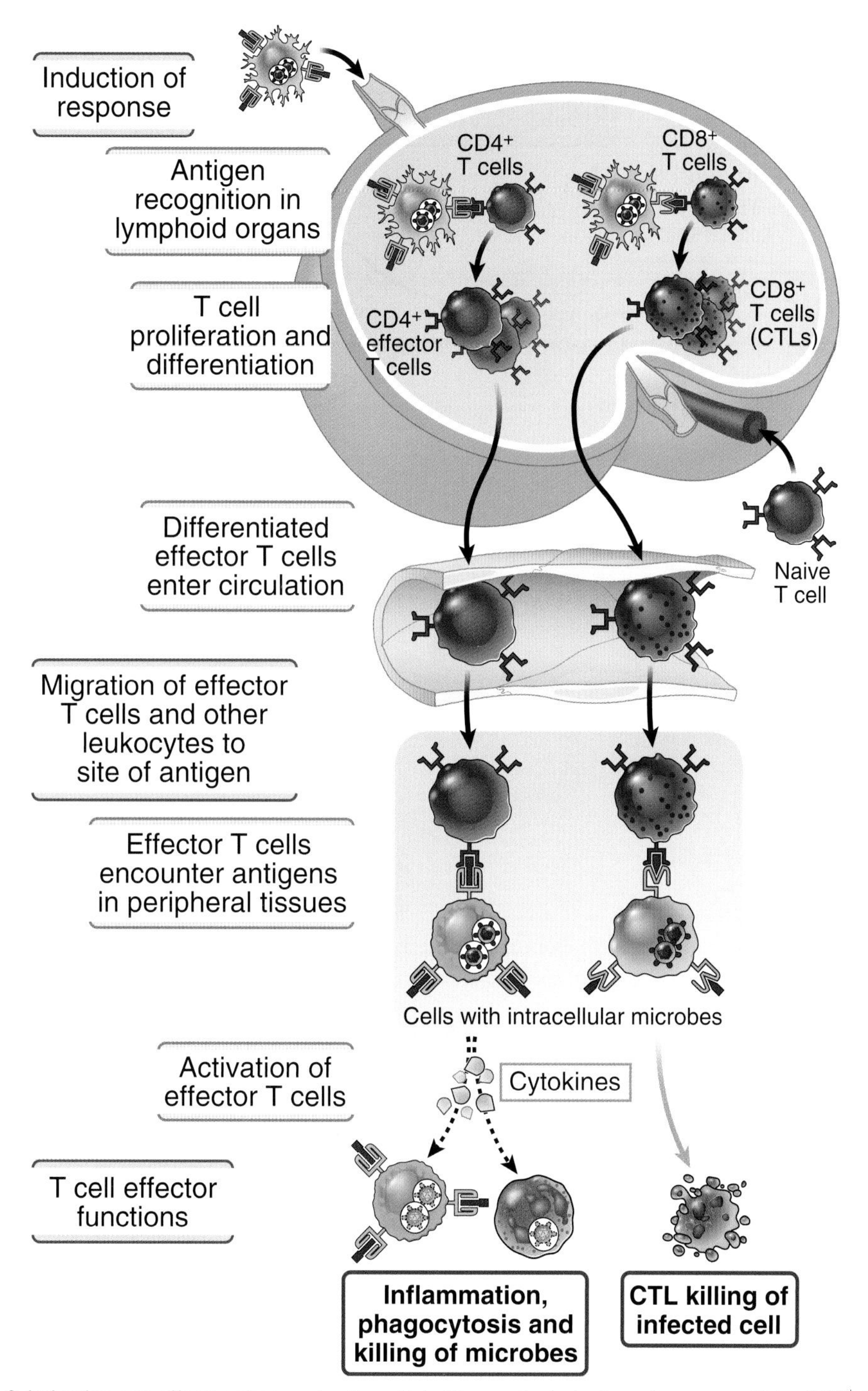

Fig. 5.2 **Induction and effector phases of cell-mediated immunity.** Induction of response: Naive $CD4^+$ T cells and $CD8^+$ T cells recognize peptides that are derived from protein antigens and presented by DCs in peripheral lymphoid organs. The T lymphocytes are stimulated to proliferate and differentiate into effector cells, many of which enter the circulation. Some of the activated $CD4^+$ T cells remain in the lymph node, migrate into follicles, and help B cells to produce antibodies (shown in Fig. 5.13). Migration of effector T cells and other leukocytes to site of antigen: effector T cells and other leukocytes migrate through blood vessels in peripheral tissues by binding to endothelial cells that have been activated by cytokines produced in response to infection in these tissues. T cell effector functions: $CD4^+$ T cells recruit and activate phagocytes to destroy microbes, and $CD8^+$ cytotoxic T lymphocytes (CTLs) kill infected cells.

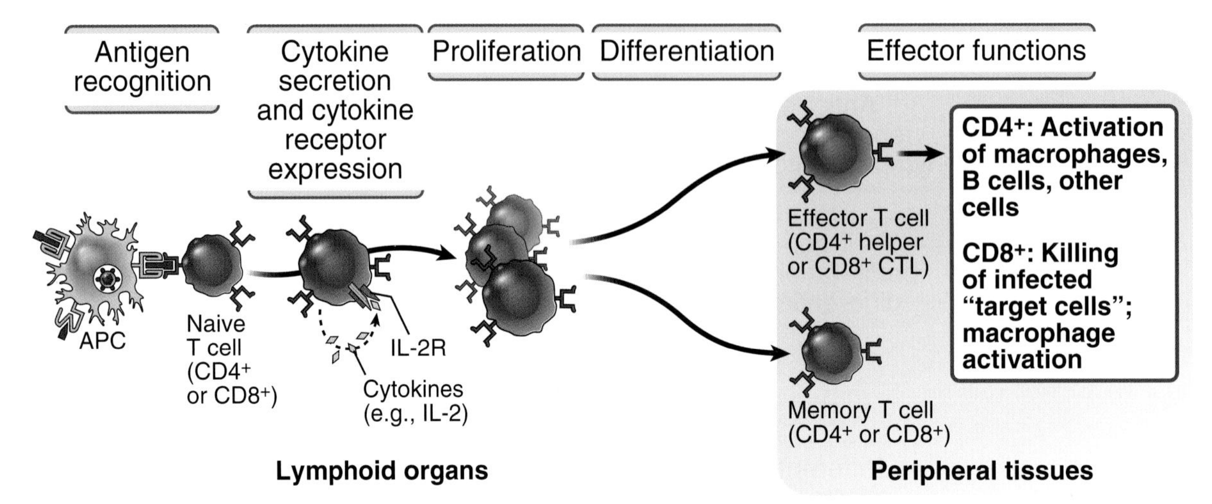

Fig. 5.3 **Steps in the activation of T lymphocytes.** Naive T cells recognize MHC-associated peptide antigens displayed on antigen-presenting cells and other signals (not shown). The T cells respond by producing interleukin-2 (IL-2) and expressing receptors for IL-2, leading to an autocrine pathway of cell proliferation. The result is expansion of the clone of T cells that are specific for the antigen. Some of the progeny differentiate into effector cells, which serve various functions in cell-mediated immunity, and some differentiate into memory cells, which survive for long periods. Other changes associated with activation, such as the expression of various surface molecules, are not shown. *APC,* Antigen-presenting cell; *CTL,* cytotoxic T lymphocyte; *IL-2R,* interleukin-2 receptor.

ANTIGEN RECOGNITION AND COSTIMULATION

The initiation of T cell responses requires multiple receptors on the T cells recognizing their specific ligands on APCs (Fig. 5.4).

- The T cell receptor (TCR) recognizes MHC-associated peptide antigens.
- CD4 or CD8 coreceptors on the T cells bind to MHC molecules on the APC and help the TCR complex deliver activating signals.
- Adhesion molecules strengthen the binding of T cells to APCs.
- Molecules called costimulators, which are expressed on APCs after encounter with microbes, bind to their receptors on the naive T cells, and promote responses, especially to infectious pathogens.
- Cytokines secreted by various cell types bind to receptors on the T cells and amplify the T cell response and direct it along various differentiation pathways.

The roles of these molecules in T cell responses to antigens are described next. Cytokines are discussed mainly in Chapter 6.

Recognition of Peptide-MHC Complexes

The TCR and the CD4 or CD8 coreceptor together recognize complexes of peptide antigens and MHC molecules on APCs, and this recognition provides the initiating, or first, signal for T cell activation (Fig. 5.5). The TCRs expressed on all $CD4^+$ and $CD8^+$ T cells consist of an α chain and a β chain, both of which participate in antigen recognition (see Fig. 4.7). (A small subset of T cells expresses TCRs composed of γ and δ chains, which do not recognize MHC-associated peptide antigens.) The TCR of a T cell specific for a foreign (e.g., microbial) peptide recognizes the displayed peptide and simultaneously recognizes residues of the MHC molecule located around the peptide-binding cleft. Every mature MHC-restricted T cell expresses either CD4 or CD8, both of which are called coreceptors because they bind to the same MHC molecules that the TCR binds and are required for initiation of signaling from the TCR complex. At the time when the TCR is recognizing the peptide-MHC complex, CD4 or CD8 binds the class II or class I MHC molecule, respectively, at a site of the MHC

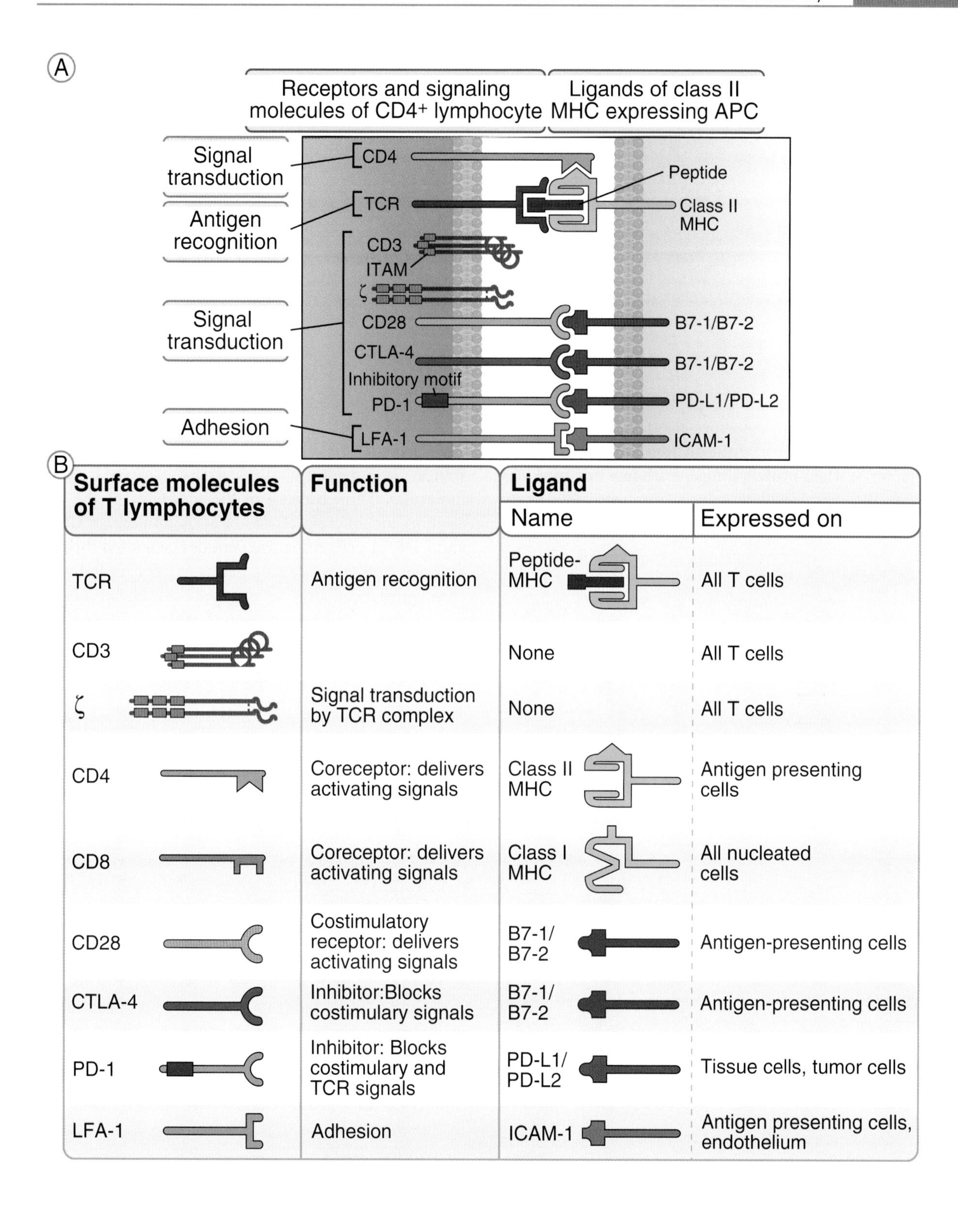

Surface molecules of T lymphocytes	Function	Ligand	
		Name	Expressed on
TCR	Antigen recognition	Peptide-MHC	All T cells
CD3	Signal transduction by TCR complex	None	All T cells
ζ		None	All T cells
CD4	Coreceptor: delivers activating signals	Class II MHC	Antigen presenting cells
CD8	Coreceptor: delivers activating signals	Class I MHC	All nucleated cells
CD28	Costimulatory receptor: delivers activating signals	B7-1/ B7-2	Antigen-presenting cells
CTLA-4	Inhibitor:Blocks costimulary signals	B7-1/ B7-2	Antigen-presenting cells
PD-1	Inhibitor: Blocks costimulary and TCR signals	PD-L1/ PD-L2	Tissue cells, tumor cells
LFA-1	Adhesion	ICAM-1	Antigen presenting cells, endothelium

Fig. 5.4 Receptors and ligands involved in T cell activation and inhibition. A, Major surface molecules of $CD4^+$ T cells involved in the activation of these cells and their corresponding ligands on antigen-presenting cells. $CD8^+$ T cells use most of the same molecules, except that the TCR recognizes peptide-class I MHC complexes, and the coreceptor is CD8, which recognizes class I MHC. CD3 is composed of three polypeptide chains, δ, ε, and γ, arranged in two pairs ($\delta\varepsilon$ and $\gamma\varepsilon$); we show CD3 as three chains. Immunoreceptor tyrosine-based activation motifs *(ITAMs)* are the regions of cytosolic tails of signaling proteins that are phosphorylated on tyrosine residues and become docking sites for other tyrosine kinases (see Fig. 5.10). Immunoreceptor tyrosine-based inhibitory motifs are the regions of signaling proteins that are sites for tyrosine phosphatases that counteract actions of ITAMs. **B,** Important properties of major surface molecules of T cells involved in functional responses. Cytokines and cytokine receptors are not listed here. The functions of most of these molecules are described later in this chapter. LFA-1 is an integrin involved in leukocyte binding to endothelium and T cell binding to APCs. *APC,* Antigen-presenting cell; *ICAM-1,* intercellular adhesion molecule 1; *LFA-1,* leukocyte function–associated antigen 1; *MHC,* major histocompatibility complex; *PD-1,* programmed death-1; *TCR,* T cell receptor.

molecule separate from the peptide-binding cleft and thus brings signaling enzymes close to the CD3 and ζ tails to initiate signal transduction. As discussed in Chapter 3, when protein antigens are ingested by APCs from the extracellular milieu into vesicles, these antigens are processed into peptides that are displayed by class II MHC molecules. By contrast, protein antigens present in the cytosol are processed by proteasomes into peptides displayed by class I MHC molecules. Thus, because of the specificity of the coreceptors for different classes of MHC molecules, $CD4^+$ and $CD8^+$ T cells recognize peptides generated through different protein processing pathways. The TCR and its coreceptor need to be engaged simultaneously to initiate the T cell response, and multiple TCRs likely need to be triggered for T cell activation to occur. Once these conditions are achieved, the T cell begins its activation program.

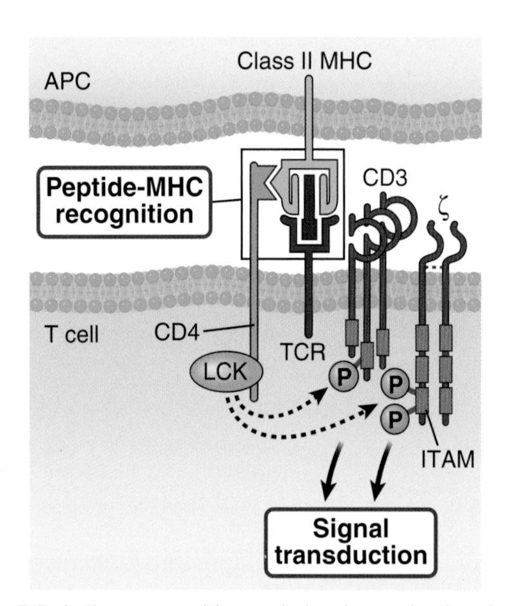

Fig. 5.5 Antigen recognition and signal transduction during T cell activation. Different T cell molecules recognize antigen and deliver biochemical signals to the interior of the cell as a result of antigen recognition. The CD3 and ζ proteins are noncovalently attached to the T cell receptor *(TCR)* α and β chains by interactions between charged amino acids in the transmembrane domains of these proteins (not shown). The figure illustrates a $CD4^+$ T cell; the same interactions are involved in the activation of $CD8^+$ T cells, except that the coreceptor is CD8 and the TCR recognizes a peptide–class I MHC complex. *APC,* Antigen-presenting cell; *ITAM,* immunoreceptor tyrosine-based activation motifs; *MHC,* major histocompatibility complex.

The biochemical signals that lead to T cell activation are triggered by a set of proteins linked to the TCR that are part of the TCR complex and by the CD4 or CD8 coreceptor (see Fig. 5.5). In lymphocytes, antigen recognition and subsequent signaling are performed by different sets of molecules. The TCR $\alpha\beta$ heterodimer recognizes antigens, but it is not able to transmit biochemical signals to the interior of the cell. The TCR is noncovalently associated with a complex of transmembrane signaling proteins, including three CD3 proteins and a protein called the ζ chain. The TCR, CD3, and ζ chain make up the TCR complex. Although the α and β TCR chains must vary among T cell clones in order to recognize diverse antigens, the signaling functions of the TCR complex are the same in all clones, and the CD3 and ζ proteins are invariant among different T cells. The mechanisms of signal transduction by these proteins of the TCR complex are discussed later in the chapter.

T cells can also be activated by molecules that bind to the TCRs of many or all clones of T cells, regardless of the peptide-MHC specificity of the TCR. For instance, some microbial toxins may bind to the TCRs of many T cell clones and also bind to MHC class II molecules on APCs outside the peptide-binding cleft. These toxins activate a large number of T cells, resulting in excessive cytokine release and systemic inflammatory disease. The toxins are called superantigens because, like conventional antigens, they bind to MHC molecules and to TCRs, but they engage many more TCRs than typical antigens do.

Role of Adhesion Molecules in T Cell Responses

Adhesion molecules on T cells recognize their ligands on APCs and stabilize the binding of the T cells to the APCs. Most TCRs bind the peptide-MHC complexes for which they are specific with low affinity. To induce a response, the adhesion of T cells to APCs must be stabilized for a sufficiently long period to achieve the necessary signaling threshold. This stabilization function is performed by adhesion molecules on the T cells that bind to ligands expressed on APCs. The most important of these adhesion molecules belong to the family of heterodimeric (two-chain) proteins called integrins. The major T cell integrin involved in binding to APCs is leukocyte function–associated antigen 1 (LFA-1), whose ligand on APCs is called intercellular adhesion molecule 1 (ICAM-1).

On resting naive T cells, which are cells that have not previously recognized and been activated by antigen, the LFA-1 integrin is in a low-affinity state. Antigen recognition by a T cell increases the affinity of that cell's LFA-1. Therefore, once a T cell sees antigen, it increases the strength of its binding to the APC presenting that antigen. Integrin-mediated adhesion is critical for the ability of T cells to bind to APCs displaying microbial antigens. Integrins also play an important role in directing the migration of effector T cells and other leukocytes from the circulation to sites of infection. This process is described in Chapter 2 and later in this chapter.

Role of Costimulation in T Cell Activation

The full activation of T cells depends on the recognition of costimulators on APCs in addition to antigen (Fig. 5.6). We have previously referred to costimulators as second signals for T cell activation. The name costimulator derives from the fact that these molecules provide stimuli to T cells that function together with stimulation by antigen.

The best-defined costimulators for T cells are two homologous proteins called B7-1 (CD80) and B7-2 (CD86), both of which are expressed on APCs and whose expression is increased when the APCs encounter microbes. These B7 proteins are recognized by a receptor called CD28, which is expressed on most T cells. Several proteins that are homologous to B7 or CD28 function to stimulate or inhibit immune responses (Fig. 5.7). The binding of B7 on the APCs to CD28 on T cells generates signals that work together with signals generated by TCR recognition of antigen presented by MHC proteins on the same APCs. CD28-mediated signaling is essential for the responses of naive T cells; in the absence of CD28:B7 interactions, antigen recognition by the TCR is insufficient for initiating T cell responses. The requirement for costimulation ensures that naive T lymphocytes are activated maximally by microbial antigens and not by harmless foreign substances or by self antigens, because, as stated previously, microbes stimulate the expression of B7 costimulators on APCs. Despite the increased expression of costimulators, most infections do not trigger harmful reactions against self antigens, mainly because numerous control mechanisms prevent autoimmunity (see Chapter 9).

A protein called inducible costimulator (ICOS), which is homologous to CD28 and also expressed on T cells, plays an important role in the development and function of follicular helper T cells during germinal center B cell responses (see Chapter 7). The CD28 family includes two inhibitory receptors that resemble CD28 structurally, called CTLA-4 and PD-1; these and their ligands are discussed later in the chapter.

Another set of molecules that participate in T cell responses are CD40 ligand (CD40L, or CD154) on activated T cells and CD40 on APCs. These molecules do not directly enhance T cell activation. Instead, CD40L expressed on an antigen-stimulated T cell binds to CD40 on APCs and activates the APCs to express more B7 costimulators and to secrete cytokines (e.g., IL-12) that enhance T cell differentiation. Thus, the CD40L-CD40 interaction promotes T cell activation by making APCs better at stimulating T cells. CD40L on effector $CD4^+$

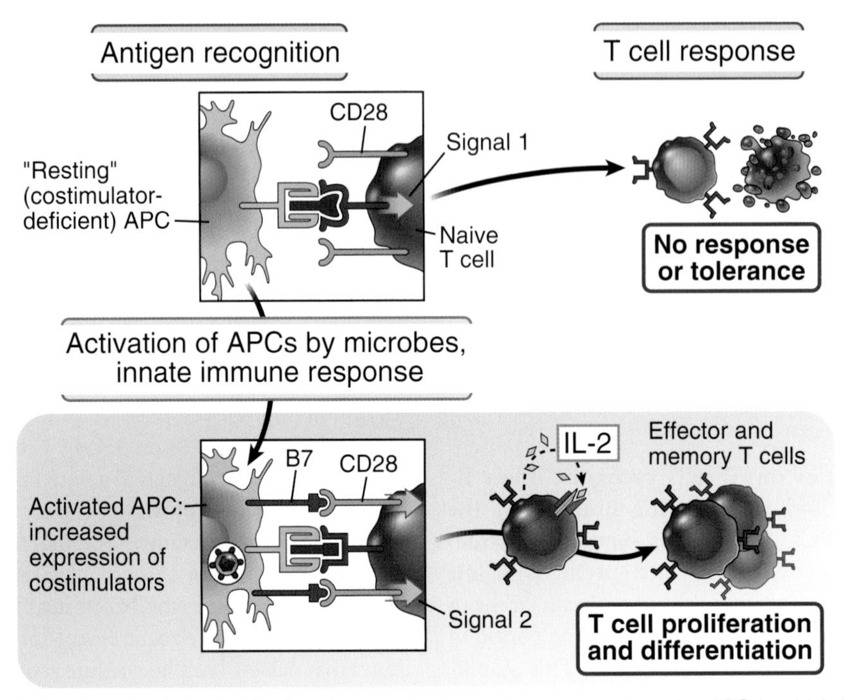

Fig. 5.6 Role of costimulation in T cell activation. Resting antigen-presenting cells (APCs), which have not been exposed to microbes or adjuvants, may present peptide antigens, but they do not express costimulators and are unable to activate naive T cells. T cells that recognize antigen without costimulation may die or become unresponsive (tolerant) to subsequent exposure to antigen. Microbes, as well as cytokines produced during innate immune responses to microbes, induce the expression of costimulators, such as B7 molecules, on the APCs. The B7 costimulators are recognized by the CD28 receptor on naive T cells, providing signal 2. In conjunction with antigen recognition (signal 1), this recognition initiates T cell responses. Activated APCs also produce cytokines that stimulate the differentiation of naive T cells into effector cells (not shown). *IL,* Interleukin.

T cells also enhances activation of macrophages and B cells, as discussed in Chapters 6 and 7.

The role of costimulation in T cell activation explains an observation mentioned in earlier chapters. Protein antigens, such as those used in vaccines, fail to elicit T cell–dependent immune responses unless these antigens are administered with substances that activate APCs, especially dendritic cells. Such substances are called **adjuvants**, and they function mainly by inducing the expression of costimulators on APCs and by stimulating the APCs to secrete cytokines that activate T cells. Most adjuvants used in experimental studies are products of microbes (e.g., killed mycobacteria) or substances that mimic microbes, and they bind to pattern recognition receptors of the innate immune system, such as Toll-like receptors and NOD-like receptors (see Chapter 2). Numerous adjuvants have been developed for human vaccines (see Chapter 8). They are postulated to work, at least in part, by also eliciting innate immune responses that activate APCs to increase expression of costimulators and secretion of T cell–activating cytokines. In mRNA vaccines, lipids that encapsulate the RNA function as adjuvants. Thus, adjuvants trick the immune system into responding to purified protein antigens in or produced by a vaccine as if these proteins were parts of infectious microbes. Some live virus vaccines, such as the attenuated viral vaccine against measles, mumps, and rubella or hybrid adenovirus vaccines against SARS-CoV-2, do not have to be given with adjuvants because the vaccine viruses express various molecules that activate the innate immune system.

The increasing understanding of costimulators has led to new strategies for inhibiting harmful immune responses. Agents that block B7:CD28 interactions are

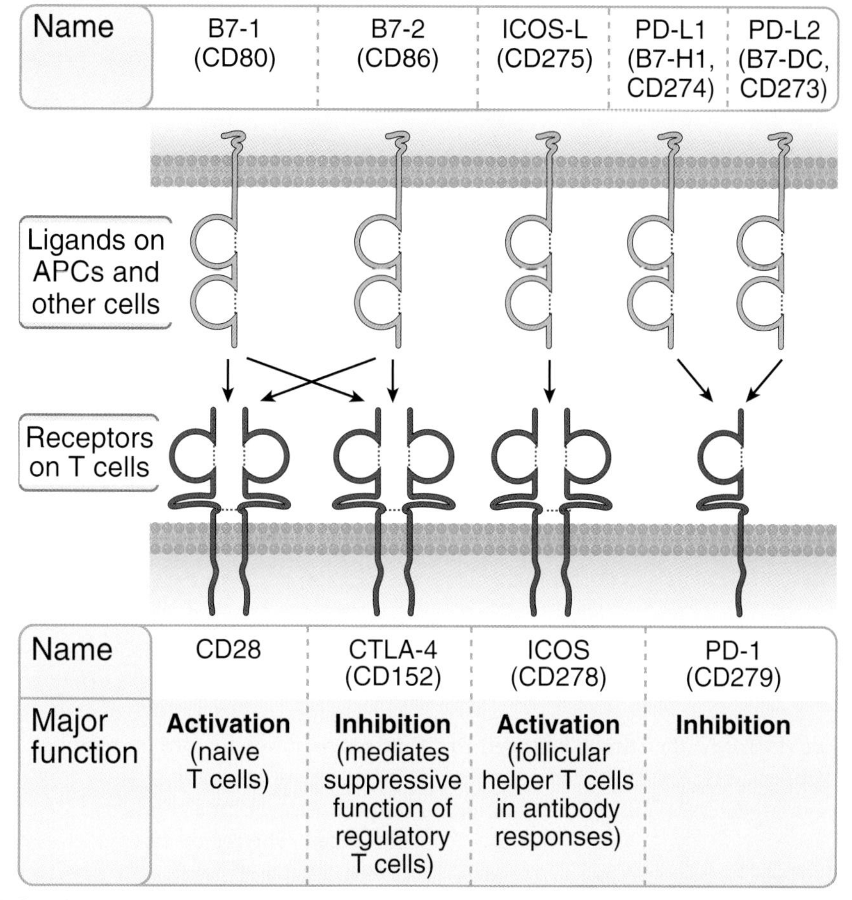

Fig. 5.7 Proteins of the B7 and CD28 families. Ligands on APCs that are homologous to B7 bind to receptors on T cells that are homologous to CD28. Different ligand-receptor pairs serve distinct roles in immune responses. CD28 and ICOS are stimulatory receptors on T cells, and CTLA-4 and PD-1 are inhibitory receptors. Their functions are discussed in the text.

used in the treatment of disorders in which T cell activation causes organ dysfunction, such as certain autoimmune diseases and graft rejection, and antibodies that block CD40:CD40L interactions are being tested as treatments for these disorders.

Stimuli for Activation of CD8$^+$ T Cells

The activation of naive CD8$^+$ T cells is stimulated by recognition of class I MHC–associated peptides and requires costimulation and helper T cells. The responses of CD8$^+$ T cells differ in several ways from responses of CD4$^+$ T lymphocytes:

- The initiation of CD8$^+$ T cell activation often requires cytosolic antigen from one cell (e.g., virus-infected or tumor cell) to be cross-presented by dendritic cells (see Fig. 3.16).
- The differentiation of naive CD8$^+$ T cells into fully active CTLs and memory cells may require the concomitant activation of CD4$^+$ helper T cells (Fig. 5.8). When virus-infected or tumor cells are ingested by dendritic cells, the APCs may present viral or tumor antigens from the cytosol in complex with class I MHC molecules and from vesicles in complex with class II MHC molecules. Thus, both CD8$^+$ T cells and CD4$^+$ T cells specific for viral or tumor antigens are activated near one another. The CD4$^+$ T cells may produce cytokines or membrane molecules that help to activate the CD8$^+$ T cells. CD4$^+$ helper cells

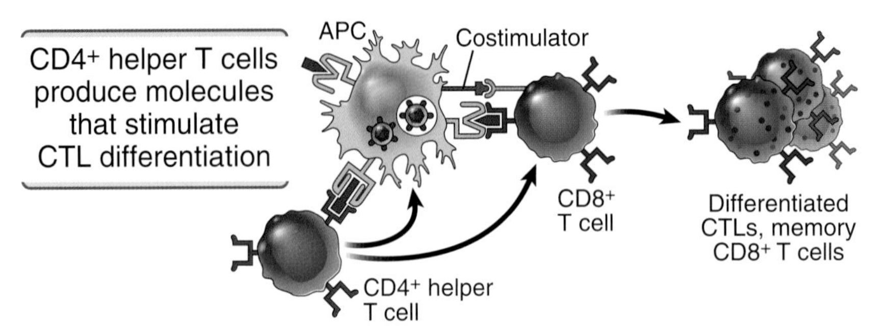

Fig. 5.8 Role of $CD4^+$ T cells in activation of $CD8^+$ T cells. Antigen-presenting cells (APCs), principally dendritic cells, may ingest and present microbial antigens to $CD8^+$ T cells (cross-presentation) and to $CD4^+$ helper T cells. The helper T cells then produce cytokines that stimulate the expansion and differentiation of the $CD8^+$ T cells. Helper cells also may activate APCs to make them potent stimulators of $CD8^+$ T cells. *CTLs,* Cytotoxic T lymphocytes.

also enhance the expression of costimulatory molecules on APCs that activate $CD8^+$ T cells. The requirement for helper T cells in $CD8^+$ T cell responses is the likely explanation for the increased susceptibility to viral infections and cancers in patients infected with the human immunodeficiency virus (HIV), which kills $CD4^+$ but not $CD8^+$ T cells.

Now that we have described the stimuli required to activate naive T lymphocytes, we next consider the biochemical pathways triggered by antigen recognition and other stimuli.

BIOCHEMICAL PATHWAYS OF T CELL ACTIVATION

Following the recognition of antigens and costimulators, T cells express proteins that are involved in their proliferation, differentiation, and effector functions (Fig. 5.9). Naive T cells that have not encountered antigen have a low level of protein synthesis. Within minutes of antigen recognition, new gene transcription and protein synthesis are seen in the activated T cells. These newly expressed proteins mediate many of the subsequent responses of the T cells. The expression of these proteins is induced by signal transduction pathways triggered by the TCR complex and costimulatory receptors.

Antigen recognition activates several biochemical signaling events, including the activation of enzymes such as kinases, recruitment of adaptor proteins, and production or activation of functional transcription factors (Fig. 5.10). These biochemical pathways are initiated when TCR complexes and the appropriate coreceptor are brought together by binding to MHC-peptide complexes on the surface of APCs. In addition, there is an orderly movement of proteins in both the APC and T cell membranes at the region of cell-to-cell contact, such that the TCR complex, CD4/CD8 coreceptors, and CD28 coalesce to the center and the integrins move to form a peripheral ring. This redistribution of signaling and adhesion molecules is required for optimal induction of activating signals in the T cell. The region of contact between the APC and T cell, including the redistributed membrane proteins, is called the immune synapse. Although the synapse was first described as the site of delivery of activating signals from membrane receptors to the cell's interior, it may serve other functions. Some effector molecules and cytokines may be secreted through this region, ensuring that they do not diffuse away but are targeted to the cell in contact with the T cell. Enzymes that degrade or inhibit signaling molecules are also recruited to the synapse, so it may be involved in terminating lymphocyte activation as well.

As discussed earlier, antigen recognition simultaneously engages the TCR and CD4 or CD8 coreceptor. The cytoplasmic tails of CD4 and CD8 have an attached protein tyrosine kinase called LCK, which is constitutively active, unlike other enzymes involved downstream in TCR signaling. Therefore, LCK is poised to initiate the signaling cascade. Several transmembrane signaling proteins are associated with the TCR, including the CD3 and ζ chains Chapter 4. CD3 and ζ contain motifs, each with two tyrosine residues, called **immunoreceptor**

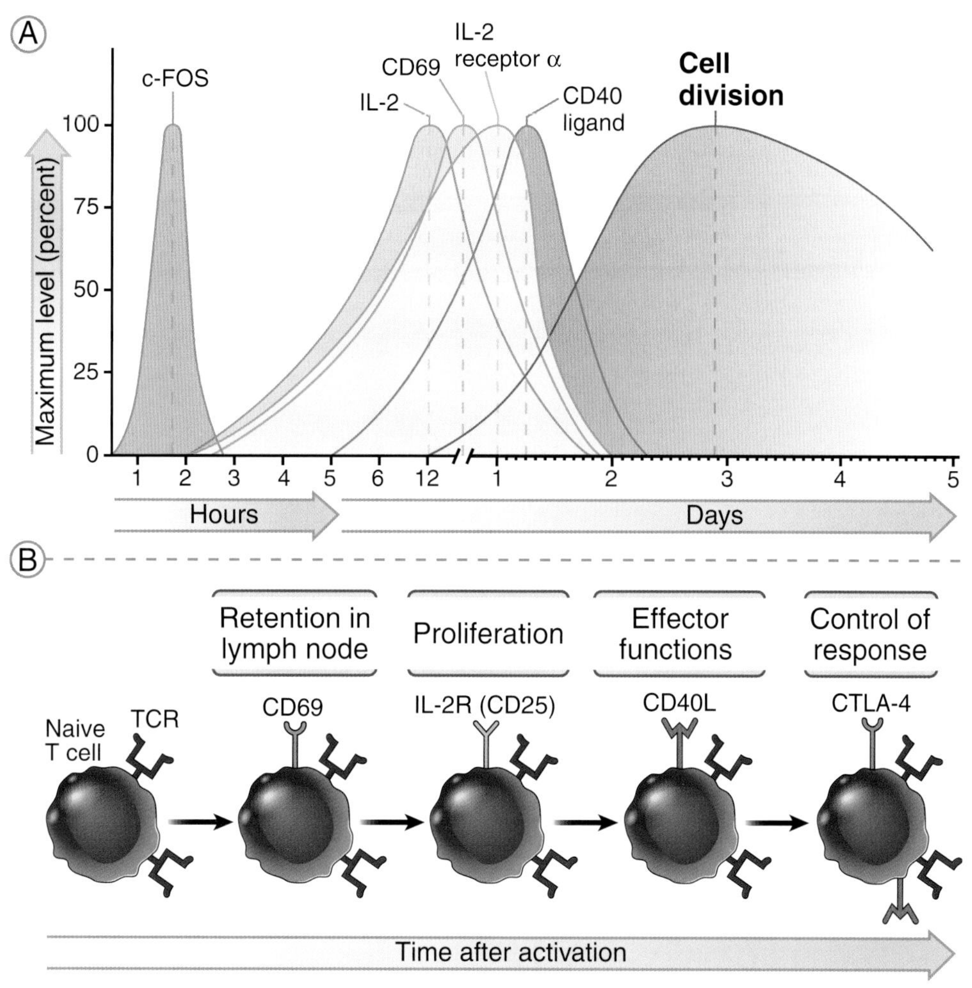

Fig. 5.9 **Proteins produced by antigen-stimulated T cells.** Antigen recognition by T cells results in the synthesis and expression of a variety of proteins, examples of which are shown. The kinetics of production of these proteins (**A**) are approximations and may vary in different T cells and with different types of stimuli. The possible effects of costimulation on the patterns or kinetics of gene expression are not shown. The functions of some of the surface proteins expressed on activated T cells are shown in (**B**). CD69 is a marker of T cell activation involved in cell migration; the interleukin-2 receptor (IL-2R) binds the cytokine IL-2 and generates signals that promote T cell survival and proliferation; CD40 ligand is an effector molecule of T cells; CTLA-4 is an inhibitor of immune responses. c-FOS (shown in **A**) is a transcription factor. *TCR,* T cell receptor.

tyrosine-based activation motifs (ITAMs), which are critical for signaling. LCK, which is brought near the TCR complex by the CD4 or CD8 molecules, phosphorylates tyrosine residues contained within the ITAMs of the CD3 and ζ proteins, and this is the event that launches signal transduction in the T cells. The importance of the coreceptors is that by binding to MHC molecules, they bring the constitutively active LCK close to its critical substrates in the TCR complex. The phosphorylated ITAMs of the ζ chain become docking sites for a tyrosine kinase called ZAP-70 (zeta-associated protein of 70 kD), which is also phosphorylated by LCK and thereby made enzymatically active. The active ZAP-70 then phosphorylates various adaptor proteins and enzymes, which assemble near the TCR complex and mediate additional signaling events.

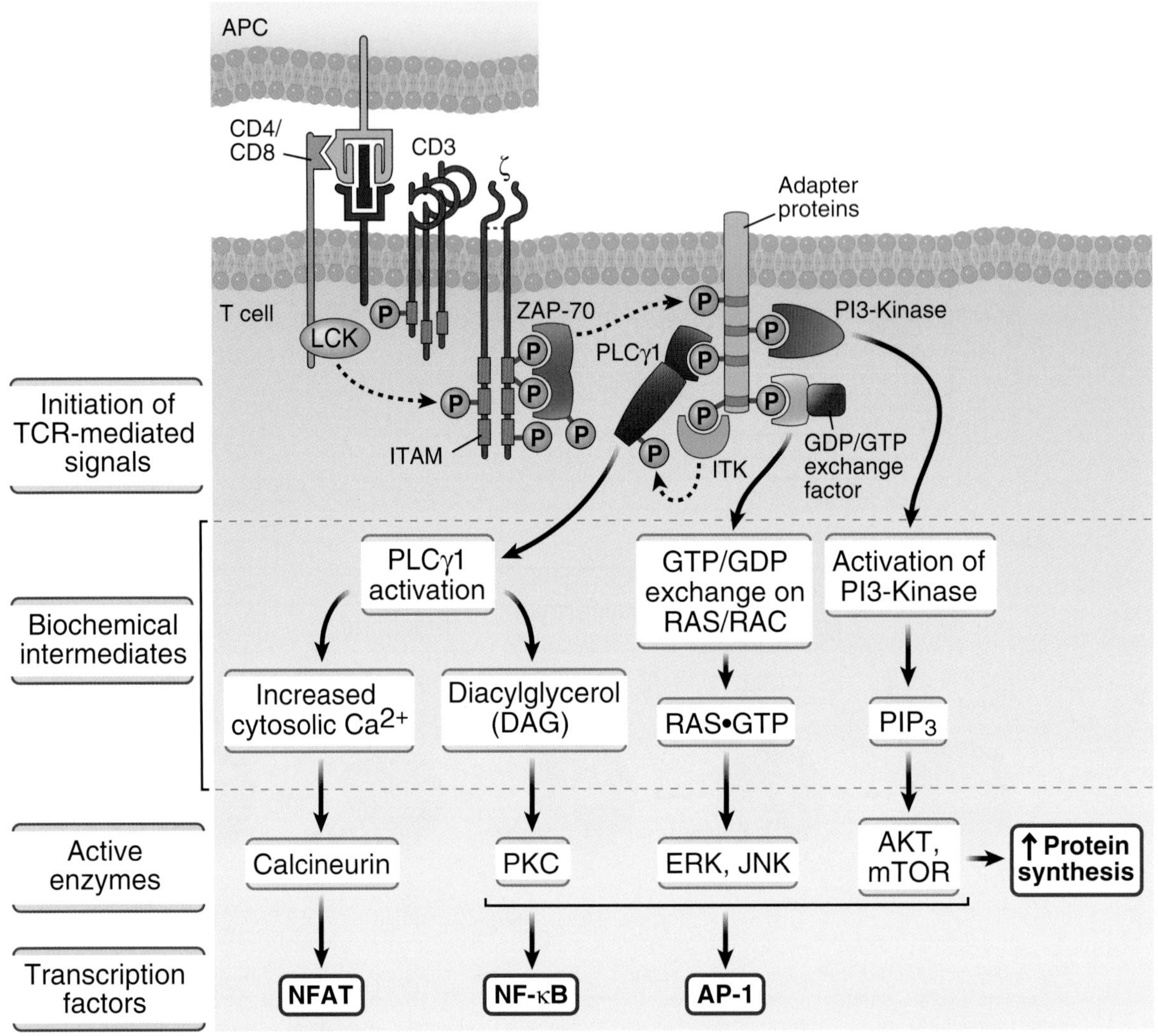

Fig. 5.10 Signal transduction pathways in T lymphocytes. Antigen recognition by T cells induces early signaling events, which include tyrosine phosphorylation of molecules of the T cell receptor *(TCR)* complex and the recruitment of adaptor proteins to the site of T cell antigen recognition. These early events lead to the activation of several biochemical intermediates, which in turn activate transcription factors that stimulate transcription of genes whose products mediate the responses of the T cells. The possible effects of costimulation on these signaling pathways are not shown. These signaling pathways are shown as independent of one another, for simplicity, but may be interconnected in more complex networks. *AP-1,* Activator protein 1; *APC,* antigen-presenting cell; *GTP/GDP,* guanosine triphosphate/diphosphate; *ITAM,* immunoreceptor tyrosine-based activation motif; *mTOR,* molecular target of rapamycin; *NFAT,* nuclear factor of activated T cells; *PKC,* protein kinase C; *PLCγ1,* γ1 isoform of phosphatidylinositol-specific phospholipase C; *PI-3,* phosphatidylinositol-3; *ZAP-70,* zeta-associated protein of 70 kD.

The signaling pathways linked to TCR complex activation lead to the production of functional transcription factors.

- **Nuclear factor of activated T cells** (NFAT) is a transcription factor present in an inactive phosphorylated form in the cytosol of resting T cells. NFAT activation and its nuclear translocation depend on the concentration of calcium (Ca^{2+}) ions in the cytosol. The signaling pathway leading to NFAT activation is initiated by phosphorylation and activation of an enzyme named phospholipase Cγ (PLCγ) by a kinase, ITK, that becomes attached to one of the adaptor proteins in the signaling complex. Activated PLCγ catalyzes the hydrolysis of a plasma membrane phospholipid called phosphatidylinositol 4,5-bisphosphate (PIP2). One by-product of PLCγ-mediated PIP2 breakdown, inositol 1,4,5-triphosphate (IP3), binds to IP3 receptors on the endoplasmic reticulum (ER) membrane and the mitochondria and initiates release of Ca^{2+} into the cytosol. In response to the loss of calcium from the endoplasmic reticulum, a plasma membrane calcium channel is opened, leading to the influx of extracellular Ca^{2+} into the cell, which causes a sustained increase in the cytosolic Ca^{2+} concentration for hours. The elevated cytosolic Ca^{2+} leads to activation of a phosphatase called calcineurin. This enzyme removes phosphates from cytoplasmic NFAT, enabling the transcription factor to migrate into the nucleus, where it binds to and activates the promoters of several genes, including those encoding the T cell growth factor IL-2 and components of the IL-2 receptor. Calcineurin inhibitors (cyclosporine and tacrolimus) are drugs that block the phosphatase activity of calcineurin and thus suppress the NFAT-dependent production of cytokines by T cells. These drugs are widely used as immunosuppressants to prevent graft rejection (see Chapter 10).

- **The RAS/RAC–MAP kinase pathways** include the guanosine triphosphate (GTP)–binding RAS and RAC proteins, several adaptor proteins, and a cascade of enzymes that eventually activate one of a family of mitogen-activated protein (MAP) kinases. These pathways are initiated by ZAP-70–dependent phosphorylation and accumulation of adaptor proteins at the plasma membrane, leading to the recruitment of RAS or RAC, and their activation by exchange of bound guanosine diphosphate (GDP) with GTP. RAS•GTP and RAC•GTP, the active forms of these proteins, initiate different enzyme cascades, leading to the activation of distinct MAP kinases. The terminal MAP kinases in these pathways, called extracellular signal–regulated kinase (ERK) and c-JUN amino-terminal (N-terminal) kinase (JNK), respectively, induce the expression of a protein called c-FOS and the phosphorylation of another protein called c-JUN. c-FOS and phosphorylated c-JUN combine to form the transcription factor **activator protein 1** (AP-1), which enhances the transcription of several T cell genes.
- Another major pathway involved in TCR signaling consists of activation of the θ isoform of the serine-threonine kinase called protein kinase C (PKCθ), which leads to activation of the transcription factor **NF-κB.** PKC is activated by diacylglycerol, which, like IP3, is generated by PLC-mediated hydrolysis of PIP2 in the membrane. PKCθ acts through adaptor proteins to activate NF-κB.
- TCR signal transduction also involves a lipid kinase called **PI-3 kinase**, which phosphorylates the membrane phospholipid PIP2 to generate phosphatidyl inositol (3,4,5)-trisphosphate (PIP3). PIP3 is required for the activation of a number of targets, including a serine-threonine kinase called AKT, which has many roles, including stimulating expression of antiapoptotic proteins and thus promoting survival of antigen-stimulated T cells. AKT activates molecular target of rapamycin (mTOR), a serine-threonine kinase that is involved in stimulating protein translation and promoting cell survival and growth. Rapamycin (also called sirolimus) is a drug that binds to and inactivates mTOR, and is used to treat graft rejection.

The various transcription factors that are induced or activated in T cells, including NFAT, AP-1, and NF-κB, stimulate transcription and subsequent production of cytokines, cytokine receptors, cell cycle inducers, and effector molecules such as CD40L (see Fig. 5.9). All of these signals are initiated by antigen recognition, because binding of the TCR and coreceptors to peptide-MHC complexes is necessary to bring together critical enzymes and substrates in T cells.

As discussed earlier, recognition of costimulators, such as B7 molecules, by their receptor CD28 is essential for full T cell responses. The biochemical signals transduced by CD28 on binding to B7 costimulators include the PI-3 kinase/AKT and the MAP-kinase pathways. CD28 engagement likely amplifies some TCR signaling pathways that are triggered by antigen recognition (signal 1) and may induce other signals that complement TCR signals. In this way,

CD28 signals increase the production of survival factors, IL-2, and cell cycle inducers, all of which promote survival and proliferation of activated T cells and their differentiation into effector and memory cells.

Lymphocyte activation is associated with profound changes in cellular metabolism. In naive (resting) T cells, low levels of glucose are taken up and used to generate energy in the form of adenosine triphosphate (ATP) by mitochondrial oxidative phosphorylation. Upon activation, glucose uptake increases markedly, and the cells switch to aerobic glycolysis. This process generates less ATP but facilitates the synthesis of more amino acids, lipids, and other molecules that provide building blocks for organelles and for producing new cells. As a result, it is possible for activated T cells to more efficiently manufacture the cellular constituents that are needed for their rapid increase in size and for producing daughter cells.

Having described the stimuli and biochemical pathways in T cell activation, we now discuss how T cells respond to antigens and differentiate into effector cells capable of combating microbes.

FUNCTIONAL RESPONSES OF T LYMPHOCYTES TO ANTIGEN AND COSTIMULATION

The recognition of antigen and costimulators by naive T cells initiates a set of responses that culminate in the expansion of the antigen-specific clones of lymphocytes and the differentiation of the naive T cells into effector cells and memory cells (see Fig. 5.3). Many of these changes in T cells are mediated by cytokines that are secreted by the T cells and act on the T cells themselves and on many other cells involved in immune defenses.

Secretion of Cytokines and Expression of Cytokine Receptors

In response to antigen and costimulators, T lymphocytes, especially $CD4^+$ T cells, rapidly secrete the cytokine IL-2. We have already discussed cytokines in innate immune responses, which are produced mainly by dendritic cells and macrophages (see Chapter 2). In adaptive immunity, cytokines are mainly secreted by $CD4^+$ T cells. Most of the cytokines of adaptive immunity, other than IL-2, are produced by effector T cells, serve diverse roles in host defense, and are described in Chapter 6 when we discuss the effector mechanisms of cell-mediated immunity.

IL-2 is produced within 1 to 2 hours after antigen stimulation of $CD4^+$ T cells. Activation also transiently increases the expression of the high-affinity IL-2 receptor, thus rapidly enhancing the ability of the T cells to bind and respond to IL-2 (Fig. 5.11). The receptor for IL-2 is a three-chain molecule. Naive T cells express two signaling chains, β and γ, which constitute the low-affinity receptor for IL-2, but these cells do not express the α chain (CD25) that enables the receptor to bind IL-2 with high affinity. Within hours after activation by antigens and costimulators, the T cells produce the α chain of the receptor, and now the complete IL-2 receptor is able to bind IL-2 strongly. Thus, IL-2 produced by antigen-stimulated T cells preferentially binds to and acts on the same T cells, an example of autocrine cytokine action.

The principal functions of IL-2 are to stimulate the survival and proliferation of T cells, resulting in an increase in the number of the antigen-specific T cells; because of these actions, IL-2 was originally called T cell growth factor. The high-affinity IL-2 receptor is constitutively expressed on regulatory T cells, so these cells are very sensitive to IL-2. In fact, IL-2 is essential for the maintenance of regulatory T cells and thus for controlling immune responses, as we discuss in Chapter 9. Activated $CD8^+$ T cells and natural killer (NK) cells express the low-affinity βγ receptor and respond to higher concentrations of IL-2.

Clonal Expansion

T lymphocytes activated by antigen and costimulation begin to proliferate within 1 or 2 days, resulting in expansion of antigen-specific clones (Fig. 5.12). This expansion quickly provides a large pool of antigen-specific lymphocytes from which effector cells can be generated to combat infection.

The magnitude of clonal expansion is remarkable, especially for $CD8^+$ T cells. Before infection, the frequency of $CD8^+$ T cells specific for any one microbial protein antigen is in the range of 1 in 10^5 to 1 in 10^6

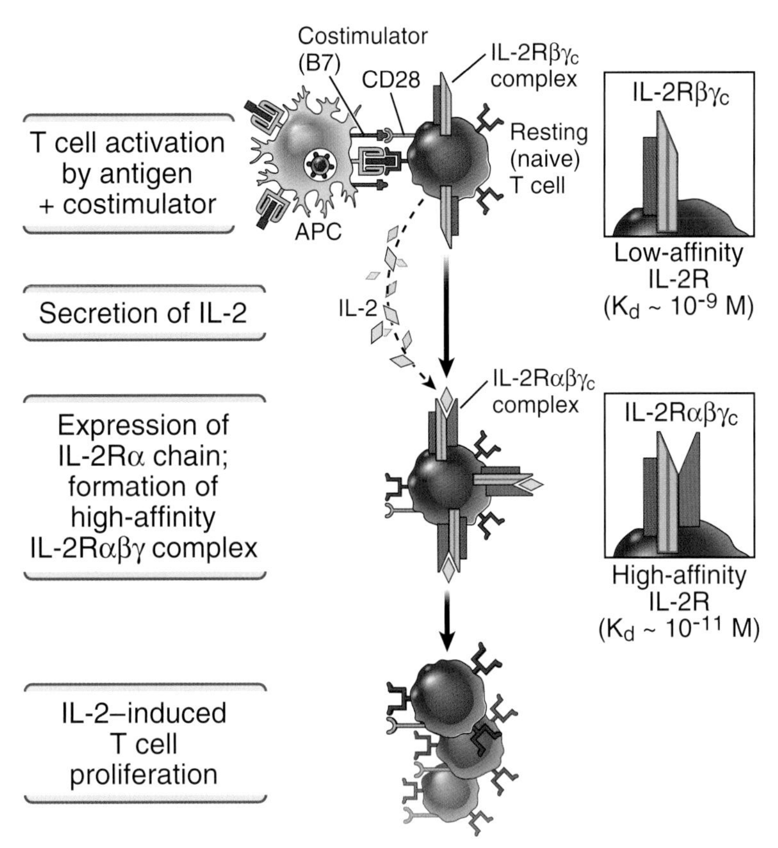

Fig. 5.11 **Role of interleukin-2 and IL-2 receptors in T cell proliferation.** Naive T cells express the low-affinity IL-2 receptor (IL-2R) complex, made up of the β and γc chains (γc designates common γ chain, so called because it is a component of receptors for several cytokines). On activation by antigen recognition and costimulation, the cells produce IL-2 and express the α chain of the IL-2R (CD25), which associates with the β and γc chains to form the high-affinity IL-2 receptor. Binding of IL-2 to its receptor initiates proliferation of the T cells that recognized the antigen. *APC,* Antigen-presenting cell.

lymphocytes in the body. At the peak of some viral infections, possibly within a week after the infection, as many as 10% to 20% of all the lymphocytes in the lymphoid organs may be specific for that virus. This means that the numbers of cells in antigen-specific clones have increased by more than 10,000-fold, with an estimated doubling time of approximately 6 hours. This enormous expansion of T cells specific for a microbe is not accompanied by a detectable increase in the numbers of bystander cells that do not recognize that microbe.

The magnitude of expansion of $CD4^+$ T cells appears to be 100-fold to 1000-fold less than that of $CD8^+$ cells. This difference may reflect differences in the functions of the two types of T cells. $CD8^+$ CTLs are effector cells that kill infected and tumor cells by direct contact, and many CTLs may be needed to kill large numbers of infected or tumor cells. By contrast, each $CD4^+$ effector cell secretes cytokines that activate numerous other effector cells, so a relatively small number of cytokine producers may be sufficient.

Differentiation of Naive T Cells into Effector Cells

Some of the progeny of antigen-stimulated, proliferating T cells differentiate into effector cells whose function is to eradicate infections. Differentiation is the result of changes in gene expression, such as the activation of genes encoding cytokines (in $CD4^+$ T cells) or cytotoxic proteins (in $CD8^+$ CTLs). It begins in concert with clonal expansion, and differentiated effector cells appear within 3 or 4 days

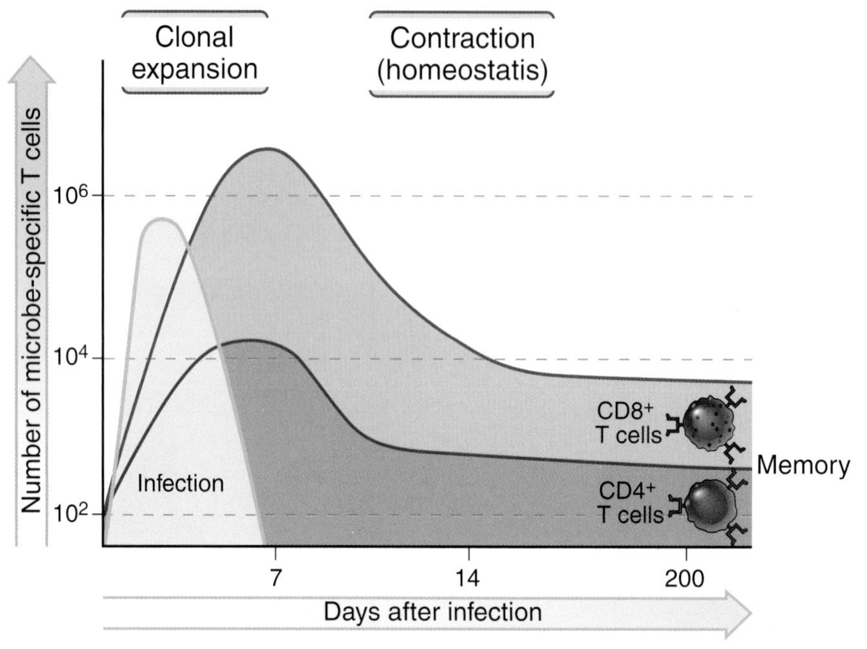

Fig. 5.12 Expansion and decline of T cell responses. The numbers of $CD4^+$ and $CD8^+$ T cells specific for various antigens in inbred mice and the clonal expansion and contraction during immune responses are illustrated. The numbers are approximations based on studies of model microbial and other antigens in inbred mice; in humans, the numbers of lymphocytes are approximately 1000-fold greater.

after exposure to microbes. Effector cells of the $CD4^+$ lineage acquire the capacity to produce different sets of cytokines. The subsets of T cells that are distinguished by their cytokine profiles are named Th1, Th2, and Th17 (Fig. 5.13). Many of these cells leave the secondary lymphoid organs where they are generated and migrate to sites of infection, where their cytokines recruit other leukocytes that destroy or help contain the inciting infectious agents. The development and functions of these effector cells are described in Chapter 6, when we discuss cell-mediated immunity. Other differentiated $CD4^+$ T cells remain in the lymphoid organs and migrate into lymphoid follicles, where they further differentiate into T follicular helper (Tfh) cells and help B lymphocytes to produce high-affinity antibodies (see Chapter 7). As we discuss in Chapters 6 and 7, $CD4^+$ helper T cells activate phagocytes and B lymphocytes through the actions of the plasma membrane protein CD40L and secreted cytokines. Effector cells of the $CD8^+$ lineage acquire the ability to kill infected and tumor cells; their development and function are also described in Chapter 6.

Development of Memory T Lymphocytes

A fraction of antigen-activated T lymphocytes differentiates into long-lived memory cells. These cells are a pool of lymphocytes that are induced by microbes and are ready to respond rapidly if the microbe returns. The factors that determine whether the progeny of antigen-stimulated lymphocytes will differentiate into effector cells or memory cells are not well defined. Memory cells have several important characteristics.

- Memory cells survive even after the infection is eradicated and antigen is no longer present. Certain cytokines, including IL-7 and IL-15, which are produced by stromal cells and myeloid cells in tissues, may serve to keep memory cells alive and cycling slowly.
- Memory T cells can be rapidly induced to produce cytokines or kill infected cells on encountering the antigen that they recognize. These cells do not perform any effector functions until they encounter antigen, but once activated, they respond much more rapidly than do naive lymphocytes and produce larger (secondary) responses than those of newly activated T cells (primary responses).

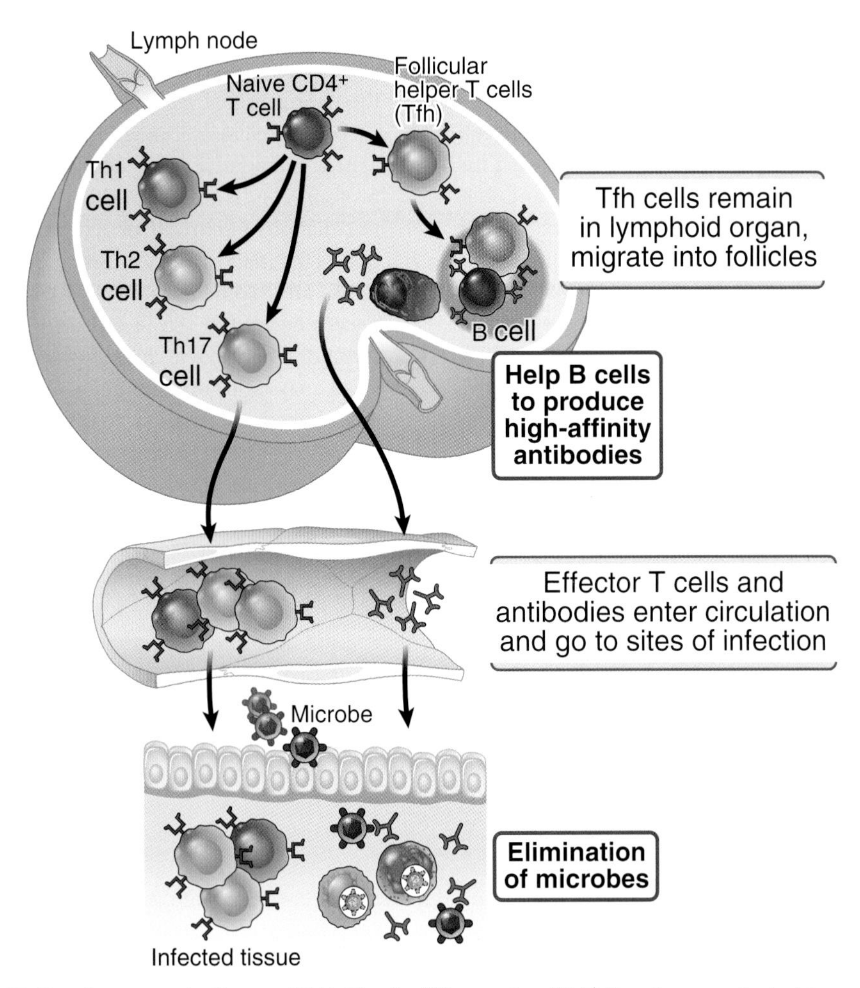

Fig. 5.13 Development of effector CD4$^+$ T cells. When naive CD4$^+$ T cells are activated in secondary lymphoid organs, they proliferate and differentiate into effector cells. Some of the effectors (the Th1, Th2, and Th17 populations) mostly exit the lymphoid organ and function to eradicate microbes in peripheral tissues. Other differentiated cells, called follicular helper T (Tfh) cells, remain in the lymphoid organ and help B cells to produce potent antibodies.

- Memory T cells are found in secondary lymphoid organs, in various peripheral tissues, especially mucosa and skin, and in the circulation. They can be distinguished from naive and effector cells by several criteria (see Chapter 1). A subset of memory T cells, called central memory cells, populate lymphoid organs and are responsible for rapid clonal expansion after reexposure to antigen. Another subset, called effector memory cells, localize in mucosal and other peripheral tissues and mediate rapid effector functions on encountering antigen at these sites. These cells retain the ability to exit the tissue and recirculate. A third subset, called tissue-resident memory cells, reside in the skin and mucosal tissues and do not readily enter the circulation. They mediate rapid secondary responses to antigens encountered in tissues.

Memory T cells likely can be activated in both lymphoid and nonlymphoid tissues, and their activation, unlike that of naive T cells, does not require high levels of costimulation or antigen presentation by dendritic cells. In fact, various other APCs, including B cells and macrophages, may be capable of activating memory T cells.

REGULATION OF T CELL RESPONSES BY INHIBITORY RECEPTORS (COINHIBITORS)

Immune responses are influenced by a balance between engagement of activating and inhibitory receptors. This idea is established for B and T lymphocytes and NK cells. In T cells, the main activating receptors are the TCR complex and costimulatory receptors such as CD28, and the best-defined inhibitory receptors, also called coinhibitors, are cytotoxic T-lymphocyte antigen 4 (CTLA-4) and programmed cell death protein 1 (PD-1). The functions and mechanisms of action of these inhibitors are complementary (Fig. 5.14).

- *CTLA-4.* CTLA-4 is a B7-binding protein expressed transiently on activated $CD4^+$ T cells and constitutively on regulatory T cells (discussed in Chapter 9). It functions to suppress the activation of responding T cells. CTLA-4 works by blocking and removing B7 molecules from the surface of APCs, thus reducing costimulation by CD28 and preventing the activation of T cells. The choice between B7 engagement of CTLA-4 or CD28 is determined by the affinity of these receptors for B7 and the level of B7 expression. CTLA-4 has a higher affinity for B7 molecules than does CD28, so it binds B7 tightly and prevents the binding of CD28. This competition is especially effective when B7 levels are low (as would be expected when APCs are only displaying self and maybe some tumor antigens but not microbial antigens); in these situations, the receptor that is preferentially engaged is the high-affinity blocking receptor CTLA-4. However, when B7 levels are high (as in infections), not all the ligands will be occupied by CTLA-4 and some B7 will be available to bind to the low-affinity activating receptor CD28, leading to T cell costimulation.
- *PD-1.* PD-1 is expressed on $CD8^+$ and $CD4^+$ T cells after antigen stimulation. Its cytoplasmic tail has inhibitory signaling motifs with tyrosine residues that are phosphorylated upon recognition of its ligands PD-L1 or PD-L2, which are homologous to the B7 molecules described earlier (see Fig. 5.7). Once phosphorylated, the tyrosines in the PD-1 tail bind a tyrosine phosphatase that inhibits kinase-dependent activating signals from CD28 and the TCR complex. Because the expression of PD-1 on T cells is increased upon chronic T cell activation and expression of the PD-1 ligands is increased by cytokines produced during prolonged inflammation, this pathway is most active in situations of chronic or repeated antigenic stimulation. This may happen in responses to chronic infections and some tumors, when PD-1–expressing T cells encounter the ligand on infected cells, tumor cells, or APCs. The major physiologic function of PD-1 may be to limit responses to infections, such as viral infections, enough to prevent the immunopathology that often results from strong T cell activation.

An important therapeutic application of our understanding of these inhibitory receptors is the treatment of cancer patients with antibodies that block these receptors, a form of cancer immunotherapy called checkpoint blockade. Such treatment leads to enhanced antitumor immune responses and tumor regression in many patients (see Chapter 10). However, patients treated with antibodies to inhibitory receptors often develop autoimmune reactions, consistent with the idea that the inhibitory receptors are constantly functioning to keep autoreactive T cells in check. Rare patients with mutations in one of their two copies of the *CTLA4* gene, which reduce expression of the receptor, also develop multiorgan inflammation (and a profound, as yet unexplained, defect in antibody production).

Several receptors on T cells other than CTLA-4 and PD-1 have been shown to inhibit immune responses and are currently being tested as targets of checkpoint blockade therapy. Some of these receptors are members of the tumor necrosis factor (TNF) receptor family or other protein families. Their physiologic role is not clearly established.

MIGRATION OF T LYMPHOCYTES IN CELL-MEDIATED IMMUNE REACTIONS

As we discussed at the beginning of this chapter, T cell responses are initiated primarily in secondary lymphoid organs, and the effector phase occurs mainly in

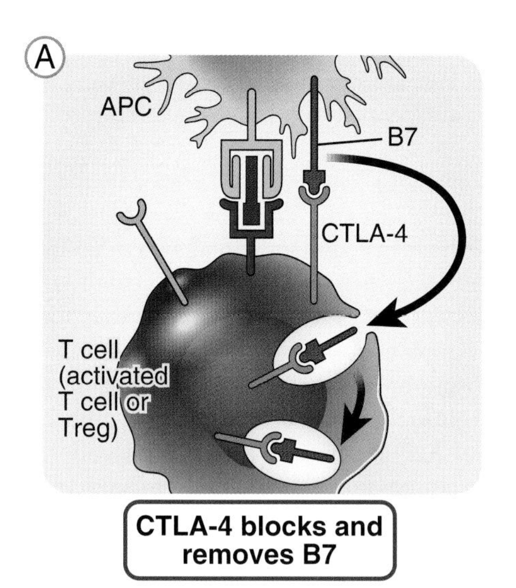

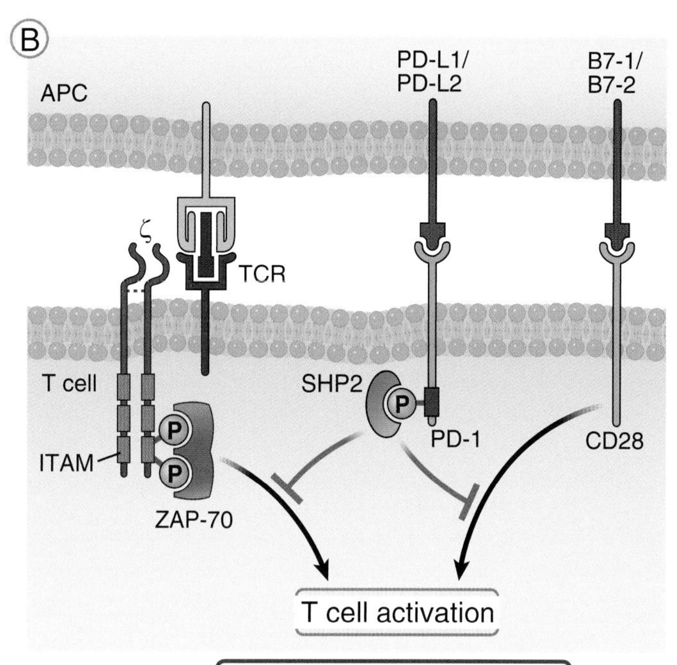

C

	CTLA-4	PD-1
Major site of action	Secondary lymphoid organs	Peripheral tissues (and lymphoid organs?)
Stage of immune response that is inhibited	Induction (priming)	Effector phase
Cell type that is inhibited	$CD4^+$ same as or more than $CD8^+$	$CD8^+ \geq CD4^+$
Cellular expression	Tregs, activated T cells	Mainly activated T cells
Main signals inhibited	Competitive inhibitor of CD28 costimulation by binding to B7 with high affinity and removing B7 from APCs	Signaling inhibitor of CD28 and TCR: inhibits kinase-depending signals by activating phosphatase
Role in Treg-mediated suppression of immune responses	Yes	No

Fig. 5.14 Functions of CTLA-4 and PD-1. **A,** CTLA-4 blocks and removes B7 molecules on antigen-presenting cells. **B,** PD-1 activates a phosphatase (SHP2) that inhibits kinase-dependent signals from CD28 and the TCR complex. **C,** Some of the major differences between these checkpoint molecules are summarized. *ITAM,* immunoreceptor tyrosine-based activation motifs; *Treg,* T regulatory cell.

peripheral tissue sites of infection (see Fig. 5.2). **Thus, T cells at different stages of their lives have to migrate in different ways:**

- Naive T cells migrate between blood and secondary (peripheral) lymphoid organs throughout the body, until they encounter dendritic cells within the lymphoid organ that display the antigens the T cells recognize (see Chapter 3).
- After the naive T cells are activated and differentiate into effector cells, these cells migrate back to the sites of infection, where they function to kill microbes.

The migration of naive and effector T cells is controlled by three families of proteins—selectins, integrins, and chemokines—that regulate the migration of all leukocytes, as described in Chapter 2 (see Fig. 2.16). The routes of migration of naive and effector T cells differ significantly because of selective expression of different adhesion molecules and chemokine receptors on naive T cells versus effector T cells, together with the selective expression of endothelial adhesion molecules and chemokines in lymphoid tissues and sites of inflammation (Fig. 5.15).

Naive T cells express the adhesion molecule L-selectin (CD62L) and the chemokine receptor CCR7, which mediate the selective migration of the naive cells into lymph nodes through specialized blood vessels called high endothelial venules (HEVs). HEVs are located in the T cell zones of lymph nodes and mucosal lymphoid tissues and are lined by specialized endothelial cells, which express carbohydrate ligands that bind L-selectin. HEVs also display chemokines that are made only in lymphoid tissues and are specifically recognized by CCR7. The migration of naive T cells proceeds in a multistep sequence like that of migration of all leukocytes through blood vessels (see Fig. 2.18):

- Naive T cells in the blood engage in L-selectin–mediated rolling interactions with the HEV, allowing chemokines to bind to CCR7 on the T cells.
- CCR7 transduces intracellular signals that activate the integrin LFA-1 on the naive T cell, increasing the binding affinity of the integrin.
- The increased affinity of the integrin for its ligand, ICAM-1, on the HEV results in firm adhesion and arrest of the rolling T cells.
- The T cells then exit the vessel through endothelial junctions and are retained in the T cell zone of the lymph node because of the chemokines produced there.

Thus, many naive T cells that are carried by the blood into an HEV migrate to the T cell zone of the lymph node stroma. This happens constantly in all lymph nodes and mucosal lymphoid tissues in the body. Effector T cells do not express CCR7 or L-selectin, and thus they are not drawn into lymph nodes.

The phospholipid sphingosine 1-phosphate (S1P) plays a key role in the egress of T cells from lymph nodes. The levels of S1P are higher in the blood and lymph than inside lymph nodes. S1P binds to and induces internalization of its receptor, which keeps the expression of the receptor on circulating naive T cells low. When a naive T cell enters the node, it is exposed to lower concentrations of S1P, and expression of the receptor begins to increase. If the T cell does not recognize any antigen, the cell leaves the node through efferent lymphatic vessels, following the gradient of S1P into the lymph. If the T cell does encounter specific antigen and is activated, the surface expression of the S1P receptor is suppressed for several days by CD69, which is transiently expressed following T cell activation. As a result, recently activated T cells stay in the lymph node long enough to undergo clonal expansion and differentiation. When this process is completed, CD69 levels decline and the S1P receptor is reexpressed on the cell surface; at the same time, the cells lose expression of L-selectin and CCR7, which previously attracted the naive T cells to the lymph nodes. Therefore, activated T cells are drawn out of the nodes into the draining lymph, which then transports the cells to the circulation. The net result of these changes is that differentiated effector T cells leave the lymph nodes and enter the circulation. The importance of the S1P pathway has been highlighted by the development of a drug (fingolimod) that binds to the S1P receptor and blocks the exit of T cells from lymph nodes. This drug is approved for the treatment of the inflammatory disease multiple sclerosis.

Effector T cells migrate to sites of infection because they express adhesion molecules and chemokine

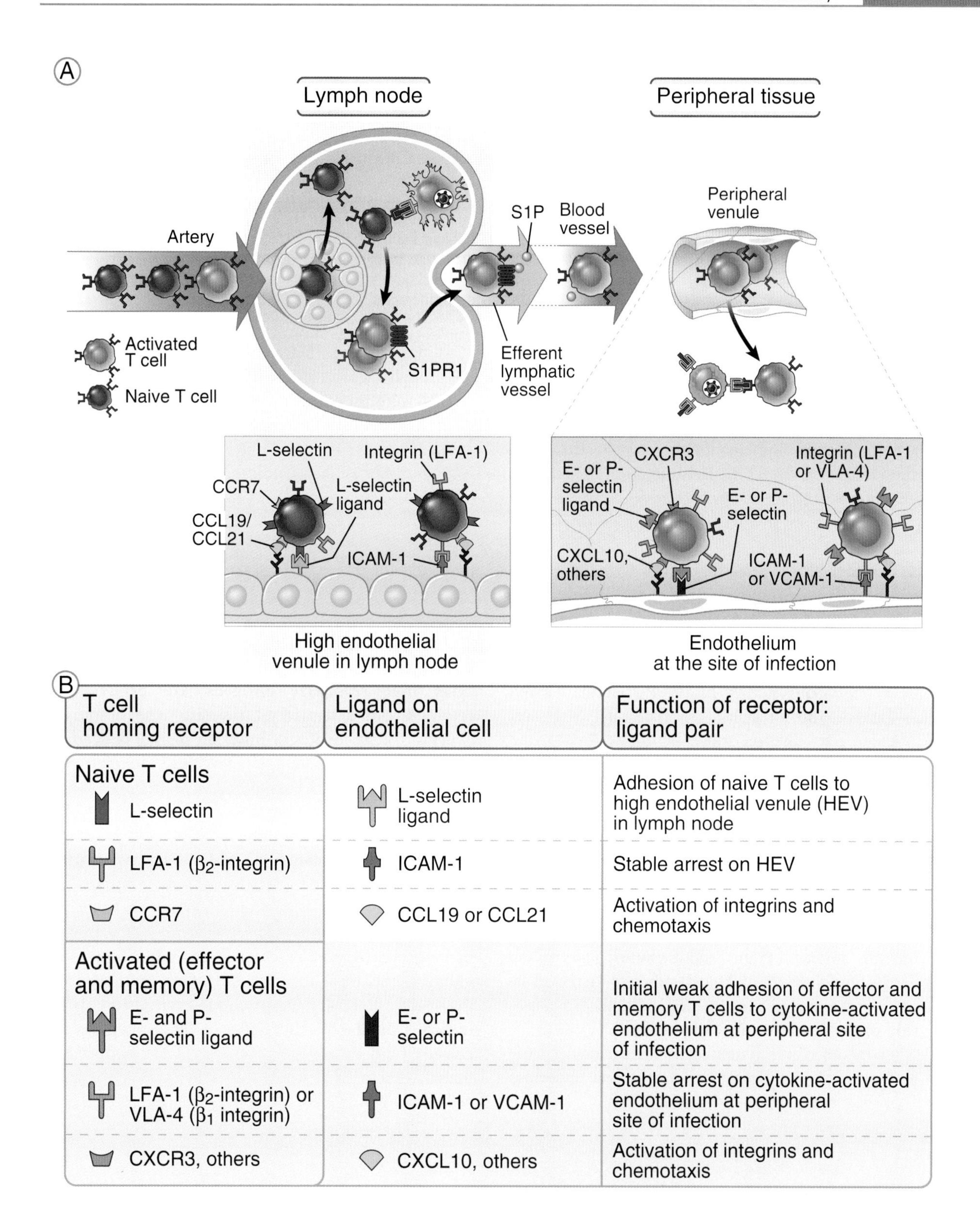

T cell homing receptor	Ligand on endothelial cell	Function of receptor: ligand pair
Naive T cells		
L-selectin	L-selectin ligand	Adhesion of naive T cells to high endothelial venule (HEV) in lymph node
LFA-1 (β_2-integrin)	ICAM-1	Stable arrest on HEV
CCR7	CCL19 or CCL21	Activation of integrins and chemotaxis
Activated (effector and memory) T cells		
E- and P-selectin ligand	E- or P-selectin	Initial weak adhesion of effector and memory T cells to cytokine-activated endothelium at peripheral site of infection
LFA-1 (β_2-integrin) or VLA-4 (β_1 integrin)	ICAM-1 or VCAM-1	Stable arrest on cytokine-activated endothelium at peripheral site of infection
CXCR3, others	CXCL10, others	Activation of integrins and chemotaxis

Fig. 5.15 Migration of naive and effector T lymphocytes. **A,** Naive T lymphocytes home to lymph nodes as a result of L-selectin, integrin, and chemokine receptor CCR7 binding to their ligands on high endothelial venules (HEVs). Chemokines expressed in lymph nodes bind to CCR7 on naive T cells, enhancing integrin-dependent adhesion and migration through the HEV. The phospholipid sphingosine 1-phosphate (S1P) plays a role in the exit of T cells from lymph nodes, by binding to its receptor, called S1PR1 (type 1 sphingosine 1-phosphate receptor). Activated T lymphocytes, including the majority of effector cells, home to sites of infection in peripheral tissues, and this migration is mediated by E-selectin and P-selectin, integrins, and chemokines secreted at inflammatory sites. Follicular helper T (Tfh) cells (not shown) are effector cells that remain in lymphoid organs, because they express a chemokine receptor (CXCR5) that draws them into lymphoid follicles, where they can interact with resident B lymphocytes. **B,** This table summarizes the functions of the principal T cell homing receptors and chemokine receptors and their ligands. *ICAM-1,* Intercellular adhesion molecule 1; *LFA-1,* leukocyte function–associated antigen 1; *VCAM-1,* vascular cell adhesion molecule 1; *VLA-4,* very late antigen 4.

receptors that bind to ligands expressed or displayed on vascular endothelium at sites of infection. The process of differentiation of naive T lymphocytes into effector cells is accompanied by changes in the types of adhesion molecules and chemokine receptors expressed on these cells (see Fig. 5.15). The migration of activated T cells into peripheral tissues is controlled by the same kinds of interactions involved in the migration of other leukocytes into tissues (see Chapter 2):

- Activated T cells express high levels of the glycoprotein ligands for E- and P-selectins and the integrins LFA-1 and VLA-4 (very late antigen 4). Innate immune cytokines produced in response to infection, such as TNF and IL-1, act on the endothelial cells to increase expression of E- and P-selectins, as well as ligands for integrins, especially ICAM-1 and vascular cell adhesion molecule 1 (VCAM-1), the ligand for the VLA-4 integrin.
- Effector T cells that are passing through the blood vessels at the infection site bind first to the endothelial selectins with low affinity, leading to rolling interactions.
- Effector T cells also express receptors for chemokines that are produced by macrophages and endothelial cells at these inflammatory sites and are displayed on the surface of the endothelium. The rolling T cells bind these chemokines, leading to increased affinity of the integrins for their ligands and firm adhesion of the T cells to the endothelium.
- After the effector T lymphocytes are arrested on the endothelium, they engage other adhesion molecules at the junctions between endothelial cells, crawling through these junctions into the tissue. Chemokines that were produced by macrophages and other cells in the tissues stimulate the motility of the transmigrating T cells.

The result of these molecular interactions between the T cells and endothelial cells is that effector T cells migrate out of the blood vessels to the area of infection. Naive T cells do not express high levels of ligands for E- or P-selectin or receptors for chemokines produced at inflammatory sites. Therefore, naive T cells do not migrate into sites of infection or tissue injury.

The homing of effector T cells to an infected tissue is independent of antigen recognition, but lymphocytes that recognize antigens are preferentially retained and activated at the site. The homing of effector T cells to sites of infection mainly depends on adhesion molecules and chemokines. Therefore, any effector T cell present in the blood, regardless of antigen specificity, can enter the site of any infection. This nonselective migration presumably maximizes the chances of effector lymphocytes entering tissues where they may encounter the microbes they recognize. The effector T cells that leave the circulation and that specifically recognize microbial antigen presented by local tissue APCs become reactivated and contribute to the killing of the microbe in the APC. One consequence of this reactivation is an increase in the expression of VLA integrins on the T cells. Some of these integrins specifically bind to molecules present in the extracellular matrix, such as hyaluronic acid and fibronectin.

Therefore, the antigen-stimulated lymphocytes adhere firmly to the tissue matrix proteins near the antigen, which may serve to keep the cells at the inflammatory sites. This selective retention contributes to accumulation of more and more T cells specific for microbial antigens in the region of the infection.

As a result of this sequence of T cell migration events, the effector phase of T cell–mediated immune responses may occur at any site of infection. Whereas the activation of naive T cells requires antigen presentation and costimulation by dendritic cells, differentiated effector cells are less dependent on costimulation. Therefore, the proliferation and differentiation of naive T cells are confined to lymphoid organs, where dendritic cells (which express abundant costimulators) display antigens, but the functions of effector T cells may be reactivated by any host cell displaying microbial peptides bound to MHC molecules, not just dendritic cells.

Elucidation of the molecular interactions involved in leukocyte migration has spurred many attempts to develop agents to block the process of cell migration into tissues. Antibodies against integrins are effective in the inflammatory diseases multiple sclerosis and inflammatory bowel disease. The clinical utility of these drugs is limited by the increased risk of new infection or reactivation of latent infections, because the immune surveillance function of the T cells is impaired when their migration into tissues is blocked. A small-molecule inhibitor of the S1P pathway is used for treating multiple sclerosis, as mentioned previously. Small molecules that bind to and block chemokine receptors have also been developed, and some have shown efficacy in inflammatory bowel disease.

DECLINE OF THE IMMUNE RESPONSE

Because of the remarkable expansion of antigen-specific lymphocytes at the peak of an immune response, it is predictable that once the response is over, the system returns to its steady state, called homeostasis, so that it is prepared to respond to the next infectious pathogen (see Fig. 5.12). During the response, the survival and proliferation of T cells are maintained by antigen, costimulatory signals from CD28, and cytokines such as IL-2. Once an infection is cleared and the stimuli for lymphocyte activation disappear, many of the cells that had proliferated in response to antigen are deprived of these survival signals. As a result, these cells die by apoptosis (programmed cell death). The response subsides within 1 or 2 weeks after the infection is eradicated, and the only sign that a T cell–mediated immune response had occurred is the pool of surviving memory lymphocytes.

To summarize, numerous mechanisms have evolved to overcome the challenges that T cells face in the generation of a useful cell-mediated immune response:

- Naive T cells need to find the antigen. This problem is solved by APCs that capture the antigen and concentrate it in specialized lymphoid organs in the regions through which naive T cells recirculate.
- The correct type of T lymphocytes (i.e., $CD4^+$ helper T cells or $CD8^+$ CTLs) must respond to antigens from the endosomal and cytosolic compartments. This selectivity is determined by the specificity of the CD4 and CD8 coreceptors for class II and class I MHC molecules and by the segregation of extracellular (vesicular) and intracellular (cytosolic) protein antigens for display by class II and class I MHC molecules, respectively.
- T cells should respond to microbial antigens but not to harmless proteins. This preference for microbes is maintained because T cell activation requires costimulators that are induced on APCs by microbes.
- Antigen recognition by a small number of T cells must lead to a response that is large enough to be effective. This is accomplished by robust clonal expansion after stimulation and by several amplification mechanisms induced by microbes and activated T cells themselves that enhance the response.
- The response must be optimized to combat different types of microbes. This is accomplished largely by the development of specialized subsets of effector T cells.

SUMMARY

- T lymphocytes are the mediators of the cell-mediated arm of the adaptive immune response, which combats microbes that are ingested by phagocytes and live within these cells or microbes that infect host cells. T lymphocytes also mediate defense against some extracellular microbes, help B lymphocytes to produce antibodies, and destroy cancer cells.
- The responses of T lymphocytes consist of sequential steps: recognition of host cell–associated microbes by naive T cells, expansion of the antigen-specific clones by proliferation, and differentiation of some of the progeny into effector cells and memory cells.
- T cells use their antigen receptors to recognize peptide antigens displayed by major histocompatibility complex (MHC) molecules on antigen-presenting cells (APCs), which accounts for the specificity of the ensuing response, and also recognize polymorphic residues of the MHC molecules, accounting for the MHC restriction of T cell responses.
- Antigen recognition by the T cell receptor (TCR) triggers signals that are delivered to the interior of the cells by molecules associated with the TCR (CD3 and ζ chains) and by the coreceptors CD4 and CD8.
- The binding of T cells to APCs is enhanced by adhesion molecules, notably the integrins, whose affinity for their ligands is increased by antigen recognition by the TCR.
- APCs exposed to microbes or to cytokines produced as part of the innate immune reactions to microbes express costimulators that bind to receptors on T cells and deliver necessary second signals for T cell activation.
- The biochemical signals triggered in T cells by antigen recognition and costimulation result in the activation of various transcription factors that stimulate the expression of genes encoding cytokines, cytokine receptors, and other molecules involved in T cell responses.
- The TCR signaling pathways involve protein tyrosine kinases, which phosphorylate proteins that become docking sites for additional kinases and other signaling molecules. The signaling pathways include the calcineurin/NFAT, RAS-MAP kinase, and PI-3 kinase/mTOR pathways.
- In response to antigen recognition and costimulation, T cells secrete cytokines that induce proliferation of the antigen-stimulated T cells and mediate the effector functions of T cells.
- T cells proliferate following activation by antigen and costimulators, resulting in expansion of the antigen-specific clones. The survival and proliferation of activated T cells are driven by the growth factor IL-2.
- Some of the T cells differentiate into effector cells that are responsible for eradicating infections. $CD4^+$ effector cells produce surface molecules, notably CD40L, and secrete various cytokines that activate other leukocytes to destroy microbes, and $CD8^+$ effector cells are able to kill infected and tumor cells.
- Other activated T cells differentiate into memory cells, which survive even after the antigen is eliminated and are capable of rapid responses to subsequent encounter with the antigen.
- Naive T cells migrate to peripheral lymphoid organs, mainly lymph nodes draining sites of microbe entry, whereas many of the effector T cells generated in lymphoid organs are able to migrate to any site of infection.
- The pathways of migration of naive and effector T cells are controlled by adhesion molecules and chemokines. The migration of T cells is independent of antigen, but cells that recognize microbial antigens in tissues are retained at these sites.

REVIEW QUESTIONS

1. What are the components of the TCR complex? Which of these components are responsible for antigen recognition and which for signal transduction?
2. What are some of the molecules in addition to the TCR that T cells use to initiate their responses to antigens, and what are the functions of these molecules?
3. What is costimulation? What is the physiologic significance of costimulation? What are some of the ligand-receptor pairs involved in costimulation?
4. Summarize the links among antigen recognition, the major biochemical signaling pathways in T cells, and the production of transcription factors.

5. What is the principal growth factor for T cells? Why do antigen-specific T cells expand more than other (bystander) T cells on exposure to an antigen?
6. What are the mechanisms by which $CD4^+$ effector T cells activate other leukocytes?
7. What are the major properties of memory T lymphocytes?
8. What proteins of the CD28 family function to inhibit T cell responses, and how do they work?
9. Why do naive T cells migrate preferentially to lymphoid organs and differentiated effector T cells (which have been activated by antigen) migrate preferentially to tissues that are sites of infection?

Answers to and discussion of the Review Questions may be found on p. 322.

6

Effector Mechanisms of T Cell–Mediated Immunity

Functions of T Cells in Host Defense

CHAPTER OUTLINE

Host defense in which T lymphocytes serve as the effector cells is called cell-mediated immunity. T cells are essential for eliminating microbes that survive and replicate inside cells and for eradicating infections by some extracellular microbes, often by recruiting other leukocytes to clear the infectious pathogens. T cells also kill tumor cells that produce mutated proteins that are recognized as foreign antigens (see Chapter 10). In this chapter, we focus on the role of T cell responses in defense against pathogenic microbes. Cell-mediated immune responses begin with the activation of naive T cells to proliferate and to differentiate into effector cells. The majority of these effector T cells then migrate to sites of infection, where they function to eliminate the microbes. Some CD4$^+$ effector cells stay in lymphoid organs and help B lymphocytes to produce high-affinity antibodies (humoral immunity, see Chapter 7). In Chapter 3, we described the function of major histocompatibility complex (MHC) molecules in displaying the antigens of intracellular microbes for recognition by T lymphocytes, and in Chapter 5, we discussed the events that occur in the activation of naive T lymphocytes. In this chapter, we address the following questions:

- What types of effector T cells are involved in the elimination of microbes?
- How do specialized effector T cells differentiate from naive T cells, and how do these effector cells eradicate infections by diverse microbes?
- What are the roles of macrophages and other leukocytes in the destruction of infectious pathogens?

TYPES OF T CELL–MEDIATED IMMUNE REACTIONS

Two main types of cell-mediated immune reactions eliminate different types of microbes: CD4$^+$ helper T cells express molecules that recruit and activate other leukocytes to phagocytose (ingest) and destroy

microbes, and CD8$^+$ cytotoxic T lymphocytes (CTLs) directly kill infected cells (Fig. 6.1). Microbial infections may occur anywhere in the body, and some infectious pathogens are able to infect and live within host cells. Pathogenic microbes that infect and survive inside host cells include (1) many bacteria, fungi, and protozoa that invade or are ingested by phagocytes but resist the killing mechanisms of these phagocytes and thus survive in vesicles or in the cytosol, and (2) viruses that infect phagocytic and nonphagocytic cells and replicate in these cells (see Fig. 5.1). CD4$^+$ and CD8$^+$ T cells recognize microbial antigens in different cellular compartments and differ in the nature of the reactions they elicit.

- CD4$^+$ T cells recognize antigens of microbes that have been internalized into endocytic vesicles in macrophages. These microbes may have evolved to resist the killing mechanisms of macrophages, which are largely restricted to vesicles. The T cells secrete cytokines and express membrane molecules that stimulate the microbicidal mechanisms of the phagocytes, enabling them to kill the microbes.
- CD4$^+$ T cells also secrete cytokines that recruit other leukocytes (i.e., neutrophils, eosinophils) to destroy extracellular microbes by phagocytosis and other mechanisms.
- By contrast, CD8$^+$ T cells recognize microbial antigens that are present in the cytosol of infected cells and destroy these cells.

Cell-mediated immunity against pathogens was discovered as a form of immunity to an infection by bacteria that survive in phagocytes that could be transferred from immune animals to naive animals by cells (now known to be T lymphocytes) but not by serum antibodies (Fig. 6.2). It was known from early studies that lymphocytes were responsible for the specificity of cell-mediated immunity against different microbes, but the elimination of the microbes was a function of activated macrophages. As already mentioned, CD4$^+$ T cells are mainly responsible for this classical type of cell-mediated immunity, whereas CD8$^+$ T cells can eradicate infections without a requirement for phagocytes.

T cell–mediated immune reactions consist of multiple steps (see Fig. 5.2). Naive T cells are stimulated by microbial antigens in secondary (peripheral) lymphoid organs, giving rise to effector T cells whose function is to

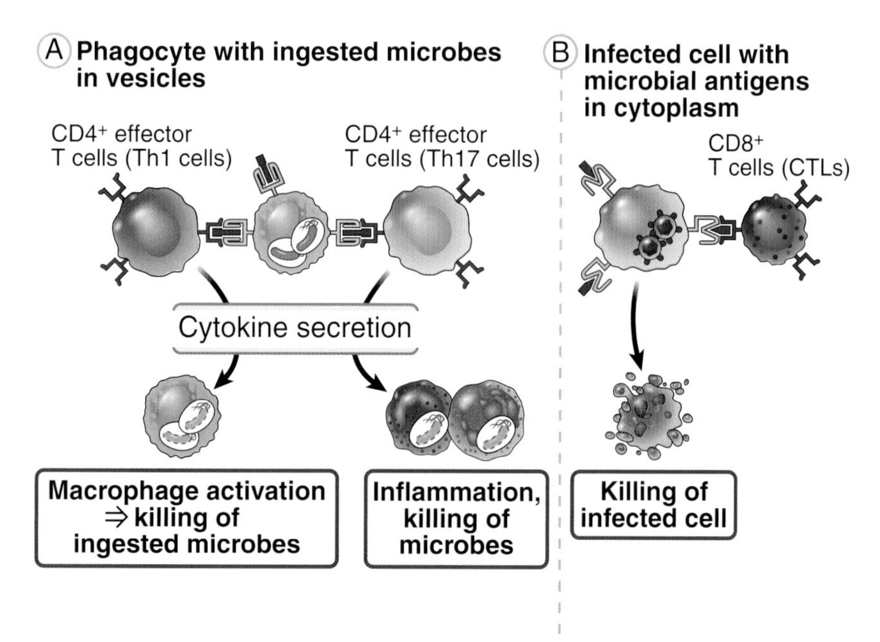

Fig. 6.1 Cell-mediated immunity. **A,** Effector T cells of the CD4$^+$ Th1 and Th17 subsets recognize microbial antigens and secrete cytokines that recruit leukocytes (inflammation) and activate phagocytes to kill the microbes. Effector cells of the Th2 subset (not shown) recruit eosinophils, which destroy helminthic parasites. **B,** CD8$^+$ cytotoxic T lymphocytes (CTLs) kill infected cells with microbial antigens in the cytosol. CD8$^+$ T cells also produce cytokines that induce inflammation and activate macrophages (not shown).

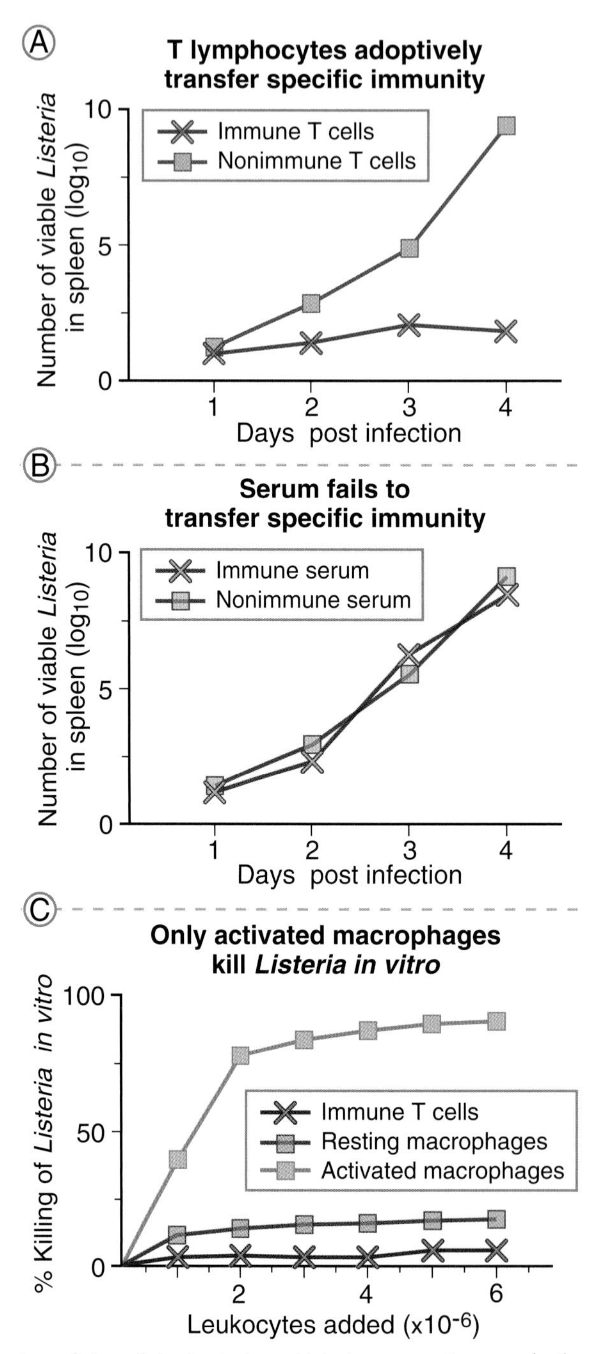

Fig. 6.2 Cell-mediated immunity to an intracellular bacterium, *Listeria monocytogenes*. In these experiments, a sample of lymphocytes or serum (a source of antibodies) was taken from a mouse that had previously been exposed to a sublethal dose of *Listeria* organisms (immune mouse) and transferred to a normal (naive) mouse, and the recipient of the adoptive transfer was challenged with the bacteria. The number of bacteria were measured in the spleen of the recipient mouse to determine if the transfer had conferred immunity. Protection against bacterial challenge (seen by reduced recovery of live bacteria) was induced by the transfer of immune lymphoid cells, now known to be T cells (**A**), but not by the transfer of serum (**B**) or nonimmune cells or serum. The bacteria were killed in vitro by activated macrophages but not by T cells (**C**). Therefore, protection depends on antigen-specific T lymphocytes (induced by the initial sublethal infection), but bacterial killing is the function of activated macrophages.

eradicate the infections. The differentiated effector T cells then migrate to the site of infection. Phagocytes at these sites ingest microbes or microbial proteins into intracellular vesicles, where they are proteolytically processed into peptide fragments, which are then bound to class II MHC molecules and displayed at the cell surface for recognition by $CD4^+$ T cells. Peptide antigens derived by proteasomal processing of microbial proteins in the cytosol of infected cells are bound to class I MHC molecules and displayed on the cell surface for recognition by $CD8^+$ T cells. Antigen recognition activates the effector T cells to perform their task of eliminating the infectious pathogens. Thus, in cell-mediated immunity, T cells recognize protein antigens at two stages. First, naive T cells recognize antigens in secondary lymphoid organs and respond by proliferating and by differentiating into effector cells (see Chapter 5). Second, effector T cells recognize the same antigens at the site of infection anywhere in the body, and the T cells respond by eliminating these microbes.

This chapter describes how $CD4^+$ and $CD8^+$ effector T cells develop in response to microbes and eliminate these microbes. Because $CD4^+$ helper T lymphocytes and $CD8^+$ CTLs use distinct mechanisms to combat infections, we discuss the development and functions of the effector cells of these lymphocyte classes individually. We conclude by describing how the two classes of lymphocytes may cooperate to eliminate intracellular microbes.

DEVELOPMENT AND FUNCTIONS OF $CD4^+$ EFFECTOR T LYMPHOCYTES

In Chapter 5, we introduced the concept that effector cells of the $CD4^+$ lineage could be distinguished on the basis of the cytokines they produce. These subsets of $CD4^+$ T cells differ in their functions and serve distinct roles in cell-mediated immunity.

Subsets of $CD4^+$ Helper T Cells Distinguished by Cytokine Profiles

Analysis of cytokine production by helper T cells has revealed that functionally distinct subsets of $CD4^+$ T cells exist that produce different cytokines and that eliminate different types of pathogens. The existence of these subsets is an illustration of specialized immune responses that are optimized to combat diverse microbes. For example, some microbes such as mycobacteria infect phagocytes but resist intracellular killing by innate immune activation of the phagocyte. The T cell response to such microbes results in further activation of the phagocytes, enabling them to kill the ingested microbes. By contrast, the immune response to helminths is dominated by the production of immunoglobulin E (IgE) antibodies and the activation of eosinophils, both of which are stimulated by T cells and help eliminate helminthic infections. Extracellular bacterial and fungal infections are controlled by phagocytosis and killing of the pathogens by neutrophils, and cytokines produced by T cells enhance these neutrophilic responses. These different types of immune responses are mediated by subpopulations of $CD4^+$ effector T cells that produce different cytokines.

$CD4^+$ helper T cells may differentiate into three subsets of effector cells that produce distinct sets of cytokines that function to defend against different types of microbial infections in tissues and a fourth subset that activates B cells in secondary lymphoid organs (Fig. 6.3). The two subsets that were defined first are called Th1 cells and Th2 cells (for type 1 helper T cells and type 2 helper T cells, respectively); the third population, which was identified later, is called Th17 cells because its signature cytokine is interleukin-17 (IL-17). The T cells that help B lymphocytes, called T follicular helper (Tfh) cells, are described in Chapter 7 and will not be considered further in this chapter. The discovery of these subpopulations has been an important milestone in understanding immune responses and provides models for studying the process of cell differentiation. However, it should be noted that some activated $CD4^+$ T cells may produce mixtures of cytokines and therefore cannot be readily classified into these subsets, and there may be plasticity in these populations so that one subset may convert into another under some conditions. Despite these caveats, considering the functions of $CD4^+$ effector cells in the context of the major subsets is helpful for understanding the mechanisms of cell-mediated immunity.

The cytokines produced in adaptive immune responses include those made by the helper T cell subsets, as well as cytokines produced by $CD4^+$ regulatory T cells and $CD8^+$ T cells. These cytokines of adaptive immunity share some general properties, but they each have different biologic activities and play unique roles in the effector phase or regulation of these responses (Fig. 6.4). The functions of the $CD4^+$ T cell subsets reflect the actions of the cytokines they produce. Similar sets of cytokines may be produced early in immune responses by innate lymphoid cells, such as ILC1, ILC2, and ILC3 (see Chapter 2), and later, in

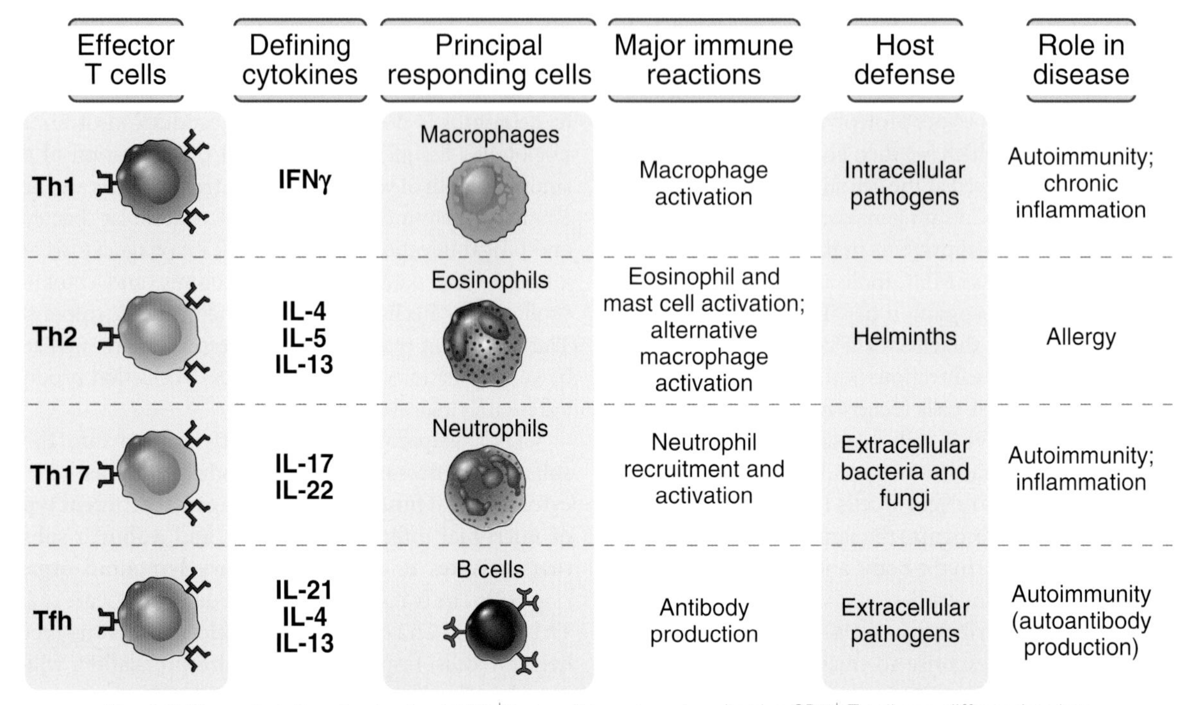

Fig. 6.3 **Characteristics of subsets of CD4$^+$ helper T lymphocytes.** A naive CD4$^+$ T cell may differentiate into subsets that produce different cytokines that recruit and activate different cell types (referred to here as responding cells) and combat different types of infections in host defense. These subsets also are involved in various kinds of inflammatory diseases. The table summarizes the major differences among Th1, Th2, Th17, and Tfh subsets of helper T cells. *IFN,* Interferon; *IL,* interleukin.

much larger amounts, by Th1, Th2, and Th17 cells, respectively. These combined innate and adaptive responses with similar cytokine profiles and functions are sometimes grouped under type 1, type 2, and type 3 immunity, respectively.

Each subset of CD4$^+$ T cells develops in response to the types of microbes that subset is best at eradicating. Different microbes elicit the production of different cytokines from dendritic cells and other cell types and these cytokines drive the differentiation of antigen-activated T cells to one or another subset. We next discuss the functions and development of each of the major subsets of CD4$^+$ effector T cells.

Th1 Cells

Th1 cells stimulate phagocyte-mediated killing of internalized microbes (Fig. 6.5). The signature cytokine of Th1 cells is interferon-γ (IFN-γ), the most potent macrophage-activating cytokine known. (Despite its similar name, IFN-γ is a much less potent antiviral cytokine than the type I IFNs [see Chapter 2]).

Th1 cells, acting through CD40 ligand and IFN-γ, increase the ability of macrophages to kill phagocytosed microbes (Fig. 6.6). Macrophages ingest and attempt to destroy microbes as part of the innate immune response (see Chapter 2). The efficiency of this process is greatly enhanced by the interaction of Th1 cells with the macrophages. The Th1 cells are induced to express CD40 ligand (CD40L, or CD154) and to secrete IFN-γ. Binding of CD40L to CD40 on macrophages functions together with IFN-γ binding to its receptor on the same macrophages to trigger biochemical signaling pathways that lead to the generation of reactive oxygen species (ROS) and nitric oxide (NO) and enhanced production of lysosomal proteases. All these molecules are potent destroyers of microbes. The net result of CD40- and IFN-γ–mediated activation is that macrophages become strongly microbicidal and can kill most ingested microbes. This pathway of macrophage activation by CD40L and IFN-γ is called **classical macrophage activation**, in contrast to Th2-mediated alternative macrophage activation, discussed later. Classically

A General properties of T cell cytokines

Property	Significance
Produced transiently in response to antigen	Provides cytokine when needed
Usually acts on same cell that produces the cytokine (autocrine) or nearby cells (paracrine)	Most effects are at tissue sites of infection and inflammation
Pleiotropism: each cytokine has multiple biological actions	Provides diversity of actions but may limit clinical utility of cytokines because of unwanted effects
Redundancy: multiple cytokines may share the same or similar biological activities	Therapuetic blocking of any one cytokine may not achieve a desired effect

B Biologic actions of selected T cell cytokines

Cytokine	Principal action	Cellular source(s)
IL-2	T cell proliferation; regulatory T cell development and maintenance	Activated T cells
Interferon-γ (IFN-γ)	Activation of macrophages (classical pathway)	$CD4^+$ Th1 and $CD8^+$ T cells, natural killer (NK) cells, ILC1
IL-4	B cell switching to IgE; alternative macrophage activation	$CD4^+$ Th2 cells, mast cells
IL-5	Activation of eosinophils	$CD4^+$ Th2 cells, mast cells, ILC2
IL-13	B cell switching to IgE; alternative macrophage activation	$CD4^+$ Th2 cells, mast cells, ILC2
IL-17	Stimulation of acute inflammation	$CD4^+$ Th17 cells, ILC3, other cells
IL-21	B cell activation; Tfh differentiation	$CD4^+$ Tfh cells
IL-22	Maintenance of epithelial barrier function	$CD4^+$ Th17 cells, NK cells, ILC3

Fig. 6.4 Properties of the major cytokines produced by $CD4^+$ helper T lymphocytes. **A,** General properties of cytokines produced during adaptive immune responses. **B,** Functions of cytokines involved in T cell–mediated immunity. Note that IL-2, which is produced by T cells early after activation and is the first identified T cell cytokine, was discussed in Chapter 5 in the context of T cell activation. Transforming growth factor β (TGF-β) functions mainly as an inhibitor of immune responses; its role is discussed in Chapter 9. The cytokines of innate immunity are shown in Fig. 2.14; several of these are also made by T cells and thus function in adaptive immunity as well. More information about these cytokines and their receptors is provided in Appendix III. *IgE,* Immunoglobulin E; *IL,* interleukin; *ILC,* innate lymphoid cell (the different types are indicated by numbers); *Tfh,* T follicular helper (cell); *Th,* T helper (cell).

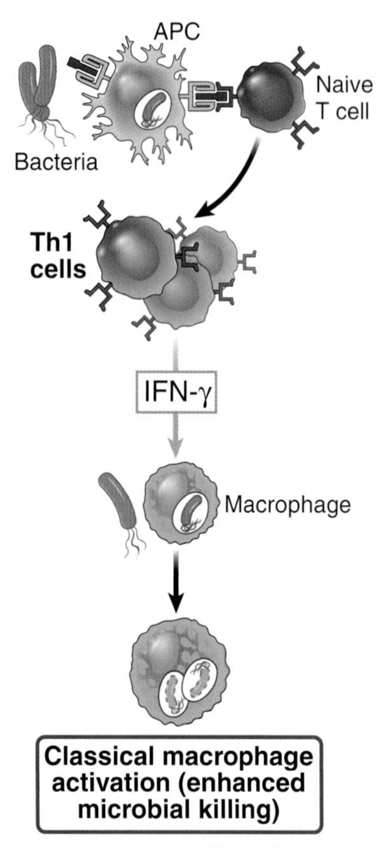

Fig. 6.5 Functions of Th1 cells. Th1 cells produce the cytokine interferon-γ (IFN-γ), which activates macrophages to kill phagocytosed microbes (classical pathway of macrophage activation). In mice, IFN-γ stimulates the production of IgG antibodies, but this has not been established in humans. *APC,* Antigen-presenting cell; *Th1,* T helper 1 (cell).

activated macrophages, often called M1 macrophages, also secrete cytokines that stimulate inflammation and express increased levels of MHC molecules and costimulators, which amplify the T cell response. $CD8^+$ T cells secrete IFN-γ as well and may contribute to macrophage activation and killing of internalized microbes.

The critical role of Th1 cells in defense against intracellular microbes is demonstrated by the observation that individuals with inherited defects in the development or function of this subset are susceptible to damaging infections with such microbes, especially nontuberculous mycobacterial species that are widespread in the environment but do not cause disease in immunocompetent individuals.

Essentially the same reaction, consisting of leukocyte recruitment and activation, may be elicited by injecting a microbial (or other) protein into the skin of an individual who has been immunized with the protein or previously infected with the microbe. This reaction is called **delayed-type hypersensitivity** (DTH), and it is described in Chapter 11, in which we discuss injurious immune reactions.

Differentiation of Th1 Cells

The differentiation of naive $CD4^+$ T cells to Th1 effector cells is driven by a combination of antigen-induced T cell receptor (TCR) signaling and the cytokines IL-12 and IFN-γ (Fig. 6.7). In response to many bacteria (especially bacteria that can live within phagocytes) and viruses, dendritic cells and macrophages produce IL-12, and natural killer (NK) cells produce IFN-γ. Therefore, when naive T cells recognize the antigens of these microbes, the T cells are also exposed to IL-12 and IFN-γ. IL-12 and IFN-γ activate the transcription factors STAT4 and STAT1, respectively, and antigen-induced signals in combination with the cytokines induce expression of a transcription factor called T-BET that is essential for Th1 development and function. These transcription factors work together to stimulate the expression of IFN-γ and other proteins that are involved in the migration of Th1 cells to sites of infection. IFN-γ not only activates macrophages to kill ingested microbes but also promotes more Th1 development and inhibits the development of Th2 and Th17 cells. Thus, IFN-γ increasingly polarizes the response to the Th1 subset.

Th2 Cells

Th2 cells promote eosinophil-mediated destruction of helminthic parasites and are involved in tissue repair (Fig. 6.8). The signature cytokines of Th2 cells—IL-4, IL-5, and IL-13—function cooperatively in eradicating worm infections. Helminths are too large to be phagocytosed, so mechanisms other than macrophage activation are needed for their destruction. When Th2 cells encounter the antigens of helminths, the T cells secrete their cytokines. Eosinophils are activated by IL-5 produced by the Th2 cells and release their granule contents, which are toxic to the parasites. IL-13 stimulates mucus secretion and intestinal peristalsis, increasing the expulsion of parasites from the intestines. Th2 responses are often accompanied by Tfh cells that

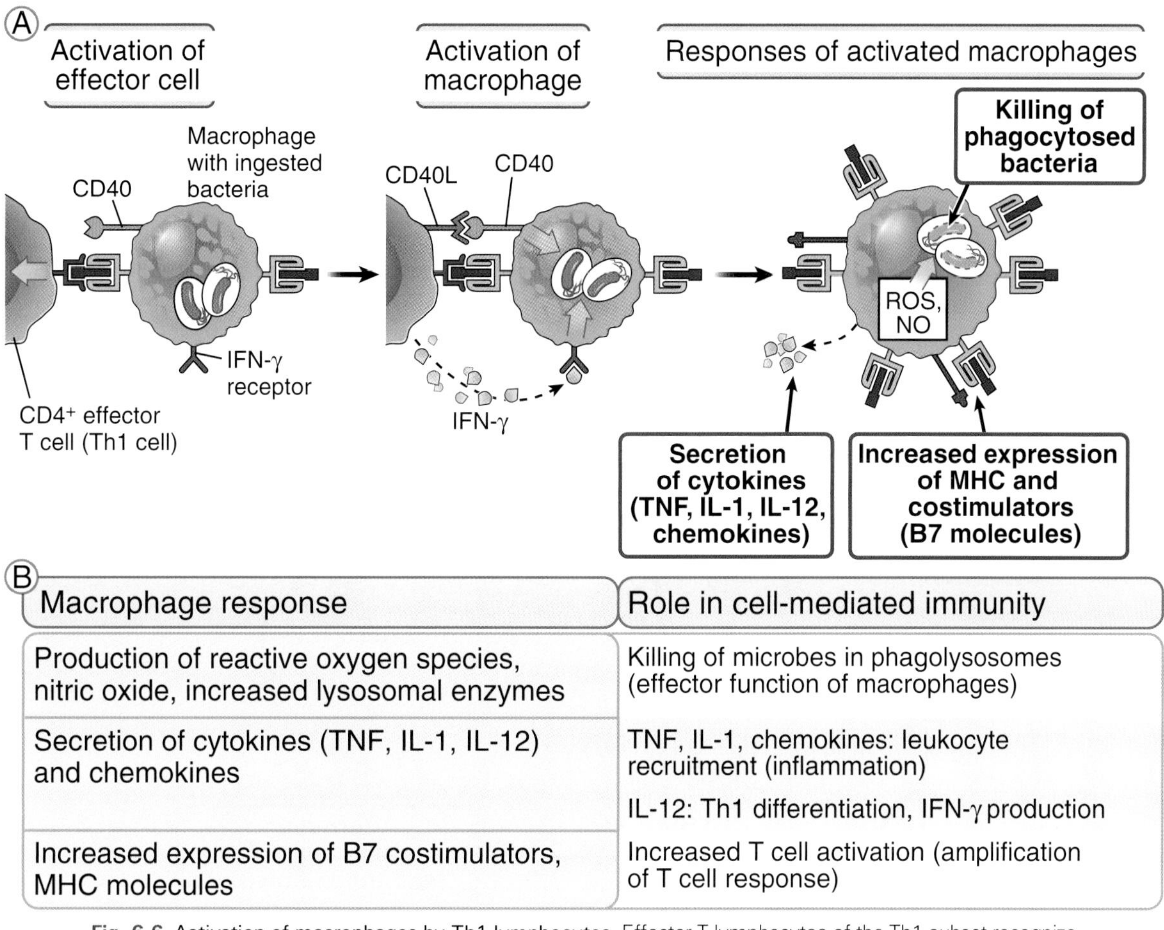

Fig. 6.6 Activation of macrophages by Th1 lymphocytes. Effector T lymphocytes of the Th1 subset recognize the antigens of ingested microbes on macrophages. In response to this recognition, the T lymphocytes express CD40L, which engages CD40 on the macrophages, and the T cells secrete interferon-γ (IFN-γ), which binds to IFN-γ receptors on the macrophages. This combination of signals activates the macrophages to produce microbicidal substances that kill the ingested microbes. This is known as the classical pathway of macrophage activation. The activated macrophages also secrete tumor necrosis factor (TNF); interleukin-1 (IL-1); and chemokines, which induce inflammation; and IL-12, which promotes Th1 responses. Activated macrophages express more major histocompatibility complex (MHC) molecules and costimulators, which further enhance T cell responses. **A,** Illustration shows a $CD4^+$ T cell recognizing class II MHC–associated peptides and activating the macrophage. **B,** Summary of macrophage responses and their roles in cell-mediated immunity. *NO,* nitric oxide; *ROS,* reactive oxygen species

produce IL-4 and IL-13, which drive B cell IgE class switching. IgE binds to mast cells and is responsible for their activation, leading to the secretion of chemical mediators that stimulate inflammation and proteases that destroy toxins.

Th2 cytokines promote epithelial barrier tissue repair following injury, in part by stimulating the alternative pathway of macrophage activation (Fig. 6.9). IL-4 and IL-13 inhibit the classical pathway of macrophage activation, which generates microbicidal and proinflammatory macrophages, described earlier. At the same time, Th2 cytokines induce alternative macrophage activation, so called to distinguish it from classical activation. Alternatively activated macrophages produce growth factors that promote collagen synthesis by fibroblasts and angiogenesis and thus may play a role in tissue repair following injury, but they may contribute to fibrosis in a variety of disease states.

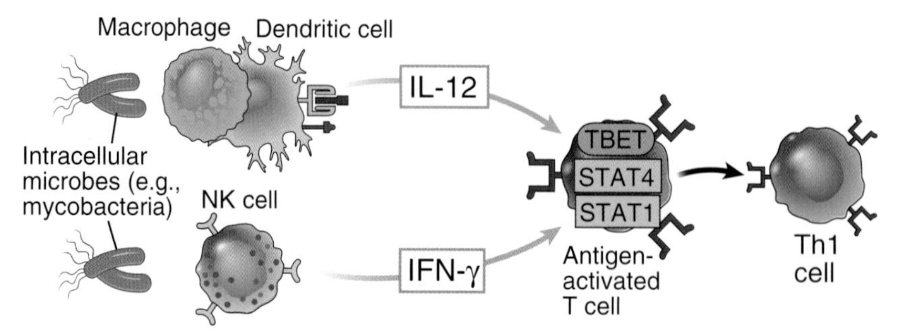

Fig. 6.7 Differentiation of Th1 effector cells. Dendritic cells, macrophages, and NK cells that respond to different types of microbes secrete cytokines that induce the differentiation of antigen-activated $CD4^+$ T cells into the Th1 subset. The transcription factors that are involved in this process are indicated in boxes in the antigen-activated T cells. *IFN-γ,* Interferon-γ; *IL,* interleukin; *NK,* natural killer.

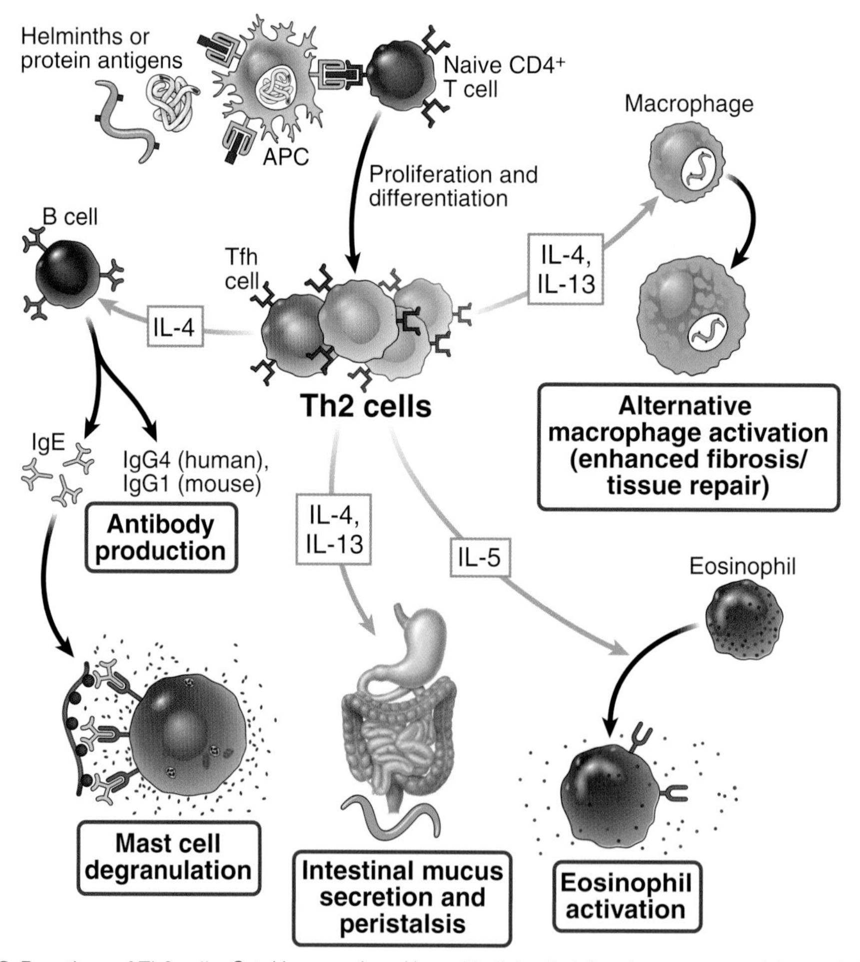

Fig. 6.8 Functions of Th2 cells. Cytokines produced by epithelial cells (often in response to injury or infection) act on antigen-presenting cells *(APCs)* or on naive T cells to stimulate the differentiation of the T cells into Th2 cells. Th2 cells produce the cytokines interleukin-4 (IL-4), IL-5, and IL-13. IL-4 (and IL-13) act on B cells to stimulate production mainly of IgE antibodies, which bind to mast cells. Help for antibody production may be provided by Tfh cells that produce Th2 cytokines and reside in lymphoid organs and not by classical Th2 cells. IL-5 activates eosinophils, a response that is important in the destruction of helminths. IL-4 and IL-13 induce alternative macrophage activation. *APC,* Antigen-presenting cell; *Ig,* immunoglobulin.

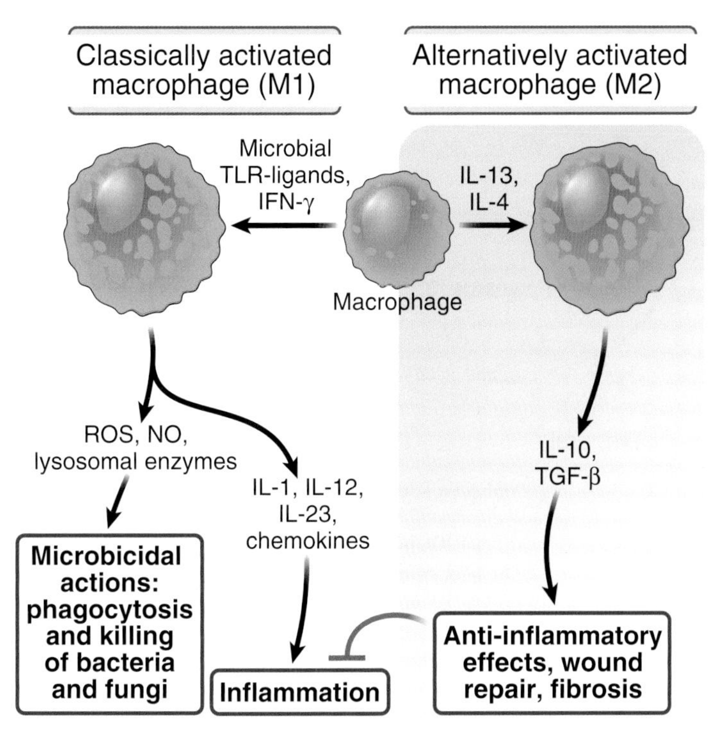

Fig. 6.9 Classical and alternative macrophage activation. Classically activated (M1) macrophages are induced by microbial products binding to TLRs and cytokines, particularly interferon-γ (IFN-γ), and are microbicidal and proinflammatory. Alternatively activated (M2) macrophages are induced by interleukin-4 (IL-4) and IL-13 (produced by certain subsets of T lymphocytes and other leukocytes) and are important in tissue repair and fibrosis. The M1 and M2 populations may represent extreme phenotypes, and there may be macrophage populations with intermediate characteristics. Various mixtures of activated macrophages are likely induced in most infections. *NO,* Nitric oxide; *ROS,* reactive oxygen species; *TGF-β,* transforming growth factor β; *TLR,* Toll-like receptor.

Th2 cells are involved in allergic reactions to environmental antigens. The antigens that elicit such reactions are called allergens. They induce Th2 responses in genetically susceptible individuals and subsequent exposure to the allergens triggers mast cell and eosinophil activation. Allergies are the most common type of immune disorder; we will return to these diseases in Chapter 11 when we discuss hypersensitivity reactions. Antagonists of IL-5 are approved for the treatment of allergic asthma, and an antibody against the IL-4/IL-13 receptor is approved for asthma and for the allergic skin disease atopic dermatitis. Antibodies that block the Th2-inducing cytokines TSLP and IL-33 are also approved for the treatment of asthma.

The relative activation of Th1 and Th2 cells in response to an infectious microbe may determine the outcome of the infection (Fig. 6.10). For example, the protozoan parasite *Leishmania major* lives inside the phagocytic vesicles of macrophages, and its elimination requires the activation of the macrophages by *L. major*–specific Th1 cells. Most inbred strains of mice make an effective Th1 response to the parasite and are thus able to eradicate the infection. However, in some inbred mouse strains, the response to *L. major* is dominated by Th2 cells, and these mice succumb to the infection. *Mycobacterium leprae*, the bacterium that causes leprosy, is a pathogen for humans that also lives inside macrophages and is eliminated by cell-mediated immune mechanisms. Some people infected with *M. leprae* are unable to eradicate the infection, which, if left untreated, will progress to a destructive form of the disease, called lepromatous leprosy. By contrast, in other patients, the bacteria induce strong cell-mediated immune responses with activated T cells and macrophages around the

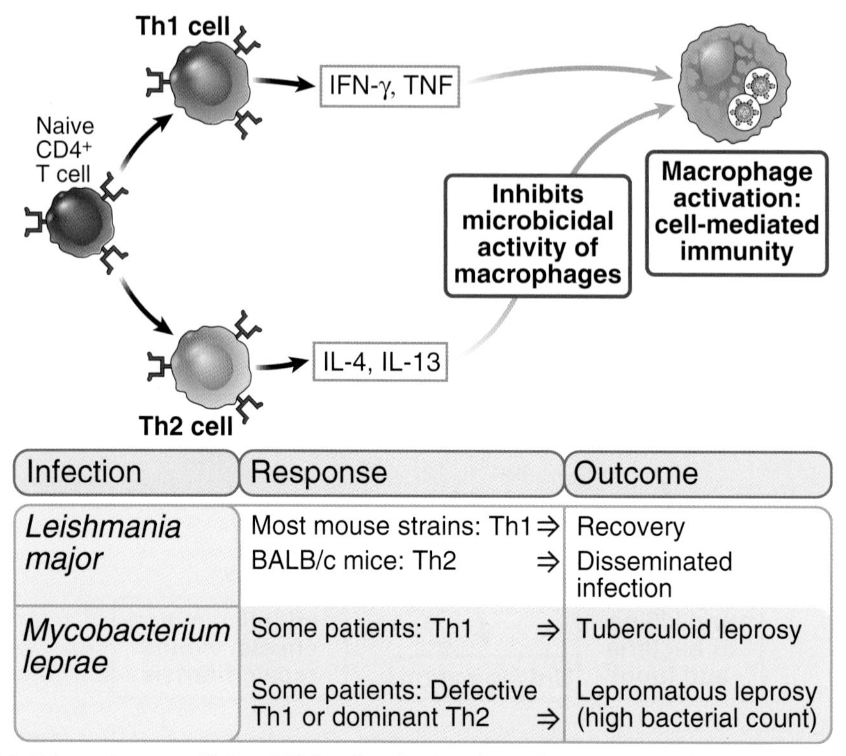

Infection	Response	Outcome
Leishmania major	Most mouse strains: Th1 ⇒	Recovery
	BALB/c mice: Th2 ⇒	Disseminated infection
Mycobacterium leprae	Some patients: Th1 ⇒	Tuberculoid leprosy
	Some patients: Defective Th1 or dominant Th2 ⇒	Lepromatous leprosy (high bacterial count)

Fig. 6.10 Balance between Th1 and Th2 cell activation determines outcome of intracellular infections. Naive CD4$^+$ T lymphocytes may differentiate into Th1 cells, which activate phagocytes to kill ingested microbes, and Th2 cells, which inhibit classical macrophage activation. The balance between these two subsets may influence the outcome of infections, as illustrated by *Leishmania* infection in mice and leprosy in humans. *IFN,* Interferon; *IL,* interleukin; *TNF,* tumor necrosis factor.

infection site and few surviving microbes; this form of less injurious infection is called tuberculoid leprosy. The tuberculoid form is associated with the activation of *M. leprae*–specific Th1 cells, whereas the destructive lepromatous form is associated with a defect in Th1 cell activation and sometimes a strong Th2 response. The same principle—that the T cell cytokine response to an infectious pathogen is an important determinant of the outcome of the infection—may be true for other infectious diseases.

Differentiation of Th2 Cells

Differentiation of naive CD4$^+$ T cells to Th2 cells is initiated by cytokines secreted by damaged epithelial cells, including IL-25, IL-33, and thymic stromal lymphopoietin (TSLP), and then further stimulated by IL-4, which may be produced by mast cells, other tissue cells, and T cells themselves at sites of helminth infection (see Fig. 6.11). IL-4 activates the transcription factor STAT6 and antigen-induced signals in combination with IL-4 induce expression of a transcription factor GATA-3, which is required for Th2 differentiation. Analogous to Th1 cells, these transcription factors stimulate the expression of Th2 cytokines and proteins involved in cell migration and thus promote Th2 responses. IL-4 produced by Th2 cells enhances further Th2 differentiation, thus amplifying the Th2 response.

Th17 Cells

Th17 cells induce inflammatory reactions that destroy bacteria and fungi (Fig. 6.12). The major cytokines produced by Th17 cells are IL-17 and IL-22. This T cell subset was discovered during studies of inflammatory diseases, many years after Th1 and Th2 subsets were described, and its role in host defense was established later.

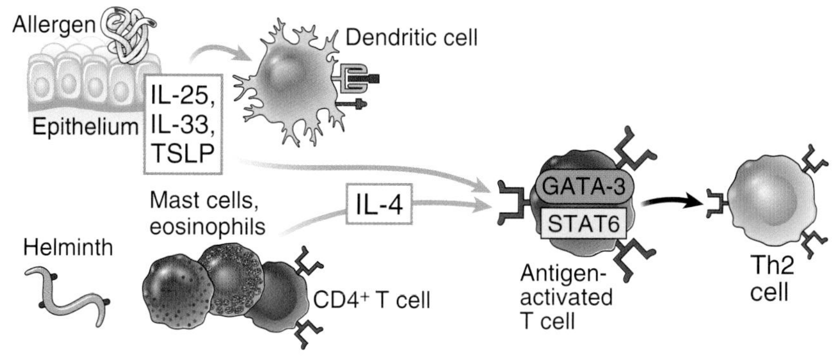

Fig. 6.11 **Differentiation of Th2 effector cells.** Cytokines released from damaged epithelium, dendritic cells, and other cells responding to helminths induce the differentiation of antigen-activated CD4$^+$ T cells into the Th2 subset. The transcription factors involved in this process are indicated in boxes in the antigen-activated T cells. *IL,* interleukin; *TSLP,* thymic stromal lymphopoietin.

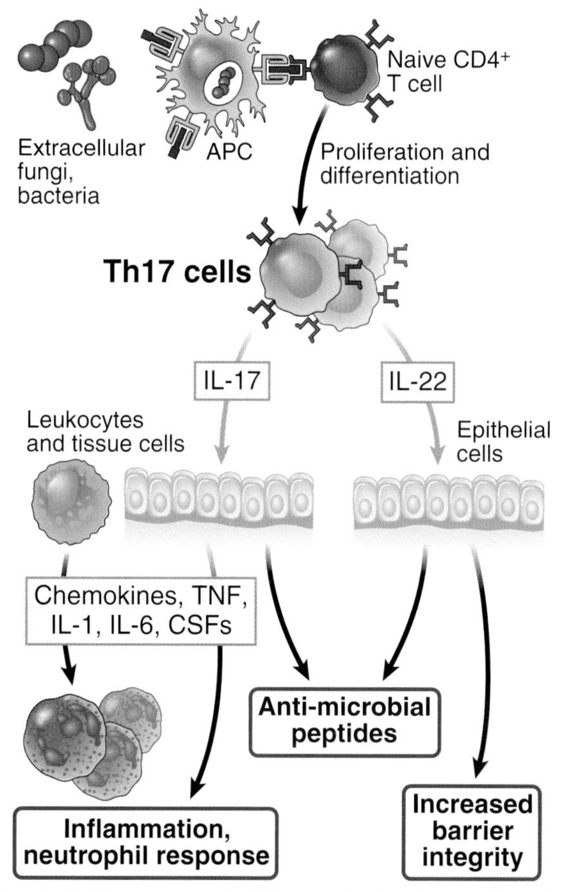

Fig. 6.12 **Functions of Th17 cells.** Th17 cells produce the cytokine interleukin-17 (IL-17), which induces production of chemokines and other cytokines from various cells, and these recruit neutrophils (and monocytes, not shown) into the site of inflammation. Some of the cytokines made by Th17 cells, notably IL-22, function to maintain epithelial barrier function in the intestinal tract and other tissues. *APC,* Antigen-presenting cell; *CSFs,* colony-stimulating factors; *TNF,* tumor necrosis factor.

The major function of Th17 cells is to stimulate the recruitment of neutrophils and, to less extent, monocytes. IL-17 secreted by Th17 cells induces the production of chemokines from other cells, and these chemokines are responsible for leukocyte recruitment. Bacteria and fungi that survive outside cells are usually rapidly destroyed following ingestion by phagocytes, especially neutrophils, so the recruited leukocytes are effective at eliminating these microbes. Th17 cells also stimulate the production of antimicrobial substances, called defensins, that function like locally produced endogenous antibiotics. IL-22 produced by Th17 cells induces epithelial cell defensin production, helps to maintain the integrity of epithelial barriers, and may promote repair of damaged epithelia. Rare individuals who have inherited defects in Th17 responses are prone to developing chronic mucocutaneous candidiasis and bacterial abscesses in the skin.

Th17 cells also induce inflammation and are implicated in numerous chronic inflammatory diseases. Antibodies that block IL-17 or the Th17-inducing cytokine IL-23 are very effective treatments for psoriasis, an inflammatory skin disease. An antibody that neutralizes IL-12 and IL-23 (by binding to a protein shared by these two-chain cytokines), and thus inhibits the development of both Th1 and Th17 cells, is used for the treatment of inflammatory bowel disease and psoriasis.

Differentiation of Th17 Cells

The development of Th17 cells from naive CD4$^+$ cells is driven by cytokines secreted by dendritic cells and macrophages in response to fungi and extracellular bacteria (Fig. 6.13). Recognition of fungal glycans and

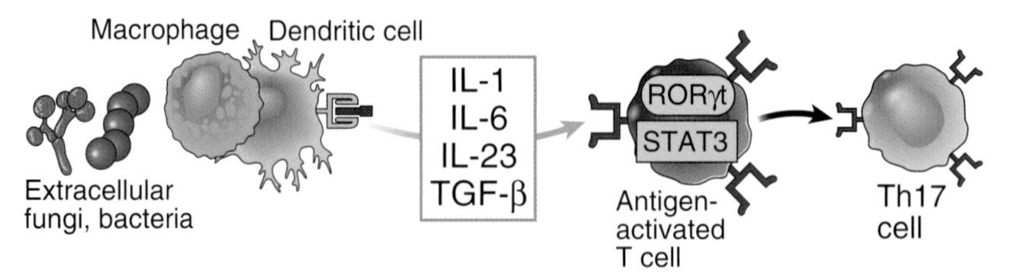

Fig. 6.13 Differentiation of Th17 effector cells. In response to some fungi and bacteria, dendritic cells and macrophages secrete cytokines that induce the differentiation of antigen-activated CD4$^+$T cells into Th17 cells. The transcription factors involved in this process are indicated in boxes in the antigen-activated T cells. *IL,* Interleukin; *TGF-β,* transforming growth factor β.

bacterial peptidoglycans and lipopeptides by innate immune receptors on dendritic cells stimulates the secretion of several cytokines, including IL-1, IL-6, and IL-23. IL-6 and IL-23 activate the transcription factor STAT3. Signals induced by these cytokines and another cytokine called transforming growth factor β (TGF-β), in combination with TCR signals, induce the expression of the transcription factor RORγT, which works together with STAT3 to promote Th17 differentiation. Interestingly, TGF-β is an inhibitor of many immune responses, but when present together with IL-6 or IL-1, it promotes the development of Th17 cells.

DIFFERENTIATION AND FUNCTIONS OF CD8$^+$ CYTOTOXIC T LYMPHOCYTES

Viruses typically infect various tissue cells that have no intrinsic microbicidal activity, so they cannot be eradicated by activating the infected cells. Microbes that survive in the cytosol (e.g., viruses) or escape from phagosomes into the cytosol (e.g., some ingested bacteria) are resistant to microbicidal mechanisms, which are largely confined to vesicles. Eradication of such cytosolic pathogens requires CD8$^+$ CTLs. CTLs also serve an important role in defense against cancers (see Chapter 10).

Naive CD8$^+$ T lymphocytes activated by antigen, costimulatory molecules, and cytokines differentiate into CTLs that are able to kill infected cells expressing the antigen. Naive CD8$^+$ T cells can recognize antigens but are not capable of killing antigen-expressing cells. The differentiation of naive CD8$^+$ T cells into fully active CTLs is accompanied by the synthesis of molecules involved in cell killing, giving these effector T cells the functional capacity that is the basis for their designation as cytotoxic. CD8$^+$ T lymphocytes recognize class I MHC–associated peptides on infected cells and tumor cells. The sources of peptides that bind to class I MHC in a cell are microbial protein antigens synthesized in the cytosol of that cell and protein antigens made by phagocytosed microbes that escape from phagocytic vesicles into the cytosol (see Chapter 3). Dendritic cells capture the antigens of infected cells and tumors in tissues, transfer these antigens into the cytosol, migrate to lymph nodes and present the antigens on class I MHC molecules to naive CD8$^+$ T cells, by the process known as cross-presentation (see Fig. 3.16). The differentiation of naive CD8$^+$ T cells into functional CTLs and memory cells requires not only antigen recognition but also costimulation and, in some situations, help from CD4$^+$ T cells (see Fig. 5.7).

CD8$^+$ CTLs recognize class I MHC–peptide complexes on the surface of infected cells and kill these cells, thus eliminating the reservoir of infection. The T cells recognize MHC-associated peptides by their TCR and the CD8 coreceptor. These infected cells are called targets of CTLs, because they are destroyed by the CTLs. The TCR and CD8, as well as other signaling proteins, cluster in the CTL membrane at the site of contact with the target cell and are surrounded by the integrin leukocyte function–associated antigen 1 (LFA-1). These molecules bind their ligands on the target cell, forming an immune synapse (see Chapter 5).

Antigen recognition by CTLs results in the activation of signal transduction pathways that lead to the exocytosis of the contents of the CTL's granules into the synapse between the CTL and the target cell (Fig. 6.14). CTLs kill target cells mainly as a result of delivery of granule proteins into the target cells. Two types of granule proteins critical for killing are granzymes

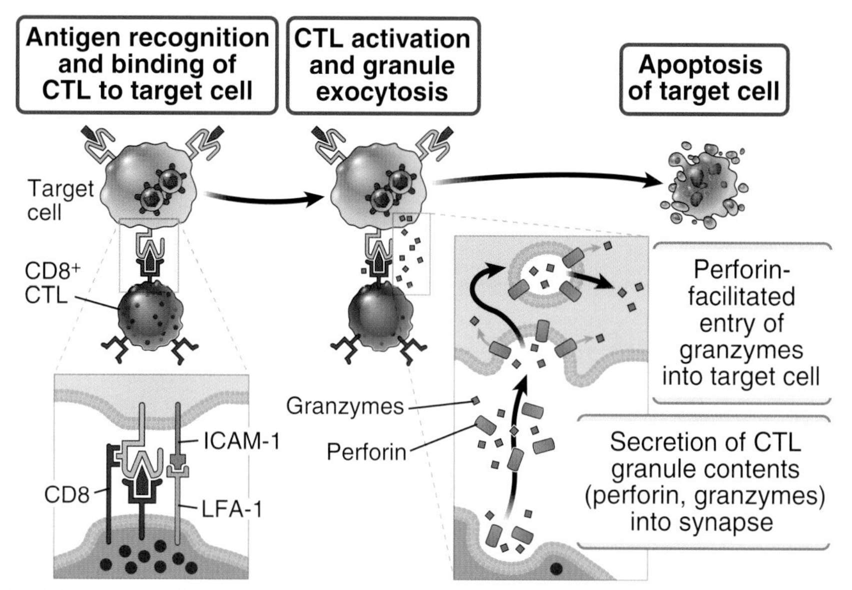

Fig. 6.14 Mechanisms of killing of infected cells by CD8$^+$ cytotoxic T lymphocytes (CTLs). CTLs recognize class I major histocompatibility complex (MHC)–associated peptides of cytoplasmic microbes in infected cells and form tight adhesions (conjugates) with these cells. Adhesion molecules such as the integrin LFA-1 and its ligand ICAM-1 stabilize the binding of the CTLs to infected cells. The CTLs are activated to release (exocytose) their granule contents (perforin and granzymes) into the synapse between the two cells. Granzymes are delivered to the cytosol of the target cell by a perforin-dependent mechanism. Granzymes then induce apoptosis. Note that the cytotoxic activity of CTLs does not require costimulation or T cell help. *ICAM-1,* Intercellular adhesion molecule 1; *LFA-1,* leukocyte function–associated antigen 1.

(granule enzymes) and perforin. **Perforin** disrupts the integrity of the target cell membranes, thereby facilitating the delivery of granzymes into the cytosol. **Granzymes** cleave and thereby activate enzymes called caspases (cysteine proteases that cleave proteins after aspartic acid residues) that are present in the cytosol of target cells, and the active caspases induce apoptotic death of the cell. Another CTL granule protein, **granulysin**, is delivered into the cytosol of infected target cells, where it kills microbes by inserting into their cell membranes.

Activated CTLs also express a membrane protein called FAS ligand, which binds to a death-inducing receptor, called FAS (CD95), on target cells. Engagement of FAS activates caspases and induces target cell apoptosis. This pathway does not require granule exocytosis and probably plays only a minor role in killing by CD8$^+$ CTLs.

The net result of these effector mechanisms of CTLs is that the infected cells are killed. Because all nucleated cells express class I MHC, the CTLs can be activated by, and are able to kill, any infected cell in any tissue. Cells that have undergone apoptosis are rapidly phagocytosed and eliminated without eliciting an inflammatory reaction. CTLs themselves are not injured during the process of killing other cells, so each CTL can kill a target cell, detach, and go on to kill additional targets.

In addition to their cytotoxic activity, CD8$^+$ effector cells secrete IFN-γ. This cytokine is responsible for activation of macrophages in infections and in disease states in which excessive activation of CD8$^+$ T cells may be a feature. It also plays a role in defense against some tumors.

Although we have described the effector functions of CD4$^+$ T cells and CD8$^+$ T cells separately, these types of T lymphocytes may function cooperatively to destroy intracellular microbes (Fig. 6.15). If microbes are phagocytosed and remain sequestered in macrophage vesicles, CD4$^+$ T cells may be adequate to eradicate these infections by secreting IFN-γ and activating the microbicidal mechanisms of the macrophages. However, if the microbes are able to escape from vesicles into the cytoplasm, they become insusceptible to the killing mechanisms of activated macrophages, and their

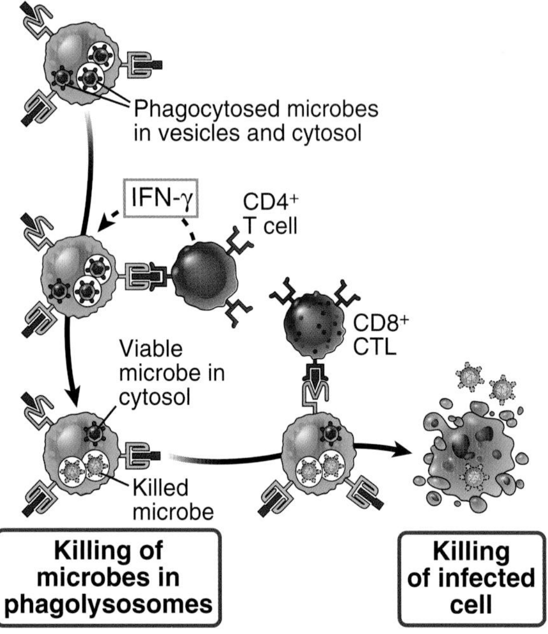

Fig. 6.15 Cooperation between CD4$^+$ and CD8$^+$ T cells in eradication of intracellular infections. In a macrophage infected by an intracellular bacterium, some of the bacteria are sequestered in vesicles (phagosomes), and others may escape into the cytosol. CD4$^+$ T cells recognize antigens derived from the vesicular microbes and activate the macrophage to kill the microbes in the vesicles. CD8$^+$ T cells recognize antigens derived from the cytosolic bacteria and are needed to kill the infected cell, thus eliminating the reservoir of infection. *CTL,* Cytotoxic T lymphocyte; *IFN,* interferon.

elimination requires destruction of the infected cells by CD8$^+$ CTLs.

RESISTANCE OF PATHOGENIC MICROBES TO CELL-MEDIATED IMMUNITY

Different microbes have developed diverse mechanisms to resist T lymphocyte–mediated host defense (Fig. 6.16). Many intracellular bacteria, such as *Mycobacterium tuberculosis, Legionella pneumophila,* and *Listeria monocytogenes,* inhibit the fusion of phagosomes with lysosomes or create pores in phagosome membranes, allowing these organisms to escape into the cytosol. Thus, these microbes cannot be killed by the microbicidal mechanisms of phagocytes and survive and even replicate inside phagocytes. Many viruses inhibit class I MHC–associated antigen presentation by inhibiting production or expression of class I molecules, by blocking transport of antigenic peptides from the cytosol into the endoplasmic reticulum (ER), and by removing newly synthesized class I molecules from the ER. All these viral mechanisms reduce the loading of class I MHC molecules by viral peptides. The result of this defective loading is reduced surface expression of class I MHC molecules, because empty class I molecules are unstable and are not expressed on the cell surface. It is interesting that NK cells are activated by class I–deficient cells (see Chapter 2). Thus, host defenses have evolved to combat immune evasion mechanisms of microbes: CTLs recognize class I MHC–associated viral peptides, viruses inhibit class I MHC expression, and NK cells recognize the absence of class I MHC molecules on infected or stressed cells.

Other viruses produce inhibitory cytokines or soluble (decoy) cytokine receptors that bind and neutralize cytokines such as IFN-γ, reducing the amount of cytokines available to trigger cell-mediated immune reactions. Some viruses evade elimination and establish chronic infections by stimulating expression of inhibitory receptors, including programmed cell death protein 1 (PD-1; see Chapter 5) on CD8$^+$ T cells, thus inhibiting the effector functions of CTLs. This phenomenon, in which the T cells mount an initial response against the virus but the response is prematurely terminated, is called **T cell exhaustion** (Fig. 6.17). It typically occurs as a reaction to chronic antigenic stimulation, as in chronic viral infections or tumors, and is a mechanism by which the repeatedly stimulated T cell terminates its own response. Still other viruses directly infect and kill immune cells, the best example being human immunodeficiency virus (HIV), which is able to survive in infected persons by killing CD4$^+$ T cells.

The outcome of infections is influenced by the strength of host defenses and the ability of pathogens to resist these defenses. The same principle is evident when the effector mechanisms of humoral immunity are considered. One approach for tilting the balance between the host and microbes in favor of protective immunity is to vaccinate individuals to enhance adaptive immune responses. The principles underlying vaccination strategies are described at the end of Chapter 8, after the discussion of humoral immunity.

As we will discuss in Chapter 10, tumors, like infectious pathogens, develop several mechanisms for evading or resisting CD8$^+$ T cell–mediated immunity. These mechanisms include inhibiting expression of

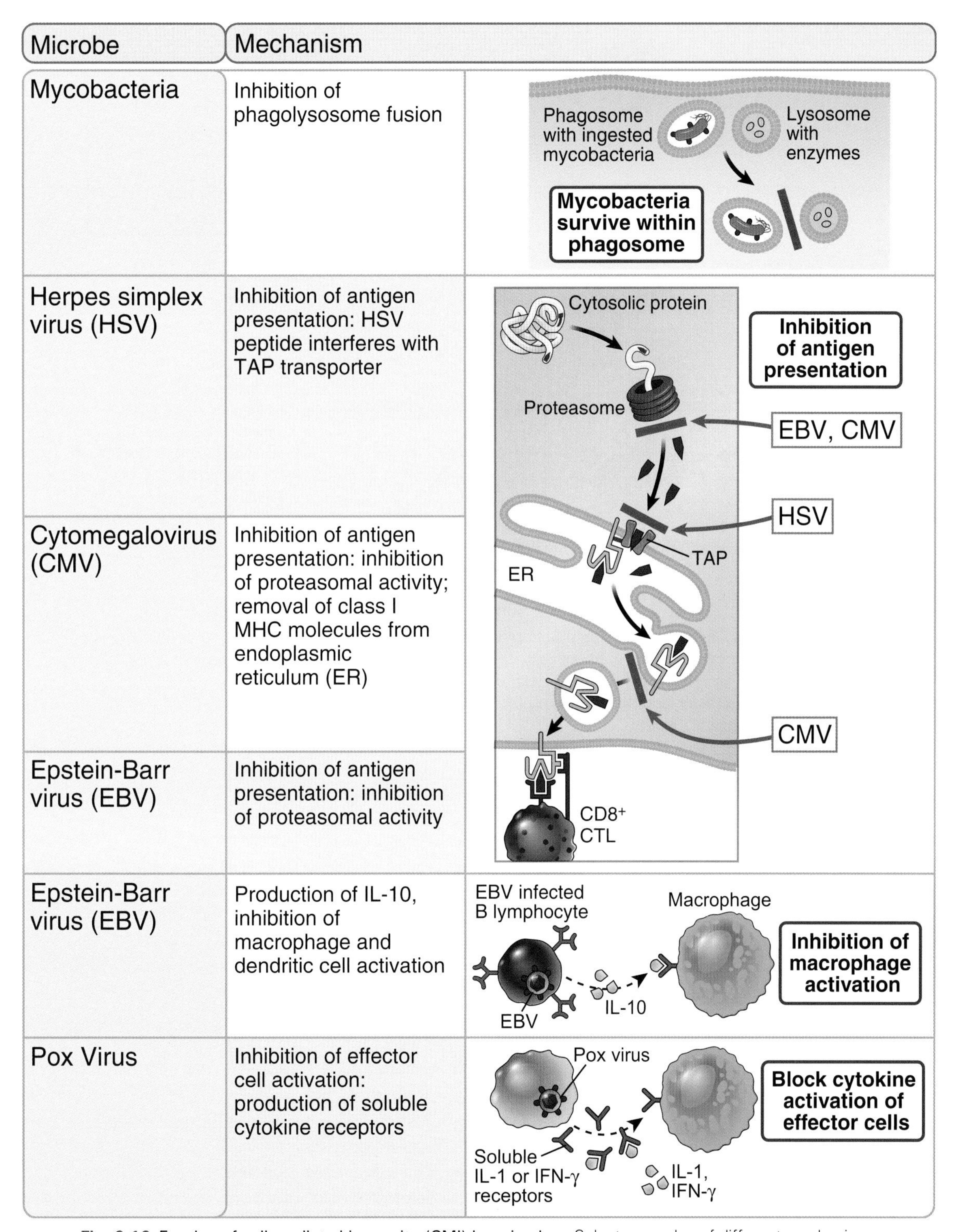

Fig. 6.16 Evasion of cell-mediated immunity (CMI) by microbes. Select examples of different mechanisms by which bacteria and viruses resist the effector mechanisms of CMI. *CTL*, Cytotoxic T lymphocyte; *ER*, endoplasmic reticulum; *IFN*, interferon; *IL*, interleukin; *TAP*, transporter associated with antigen processing.

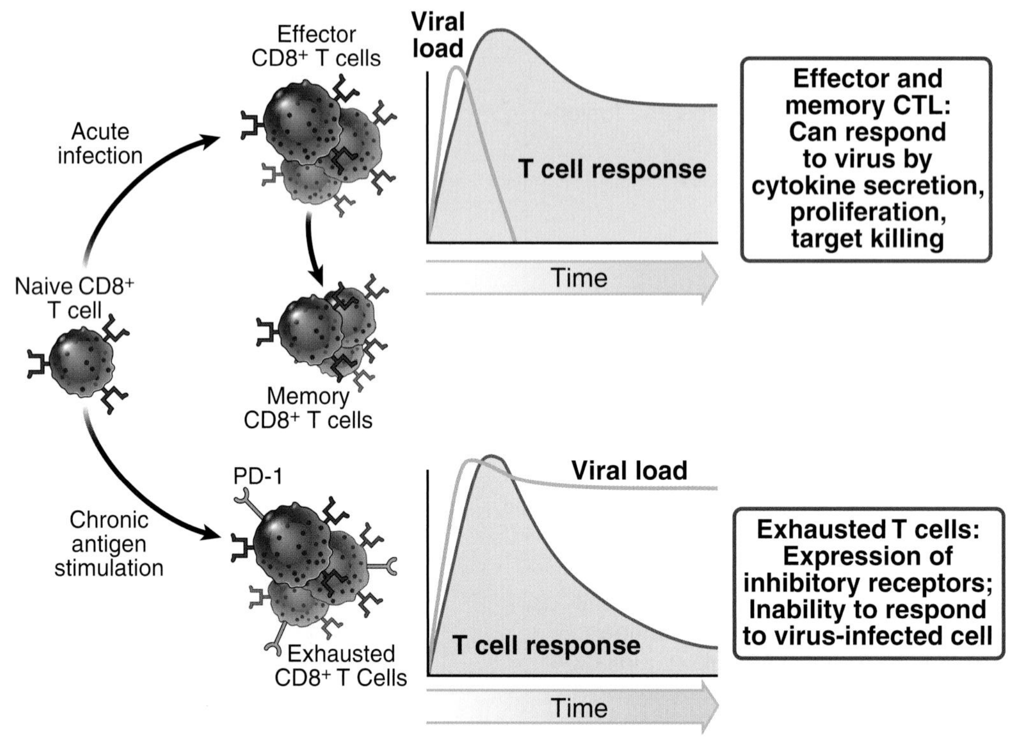

Fig. 6.17 T cell activation and exhaustion. A, In an acute viral infection, virus-specific $CD8^+$ T cells proliferate, differentiate into effector CTLs and memory cells, and clear the virus. **B,** In some chronic viral infections, $CD8^+$ T cells mount an initial response but begin to express inhibitory receptors (e.g., PD-1, CTLA-4) and are inactivated, leading to persistence of the virus. This process is called *exhaustion* because the T cells do make a response, but this is short lived.

class I MHC molecules and inducing T cell exhaustion. Blocking some of these evasion mechanisms provides effective strategies for unleashing antitumor immunity (see Chapter 10).

SUMMARY

- Cell-mediated immunity is the arm of adaptive immunity that eradicates infections by cell-associated microbes. This form of host defense uses two types of T cells: $CD4^+$ helper T cells recruit and activate phagocytes and other leukocytes to kill ingested and some extracellular microbes, and $CD8^+$ CTLs eliminate the reservoirs of infection by killing cells harboring microbes in the cytosol.
- $CD4^+$ T cells can differentiate into subsets of effector cells that make different cytokines and perform distinct functions.
- Effector cells of the Th1 subset recognize the antigens of microbes that have been ingested by macrophages. These T cells secrete IFN-γ and express CD40 ligand, which function cooperatively to activate macrophages.
- Classically activated macrophages produce substances, including ROS, NO, and lysosomal enzymes, that kill ingested microbes. Macrophages also produce cytokines that induce inflammation.
- Th2 cells stimulate eosinophilic inflammation and trigger the alternative pathway of macrophage activation, and Tfh cells induced in parallel trigger IgE production. IgE and eosinophils are important in host defense against helminthic parasites and in allergic diseases.
- The balance between activation of Th1 and Th2 cells determines the outcomes of many infections, with Th1 cells promoting and Th2 cells suppressing defense against intracellular microbes.

- Th17 cells stimulate the recruitment of neutrophils and monocytes that destroy extracellular bacteria and fungi. Th17 responses also underlie some chronic inflammatory diseases.
- $CD8^+$ T cells differentiate into CTLs that kill infected cells, mainly by inducing apoptosis of the cells. $CD4^+$ and $CD8^+$ T cells often function cooperatively to eradicate intracellular infections. $CD8^+$ CTLs also kill cancer cells and are the principal mediators of antitumor immunity.
- Many pathogenic microbes have evolved mechanisms to resist cell-mediated immunity. These mechanisms include inhibiting phagolysosome fusion, escaping from the vesicles of phagocytes, inhibiting the assembly of class I MHC–peptide complexes, producing inhibitory cytokines or decoy cytokine receptors, and inactivating T cells, thus prematurely terminating T cell responses.

REVIEW QUESTIONS

1. What are the types of T lymphocyte–mediated immune reactions that eliminate microbes that are sequestered in the vesicles of phagocytes and microbes that live in the cytoplasm of infected host cells?
2. What are the major subsets of $CD4^+$ effector T cells, how do they differ, and what are their roles in defense against different types of infectious pathogens?
3. What are the mechanisms by which T cells activate macrophages, and what are the responses of macrophages that result in the killing of ingested microbes?
4. How do $CD8^+$ CTLs kill cells infected with viruses?
5. What are some of the mechanisms by which intracellular microbes resist the effector mechanisms of cell-mediated immunity?

Answers to and discussion of the Review Questions may be found on p. 322.

7

Humoral Immune Responses

Activation of B Lymphocytes and Production of Antibodies

CHAPTER OUTLINE

Humoral immunity is mediated by antibodies and is the arm of the adaptive immune response that functions to neutralize and eliminate extracellular microbes and microbial toxins. It is the principal type of immune response against extracellular microbial protein antigens, as well as against nonprotein antigens, such as microbial capsular polysaccharides and lipids that cannot be recognized by most types of T cells. Antibodies are produced by B lymphocytes and their progeny. Naive B lymphocytes recognize antigens but do not secrete antibodies, and antigen-induced activation of these cells stimulates their differentiation into antibody-secreting plasma cells and memory cells.

This chapter describes the process and mechanisms of B cell activation and antibody production, focusing on the following questions.

- How are antigen receptor—expressing naive B lymphocytes activated and converted to antibody-secreting plasma cells?
- How do helper T cells stimulate the production of diverse and potent antibodies in response to different types of microbes?
- How are the responses of B cells controlled?

Chapter 8 describes how the antibodies that are produced during humoral immune responses function to defend individuals against microbes and toxins.

PHASES AND TYPES OF HUMORAL IMMUNE RESPONSES

The activation of B lymphocytes results in their proliferation, leading to expansion of antigen-specific clones, and their differentiation into plasma cells, which secrete antibodies (Fig. 7.1). Naive B lymphocytes express two classes of membrane-bound antibodies, immunoglobulins M and D (IgM and IgD), that function as receptors for antigens. These naive B cells are activated by antigen binding to membrane immunoglobulin (Ig) and by other signals discussed later in the chapter. The antibodies secreted in response to an antigen have the same specificity as the surface receptors on naive B cells that recognize that antigen in order to initiate the response. One activated B cell may generate a few thousand **plasma cells,** each of which can produce copious amounts of antibody, in the range of several thousand molecules per hour. In this way, humoral immunity can keep pace with rapidly proliferating microbes. During their differentiation, some B cells may produce antibodies of different heavy-chain classes (or isotypes) that mediate different effector functions and are specialized to combat different types of microbes. This process is called heavy-chain class switching or isotype switching. The affinity of antibodies specific for microbial proteins increases during the course of a

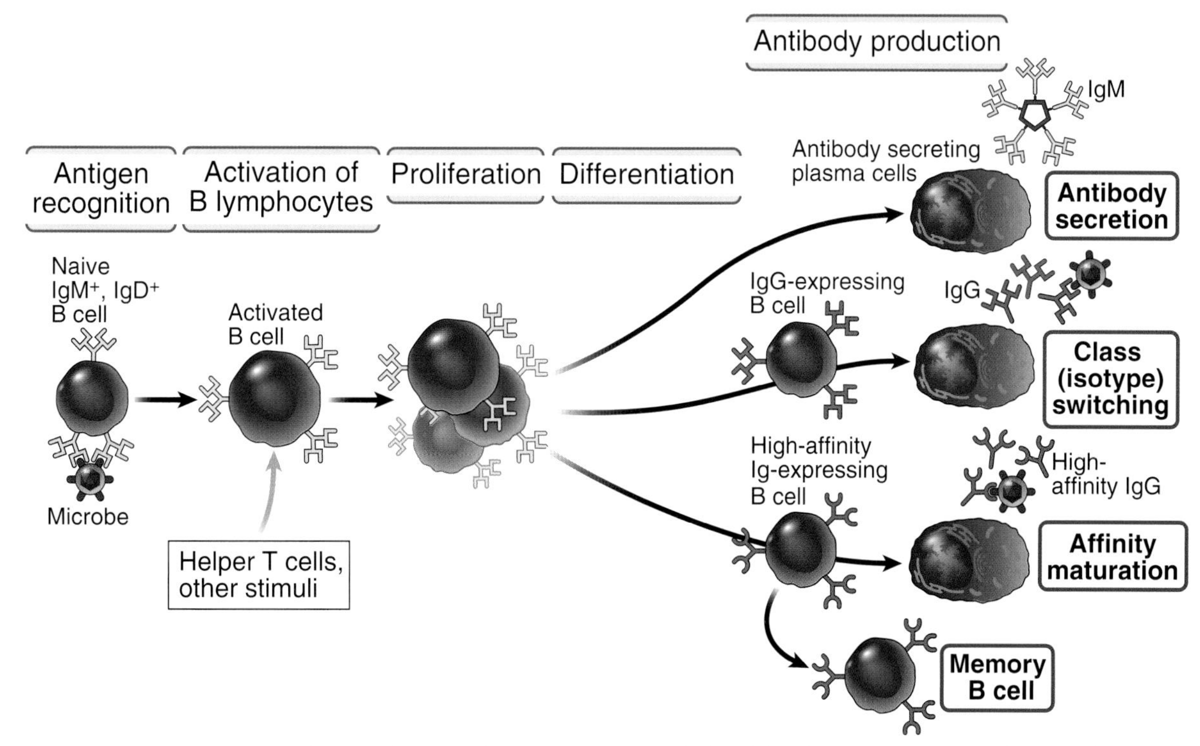

Fig. 7.1 **Sequence of events in humoral immune responses.** Naive B lymphocytes recognize antigens, and under the influence of helper T cells and other stimuli (not shown), the B cells are activated to proliferate, giving rise to clonal expansion, and to differentiate into antibody-secreting plasma cells. Some of the activated B cells undergo heavy-chain isotype switching and affinity maturation, and some become long-lived memory cells. *Ig,* Immunoglobulin.

response to microbes. This process is called affinity maturation, and it leads to the production of antibodies with improved capacity to bind to and neutralize microbes and their toxins.

Antibody responses to different types of antigens are classified as T-dependent or T-independent, based on the requirement for T cell help (Fig. 7.2). B lymphocytes recognize and are activated by a wide variety of chemically distinct antigens, including proteins, polysaccharides, lipids, nucleic acids, and small chemicals. Helper T lymphocytes play an important role in B cell activation by protein antigens. (The designation *helper* came from the discovery that some T cells stimulate, or help, B lymphocytes to produce antibodies.) In order for T cells to help B cells, each cell has to recognize the same protein antigen, as explained later. Therefore, T cells help B cell responses only to protein antigens because T cells can only recognize peptides derived from proteins and displayed by major histocompatibility complex (MHC) molecules (see Chapter 3). In the absence of T cell help, most protein antigens elicit weak or no antibody responses. Therefore, protein antigens and the antibody responses to these antigens are called T-dependent. Polysaccharides, nucleic acids, lipids, and other multivalent antigens (which contain the same antigenic epitope repeated multiple times in tandem) can stimulate antibody production without the involvement of helper T cells. Therefore, these multivalent nonprotein antigens and the antibody responses to them are called T-independent. The antibodies produced in response to proteins exhibit class switching and affinity maturation because helper T cells stimulate these processes. T-dependent antigens also stimulate the generation of long-lived plasma cells and memory B cells. Thus, the most specialized and long-lived antibody responses are elicited by protein antigens and are generated under the influence of helper T cells, whereas T-independent responses are relatively short-lived and

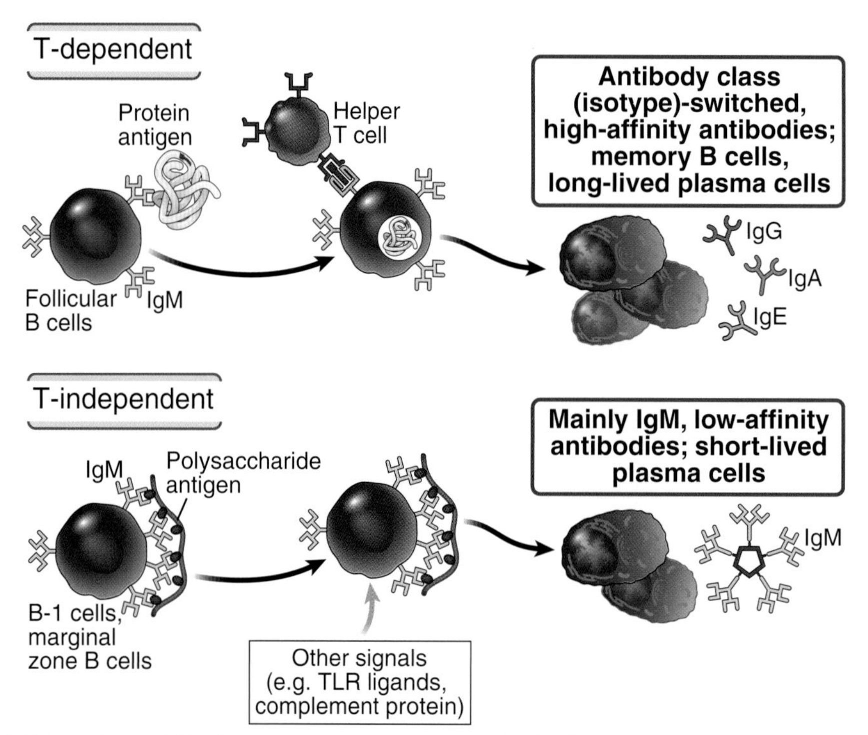

Fig. 7.2 T-dependent and T-independent antibody responses. Antibody responses to protein antigens require T cell help, and the antibodies produced typically show heavy-chain class switching and are of high affinity. Nonprotein (e.g., polysaccharide) antigens are able to activate B cells without T cell help. Most T-dependent responses are made by follicular B cells, whereas marginal zone B cells and B-1 cells play greater roles in T-independent responses. *Ig,* Immunoglobulin; *TLR,* Toll-like receptor.

require the direct activation of B cells by antigens in concert with signals generated by innate immune mechanisms but without a role for T cells.

Different subsets of B cells respond preferentially to T-dependent and T-independent antigens (see Fig. 7.2). The majority of B cells are called **follicular B cells** because they reside in and recirculate through the follicles of lymphoid organs (see Chapter 1). These follicular B cells make the bulk of T-dependent, class-switched, and high-affinity antibody responses to protein antigens and give rise to long-lived plasma cells. **Marginal-zone B cells**, which are located in the peripheral region of the splenic white pulp and also in the outer rim of follicles in lymph nodes, respond largely to blood-borne polysaccharide and lipid antigens; **B-1 cells**, present in the mucosal tissues and peritoneum, also mainly respond to multivalent polysaccharide and lipid antigens. Marginal-zone B cells and B-1 cells express antigen receptors of limited diversity and make predominantly T-independent IgM responses. IgM antibodies may be produced spontaneously by B-1 cells, without overt immunization. These antibodies, called **natural antibodies**, may help to clear some cells that die by apoptosis during normal cell turnover and may also provide protection against some bacterial pathogens.

Antibody responses generated during the first exposure to an antigen, called primary responses, differ quantitatively and qualitatively from responses to subsequent exposures, called secondary responses (Fig. 7.3). The amounts of antibody produced in the secondary immune response are greater than the amounts produced in primary responses. In secondary responses to protein antigens, there is increased heavy-chain class switching and affinity maturation, because repeated stimulation by a protein antigen leads to an increase in the number and activity of antigen-specific helper T lymphocytes.

With this introduction, we now discuss the steps in B cell activation and antibody production, beginning with the responses of B cells to the initial encounter with antigen.

STIMULATION OF B LYMPHOCYTES BY ANTIGEN

Humoral immune responses are initiated when antigen-specific B lymphocytes in the spleen, lymph nodes, and mucosal lymphoid tissues recognize antigens. Some of the antigens in tissues or in the blood are transported to and concentrated in the B cell–rich follicles and marginal zones of these secondary lymphoid organs. In lymph nodes, macrophages lining the subcapsular sinus may capture antigens and take them to the adjacent follicles, where the bound antigens are displayed to B cells. B lymphocytes specific for an antigen use their membrane Ig as receptors that recognize the intact antigen directly, without any need for processing of the antigen. Because B cells are capable of recognizing the native antigen, the antibodies that are subsequently secreted (which have the same specificity as the B cell antigen receptors) are able to bind to the native microbe or microbial product.

The recognition of antigen triggers signaling pathways that initiate B cell activation. As with T lymphocytes, B cell activation is enhanced by signals produced during innate immune reactions to microbes. In the following sections, we describe the mechanisms of B cell activation by antigen and other stimuli, followed by a discussion of the functional consequences of antigen recognition.

Antigen-Induced Signaling in B Cells

Antigen-induced clustering of membrane Ig receptors triggers biochemical signals that activate B cells (Fig. 7.4). The process of B lymphocyte activation is, in principle, similar to the activation of T cells (see Fig. 5.9). In B cells, antigen receptor–mediated signal transduction requires the bringing together (cross-linking) of two or more membrane Ig molecules. Receptor cross-linking occurs when two or more antigen molecules in an aggregate, or repeating epitopes of one antigen molecule, bind to adjacent membrane Ig molecules of a B cell. Polysaccharides, lipids, and other nonprotein antigens often contain multiple identical epitopes in each molecule and are therefore able to bind to numerous Ig receptors on a B cell at the same time. Even protein antigens may be expressed in an array on the surface of microbes and are thus able to cross-link antigen receptors of a B cell.

Signals initiated by antigen receptor cross-linking are transduced by receptor-associated proteins. Membrane IgM and IgD, the antigen receptors of naive B lymphocytes, have highly variable extracellular antigen-binding regions (see Chapter 4). However, these membrane receptors have short cytoplasmic tails,

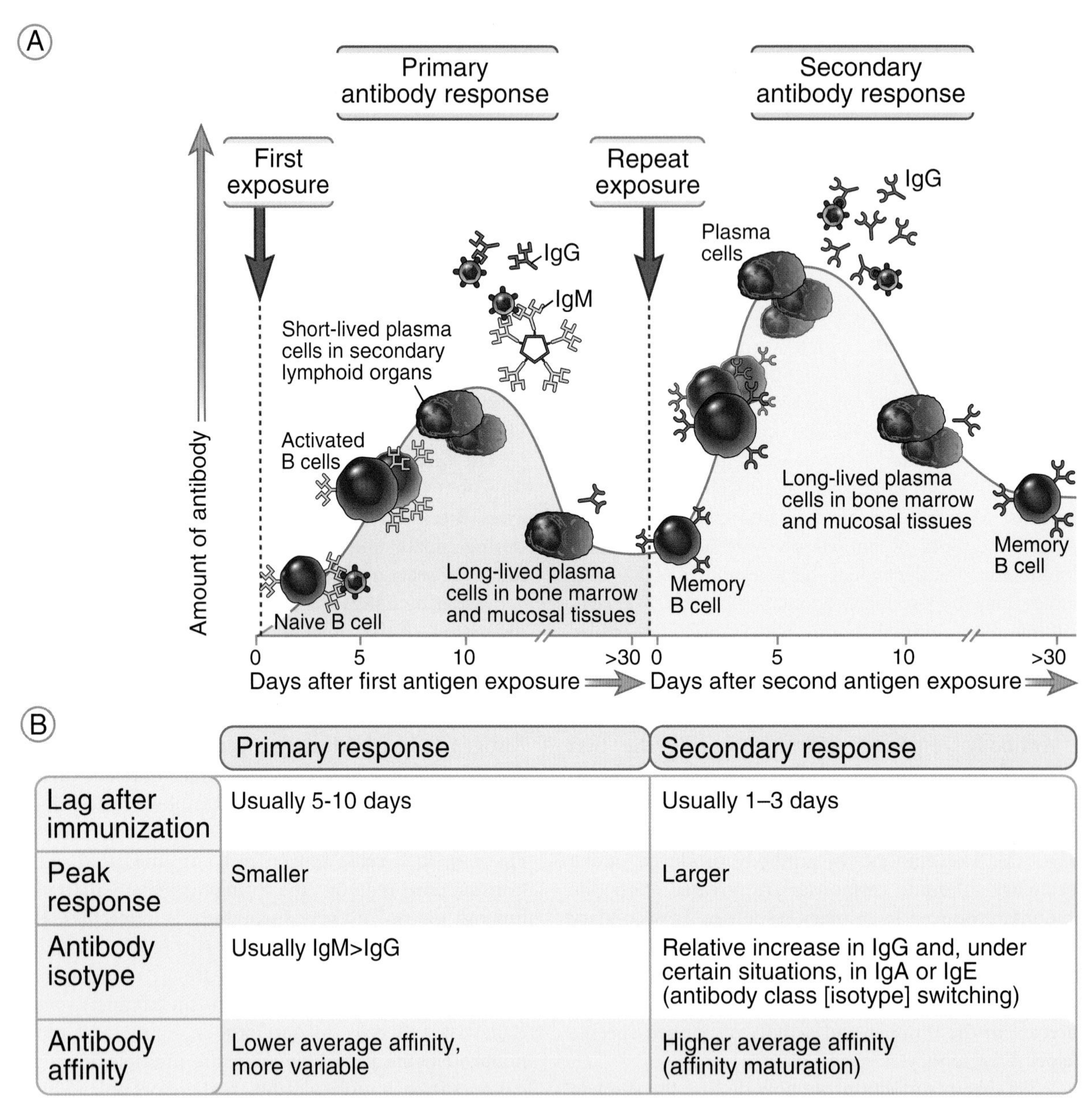

	Primary response	Secondary response
Lag after immunization	Usually 5-10 days	Usually 1–3 days
Peak response	Smaller	Larger
Antibody isotype	Usually IgM>IgG	Relative increase in IgG and, under certain situations, in IgA or IgE (antibody class [isotype] switching)
Antibody affinity	Lower average affinity, more variable	Higher average affinity (affinity maturation)

Fig. 7.3 Features of primary and secondary antibody responses. Primary and secondary antibody responses differ in several respects, illustrated schematically in **(A)** and summarized in **(B)**. In a primary response to infection or vaccination, naive B cells in secondary lymphoid organs are activated to proliferate and differentiate into antibody-secreting plasma cells and memory cells. Some plasma cells may migrate to and survive in the bone marrow for long periods. In a secondary response, memory B cells are activated to produce larger amounts of antibodies, often with more heavy-chain class switching and affinity maturation. These features of secondary responses are seen mainly in responses to protein antigens, because these changes in B cells are stimulated by helper T cells, and only proteins activate T cells (not shown). The kinetics of the responses may vary with different antigens and types of immunization. *Ig,* Immunoglobulin.

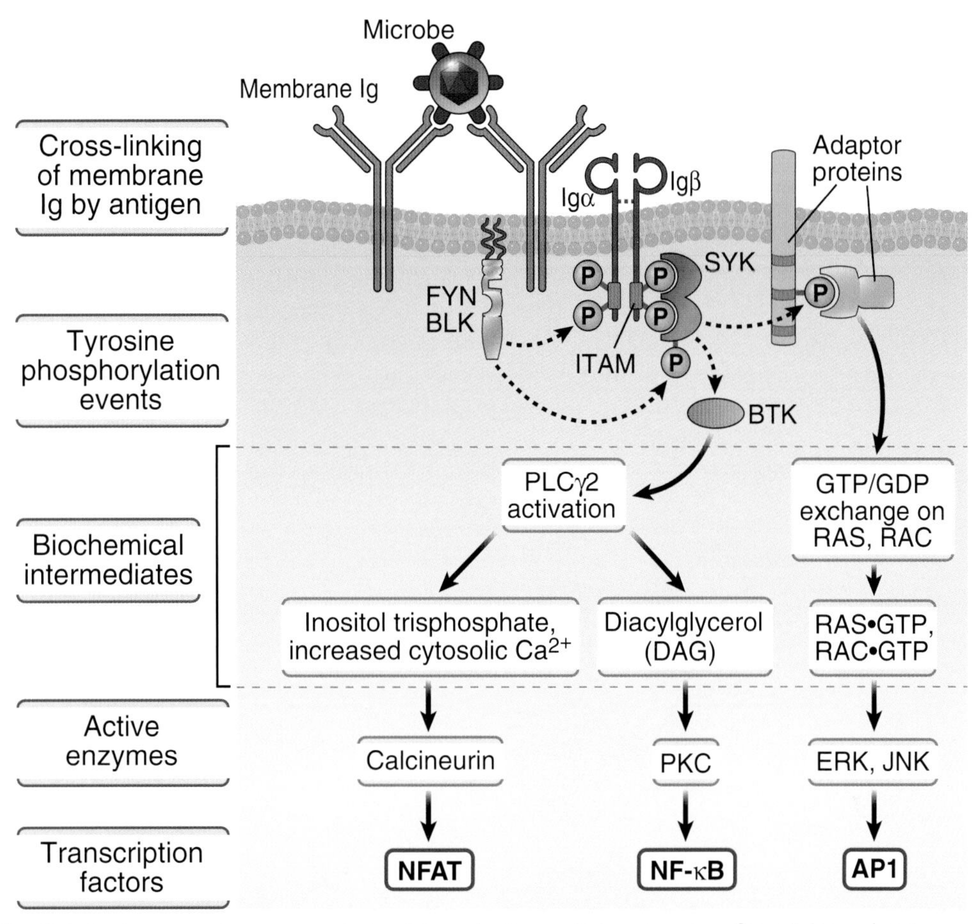

Fig. 7.4 Antigen receptor–mediated signal transduction in B lymphocytes. Cross-linking of antigen receptors on B cells by antigen triggers biochemical signals that are transduced by the immunoglobulin (Ig)-associated proteins Igα and Igβ. These signals induce early tyrosine phosphorylation events, activation of various biochemical intermediates and enzymes, and activation of transcription factors. Similar signaling events are seen in T cells after antigen recognition. Maximal signaling requires cross-linking of at least two Ig receptors by antigens. *AP-1,* Activator protein 1; *GDP,* guanosine diphosphate; *GTP,* guanosine triphosphate; *ITAM,* immunoreceptor tyrosine-based activation motif; *NFAT,* nuclear factor of activated T cells; *NF-κB,* nuclear factor-κB; *PKC,* protein kinase C; *PLC,* phospholipase C.

so although they recognize antigens, they do not themselves transduce signals. The receptors are noncovalently associated with two proteins, called Igα and Igβ, to form the **B cell receptor (BCR) complex**, analogous to the T cell receptor (TCR) complex of T lymphocytes. The cytoplasmic domains of Igα and Igβ each contain a conserved immunoreceptor tyrosine-based activation motif (ITAM), similar to those found in signaling subunits of many other activating receptors in the immune system (e.g., CD3 and ζ proteins of the TCR complex; see Chapter 5). When two or more antigen receptors of a B cell are brought together by antigen-induced cross-linking, the tyrosines in the ITAMs of Igα and Igβ are phosphorylated by tyrosine kinases, including LYN, FYN, and BLK, which are tethered to the plasma membrane and can access the tails of Igα and Igβ after BCR cross-linking. The phosphorylated ITAMs recruit the SYK tyrosine kinase (equivalent to ZAP-70 in T cells), which is activated and in turn phosphorylates tyrosine residues on adaptor proteins. These phosphorylated proteins then recruit and activate a number of downstream molecules, mainly enzymes that initiate signaling cascades that activate transcription factors.

The net result of receptor-induced signaling in B cells is the activation of transcription factors that switch on the expression of genes whose protein products are involved in B cell proliferation and differentiation. Some of the important proteins are described later.

Role of Innate Immune Signals in B Cell Activation

Antigen-induced signals are augmented by two additional types of signals that are produced during innate immune responses to microbes. This ensures that B cells respond preferentially to microbes and not to harmless antigens (similar to costimulation in T cells; see Chapter 5). The requirement for these additional stimuli illustrates a fundamental tenet of the two-signal hypothesis that was introduced in Chapter 2, that microbes or innate immune responses to microbes provide signals in addition to antigen that are necessary for lymphocyte activation.

B lymphocytes express a receptor for a complement protein that promotes the activation of these cells (Fig. 7.5A). The complement system, introduced in Chapter 2, is a collection of plasma proteins that are

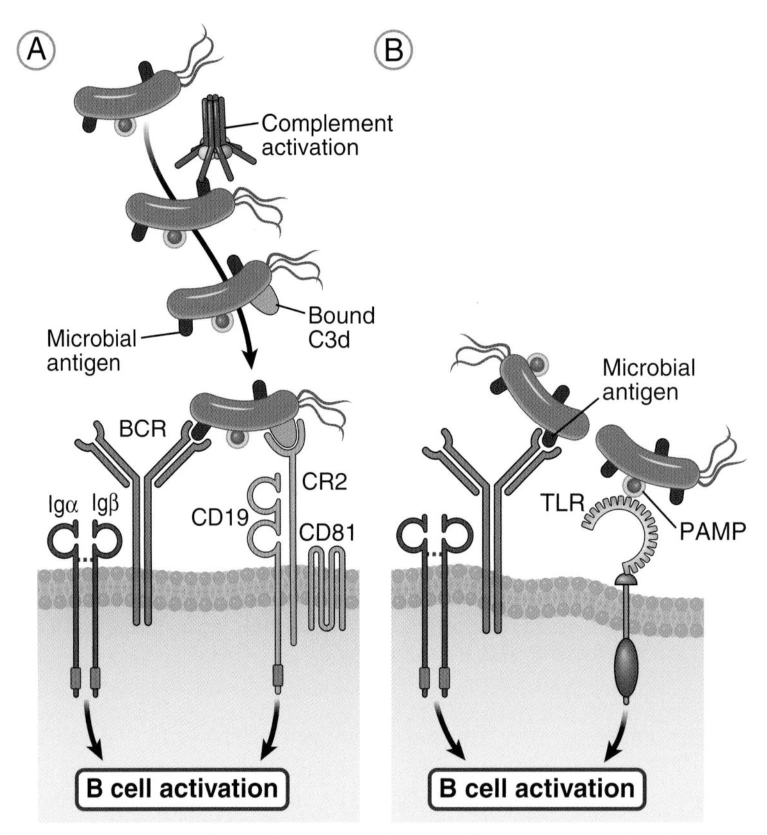

Fig. 7.5 Role of innate immune signals in B cell activation. Signals generated during innate immune responses to microbes and some antigens cooperate with recognition of antigen by antigen receptors to initiate B cell responses. **A,** Activation of complement by microbes leads to the binding of a complement breakdown product, C3d, to the microbes. The B cell simultaneously recognizes a microbial antigen (by the immunoglobulin receptor) and bound C3d by CR2 (type 2 complement receptor). CR2 is attached to a complex of proteins (CD19, CD81) that are involved in delivering activating signals to the B cell. **B,** Molecules derived from microbes (so-called pathogen-associated molecular patterns [PAMPs]; see Chapter 2) may activate Toll-like receptors (TLRs) of B cells at the same time as microbial antigens are being recognized by the antigen receptor. *BCR,* B cell receptor.

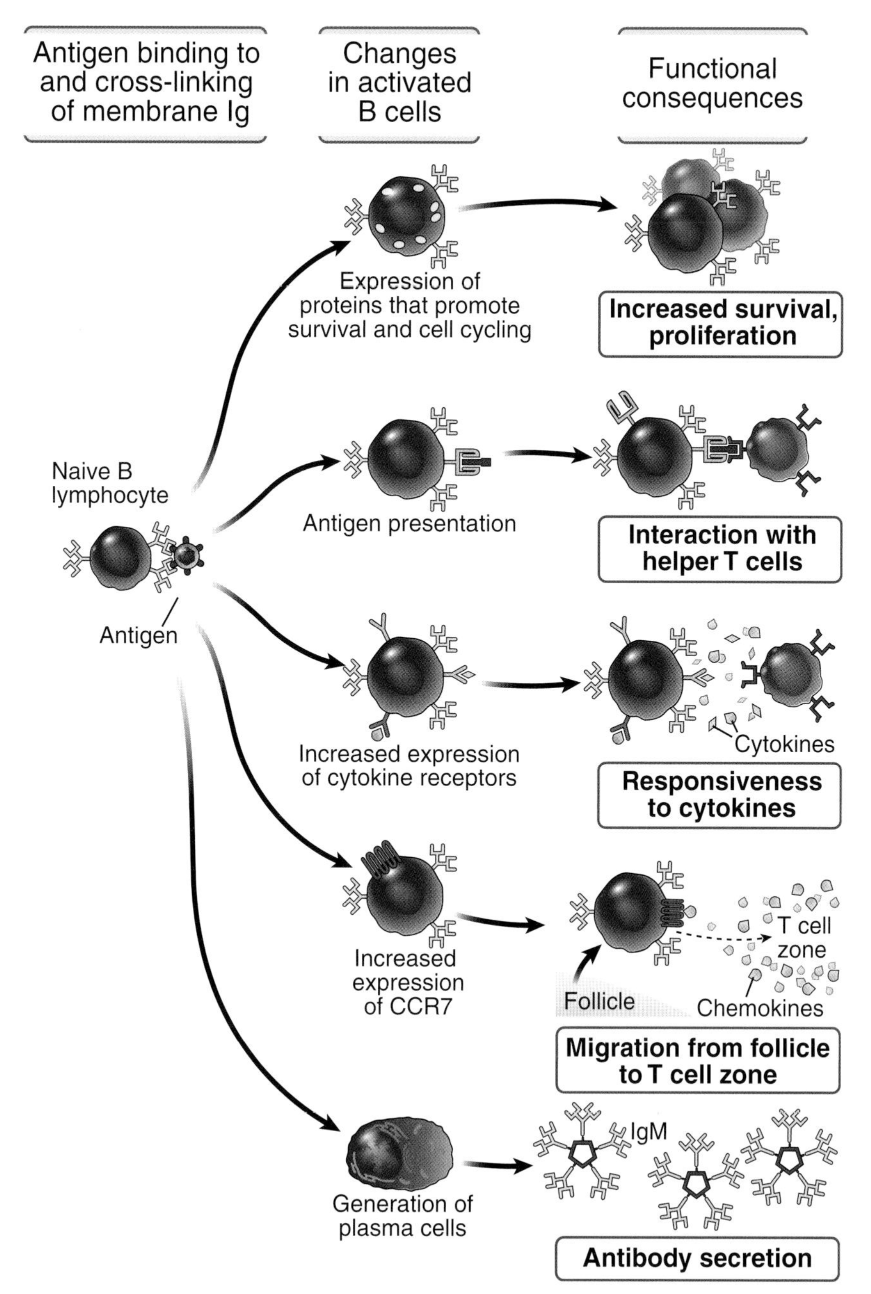

Fig. 7.6 **Functional consequences of antigen receptor–mediated B cell activation.** The activation of B cells by antigen in secondary lymphoid organs initiates the process of B cell proliferation and IgM secretion and prepares the B cell for interaction with helper T cells.

activated by microbes and by antibodies attached to microbes and function as effector mechanisms of host defense (see Chapter 8). When the complement system is activated by a microbe as part of the innate immune response, the microbe becomes coated with proteolytic fragments of the most abundant complement protein, C3. One of these fragments is called C3d. B lymphocytes express a receptor for C3d called complement receptor type 2 (CR2, or CD21). B cells that are specific for a microbe's antigens recognize the antigens by their BCRs and simultaneously recognize the bound C3d via CR2. Engagement of CR2 greatly enhances antigen-dependent activation of B cells by stimulating tyrosine phosphorylation of ITAMs.

Microbial products also directly activate B cells by engaging innate pattern recognition receptors (see Fig. 7.5B). B lymphocytes, similar to dendritic cells and other leukocytes, express numerous Toll-like receptors (TLRs; see Chapter 2). Pathogen-associated molecular patterns of microbes bind to TLRs on the plasma membrane or in endosomes of B cells, which triggers signaling pathways that work in concert with signals from the antigen receptor. This combination of signals stimulates B cell proliferation, differentiation, and Ig secretion, thus promoting antibody responses against microbes.

Functional Consequences of B Cell Activation by Antigen

B cell activation by multivalent antigen may initiate the proliferation and differentiation of the cells and prepares them to interact with helper T lymphocytes if the antigen is a protein (Fig. 7.6). The activated B lymphocytes begin to synthesize more IgM and to produce some of this IgM in a secreted form. Thus, antigen stimulation induces the early phase of the humoral immune response. This response is greatest when the antigen is multivalent, cross-links many antigen receptors, and activates complement and innate immune receptors strongly; all these features are typically seen with polysaccharides and other T-independent microbial antigens, as discussed later, but not most soluble proteins. Therefore, by themselves, protein antigens typically do not stimulate high levels of B cell proliferation and differentiation. However, protein antigens induce changes in B cells that enhance their ability to interact with helper T lymphocytes.

When a protein antigen binds specifically to the BCR of a B cell in the follicle of a secondary lymphoid organ, the antigen is efficiently endocytosed, leading to degradation of the antigen and display of peptides bound to class II MHC molecules, which can be recognized by helper T cells.

The next section describes the interactions of helper T cells with B lymphocytes in antibody responses to T-dependent protein antigens. Responses to T-independent antigens are discussed at the end of the chapter.

FUNCTIONS OF HELPER T LYMPHOCYTES IN HUMORAL IMMUNE RESPONSES

For a protein antigen to stimulate an antibody response, B lymphocytes and helper T lymphocytes specific for that antigen must come together in lymphoid organs and interact in a way that stimulates B cell proliferation and differentiation. We know this process works efficiently because protein antigens elicit antibody responses within 3 to 7 days after antigen exposure. The efficiency of antigen-induced T-B cell interaction raises many questions. How do B cells and T cells specific for different epitopes of the same antigen find one another, considering that naive B and T lymphocytes specific for any one antigen are rare, probably less than 1 in 100,000 of all the lymphocytes in the body? How do helper T cells specific for an antigen interact with B cells specific for an epitope of the same antigen and not with irrelevant B cells? What signals are delivered by helper T cells that stimulate not only the secretion of antibody but also the special features of the antibody response to proteins—namely, heavy-chain class switching and affinity maturation? As discussed next, the answers to these questions are now well understood.

The process of T cell–B cell interaction and T cell–dependent antibody responses is initiated by recognition of different epitopes of the same protein antigen by the two cell types and occurs in a series of sequential steps (Fig. 7.7):

- Naive $CD4^+$ T cells are activated in the T cell zone of a secondary lymphoid organ by antigen (in the form of processed peptides bound to class II MHC molecules) presented by dendritic cells (which also provide costimulatory signals) and differentiate into functional (cytokine-producing) helper T cells.
- Naive B cells are activated in the follicles of the same lymphoid organ by an exposed epitope on the same protein (in its native conformation) that has been transported to the follicle.
- The antigen-activated helper T cells and B cells migrate toward one another and interact at the edges

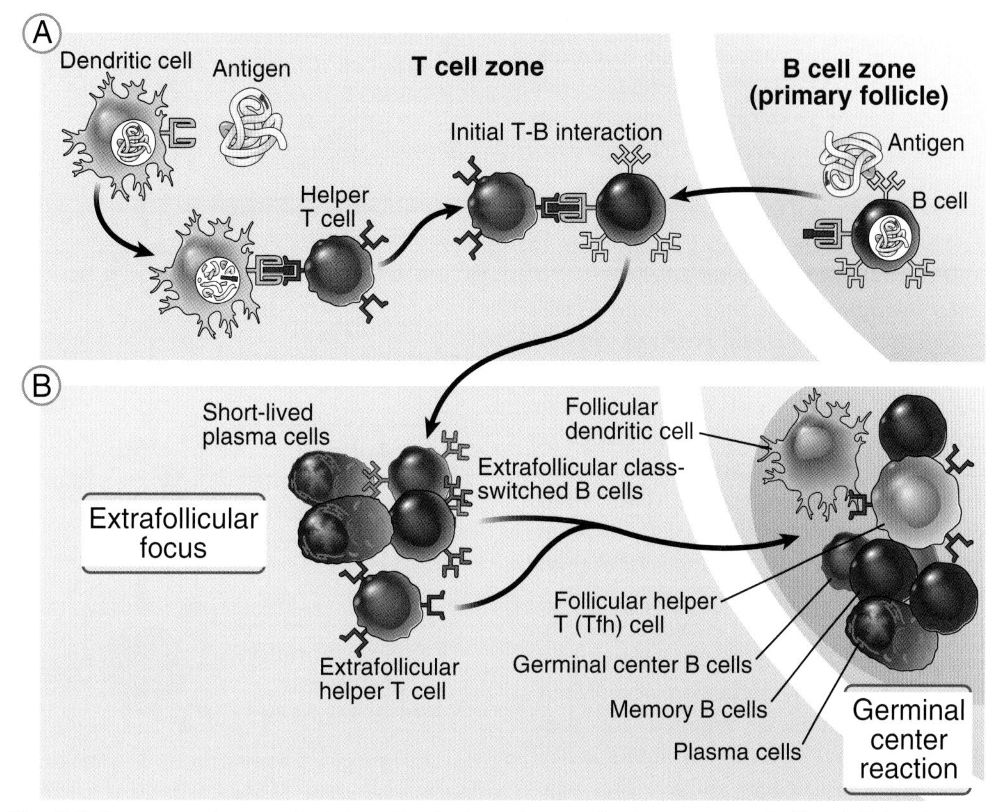

Fig. 7.7 Sequence of events in helper T cell–dependent antibody responses. **A,** T and B lymphocytes independently recognize epitopes of the same antigen in different regions of secondary lymphoid organs and are activated. The activated cells migrate toward one another and interact outside lymphoid follicles. **B,** Antibody-secreting plasma cells are initially produced in the extrafollicular focus where the antigen-activated T and B cells interact. Some of the helper T cells differentiate further into T follicular helper (Tfh) cells, and then these Tfh cells and some activated B cells migrate back into the follicle to form the germinal center, where the antibody response develops fully.

of the follicles, where the initial antibody response develops.

- Some of the antigen-specific B and T cells migrate back into follicles to form germinal centers, where more specialized antibody responses are induced.

Next we describe each of these steps in detail.

Activation and Migration of Helper T Cells and B Cells

Helper T cells that have been activated by dendritic cells in the parafollicular cortex of secondary lymphoid organs migrate toward B cell follicles and interact with antigen-stimulated B lymphocytes that have migrated out of the follicles into the parafollicular areas (see Fig. 7.7A).

- The initial activation of T cells requires antigen recognition and costimulation, as described in Chapter 5. The antigens that stimulate $CD4^+$ helper T cells are proteins derived from microbes that are internalized, processed in late endosomes and lysosomes, and displayed as peptides bound to class II MHC molecules of antigen-presenting cells (APCs) in the T cell–rich zones of peripheral lymphoid tissues. T cell activation is induced best by microbial protein antigens and, in the case of vaccines, by protein antigens that are administered with adjuvants, which stimulate the expression of costimulators on APCs. The $CD4^+$ T cells differentiate into effector cells capable of producing various cytokines and CD40 ligand, and some of these T lymphocytes migrate toward the edges of lymphoid follicles.

- B lymphocytes are activated by antigen in the follicles, as described earlier, and the activated B cells begin to move out of the follicles toward the T cells.

The directed migration of activated B and T cells toward one another depends on changes in the expression of certain chemokine receptors on the activated lymphocytes. Activated T cells reduce expression of the chemokine receptor CCR7, which recognizes chemokines produced in T cell zones, and increase expression of the chemokine receptor CXCR5, which binds a chemokine produced in B cell follicles. Activated B cells undergo precisely the opposite changes, decreasing CXCR5 and increasing CCR7 expression. As a result, antigen-stimulated B and T cells migrate toward one another and meet outside the lymphoid follicles. The next step in their interaction occurs here. Because antigen recognition is required for these changes, the cells that move toward one another are the ones that have been stimulated by antigen. This regulated migration is one mechanism for ensuring that rare antigen-specific lymphocytes can locate one another and interact productively during immune responses to the antigen.

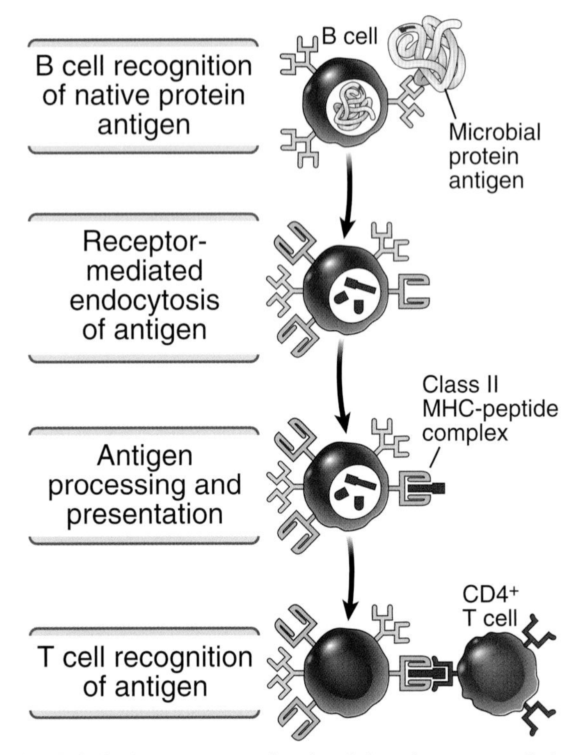

Fig. 7.8 Antigen presentation by B lymphocytes to helper T cells. B cells specific for a protein antigen bind and internalize that antigen, process it, and present peptides attached to class II major histocompatibility complex (MHC) molecules to helper T cells. The B cells and helper T cells are specific for the same antigen, but the B cells recognize native (conformational) epitopes, and the helper T cells recognize peptide fragments of the antigen bound to class II MHC molecules.

Presentation of Antigens by B Lymphocytes to Helper T Cells

The B lymphocytes that bind protein antigens by their membrane Ig antigen receptors endocytose these antigens, process them in endosomal vesicles, and display class II MHC–associated peptides for recognition by $CD4^+$ helper T cells (Fig. 7.8). The membrane Ig of B cells is a receptor that enables a B cell to specifically bind a particular antigen, even when the extracellular concentration of the antigen is low. Antigen bound by membrane Ig is endocytosed efficiently and delivered to late endosomal vesicles and lysosomes, where proteins are processed into peptides that bind to class II MHC molecules (see Chapter 3). Therefore, B lymphocytes are efficient APCs for the antigens they specifically recognize.

Any one B cell may bind a conformational epitope of a native protein antigen, internalize and process the protein, and display multiple peptides from that protein for T cell recognition. Therefore, B cells recognize one epitope of a protein antigen first, and helper T cells recognize different epitopes of the same protein later. Because B cells efficiently internalize and process the antigen for which they have specific receptors, and helper T cells recognize peptides derived from the same antigen, the ensuing interaction remains antigen specific. B cells are capable of activating previously differentiated effector T cells but are inefficient at initiating the responses of naive T cells.

The idea that a B cell recognizes one epitope of an intact antigen and displays different epitopes (peptides) for recognition by helper T cells was first demonstrated by studies using hapten-carrier conjugates. A hapten is a small chemical that is recognized by B cells but stimulates strong antibody responses only if it is attached to a carrier protein. In this situation, the B cell binds the hapten portion, ingests the conjugate, and displays peptides derived from the carrier to helper T cells. The antibody response is, of course, specific for the epitope that the B cell recognized (the hapten in this example), and the peptides derived from the carrier protein simply bring helper T cells into the reaction. This concept has been exploited to develop effective **conjugate vaccines** against microbial polysaccharides (Fig. 7.9). Some bacteria have polysaccharide-rich capsules, and the polysaccharides stimulate T-independent antibody responses, but they do not elicit high-affinity, long-lasting antibody responses, especially in infants and young children. If the polysaccharide is coupled to a carrier protein, however, high-affinity T-dependent antibody responses are induced against the polysaccharide because helper T cells specific for the carrier are engaged in the response. In this situation, the B cell recognizes the polysaccharide (equivalent to the hapten) and the T cell recognizes peptides from the attached protein (the carrier); the antibody response is specific for the polysaccharide, but it is much stronger than conventional T-independent responses because helper T cells are able to participate. Such conjugate vaccines have been very useful for inducing protective immunity against bacteria such as *Haemophilus influenzae*, meningococci, pneumococci, and typhoid.

Mechanisms of Helper T Cell–Mediated Activation of B Lymphocytes

Activated helper T lymphocytes that recognize antigen presented by B cells use CD40 ligand (CD40L) and secreted cytokines to activate the antigen-specific B cells (Fig. 7.10). The process of helper T cell–mediated B lymphocyte activation is analogous to the process of T cell–mediated macrophage activation in cell-mediated immunity (see Fig. 6.6). CD40L expressed on activated helper T cells binds to CD40 on B lymphocytes. Engagement of CD40 generates signals in the B cells that stimulate proliferation and the synthesis and secretion of antibodies. At the same time, cytokines produced by the helper T cells bind to cytokine receptors on B lymphocytes and stimulate more B cell

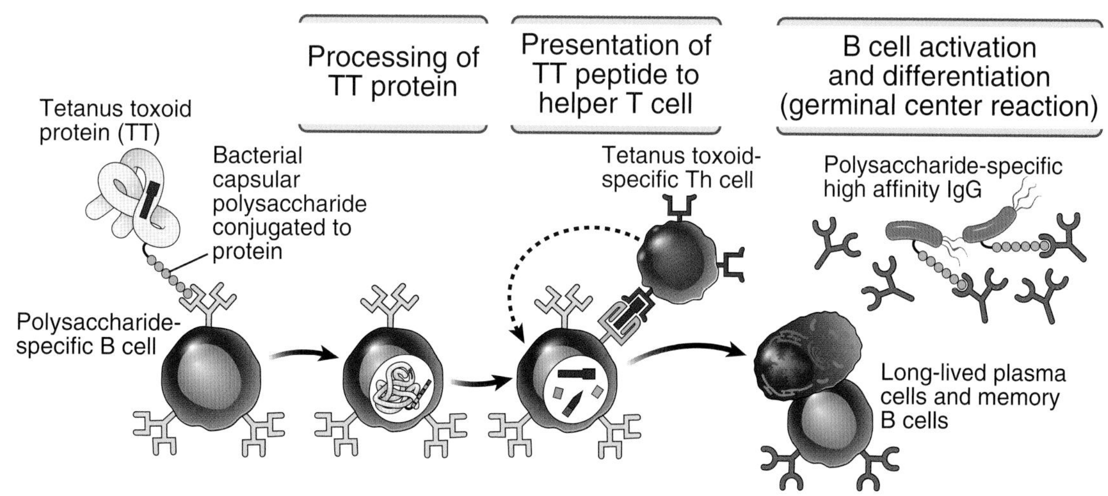

Fig. 7.9 The principle of conjugate vaccines: the hapten-carrier concept. In order to generate strong antibody responses against a microbial polysaccharide, the polysaccharide is coupled to a protein (in this case, tetanus toxoid). B cells that recognize the polysaccharide ingest it and present peptides from the protein to helper T cells, which stimulate the polysaccharide-specific B cells. Thus, isotype switching, affinity maturation, and long-lived plasma cells and memory cells (all features of responses to proteins) are induced in a response to polysaccharides. (Note that some B cells will also recognize the tetanus toxoid and antibodies will be produced against the carrier protein, but this has no bearing on the anti-polysaccharide response.) *Ig,* Immunoglobulin.

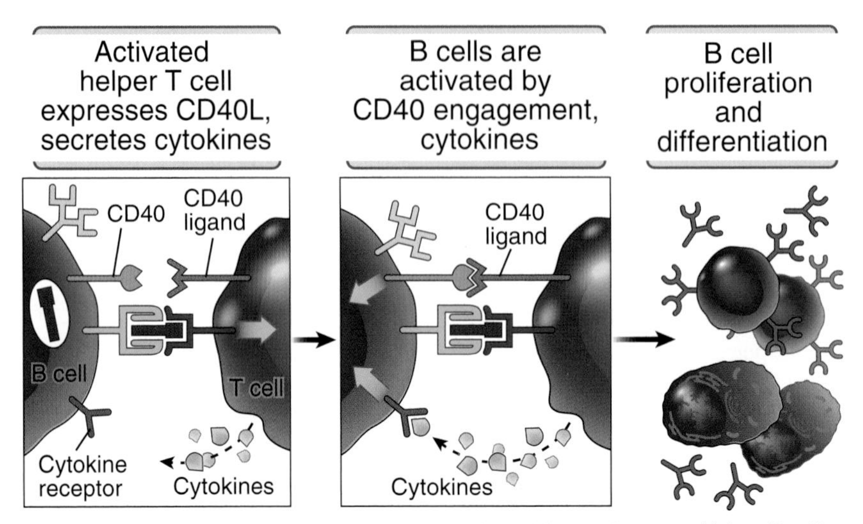

Fig. 7.10 Mechanisms of helper T cell–mediated activation of B lymphocytes. Helper T cells recognize peptide antigens presented by B cells. The helper T cells are activated to express CD40 ligand (CD40L) and secrete cytokines, both of which bind to their receptors on the same B cells and activate the B cells.

proliferation and Ig production. The requirement for the CD40L-CD40 interaction ensures that only T and B lymphocytes in physical contact engage in productive interactions. As described previously, the antigen-specific lymphocytes are the cells that physically interact, thus ensuring that the antigen-specific B cells are the cells that receive T cell help and are activated. The CD40L-CD40 interaction also stimulates heavy-chain class switching and affinity maturation, which explains why these changes are typically seen in antibody responses to T-dependent protein antigens.

Extrafollicular and Germinal Center Reactions

The initial T-B interaction, which occurs outside the lymphoid follicles, results in the production of low levels of antibodies, which may be of switched isotypes (described next) but are generally of low affinity. The plasma cells that are generated in these extrafollicular foci are typically short-lived and produce antibodies for a few weeks, and both the plasma cells and memory cells that develop have lower affinity than those produced in germinal centers.

Many of the events in fully developed antibody responses occur in germinal centers that are formed in lymphoid follicles and require the participation of a specialized type of helper T cell (Fig. 7.11). Some of the activated helper T cells express high levels of the chemokine receptor CXCR5, which draws these cells into the adjacent follicles. The $CD4^+$ T cells that migrate into B cell–rich follicles are called **follicular helper T (Tfh)** cells. The generation and function of Tfh cells are dependent on a receptor of the CD28 family called inducible costimulator (ICOS), which binds to its ligand expressed on B cells and other cells. Inherited mutations in the *ICOS* gene are the cause of some antibody deficiencies (see Chapter 12). Tfh cells and their immediate precursors in the extrafollicular focus may secrete cytokines, such as interleukin (IL)-4 and IL-13, which determine which antibody isotype is produced by class switching (see later). In addition, most Tfh cells secrete the cytokine IL-21, which has an important but incompletely understood role in Tfh cell function.

A few of the activated B cells from the extrafollicular focus migrate back into the lymphoid follicle, together with Tfh cells, and begin to divide rapidly in response to signals from the Tfh cells. It is estimated that these B cells have a doubling time of 3 to 4 hours, so one cell may produce several thousand progeny within a week. The region of the follicle containing these proliferating B cells is the **germinal center**, so named because it was once incorrectly thought that these were the sites where new lymphocytes are generated (germinated). In the germinal center, B cells undergo further class switching and somatic mutation of Ig genes; both processes are described later. Early in the germinal center reaction, some B cells of moderate affinity develop into memory cells and exit the germinal center.

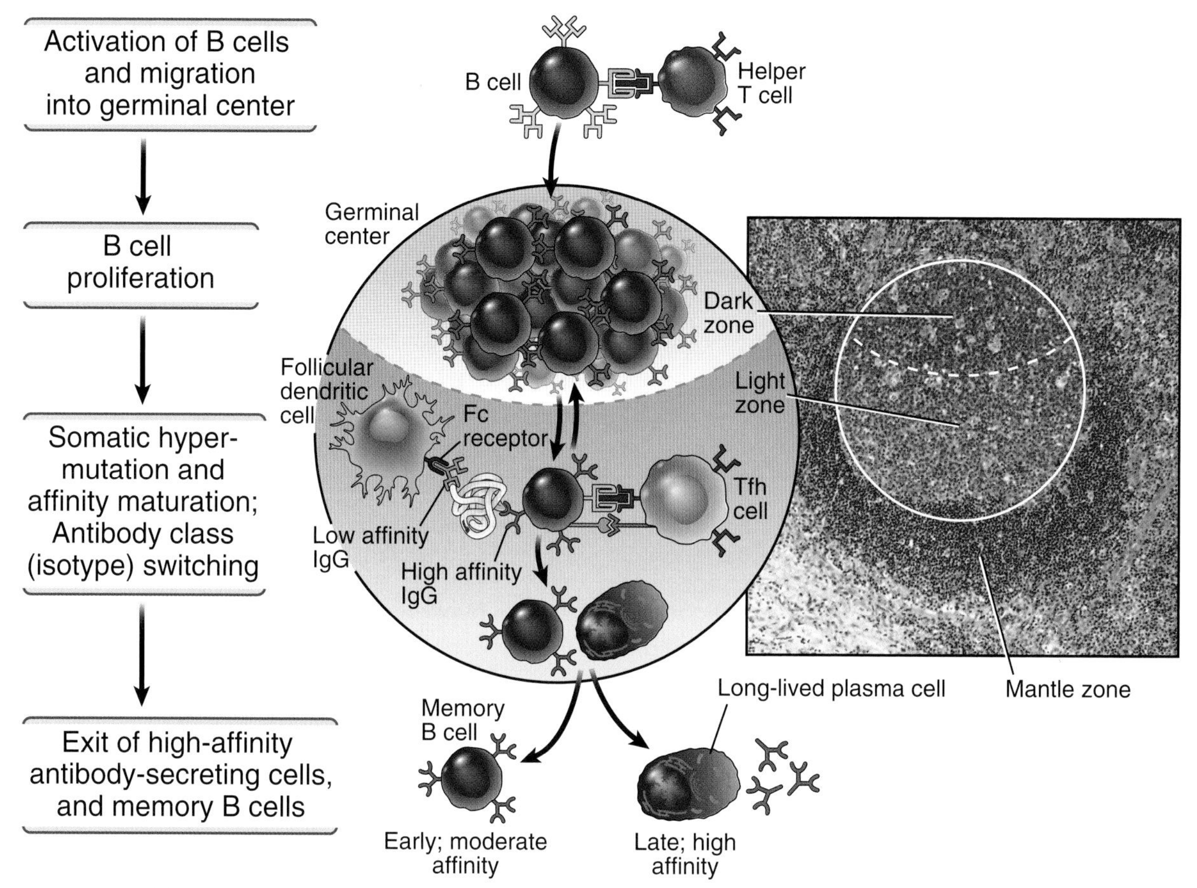

Fig. 7.11 The germinal center reaction. B cells that have been activated by T helper cells at the edge of a primary follicle migrate into the follicle and proliferate, forming the dark zone of the germinal center. Germinal center B cells in the dark zone mutate their Ig genes at an extremely high rate, a process called somatic hypermutation, and migrate into the light zone, where B cells with the highest-affinity Ig receptors are selected to survive, and they differentiate into plasma cells or memory cells, which leave the germinal center. The *right panel* shows the histology of a secondary follicle with a germinal center in a lymph node. The germinal center includes a basal dark zone and an adjacent light zone. The mantle zone is the part of the follicle outside the germinal center. *Tfh,* T follicular helper cell.

High-affinity B cells are produced later during the germinal center reaction by repeated Ig gene mutation and selection (described below) and these eventually differentiate into long-lived plasma cells and memory cells. Proliferating B cells reside in the histologically defined dark zone of the germinal center (see Fig. 7.11), while selection occurs in the less dense light zone.

Heavy-Chain Class (Isotype) Switching

Helper T cells stimulate the progeny of naive IgM- and IgD-expressing B lymphocytes to change the heavy-chain classes (isotypes) of the antibodies they produce, without changing their antigen specificities (Fig. 7.12). Different antibody isotypes perform different functions, and therefore the process of class switching broadens the functional capabilities of humoral immune responses. For example, an important defense mechanism against most bacteria and viruses when they are outside host cells is to coat (opsonize) these microbes with antibodies and cause them to be phagocytosed by neutrophils and macrophages. This reaction is best mediated by antibody classes, such as IgG1 and IgG3 (in humans), that bind to high-affinity phagocyte Fc receptors specific for the Fc portion of

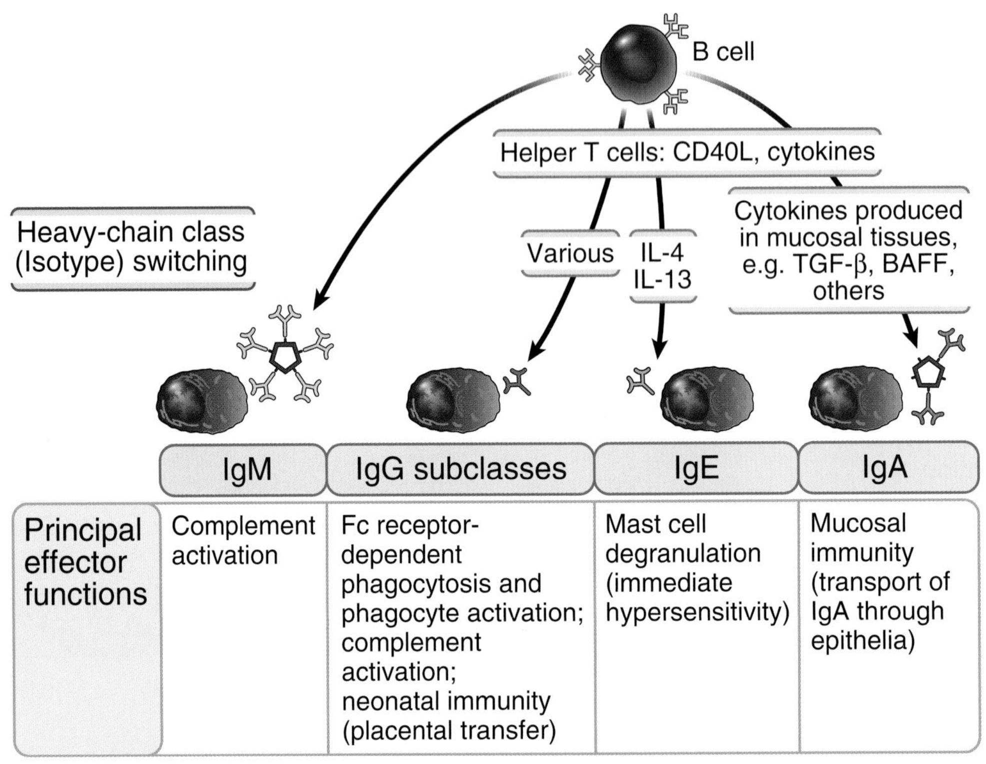

Fig. 7.12 Immunoglobulin (Ig) heavy-chain class (isotype) switching. Antigen-stimulated B lymphocytes may differentiate into IgM antibody–secreting cells, or, under the influence of CD40 ligand (CD40L) and cytokines, some of the B cells may differentiate into cells that produce different Ig heavy-chain classes (isotypes). The principal effector functions of some of these isotypes are listed; all isotypes may function to neutralize microbes and toxins. B cell–activating factor belonging to the TNF family (BAFF) is a cytokine that may be involved in switching to IgA, especially in T-independent responses. Switching to IgG subclasses is stimulated by the cytokine IFN-γ in mice, but in humans it is thought to be stimulated by other cytokines. *IL*, Interleukin; *TGF-β*, transforming growth factor β.

these γ heavy chains (see Chapter 8). Helminths stimulate the production of IgE antibodies, which bind to and activate mast cells, which have high-affinity Fc receptors for the ε heavy chain. The role of IgE and mast cells in defense against helminths is unclear. Microbes at mucosal barriers stimulate the production of IgA, which is transported into the lumens of mucosal organs and neutralizes microbes (see Chapter 8). Thus, effective host defense requires that the immune system make different antibody isotypes in response to different types and locations of microbes, even though all naive B lymphocytes specific for all these microbes express antigen receptors of the IgM and IgD isotypes.

Another functional consequence of class switching is that the IgG antibodies produced have a longer half-life than the initially produced IgM because IgG is able to bind to a specialized Fc receptor called the neonatal Fc receptor (FcRn). FcRn expressed in the placenta mediates the transfer of maternal IgG to the fetus, providing protection to the newborn, and FcRn expressed on endothelial cells and phagocytes plays a special role in protecting IgG from intracellular catabolism, thereby prolonging its half-life in the blood (see Chapter 8).

Heavy-chain class switching is induced by a combination of CD40L-mediated signals and cytokines. These signals act on antigen-stimulated B cells and induce switching in some of the progeny of these cells. In the absence of CD40 or CD40L, B cells secrete only IgM and fail to switch to other isotypes, indicating the

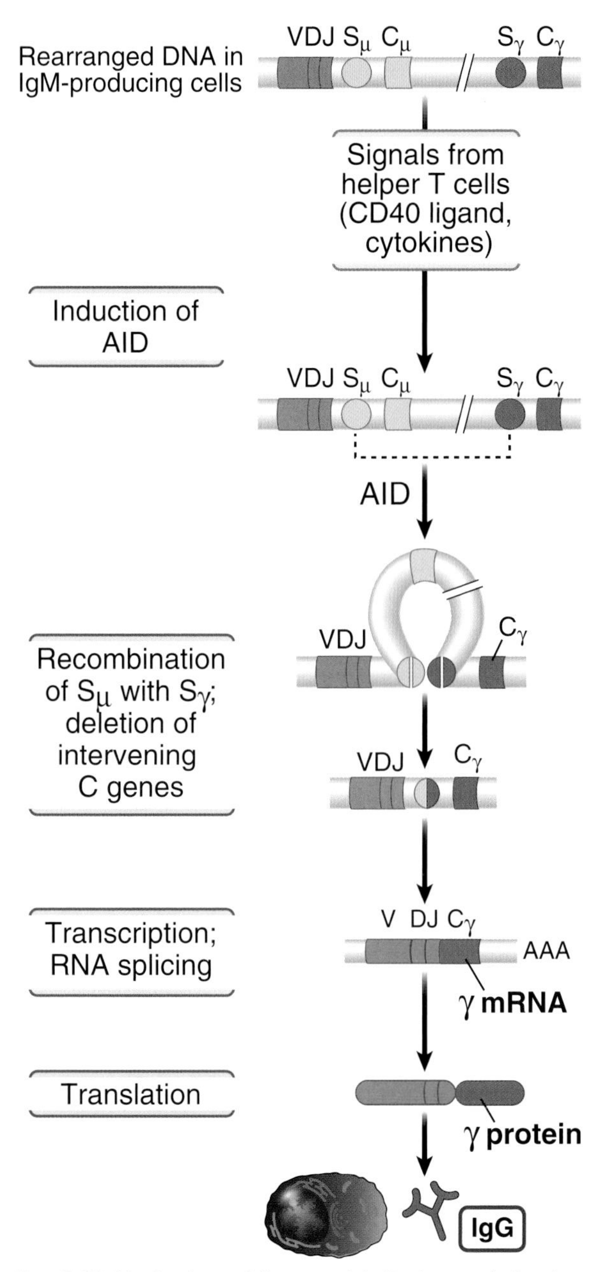

Fig. 7.13 Mechanism of immunoglobulin heavy-chain class (isotype) switching. In an IgM-producing B cell, the rearranged VDJ encoding the V region is adjacent to the μ constant region genes (Cμ). Signals from helper T cells (CD40 ligand and cytokines) may induce recombination of switch (S) regions such that the rearranged VDJ DNA is moved close to a C gene downstream of Cμ, which are Cγ genes in the example shown. The enzyme activation-induced deaminase (AID), which is induced in the B cells by signals from Tfh cells, alters nucleotides in the switch regions so that they can be cleaved by other enzymes and joined to downstream switch regions. Subsequently, when the heavy-chain gene is transcribed, the VDJ exon is spliced onto the exons of the downstream C gene, producing a heavy chain with a new constant region and thus a new class of Ig. Note that although the C region changes, the VDJ region, and thus the specificity of the antibody, is preserved. (Each C region gene consists of multiple exons, but only one is shown for simplicity.)

essential role of this ligand-receptor pair in class switching. A disease called the **X-linked hyper-IgM syndrome** is caused by mutations in the *CD40L* gene, which is located on the X chromosome, leading to production of nonfunctional forms of CD40L in males who inherit the mutation. In this disease, much of the serum antibody is IgM, because of defective heavy-chain class switching. Patients with this disease also have defective cell-mediated immunity against intracellular microbes, because CD40L is important for T cell–mediated activation of macrophages and for the amplification of T cell responses by dendritic cells (see Chapter 6).

The molecular mechanism of class switching, called switch recombination, involves taking the previously formed VDJ exon encoding the V domain of an Ig μ heavy chain and moving it adjacent to a different C region gene downstream in the Ig heavy chain locus (Fig. 7.13). IgM-producing B cells, which have not undergone switching, contain in their heavy-chain locus a rearranged VDJ exon adjacent to the exons of the first constant region gene, which is Cμ. The heavy-chain mRNA is produced by splicing a VDJ exon to the Cμ gene sexons in the initially transcribed RNA, and this mRNA is translated to produce a μ heavy chain, which combines with a light chain to give rise to an IgM antibody. Thus, the first antibody produced by B cells is IgM. In the intron 5′ of each constant region is a large guanine-cytosine (GC)–rich stretch of DNA called the switch region. Signals from CD40 and cytokine receptors stimulate transcription through one of the switch regions adjacent to the constant region exons of another antibody isotype that is downstream of Cμ. Transcription through a switch region makes the DNA there accessible to the switch machinery. During switch recombination, the switch region next to Cμ recombines with the switch region adjacent to the transcriptionally active downstream constant region, and the intervening DNA is deleted. An enzyme called activation-induced deaminase (AID), which is induced by CD40 signals, plays a key role in this process. AID converts cytosines in the transcribed switch region DNA to uracil (U), a base that is normally present only in RNA. The sequential action of DNA repair enzymes results in the removal of these aberrant uracils and the creation of nicks in the DNA. Such a process on both strands leads to double-stranded DNA breaks. When double-stranded DNA breaks in two switch regions are brought together and repaired, the intervening DNA is removed. Therefore, the VDJ exon that was previously formed by recombination events during B cell development in the bone marrow and was originally close to Cμ may now be brought immediately upstream of the constant region of a different isotype (e.g., IgG, IgA, IgE). The result is that the B cell begins to produce a new heavy-chain isotype (determined by the C region of the antibody) with the same specificity as that of the original B cell because specificity is determined by the sequence of the VDJ exon, which is not altered.

Cytokines produced by Tfh cells determine which heavy-chain isotype is produced (see Fig. 7.12). The production of opsonizing IgG antibodies, which bind to phagocyte Fc receptors, occurs as a result of class switching driven by IFN-γ in mice but the cytokines involved in IgG class switching in humans are not well established. Opsonization and phagocytosis is an important defense mechanism against many bacteria and viruses, and predictably, these microbes induce the production of IFN-γ. By contrast, switching to the IgE class is stimulated by IL-4 and IL-13 produced by Tfh cells. IgE production is associated with helminth infections, which induce strong Th2 and related Tfh cell responses. Thus, the nature of the helper T cell response to a microbe guides the subsequent antibody response, making it optimal for combating that microbe.

The antibody isotype produced is also influenced by the site of immune responses. As mentioned earlier, IgA antibody is the major isotype produced in mucosal lymphoid tissues, probably because cytokines such as transforming growth factor β (TGF-β) that promote switching to IgA are abundant in these tissues. IgA is the principal antibody isotype that can be actively secreted through mucosal epithelia (see Chapter 8).

Affinity Maturation

Affinity maturation is the process by which the affinity of antibodies produced in response to a protein antigen increases with prolonged or repeated exposure to that antigen (Fig. 7.14). Because of affinity maturation,

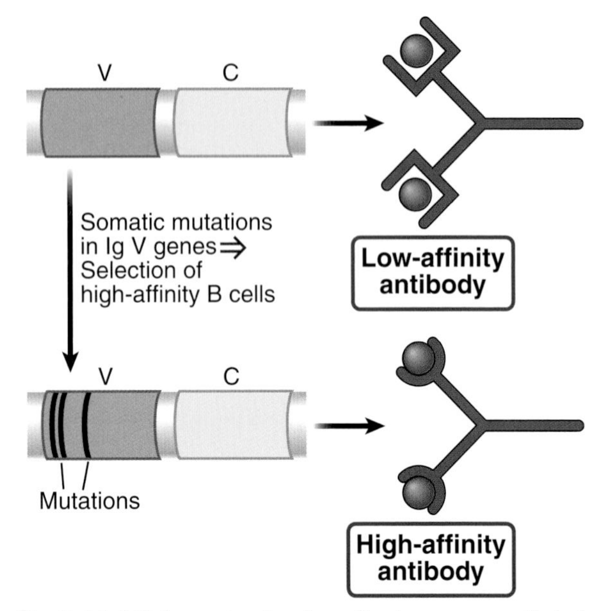

Fig. 7.14 Affinity maturation in antibody responses. Early in the immune response, low-affinity antibodies are produced. During the germinal center reaction, somatic mutation of immunoglobulin (Ig) V genes and selection of mutated B cells with high-affinity antigen receptors result in the production of antibodies with high affinity for antigen.

the ability of antibodies to bind to a microbe or microbial antigen increases if the infection is persistent or recurrent. This increase in affinity is caused by point mutations in the V regions, and particularly in the antigen-binding hypervariable regions, of the genes encoding the antibodies produced. Affinity maturation is seen only in responses to helper T cell–dependent protein antigens, indicating that T cells are critical in the process. These findings raise two intriguing questions: how are mutations in Ig genes induced in B cells, and how are the highest affinity (i.e., most useful) B cells selected to become progressively more numerous?

Affinity maturation occurs in the germinal centers of lymphoid follicles and is the result of somatic hypermutation of Ig genes in dividing B cells, followed by the selection of high-affinity B cells by antigen (Fig. 7.15). In the dark zones of germinal centers (where the proliferating B cells are concentrated), numerous point mutations are introduced into the Ig genes of the rapidly dividing B cells. The enzyme AID, which is required for class switching, also plays a critical role in somatic mutation. This enzyme, as stated earlier, converts cytosines into uracils. The uracils that are

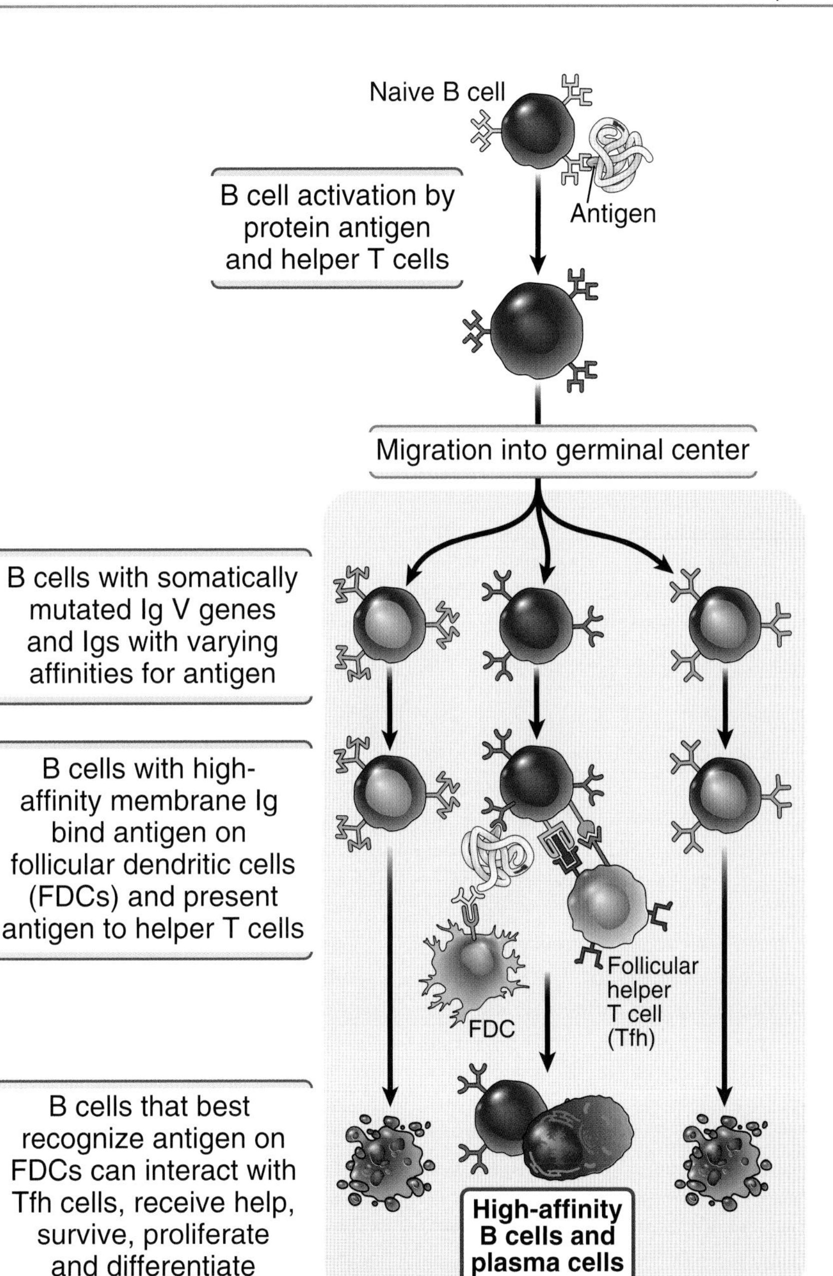

Fig. 7.15 Selection of high-affinity B cells in germinal centers. Some activated B cells migrate into follicles to form germinal centers, where they undergo rapid proliferation and accumulate mutations in their immunoglobulin (Ig) V genes. These B cells produce antibodies with different affinities for the antigen. B cells that recognize antigen bound to follicular dendritic cells (FDCs) antigen are selected to survive. FDCs display antigens by using Fc receptors to bind immune complexes or by using C3 receptors to bind immune complexes with attached C3b and C3d complement proteins (not shown). B cells bind the antigen, process it, and present it to follicular helper T (Tfh) cells in the germinal centers, and signals from the Tfh cells promote survival of the B cells. As more antibody is produced, the amount of available antigen decreases, so only the B cells that express receptors with higher affinities can bind the antigen and are selected to survive.

produced in Ig V-region DNA are frequently replaced by thymidines during DNA replication, creating C-to-T mutations, or they are removed and repaired by error-prone mechanisms that often lead to introduction of various nucleotides in the vicinity of the original mutated cytosine. The frequency of Ig gene mutations is estimated to be 1 in 10^3 base pairs per cell division, which is much greater than the mutation rate in most other genes. For this reason, Ig mutation in germinal center B cells is called somatic hypermutation. This extensive mutation results in the generation of different B cell clones whose Ig molecules may bind with varying affinities to the antigen that initiated the response. The next step in the process is the selection of B cells with the most useful antigen receptors.

Germinal center B cells undergo apoptosis unless rescued by antigen recognition and T cell help. While somatic hypermutation of Ig genes is taking place in germinal centers, the antibody secreted earlier during the immune response binds residual antigen. The antigen-antibody complexes that are formed may activate complement. These complexes are displayed by FDCs, which are a type of stromal cell developmentally related to fibroblastic reticular cells. FDCs are found only in lymphoid follicles. In spite of their similar name, FDCs are distinct from the class II MHC–expressing DCs that present antigens to T lymphocytes and are not derived from bone marrow precursors. The long cytoplasmic processes of FDCs form a meshwork around which germinal centers are formed. These cells express complement receptors and Fc receptors, which are involved in displaying antigens for the selection of germinal center B cells, as described next.

B cells that have undergone somatic hypermutation are given a chance to bind antigen. Most of the antigen in the germinal centers carries attached antibody or complement proteins, which bind to receptors on FDCs and is thus displayed by the FDCs. B cells that recognize the antigen can internalize it, process it, and present peptides to germinal center Tfh cells, which then provide critical survival signals. High-affinity B cells more effectively compete for the antigen and thus are more likely to bind the antigen and survive than B cells with Igs that have lower affinities for the antigen, akin to a process of Darwinian survival of the fittest. Selected cells return to the dark zone, and this process is repeated several times. As the immune response to a protein antigen develops, and also with repeated antigen exposure (e.g., with vaccine boosters), the amount of antibody produced increases. As a result, the amount of antigen available in the germinal center decreases. The B cells that are selected to survive must be able to bind antigen at increasingly lower concentrations, and therefore these are cells whose antigen receptors are of increasingly higher affinity.

Generation of Plasma Cells and Memory B Cells

Activated B cells in germinal centers may differentiate into memory cells or long-lived plasma cells. Memory B cells do not secrete antibody but can home to tissues and to secondary lymphoid organs. Memory B cells exit the germinal center usually after limited rounds of selection. They survive for months or years in the absence of additional antigen exposure, undergo slow cycling, and are ready to respond rapidly if the antigen is reintroduced. Therefore, memory from a T-dependent antibody response can last for a lifetime.

After repeated rounds of selection in the light zone, high-affinity B cells differentiate into antibody-secreting plasma cells. The initial antibody-secreting cells, called plasmablasts, enter the circulation and rapidly migrate to the bone marrow or mucosal tissues. In the bone marrow, they may further differentiate into long-lived plasma cells, which survive for years and continue to produce high-affinity antibodies, even after the antigen is eliminated. It is estimated that more than half of the antibodies in the blood of a normal adult are produced by these long-lived plasma cells; thus, circulating antibodies reflect each individual's history of antigen exposure. These antibodies provide a level of immediate protection if the antigen (microbe or toxin) reenters the body.

ANTIBODY RESPONSES TO T-INDEPENDENT ANTIGENS

Polysaccharides, lipids, and other nonprotein antigens elicit antibody responses without the participation of helper T cells. Recall that these nonprotein antigens cannot bind to MHC molecules, so they cannot be seen by T cells (see Chapter 3). Many bacteria contain polysaccharide-rich capsules, and defense against such bacteria is mediated primarily by antibodies that bind to capsular polysaccharides and target the bacteria for phagocytosis. Antibody responses to T-independent antigens differ from responses to proteins, and most of these differences are attributable to the roles of helper T cells in antibody responses to proteins (Fig. 7.16; see also Fig. 7.2). Extensive cross-linking of BCRs by multivalent antigens may activate the B cells strongly enough to stimulate their proliferation and differentiation without a requirement for T cell help. Polysaccharides also activate the complement system, and many T-independent antigens engage TLRs, thus providing activating signals to the B cells that enhance B cell activation in the absence of T cell help (see Fig. 7.5).

REGULATION OF HUMORAL IMMUNE RESPONSES

B cell responses are regulated by the products of B cells themselves, that is, antibodies, and by cell-intrinsic

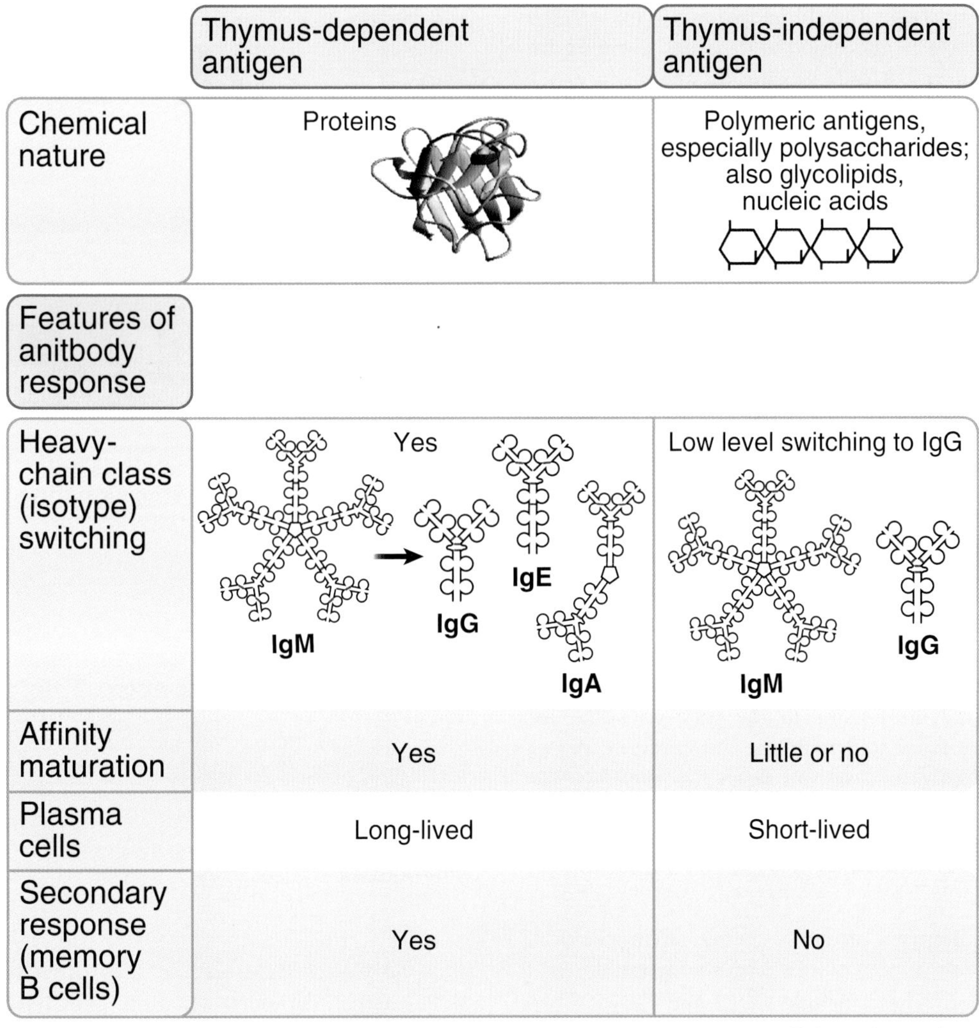

Fig. 7.16 **Features of antibody responses to T-dependent and T-independent antigens.** T-dependent antigens (proteins) and T-independent antigens (nonproteins) induce antibody responses with different characteristics, largely reflecting the influence of helper T cells in T-dependent responses to protein antigens and the absence of T cell help in T-independent responses. *Ig,* Immunoglobulin.

mechanisms, including inhibitory receptors and signaling pathways.

Antibody Feedback

After B lymphocytes differentiate into antibody-secreting cells and memory cells, a fraction of these cells survive for long periods, but most of the activated B cells probably die by apoptosis. This gradual loss of the activated B cells contributes to the physiologic decline of the humoral immune response. B cells also use a special mechanism for shutting off antibody production. As IgG antibody is produced and circulates throughout the body, the antibody binds to antigen that is still available in the blood and tissues, forming immune complexes. B cells specific for the antigen may bind the antigen part of the immune complex by their Ig receptors. At the same time, the Fc tail of the attached IgG antibody may be recognized by a special type of Fc receptor expressed on B cells (as well as on many myeloid cells) called FcγRIIB (Fig. 7.17). This Fc receptor is an inhibitory receptor, and, like other inhibitory receptors in T cells and natural killer (NK) cells, it contains a cytoplasmic immunoreceptor tyrosine-based inhibitory motif (ITIM). This ITIM is phosphorylated by the LYN tyrosine kinase, and a phosphatase is recruited to the phosphorylated ITIM. The phosphatase shuts off antigen receptor–induced kinase-dependent signals, thereby terminating B cell responses. This process, in which antibody bound to antigen inhibits further antibody production, is called **antibody feedback**. It serves to terminate further B cell activation once sufficient quantities of IgG antibodies have been produced. Inhibition by the FcγRIIB also functions to limit antibody responses against self antigens, and polymorphisms in the gene encoding this receptor are associated with the autoimmune disease systemic lupus erythematosus (see Chapter 9).

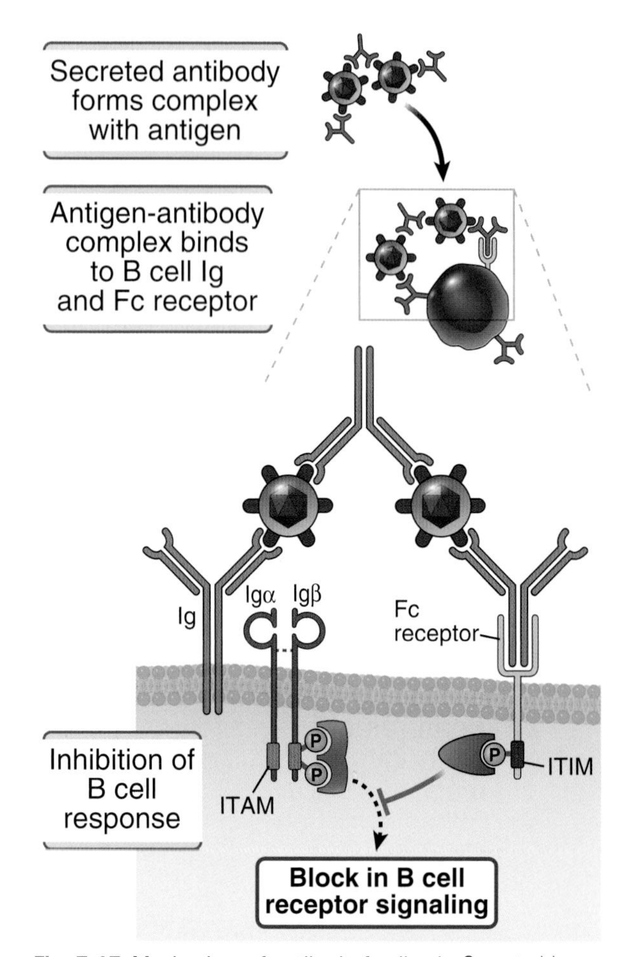

Fig. 7.17 Mechanism of antibody feedback. Secreted immunoglobulin G (IgG) antibodies form immune complexes (antigen-antibody complexes) with residual antigen (shown here as a virus but more commonly it is a soluble antigen). The complexes interact with B cells specific for the antigen, with the membrane Ig antigen receptors recognizing epitopes of the antigen and a type of Fc receptor (FcγRIIB) recognizing the bound antibody. The Fc receptors block activating signals from the antigen receptor, terminating B cell activation. The cytoplasmic domain of B cell FcγRIIB contains an ITIM that binds enzymes that inhibit antigen receptor–mediated B cell activation. *ITAM,* Immunoreceptor tyrosine-based activation motif; *ITIM,* immunoreceptor tyrosine-based inhibition motif.

B Cell Signal Attenuation by Other Inhibitory Receptors

Several inhibitory receptors other than FcγRIIB dampen B cell responses and raise the threshold for B cell activation. These include CD22 and CD72. These inhibitory receptors contain cytoplasmic ITIMs that are phosphorylated after the BCR is engaged. The phosphorylated ITIMs recruit a tyrosine phosphatase SHP-1, which dampens signaling. They have been best studied in rodents, and their contributions to human B cell biology and disease are less certain.

SUMMARY

- Humoral immunity is mediated by antibodies that bind to extracellular microbes and their toxins, which are neutralized or targeted for destruction by phagocytes and the complement system.
- Humoral immune responses are initiated by recognition of the antigens by specific membrane immunoglobulin (Ig) antigen receptors of naive B cells. The binding of multivalent antigen cross-links B cell antigen receptors of specific B cells, and biochemical signals are delivered to the inside of the B cells by Ig-associated signaling proteins. These signals induce B cell clonal expansion and IgM secretion.
- In humoral immune responses to a protein antigen, called T-dependent responses, the binding of protein antigen to specific Ig receptors of naive B cells in lymphoid follicles results in the generation of signals that prepare the B cell for interaction with activated helper T cells that express CD40L and secrete cytokines. The B cells internalize and process that antigen and present class II major histocompatibility complex (MHC)—displayed peptides to activated helper T cells specific for the displayed peptide-MHC complex. These helper T cells contribute to early B cell activation at extrafollicular sites.
- The early T-dependent humoral response occurs in extrafollicular foci and generates low levels of class-switched antibodies that are produced by short-lived plasma cells.
- Activated B cells induce the further activation of T cells and their differentiation into T follicular helper (Tfh) cells. The B cells, together with the Tfh cells, migrate into follicles and form germinal centers.
- The full T-dependent humoral response develops in germinal centers and leads to extensive class switching and affinity maturation; generation of long-lived plasma cells that secrete antibodies for many years; and development of long-lived memory B cells, which rapidly respond to re-encounter with antigen by proliferation and secretion of high-affinity antibodies.
- Heavy-chain class switching (or isotype switching) is the process by which the isotype, but not the specificity, of the antibodies produced in response to an antigen changes as the humoral response proceeds. Class switching is stimulated by the combination of CD40L and cytokines, both expressed by helper T cells. Different cytokines induce switching to different antibody isotypes, enabling the immune system to respond in the most effective way to different types of microbes.
- Affinity maturation is the process by which the affinity of antibodies for protein antigens increases with prolonged or repeated exposure to the antigens. The process is initiated by signals from Tfh cells, resulting in migration of the B cells into follicles and the formation of germinal centers. Here the B cells proliferate rapidly, and their Ig V genes undergo extensive somatic mutation. The antigen may be displayed by FDCs in the germinal centers. B cells with mutated V regions that recognize the antigen with high affinity are selected to survive, giving rise to affinity maturation of the antibody response.
- Polysaccharides, lipids, and other nonprotein antigens are called T-independent antigens because they induce antibody responses without T cell help. Most T-independent antigens contain multiple identical epitopes that are able to cross-link many Ig receptors on a B cell, providing signals that stimulate B cell responses even in the absence of helper T cell activation. Antibody responses to T-independent antigens show less heavy-chain class switching and affinity maturation than is typical for responses to T-dependent protein antigens.
- Secreted antibodies form immune complexes with residual antigen and shut off B cell activation by engaging an inhibitory Fc receptor on B cells.

REVIEW QUESTIONS

1. What are the signals that induce B cell responses to protein antigens and polysaccharide antigens?
2. What are the major differences between primary and secondary antibody responses to a protein antigen?
3. How do helper T cells specific for an antigen interact with B lymphocytes specific for the same antigen? Where in a lymph node do these interactions mainly occur?

4. What are the signals that induce heavy-chain class switching, and what is the importance of this phenomenon for host defense against different microbes?
5. What is affinity maturation? How is it induced, and how are high-affinity B cells selected to survive?
6. What are the characteristics of antibody responses to polysaccharides and lipids? What types of bacteria stimulate mostly these types of antibody responses?

Answers to and discussion of the Review Questions may be found on p. 322.

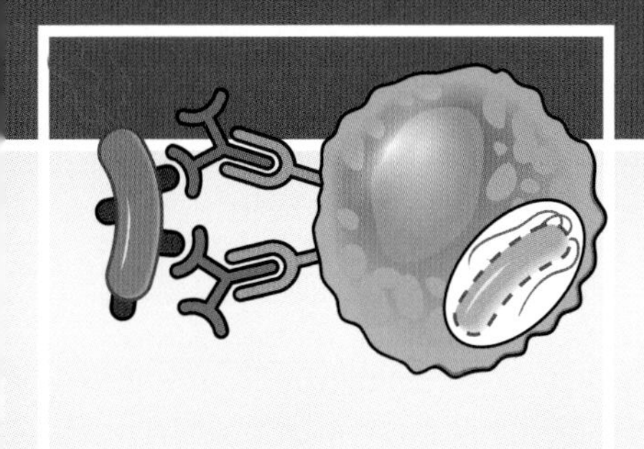

8

Effector Mechanisms of Humoral Immunity
Elimination of Extracellular Microbes and Toxins

CHAPTER OUTLINE

Humoral immunity is the type of host defense mediated by secreted antibodies that is necessary for protection against extracellular microbes and their toxins. Antibodies prevent infections by blocking microbes from binding to and entering host cells. In addition, antibodies function together with other components of the immune system (e.g., phagocytes, complement proteins) to eliminate microbes and toxins. Although antibodies are a major mechanism of adaptive immunity against extracellular microbes, they cannot reach microbes that live inside cells. However, humoral immunity is vital even for defense against microbes that live inside cells, such as viruses, because antibodies can bind to these microbes before they enter host cells or during passage from infected to uninfected cells, thus preventing spread of infection. Defects in antibody production are associated with increased susceptibility to infections by many bacteria, viruses, and parasites. All the vaccines that are currently in use work mainly by stimulating the production of antibodies.

This chapter describes how antibodies provide defense against infections, addressing the following questions:

- What are the mechanisms used by secreted antibodies to combat different types of infectious agents and their toxins?
- What is the role of the complement system in defense against microbes?
- How do antibodies combat microbes that enter the gastrointestinal and respiratory tracts?
- How do antibodies protect the fetus and newborn from infections?

Before describing the mechanisms by which antibodies function in host defense, we summarize the properties of antibody molecules that are important for these functions.

PROPERTIES OF ANTIBODIES THAT DETERMINE EFFECTOR FUNCTIONS

Several features of the production and structure of antibodies contribute in important ways to the roles of these molecules in host defense.

Antibodies function in the circulation, in tissues throughout the body, and in the lumens of mucosal organs. Antibodies are produced after stimulation of B lymphocytes by antigens in secondary (peripheral) lymphoid organs (lymph nodes, spleen, mucosal lymphoid tissues) and, in lesser amounts, at tissue sites of inflammation. Many of the antigen-stimulated B lymphocytes differentiate into antibody-secreting plasma cells, some of which remain in lymphoid organs or inflamed tissues and others migrate to and reside in the bone marrow. Different plasma cells synthesize and secrete antibodies of different heavy-chain classes (isotypes). These secreted antibodies enter the blood, from where they may reach any peripheral site of infection, or enter mucosal secretions, where they prevent infections by microbes that could enter through epithelial barriers.

Protective antibodies are produced during the first (primary) response to a microbe and in larger amounts during subsequent secondary responses (see Fig. 7.3). Antibody production usually begins within the first week after infection or vaccination. Some of the plasma cells generated in germinal center reactions migrate to the bone marrow and continue to produce antibodies for months or years. If the microbe again tries to infect the host, the continuously secreted antibodies provide immediate protection. At the same time, memory cells that had developed during the initial B cell response rapidly differentiate into antibody-producing cells upon repeat encounter with the antigen, providing a large burst of antibody for more effective defense against the infection. A goal of vaccination is to stimulate the development of long-lived plasma cells and memory cells.

Antibodies use their antigen-binding (Fab) regions to bind to and block the harmful effects of microbes and toxins, and they use their Fc regions to activate diverse effector mechanisms that eliminate these microbes and toxins (Fig. 8.1). This spatial segregation of the antigen recognition and effector functions of antibody molecules was introduced in Chapter 4. Antibodies block the infectivity of microbes and the injurious effects of microbial toxins simply by binding to the microbes and toxins. Other functions of antibodies require the participation of various components of host defense, such as phagocytes and the complement system. The Fc portions of immunoglobulin (Ig) molecules, made up of the heavy-chain constant regions, contain the binding sites for Fc receptors on phagocytes and for complement proteins. The binding of antibodies to Fc receptors and complement proteins occurs only after Ig molecules recognize and become attached to a microbe or microbial antigen. Therefore, even the Fc-dependent functions of antibodies require antigen recognition by the Fab regions. This feature of antibodies ensures that they activate effector mechanisms only when needed—that is, when they recognize their target antigens.

Heavy-chain class (isotype) switching and affinity maturation enhance the protective functions of antibodies. Class switching and affinity maturation are two changes that occur in the antibodies produced by antigen-stimulated B lymphocytes, especially during responses to protein antigens (see Chapter 7). Heavy-chain class switching results in the production of antibodies with distinct Fc regions, capable of different functions (see Fig. 7.12). By switching to different antibody classes in response to various microbes, the humoral immune system is able to engage diverse host mechanisms that are optimal for combating those microbes. Affinity maturation is induced by repeated stimulation with protein antigens, and it leads to the production of antibodies with higher and higher affinities for the antigen, compared to the antibodies initially secreted. This change increases the ability of antibodies to bind to and neutralize or eliminate microbes. The progressive increase in antibody affinity with repeated stimulation of B cells is one of the reasons for the recommended practice of giving multiple rounds of immunizations with the same antigen for generating protective immunity (e.g., booster shots of vaccines).

Antibodies of the IgG isotype survive for a longer duration in the blood than IgM and other isotypes and therefore class switching to IgG prolongs the protective functions of the humoral immune response. Most circulating proteins have half-lives of hours to days in

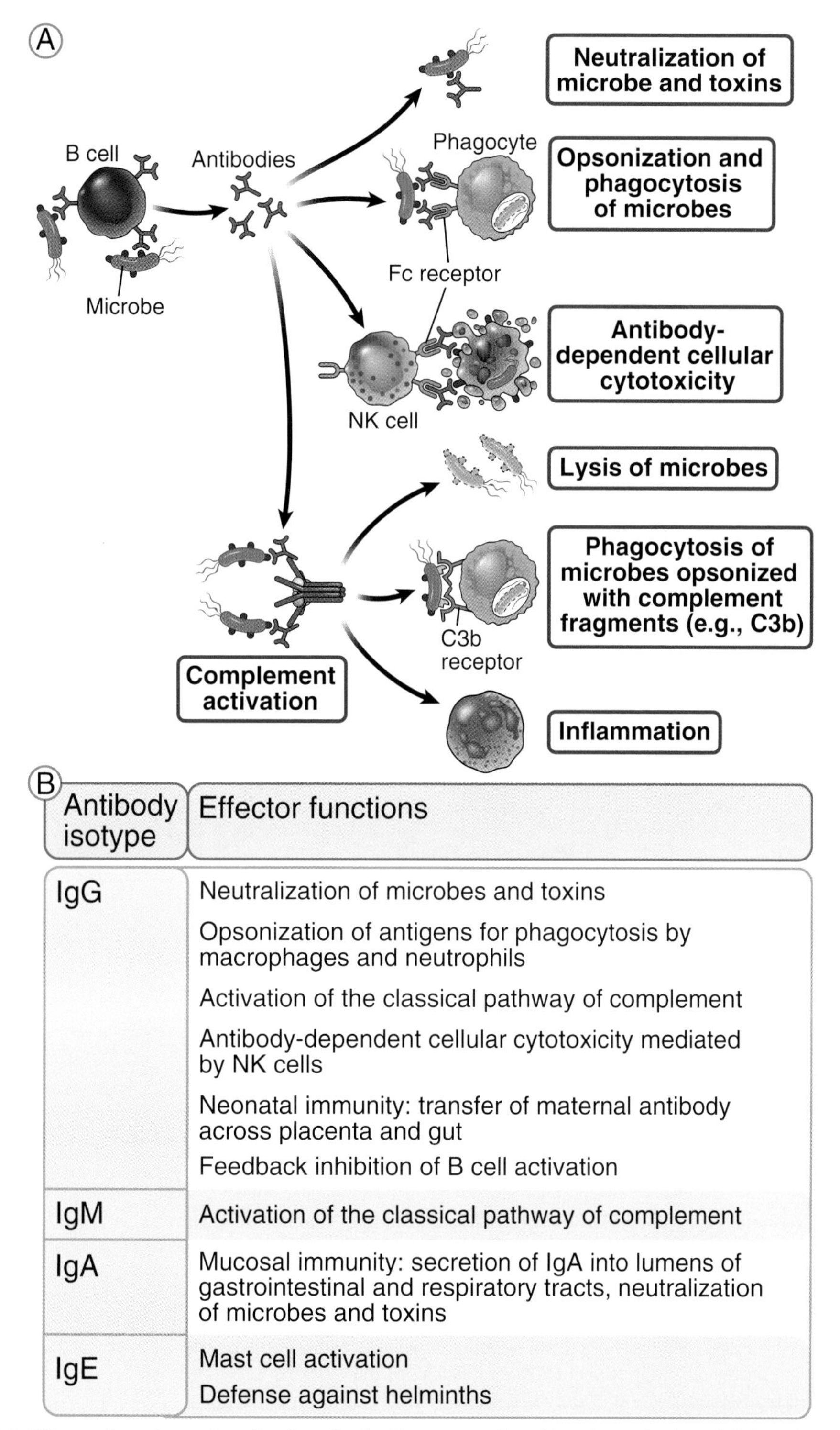

Antibody isotype	Effector functions
IgG	Neutralization of microbes and toxins Opsonization of antigens for phagocytosis by macrophages and neutrophils Activation of the classical pathway of complement Antibody-dependent cellular cytotoxicity mediated by NK cells Neonatal immunity: transfer of maternal antibody across placenta and gut Feedback inhibition of B cell activation
IgM	Activation of the classical pathway of complement
IgA	Mucosal immunity: secretion of IgA into lumens of gastrointestinal and respiratory tracts, neutralization of microbes and toxins
IgE	Mast cell activation Defense against helminths

Fig. 8.1 Effector functions of antibodies. Antibodies are produced by the activation of B lymphocytes by antigens and other signals (not shown). Antibodies of different heavy-chain classes (isotypes) perform different effector functions, as illustrated schematically in **(A)** and summarized in **(B)**. (Some properties of antibodies are listed in Fig. 4.3.) *Ig,* Immunoglobulin; *NK,* natural killer.

the blood, but IgG has an unusually long half-life of 3 to 4 weeks because of a special mechanism involving a particular Fc receptor called the neonatal Fc receptor (FcRn). This Fc receptor is expressed in placenta, endothelium, phagocytes, and a few other cell types. In the placenta, the FcRn transports antibodies from the mother's circulation to the fetus (discussed later). In other cell types, the FcRn protects IgG antibodies from intracellular catabolism (Fig. 8.2A). FcRn is found in the endosomes of endothelial cells and phagocytes, where it binds to IgG that has been taken up by the cells. Once bound to the FcRn, the IgG is recycled back into the circulation or tissue fluids, thus avoiding lysosomal degradation. This unique mechanism for protecting a blood protein is the reason why IgG antibodies have a half-life that is much longer than that of other Ig isotypes and most other plasma proteins. This property of Fc regions of IgG has been exploited to increase the half-life of other proteins by coupling the proteins to an IgG Fc region (Fig. 8.2B). One of several therapeutic agents based on this principle is the tumor necrosis factor (TNF) receptor—Fc fusion protein, which functions as an antagonist of TNF and is used to treat various inflammatory diseases. By coupling the extracellular domain of the TNF receptor to the Fc portion of a human IgG molecule using a genetic engineering approach, the half-life of the hybrid protein becomes much greater than that of the soluble receptor by itself. An engineered Fc portion of IgG that binds to and blocks the FcRn reduces levels of circulating autoantibodies (and all IgGs) and is approved for treating generalized myasthenia gravis.

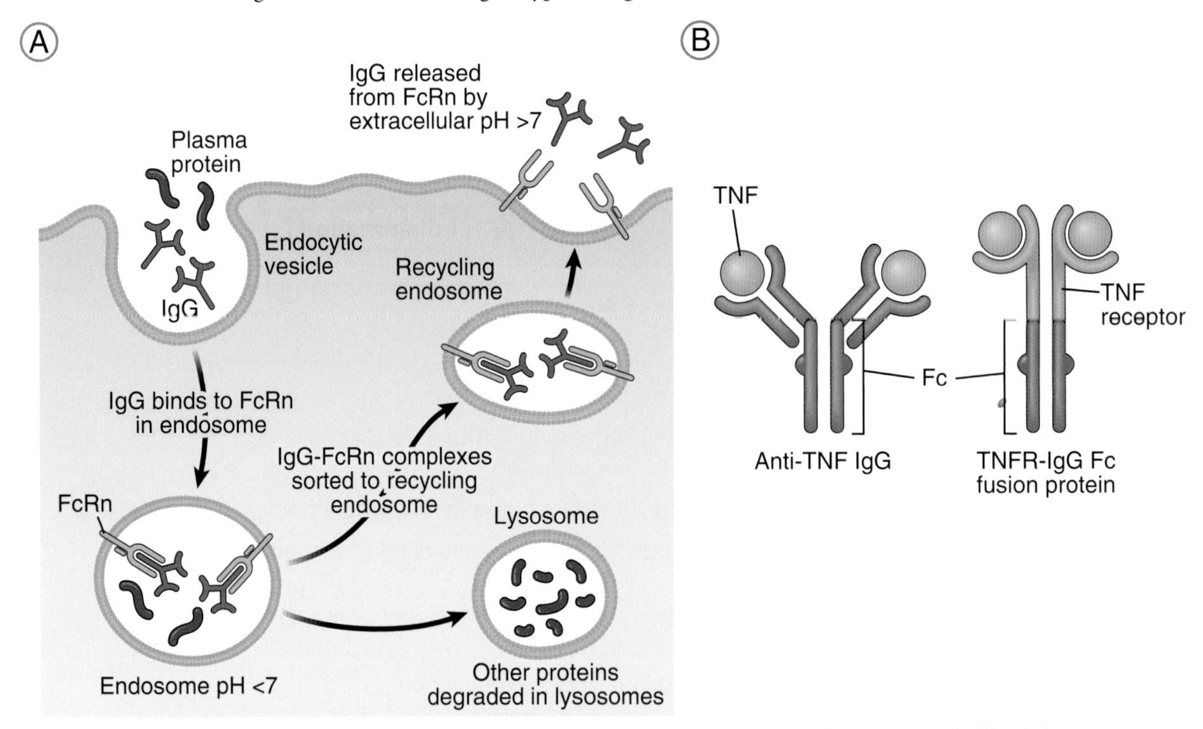

Fig. 8.2 Neonatal Fc receptor (FcRn) contributes to the long half-life of IgG molecules. A, Circulating or extravascular IgG antibodies (mainly of the IgG1, IgG2, and IgG4 subclasses) are ingested by endothelial cells and phagocytes into endosomes, where they bind the FcRn, a receptor present in the endosomal membrane. The low internal pH of endosomes favors tight IgG binding to the FcRn, protecting the IgG from lysosomal proteolysis. The FcRn-IgG complexes recycle back to the cell surface, where they are exposed to the neutral pH (~7) of the blood, which releases the bound antibody back into the circulation or tissue fluid. **B,** Fc-containing fusion proteins. A monoclonal antibody specific for the cytokine tumor necrosis factor (TNF) *(left)* can bind to and block the activity of the cytokine and remain in the circulation for a long time (weeks) due to recycling by the neonatal Fc receptor (FcRn). The soluble extracellular domain of the TNF receptor (TNFR) *(right)* can also act as an antagonist of the cytokine, and coupling the soluble receptor to an IgG Fc domain prolongs the half-life of the fusion protein in the blood by the same FcRn-dependent mechanism. Both anti-TNF IgG monoclonal antibodies and TNFR-IgG Fc fusion proteins are used as drugs to treat some inflammatory diseases. *Ig,* Immunoglobulin.

With this introduction, we proceed to a discussion of the mechanisms used by antibodies to combat infections. Much of the chapter is devoted to effector mechanisms that are not influenced by anatomic considerations; that is, they may be active anywhere in the body. At the end of the chapter, we describe the special features of antibody functions at particular anatomic locations.

NEUTRALIZATION OF MICROBES AND MICROBIAL TOXINS

Antibodies bind to and block, or neutralize, the infectivity of microbes and the interactions of microbial toxins with host cells (Fig. 8.3). Vaccine-induced immunity is related mainly to the ability of vaccines to stimulate the production of neutralizing antibodies that block initial infection. Antibodies in mucosal secretions in the gut and airways block the entry of ingested and inhaled microbes, respectively (discussed in more detail later in the chapter). After microbes enter the host, they use molecules in their envelopes or cell walls to bind to and gain entry into host cells. Antibodies may attach to these microbial surface molecules, thereby preventing the microbes from infecting host cells. Microbes, such as viruses, that are able to enter host cells may replicate inside the cells and then be released and go on to infect other neighboring cells. Antibodies can neutralize the microbes during their transit from cell to cell and thus

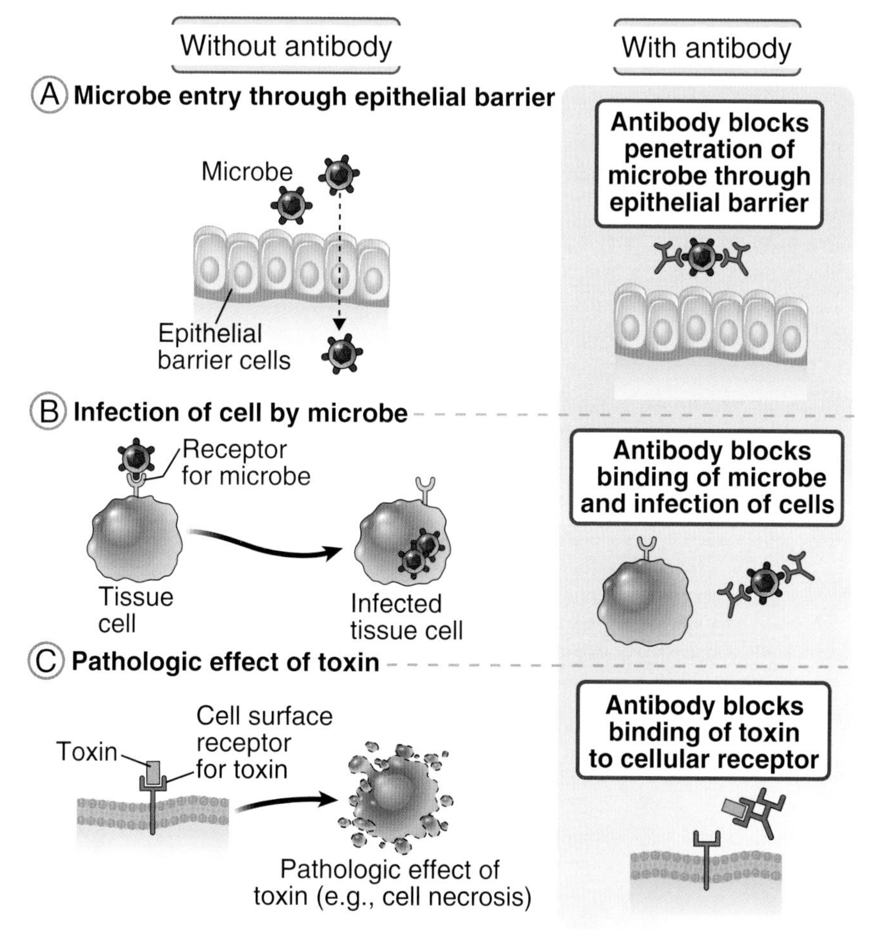

Fig. 8.3 Neutralization of microbes and toxins by antibodies. **A,** Antibodies at epithelial surfaces, such as in the gastrointestinal and respiratory tracts, block the entry of ingested and inhaled microbes, respectively. **B,** Antibodies prevent the binding of microbes to cells, thereby blocking the ability of the microbes to infect host cells. **C,** Antibodies block the binding of toxins to cells, thereby inhibiting the pathologic effects of the toxins.

also limit the spread of infection. If an infectious microbe does colonize the host, its harmful effects may be caused by endotoxins or exotoxins, which often bind to specific receptors on host cells in order to mediate their effects. Antibodies prevent binding of the toxins to host cells and thus block their harmful effects. Emil von Behring and Shibasaburo Kitasato's demonstration of this type of protection mediated by the administration of serum from immunized animals that contained antibodies against diphtheria toxin was the first formal demonstration of therapeutic immunity against a microbe or its toxin. This approach was called serum therapy, and was the basis for awarding Behring the first Nobel Prize in Physiology or Medicine in 1901.

OPSONIZATION AND PHAGOCYTOSIS

Antibodies coat microbes and promote their ingestion by phagocytes (Fig. 8.4). The process of coating particles for subsequent phagocytosis is called opsonization, and the molecules that coat microbes and enhance their phagocytosis are called opsonins. When several IgG molecules bind to a microbe, an array of their Fc regions projects away from the microbial surface. If the antibodies belong to certain isotypes (IgG1 and IgG3 in humans), their Fc regions bind to a high-affinity receptor for the Fc regions of these γ heavy chains, called FcγRI (CD64), which is expressed on neutrophils and macrophages (Fig. 8.5). The phagocyte extends its plasma membrane around the attached microbe and ingests the microbe into a vesicle called a phagosome, which fuses with lysosomes. The binding of antibody Fc tails to FcγRI also activates the phagocytes, because the FcγRI contains a signaling chain that triggers numerous biochemical pathways in the phagocytes. The signals generated lead to production of large amounts of reactive oxygen species, nitric oxide, and proteolytic enzymes in the lysosomes of the activated neutrophils and macrophages, all of which contribute to the destruction of the ingested microbe.

Antibody-mediated phagocytosis is the major mechanism of defense against encapsulated bacteria, such as pneumococci. The polysaccharide-rich capsules of these bacteria protect the organisms from phagocytosis in the absence of antibody, but opsonization by antibody promotes ingestion and destruction of the bacteria. The spleen contains large numbers of macrophages and is an important site of phagocytic clearance of opsonized bacteria. This is why patients who do not have a spleen (most often because of surgical removal after trauma or infarction in sickle cell disease) are susceptible to disseminated infections by encapsulated bacteria.

One of the Fcγ receptors, FcγRIIB, does not mediate effector functions of antibodies but rather shuts down antibody production and reduces inflammation. The role of FcγRIIB in feedback inhibition of B cell activation was discussed in Chapter 7 (see Fig. 7.16). FcγRIIB also inhibits activation of macrophages and dendritic cells and may thus serve an antiinflammatory function as well. Pooled IgG from healthy donors is given intravenously to treat various inflammatory diseases. This preparation is called

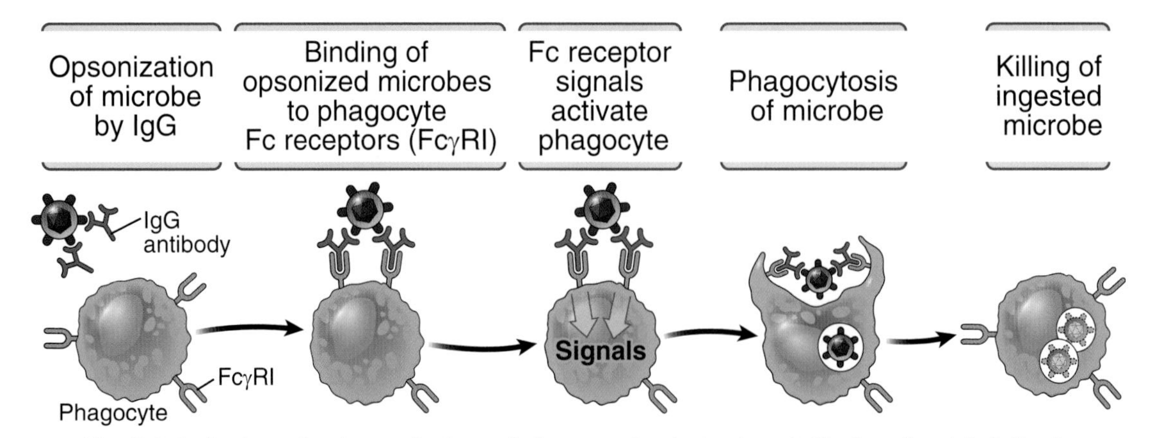

Fig. 8.4 Antibody-mediated opsonization and phagocytosis of microbes. Antibodies of certain IgG subclasses (IgG1 and IgG3) bind to microbes and are then recognized by Fc receptors on phagocytes. Signals from the Fc receptors promote the phagocytosis of the opsonized microbes and activate the phagocytes to destroy these microbes.

Fc Receptor	Affinity for Ig	Cell distribution	Function
Signaling Fc receptors			
FcγRI (CD64)	High; binds IgG1 and IgG3	Macrophages, neutrophils	Phagocytosis; activation of phagocytes
FcγRIIB (CD32)	Low	B lymphocytes, DCs, mast cells, neutrophils, macrophages	Feedback inhibition of B cells, attenuation of inflammation
FcγRIIIA (CD16)	Low	NK cells	Antibody-dependent cellular cytotoxicity (ADCC)
FcεRI	High, binds IgE	Mast cells, basophils	Activation (degranulation) of mast cells and basophils

Fig. 8.5 Leukocyte Fc receptors. The cellular distribution and functions of different types of human Fc receptors expressed on immune cells. Two other types of Fc receptors, FcRn and the poly-Ig receptor, do not deliver signals to cells but are involved in transport of Ig across cellular membranes; these are discussed in the text. *DCs,* Dendritic cells; *Ig,* immunoglobulin; *NK,* natural killer.

intravenous immune globulin (IVIG), and its beneficial effect in these diseases is partly mediated by its binding to FcγRIIB on various cells.

ANTIBODY-DEPENDENT CELLULAR CYTOTOXICITY

Natural killer (NK) cells bind to antibody-coated cells and destroy these cells (Fig. 8.6). NK cells express an Fcγ receptor called FcγRIII (CD16), which is one of several kinds of NK cell–activating receptors (see Chapter 2). FcγRIII binds to arrays of IgG antibodies already attached by their antigen binding sites to surface antigens on a cell. This Fc receptor generates signals that cause the NK cell to discharge its granule proteins, which kill the antibody-coated cell by the same mechanisms that $CD8^+$ cytotoxic T lymphocytes use to kill infected cells (see Chapter 6). This process is called antibody-dependent cellular cytotoxicity (ADCC). Cells infected with enveloped viruses typically express viral glycoproteins on their surface that can be recognized by specific antibodies, and this may facilitate ADCC-mediated destruction of the infected cells. ADCC is also one of the mechanisms by which therapeutic antibodies used to treat cancers eliminate tumor cells.

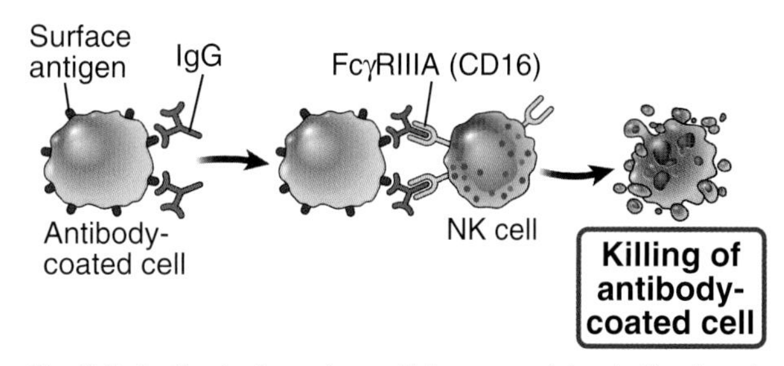

Fig. 8.6 Antibody-dependent cellular cytotoxicity. Antibodies of certain immunoglobulin G (IgG) subclasses (IgG1 and IgG3) bind to antigens on the surface of infected cells, and their Fc regions are recognized by an Fcγ receptor on natural killer (NK) cells. The NK cells are activated and kill the antibody-coated cells.

IMMUNOGLOBULIN E– AND MAST CELL–MEDIATED REACTIONS

Activation of mast cells and eosinophils contributes to allergic diseases and defense against helminthic parasites. Mast cells express the high-affinity IgE receptor, FcεRI, which binds IgE and leads to activation of the cells. This reaction is important in allergic diseases (see Chapter 11) and may contribute to the expulsion of worms. Most helminths are too large to be phagocytosed, and their thick integuments make them resistant to the microbicidal substances produced by neutrophils and macrophages. The immune response to helminthic parasites is dominated by T helper 2 (Th2) cell activation, IgE antibody production, and eosinophilia, suggesting that all may contribute to defense. However, in eosinophils, FcεRI is not expressed at high levels and it lacks a signaling chain, so IgE cannot activate eosinophils. These cells may be recruited to sites of infection by chemokines and may bind to IgG-coated parasites by FcγRI. Eosinophils can also be activated by the Th2 cytokine IL-5 independent of antibody and release their granule contents, which can destroy the thick integuments of helminths.

THE COMPLEMENT SYSTEM

The complement system is a collection of circulating and cell membrane proteins that play important roles in host defense against microbes and in antibody-mediated tissue injury. The term *complement* refers to the ability of these proteins to assist, or complement, the activity of antibodies in destroying (lysing) cells, including microbes. The complement system may be activated by microbes in the absence of antibody, as part of the innate immune response to infection, and by antibodies attached to microbes, as part of adaptive immunity (see Fig. 2.12).

The activation of the complement system involves sequential proteolytic cleavage of complement proteins, leading to the generation of effector molecules that participate in eliminating microbes in different ways. Complement protein activation, like all enzymatic cascades, is capable of achieving tremendous amplification, because even a small number of activated complement molecules produced early in the process may generate a large number of effector molecules later in the cascade. Activated complement proteins become covalently attached to the cell surfaces where the activation occurs, ensuring that complement effector functions are limited to the correct sites. Normal host cells possess several regulatory mechanisms that inhibit the activation of complement and the deposition of activated complement proteins, thus preventing complement-mediated damage to healthy cells.

Pathways of Complement Activation

There are three pathways of complement activation: the alternative and lectin pathways are initiated by microbes in the absence of antibody, and the classical pathway is initiated by certain isotypes of antibodies attached to antigens (Fig. 8.7). Several proteins in each pathway interact in a precise sequence. The most abundant complement protein in the plasma, C3, plays a central role in all three pathways. The early steps of all three pathways function to generate a large number of functionally active fragments of C3 bound to the microbe or cell where the complement pathway was initiated. (By convention, the smaller proteolytic fragment of any complement protein is given the "a" suffix, and the larger piece is the "b" fragment; C2 is an exception, for historical reasons.)

- The **alternative pathway** of complement activation is triggered by spontaneous hydrolysis of C3 in plasma at a low level. The breakdown products of C3 are unstable, and, in the absence of infection, are rapidly degraded and lost. However, when a breakdown product of C3 hydrolysis, called C3b, is deposited on the surface of a microbe, it forms stable covalent bonds with microbial proteins or polysaccharides. The microbe-bound C3b binds another protein called Factor B, which is then cleaved by a plasma protease called Factor D to generate the Bb fragment. This fragment remains attached to C3b, and the C3bBb complex functions as a proteolytic enzyme, called the alternative pathway C3 convertase, that breaks down more C3. The C3 convertase is stabilized by properdin, a positive regulator of the complement system. As a result of the enzymatic activity of the C3 convertase, many more C3b and C3bBb molecules are produced and become attached to the microbe. Some of the C3bBb molecules bind an additional C3b molecule, and the resulting C3bBb3b complexes function as a C5 convertase, to cleave the complement protein C5 and initiate the late steps of complement activation.
- The **classical pathway** of complement activation is triggered when IgM or certain subclasses of IgG (IgG1 and IgG3 in humans) bind to antigens (e.g., on a microbial cell surface). As a result of this binding, adjacent Fc regions of the antibodies become accessible to and bind the C1 complement protein (which is made up of a binding component called C1q and two proteases called C1r and C1s). The attached C1 becomes enzymatically active, resulting in the sequential cleavage of two proteins, C4 and C2. One of the C4 fragments that is generated, C4b, becomes covalently attached to the

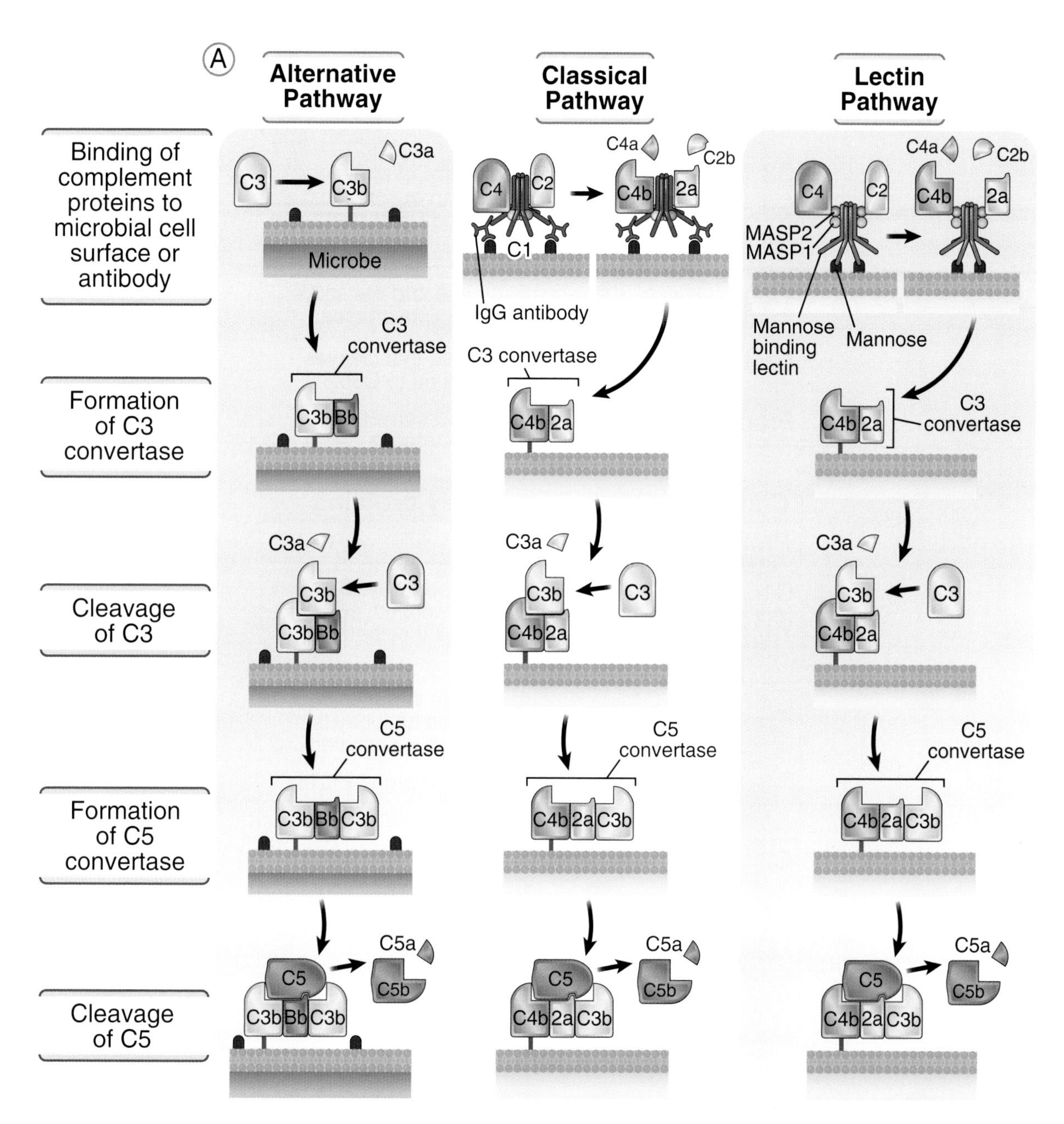

Fig. 8.7 Early steps of complement activation. **A,** The steps in the activation of the alternative, classical, and lectin pathways. Although the sequence of events is similar, the three pathways differ in their requirement for antibody and the proteins used. Note that C5 is cleaved by the C5 convertase but is not a component of the enzyme.

antibody or to the microbial surface where the antibody is bound, and then binds C2, which is cleaved by active C1 to yield the C4b2a complex. This complex is the classical pathway C3 convertase, which functions to break down C3, and the C3b that is generated again becomes attached to the microbe. Some of the C3b binds to the C4b2a complex, and the resultant C4b2a3b complex functions as a C5

(B) **Alternative pathway proteins**

Protein	Function
C3	C3b binds to the surface of a microbes, where it functions as an opsonin and as a component of C3 and C5 convertases C3a stimulates inflammation
Factor B	Bb is a serine protease and the active enzyme of C3 and C5 convertases
Factor D	Plasma serine protease that cleaves Factor B when it is bound to C3b
Properdin	Properdin stabilizes C3 convertases (C3bBb)on microbial surfaces

(C) **Classical and lectin pathway proteins**

Protein	Function
C1 ($C1qr_2s_2$)	Initiates the classical pathway; C1q binds to Fc portion of antibody; C1r and C1s are proteases that lead to C4 and C2 activation
C4	C4b covalently binds to surfaces of microbes or cells where antibody is bound and complement is activated C4b binds to C2 for cleavage by C1s C4a stimulates inflammation
C2	C2a is a serine protease functioning as an active enzyme of C3 and C5 convertases
Mannose binding lectin (MBL)	Initiates the lectin pathway; MBL binds to terminal mannose residues of microbial carbohydrates. MBL-associated proteases activate C4 and C2, as C1r and C1s do in the classical pathway.

Fig. 8.7, cont'd B, The important properties of the proteins involved in the early steps of the alternative pathway of complement activation. **C,** The important properties of the proteins involved in the early steps of the classical and lectin pathways. Note that C3, which is listed among the alternative pathway proteins **(B),** is also the central component of the classical and lectin pathways.

convertase, which cleaves the C5 complement protein.

- The **lectin pathway** of complement activation is initiated not by antibodies but by the binding of circulating lectins, such as plasma mannose-binding lectin (MBL) or ficolins, to microbial polysaccharides. Serine proteases structurally related to C1s of the classical pathway are associated with these lectins and serve to activate C4. The subsequent steps are the same as in the classical pathway.

The net result of these early steps of complement activation is that microbes acquire a coating of covalently attached C3b. Note that the alternative and lectin pathways are effector mechanisms of innate immunity, whereas the classical pathway is a mechanism of adaptive humoral immunity. These pathways differ in their initiation, but once triggered, their late steps are the same.

The late steps of complement activation lead to the formation of a cytolytic protein complex. They are initiated by proteolysis of C5 by the C5 convertase, generating C5b (Fig. 8.8). C6 binds to C5b and inserts into the target membrane and then C7, C8, and C9 bind sequentially to the initial C5b-C6 complex in the membrane. The final protein in the pathway, C9, polymerizes to form a pore in the cell membrane of the microbe where complement is activated and through which water and ions can enter, causing death of the microbe. The C5b-9 complex is called the **membrane attack complex** (MAC), and its formation is the end result of complement activation.

Functions of the Complement System

The complement system plays an important role in the elimination of microbes during innate and adaptive immune responses. The main effector

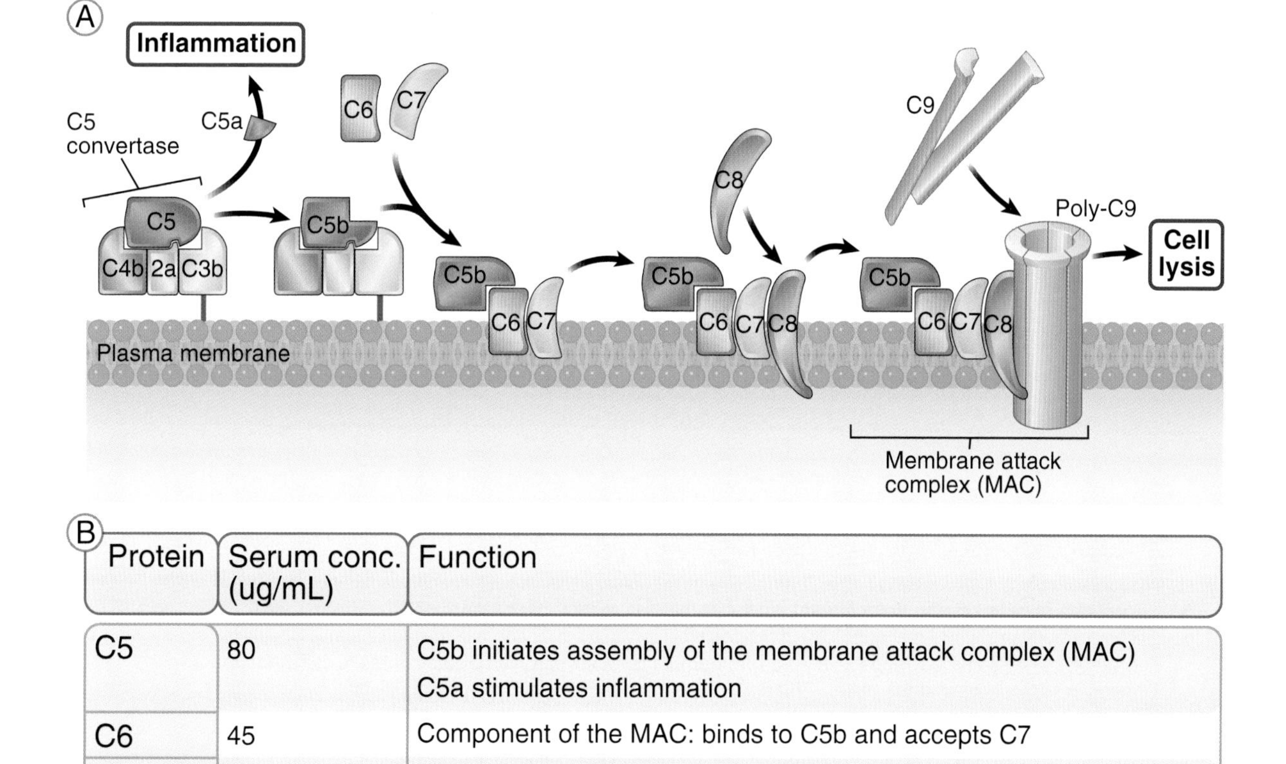

Protein	Serum conc. (ug/mL)	Function
C5	80	C5b initiates assembly of the membrane attack complex (MAC) C5a stimulates inflammation
C6	45	Component of the MAC: binds to C5b and accepts C7
C7	90	Component of the MAC: binds C5b, 6 and inserts into lipid membranes
C8	60	Component of the MAC: binds C5b, 6, 7 and initiates binding and polymerization of C9
C9	60	Component of the MAC: binds C5b, 6, 7, 8 and polymerizes to form membrane pores

Fig. 8.8 Late steps of complement activation. **A,** The late steps of complement activation start after the formation of the C5 convertase and are identical in the alternative and classical pathways. **B,** Properties of the proteins in the late steps of complement activation.

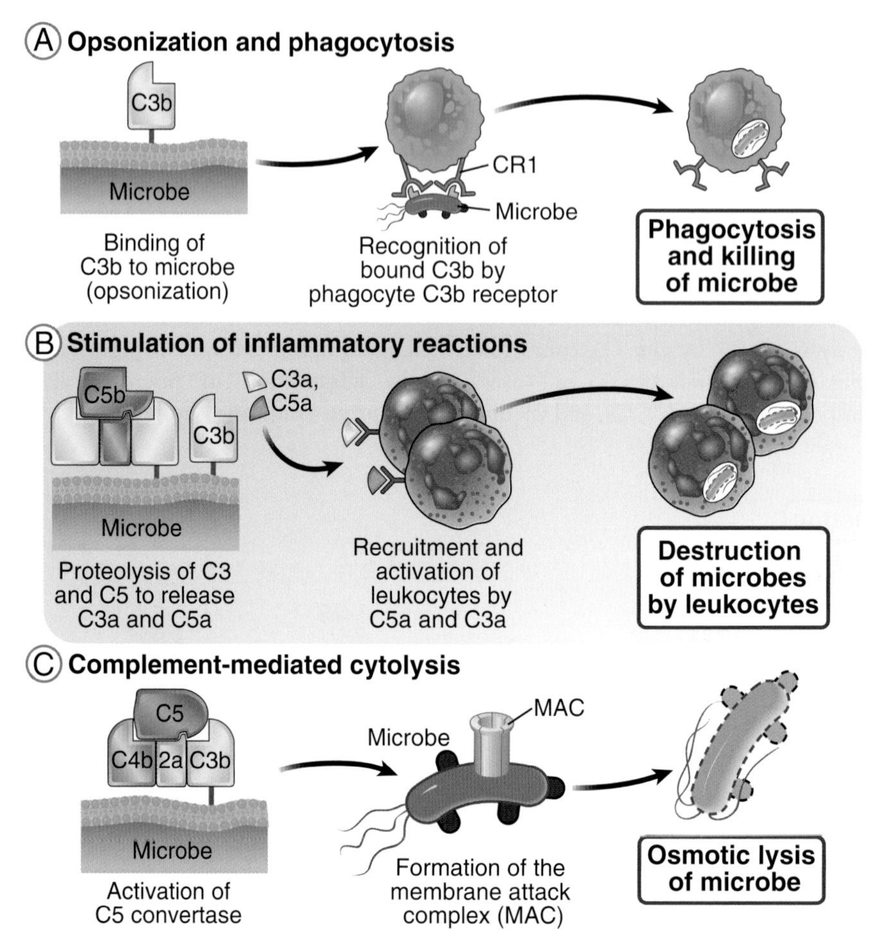

Fig. 8.9 The functions of complement. **A,** C3b opsonizes microbes and is recognized by the type 1 complement receptor (CR1) of phagocytes, resulting in ingestion and intracellular killing of the opsonized microbes. Thus, C3b is an opsonin. CR1 also recognizes C4b, which may serve the same function. Other complement products, such as the inactivated form of C3b (iC3b), also bind to microbes and are recognized by other receptors on phagocytes (e.g., type 3 complement receptor, a member of integrin family of proteins). **B,** Small peptides released during complement activation bind to receptors on neutrophils and other leukocytes and stimulate inflammatory reactions. The peptides that serve this function are mainly C5a and C3a, released by proteolysis of C5 and C3, respectively. **C,** The membrane attack complex creates pores in cell membranes and induces osmotic lysis of the cells.

functions of the complement system are illustrated in Fig. 8.9.

- **Opsonization.** Microbes coated with C3b are phagocytosed by virtue of C3b being recognized by complement receptor type 1 (CR1, or CD35), which is expressed on phagocytes. Thus, C3b functions as an opsonin.
- **Inflammation.** The small peptide fragments C3a and C5a, which are produced by proteolysis of C3 and C5, are chemotactic for neutrophils, stimulate the release of inflammatory mediators from various leukocytes, and stimulate movement of leukocytes and plasma proteins across the endothelium into tissues. In this way, complement fragments induce inflammatory reactions that also serve to eliminate microbes.
- **Cell lysis.** The MAC can induce osmotic lysis of cells, including microbes. MAC-induced lysis is

effective mostly against microbes that have thin cell walls and little or no glycocalyx, such as the *Neisseria* species of bacteria.

In addition to its antimicrobial effector functions, the complement system stimulates B cell responses and antibody production. When C3 is activated by a microbe by the alternative pathway, one of its breakdown products, C3d, is recognized by complement receptor type 2 (CR2) on B lymphocytes. Signals delivered by this receptor enhance B cell responses against the microbe. This process is described in Chapter 7 (see Fig. 7.5A) and is an example of an innate immune response to a microbe (complement activation) enhancing an adaptive immune response to the same microbe (B cell activation and antibody production). Complement proteins bound to antigen-antibody complexes are recognized by follicular dendritic cells in germinal centers, allowing the antigens to be displayed for further B cell activation and selection of high-affinity B cells (see Chapter 7). This complement-dependent antigen display is another way in which the complement system promotes antibody production.

Inherited deficiencies of complement proteins cause immune deficiencies and, in some cases, increased incidence of autoimmune disease. Deficiency of C3 results in increased susceptibility to bacterial infections that may be fatal early in life. Deficiencies of the early proteins of the classical pathway, C2 and C4, may have no clinical consequence, may result in increased susceptibility to infections, or are associated with an increased incidence of systemic lupus erythematosus, an immune complex-mediated autoimmune disease in which patients make antibodies against their own nuclear and other antigens. The increased incidence of lupus may be related to the role of complement in clearance of dead cells and immune complexes containing self antigens, such that deficiencies of the classical pathway result in increased burden of nuclear antigens. Complement deficiencies may also lead to abnormal signaling in B cells and failure of B cell tolerance. Deficiencies of C9 and MAC formation result in increased susceptibility to *Neisseria* infections. Some individuals inherit polymorphisms in the gene encoding MBL, leading to production of a protein that is functionally defective; such defects are associated with increased susceptibility to infections. Inherited deficiency of the alternative pathway protein properdin also causes increased susceptibility to bacterial infection. A monoclonal antibody that blocks the C5 molecule is approved for the treatment of diseases caused by defects in complement regulatory proteins (discussed next) and in certain antibody-mediated autoimmune diseases (e.g., myasthenia gravis).

Regulation of Complement Activation

Mammalian cells express regulatory proteins that inhibit complement activation, thus preventing complement-mediated damage to host cells (Fig. 8.10). Many such regulatory proteins have been described, and defects in these proteins are associated with clinical

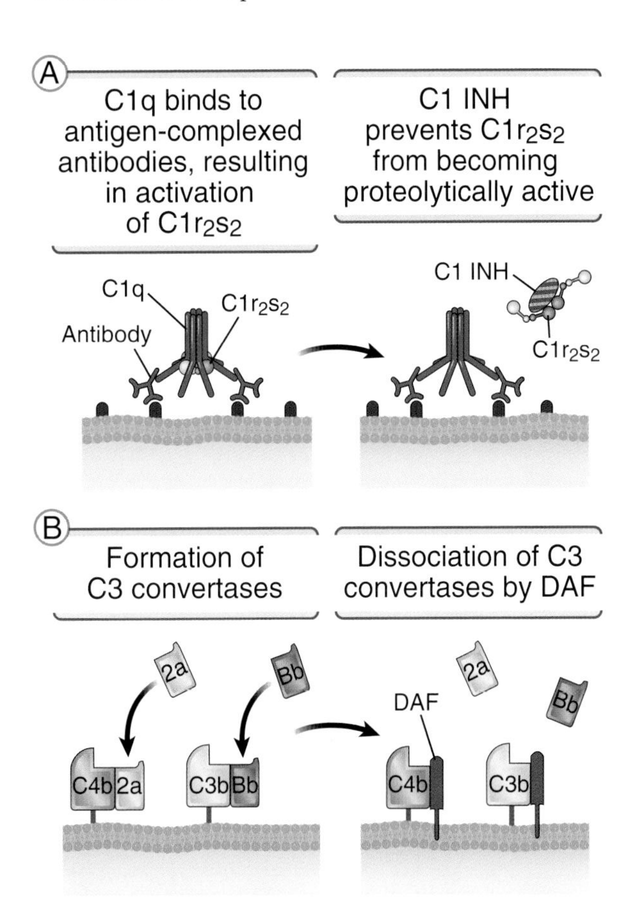

Fig. 8.10 Regulation of complement activation. **A,** C1 inhibitor (C1 INH) prevents the assembly of the C1 complex, which consists of C1q, C1r, and C1s proteins, thereby blocking complement activation by the classical pathway. **B,** The GPI-linked cell surface protein decay-accelerating factor (DAF) and the type 1 complement receptor (CR1) (not shown) interfere with the formation of the C3 convertase by blocking the binding of Bb (in the alternative pathway) or C2a (in the classical pathway).

C

Plasma proteins

Protein	Function
C1 inhibitor (C1 INH)	Inhibits C1r and C1s serine protease activity
Factor I	Proteolytically cleaves C3b and C4b
Factor H	Causes dissociation of alternative pathway C3 convertase subunits Co-factor for Factor I-mediated cleavage of C3b
C4 binding protein (C4BP)	Causes dissociation of classical pathway C3 convertase subunits Co-factor for Factor I-mediated cleavage of C4b

Membrane proteins

Protein	Distribution	Function
Membrane co-factor protein (MCP, CD46)	Leukocytes, epithelial cells, endothelial cells	Co-factor for Factor I-mediated cleavage of C3b and C4b
Decay accelerating factor (DAF)	Blood cells, endothelial cells, epithelial cells	Blocks formation of C3 convertase
CD59	Blood cells, endothelial cells, epithelial cells	Blocks C9 binding and prevents formation of the MAC
Type 1 complement receptor (CR1, CD35)	Mononuclear phagocytes, neutrophils, B and T cells, erythrocytes, eosinophils, FDCs	Causes dissociation of C3 convertase subunits Co-factor for Factor I-mediated cleavage of C3b and C4b

Fig. 8.10, cont'd C, The major regulatory proteins of the complement system and their functions. The function of DAF is shown in part B. Membrane cofactor protein (or CD46) and CR1 serve as cofactors for cleavage of C3b by a plasma enzyme called Factor I, thus destroying any C3b that may be formed (not shown). *FDCs,* Follicular dendritic cells; *MAC,* membrane attack complex.

syndromes caused by uncontrolled complement activation.

- A regulatory protein called C1 inhibitor (C1 INH) stops complement activation early, at the stage of C1 activation. Deficiency of C1 INH is the cause of a disease called **hereditary angioedema.** C1 INH is a serine protease inhibitor that functions as a major physiologic inhibitor of the cleavage of kallikrein, the precursor of the vasoactive molecule bradykinin. Therefore, C1 INH deficiency results not only in increased complement activation but also increased proteolytic activation of bradykinin, and this is the main reason for the vascular changes that lead to leakage of fluid (edema) in many tissues.

- Decay-accelerating factor (DAF) is a glycolipid-linked cell surface protein that disrupts the binding of Bb to C3b and the binding of C4b to C2a, thus blocking C3 convertase formation and terminating complement activation by both the alternative and the classical pathways. A disease called **paroxysmal nocturnal hemoglobinuria** results from the acquired deficiency in hematopoietic stem cells of an enzyme that synthesizes the glycolipid anchor for several cell-surface proteins, including the complement regulatory proteins DAF and CD59, which block MAC formation. In these patients, unregulated complement activation occurs on erythrocytes, leading to their lysis. A monoclonal antibody that blocks C5 is an effective treatment for this disease.
- A plasma enzyme called Factor I cleaves C3b into inactive fragments, with membrane cofactor protein (MCP) and the plasma protein Factor H serving as cofactors in this enzymatic process. Deficiency of the regulatory proteins Factors H and I results in increased complement activation and reduced levels of C3 because of its consumption, causing increased susceptibility to infection. Mutations in Factor H that compromise its binding to cells are associated with a rare genetic disease called atypical hemolytic uremic syndrome, in which there are clotting, vascular, and renal abnormalities. Anti-C5 antibody is also used to treat this disease. Certain genetic variants of Factor H are linked to an eye disease called age-related macular degeneration.

Complement regulatory proteins are made by vertebrate cells but not by microbes. Because microbes lack these regulatory proteins, the complement system can be activated on microbial surfaces much more effectively than on normal host cells. Even in vertebrate cells, the regulation can be overwhelmed by too much complement activation. For instance, host cells can become targets of complement if they are coated with large amounts of antibodies, as in some autoimmune diseases (see Chapter 11).

FUNCTIONS OF ANTIBODIES AT SPECIAL ANATOMIC SITES

The effector mechanisms of humoral immunity described so far may be active at any site in the body to which antibodies gain access. As mentioned previously, antibodies are produced in secondary lymphoid organs and bone marrow and readily enter the blood, from which they may go anywhere. Antibodies also serve vital protective functions at two special anatomic sites: at mucosal barriers and in the fetus.

Mucosal Immunity

Immunoglobulin A (IgA) is produced by plasma cells in mucosal tissues, transported across epithelia, and binds to and neutralizes microbes in the lumens of the mucosa-lined organs (Fig. 8.11). Microbes often are inhaled or ingested, and antibodies that are secreted into the lumens of the respiratory or gastrointestinal tract bind to these microbes and prevent them from colonizing the host. This type of immunity is called mucosal immunity (or secretory immunity). The principal class of antibody produced in mucosal tissues is IgA. In fact, IgA accounts for about two-thirds of the approximately 3 g of antibody produced daily by a healthy adult, reflecting the vast surface area of the intestines. The propensity of B cells in mucosal epithelial tissues to produce IgA is because the cytokines that induce switching to this isotype, including transforming growth factor β (TGF-β), are produced at high levels in mucosal associated lymphoid tissues. In addition, IgA-producing B cells that are generated in regional lymph nodes or spleen tend to home to mucosal tissues in response to chemokines produced in these tissues. Also, some of the IgA is produced by a subset of B cells, called B-1 cells, best studied in rodents, which are abundant in mucosal tissues; these cells undergo class switching to IgA in response to nonprotein antigens, without T cell help.

Intestinal mucosal plasma cells are located in the lamina propria, beneath the epithelial barrier, and IgA is produced in this region. To bind and neutralize microbial pathogens in the lumen before they can invade the body, the IgA must be transported across the epithelium into the lumen. Transport through the epithelium is carried out by a special Fc receptor, the poly-Ig receptor,

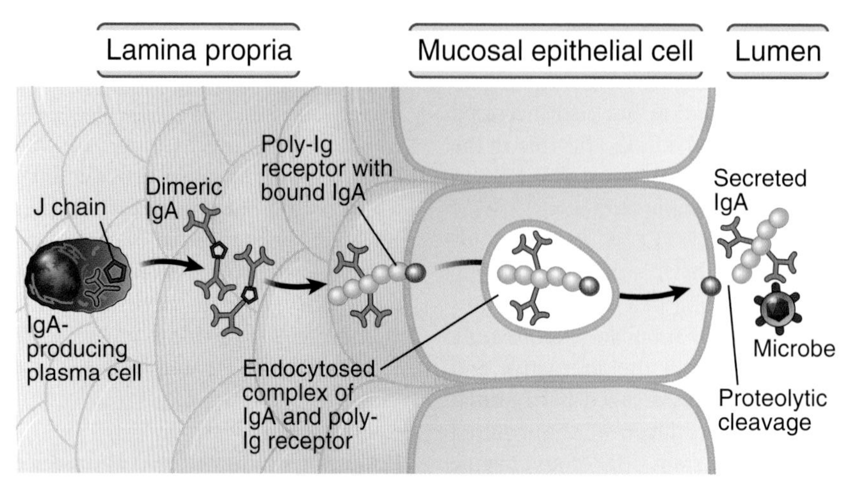

Fig. 8.11 Transport of immunoglobulin A (IgA) through epithelium. In the mucosa of the gastrointestinal and respiratory tracts, IgA is produced by plasma cells in the lamina propria and is actively transported through epithelial cells by an IgA-specific Fc receptor, called the poly-Ig receptor because it recognizes IgM as well. The J chain is required for high-affinity binding of dimeric IgA to the poly-Ig receptor. On the luminal surface, the IgA with a portion of the bound receptor is released. Here the antibody recognizes ingested or inhaled microbes and blocks their entry through the epithelium.

which is expressed on the basal surface of the epithelial cells. This receptor binds IgA, endocytoses it into vesicles, and transports it to the luminal surface. Here the receptor is cleaved by a protease, and the IgA is released into the lumen still carrying a portion of the bound poly-Ig receptor (the secretory component). The attached secretory component protects the antibody from degradation by proteases in the gut. The antibody can then recognize microbes in the lumen and block their binding to and entry through the epithelium.

The gut contains a large number of commensal bacteria that are essential for basic functions such as absorption of food and, therefore, have to be tolerated by the immune system. IgA antibodies are produced mainly against potentially harmful and proinflammatory bacteria, thus blocking their entry through the gut epithelium. Harmless commensals are tolerated by the immune system of the gut by mechanisms that are discussed in Chapter 9 and do not stimulate IgA production.

Neonatal Immunity

Maternal antibodies are transported across the placenta to the fetus and across the gut epithelium of neonates, protecting the newborn from infections. Newborn mammals have incompletely developed immune systems and are unable to mount effective immune responses against many microbes. During their early life, they are protected from infections by antibodies acquired from their mothers (Fig. 8.12). This is an example of naturally occurring passive immunity. Neonates acquire maternal antibodies by two routes. During pregnancy, maternal IgG binds to the FcRn expressed in the placenta and is transported into the fetal circulation. After birth, infants ingest maternal IgA

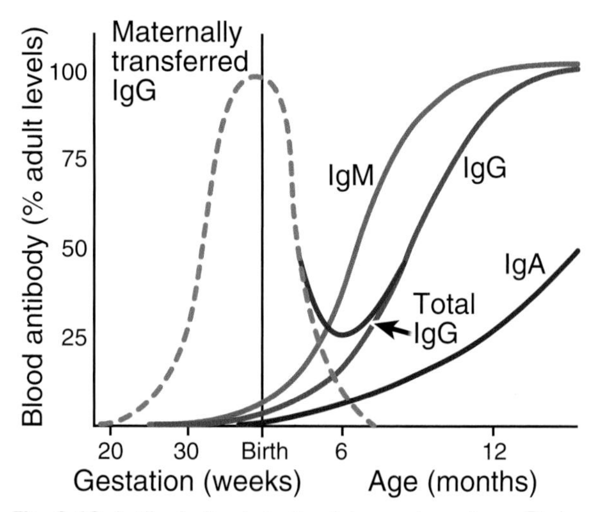

Fig. 8.12 Antibody levels in the fetus and newborn. During fetal and early neonatal life, circulating antibody is acquired from the mother though transfer across the placenta and from breast milk. The newborn starts producing antibody at about 6 months of age. Total IgG refers to IgG transferred from the mother and that produced in the newborn. *Ig,* Immunoglobulin.

antibodies secreted across the mammary gland epithelia via the poly-Ig receptor into their mothers' colostrum and milk. Ingested IgA antibodies provide mucosal immune protection to the neonate. Thus, neonates acquire the antibody profiles of their mothers and are protected from infectious microbes to which the mothers were exposed or vaccinated. Because of the long half-life of circulating IgG, the newborn is protected by the IgG from the mother for about 6 months, at which time the baby begins to make its own IgG. There is often a nadir in the blood IgG concentration at this age, and a concomitant increase in the incidence of infections.

EVASION OF HUMORAL IMMUNITY BY MICROBES

Microbes have evolved numerous mechanisms to evade humoral immunity (Fig. 8.13). Many bacteria and viruses mutate their surface molecules that are needed for entry into host cells so that they can no longer be recognized by antibodies produced in response to the original microbe. Influenza changes its major surface antigens because of mutations and reassortment of its RNA. This happens so often that every year most of the infections are caused by a new strain of the virus. HIV mutates its genome at a high rate, and therefore different strains contain many variant forms of the major antigenic surface glycoprotein of HIV, called gp120. As a result, antibodies against exposed determinants on gp120 in any one HIV subtype may not protect against other virus subtypes that appear in infected individuals. This is one reason why gp120 vaccines are not effective in protecting people from HIV infection. There are so many strains of rhinovirus that vaccines for the common cold are considered impractical. The protective antibody response to SARS-CoV-2, the causative agent of COVID-19, is directed against the spike protein, which the virus uses to gain entry into host cells. Variants of the virus have mutated the spike protein enough to evade vaccine-induced immunity,

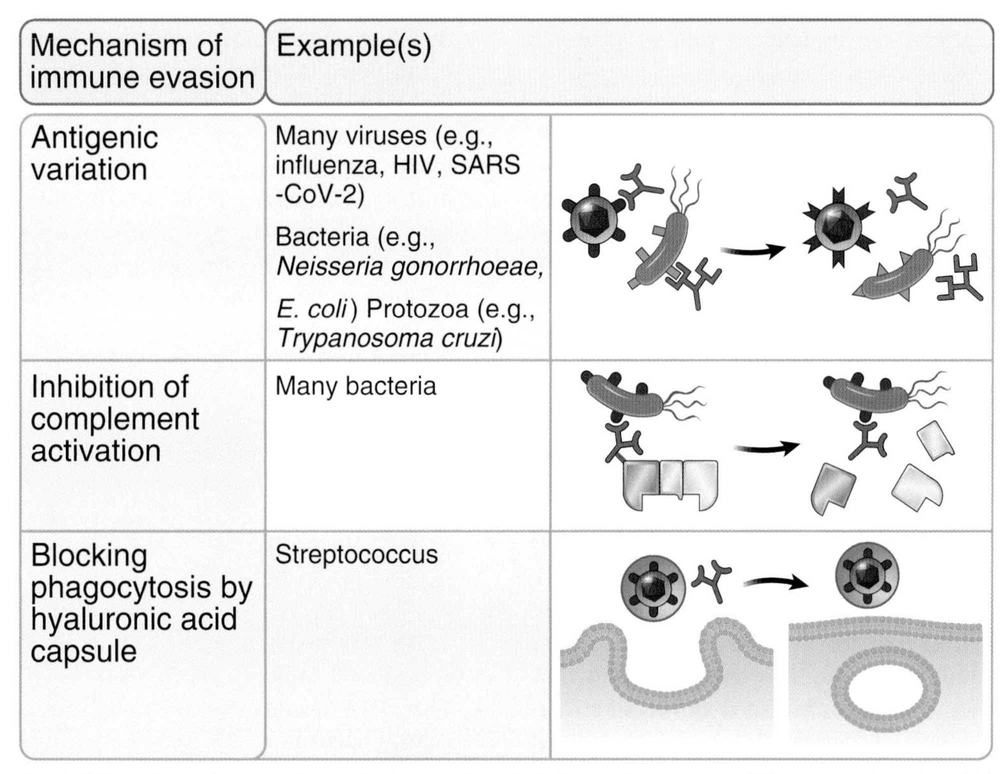

Mechanism of immune evasion	Example(s)
Antigenic variation	Many viruses (e.g., influenza, HIV, SARS-CoV-2) Bacteria (e.g., *Neisseria gonorrhoeae*, *E. coli*) Protozoa (e.g., *Trypanosoma cruzi*)
Inhibition of complement activation	Many bacteria
Blocking phagocytosis by hyaluronic acid capsule	Streptococcus

Fig. 8.13 Evasion of humoral immunity by microbes. This figure shows some of the mechanisms by which microbes evade humoral immunity, with illustrative examples.

making the first COVID-19 vaccines that were developed less effective at preventing infection, though they still offer protection from severe illness and death. Bacteria such as *Escherichia coli* vary the antigens contained in their pili and thus evade antibody-mediated defense. The trypanosome that causes sleeping sickness expresses new surface glycoproteins whenever it encounters antibodies against the original glycoprotein. As a result, infection with this protozoan parasite is characterized by waves of parasitemia, each wave consisting of an antigenically new parasite that is not recognized by antibodies produced against the parasites in the preceding wave.

Other microbes inhibit complement activation, or resist opsonization and phagocytosis by concealing surface antigens under a hyaluronic acid capsule.

VACCINATION

Now that we have discussed the mechanisms of host defense against microbes, including cell-mediated immunity in Chapter 6 and humoral immunity in this chapter, it is important to consider how these protective immune responses can be induced with prophylactic vaccines.

Vaccination is the process of stimulating protective adaptive immune responses against microbes by exposure to nonpathogenic forms or components of the microbes. The development of vaccines against infections has been one of the great successes of immunology. The only human disease to be intentionally eradicated from the earth is smallpox, and this was achieved by a worldwide program of vaccination. Polio may be the second such disease (although sporadic cases continue to appear in many countries). As mentioned in Chapter 1, many other diseases have been largely controlled by vaccination (see Fig. 1.2).

Several types of vaccines are in use and being developed (Fig. 8.14).

- Some of the most effective vaccines, such as those against yellow fever, measles, mumps, and rubella, are composed of **live attenuated viruses,** which have been selected for their lack of pathogenicity while retaining their infectivity and antigenicity. Immunization with these attenuated viruses stimulates the production of neutralizing antibodies against viral antigens that protect vaccinated individuals from subsequent infections. Killed viral vaccines (viruses are chemically inactivated in vitro) against hepatitis A, polio, and rabies are also currently in use.
- Vaccines composed of microbial proteins and polysaccharides, called **subunit vaccines**, work in the same way. Some microbial polysaccharide antigens (which cannot stimulate T cell help) are chemically coupled to proteins so that helper T cells are activated and high-affinity antibodies are produced against the polysaccharides. These are called **conjugate vaccines**, and they are excellent examples of the practical application of our knowledge of helper T cell—B cell interactions (see Fig. 7.9). Immunization with inactivated microbial toxins and with microbial proteins synthesized in the laboratory stimulates antibodies that bind to and neutralize the native toxins and the microbes, respectively. Subunit vaccines are also produced using recombinant DNA technology, such as widely used hepatitis B virus and human papilloma virus vaccines. Purified protein vaccines have to be administered with adjuvants in order to stimulate effective immune responses.
- The idea of injecting **nucleic acids** (DNA or mRNA) encoding microbial proteins has been tested for many years. Advantages of such vaccines are that the microbial proteins may be produced inside host cells and secreted and can thus elicit both humoral and cell-mediated immune responses, and nucleic acids engage Toll-like receptors and thus have an intrinsic adjuvant-like activity. Plasmid DNA vaccines have still not been successful. Initial attempts to use mRNA as vaccines were hampered by two main problems—extracellular RNA is unstable, and when RNA enters the cytosol of cells, it is recognized by innate immune receptors and elicits a type I interferon response that can cause harmful inflammation. Structural modifications in mRNA and packaging in lipid nanoparticles have alleviated both problems. The remarkable success of mRNA vaccines for SARS-CoV-2 at the end of 2020 led to reduced hospitalization and mortality and has been a major contributor to limiting the worldwide COVID-19 pandemic and its devastating health and societal consequences.
- The DNA encoding a microbial antigen can be incorporated into the genome of a replication-defective viral vector that is harmless for humans but enters host cells, where the microbial proteins

Type of vaccine	Examples	Form of protection
Live attenuated, or killed, bacteria	Pertussis, BCG, cholera	Antibody response
Live attenuated viruses	Measles, mumps, rubella, rabies, influenza A	T cell and antibody responses
Killed viruses	Hepatitis A, polio, rabies	Antibody response
Recombinant protein subunit vaccines	Human papilloma virus, hepatitis B virus	Antibody responses
Modified protein	Tetanus toxoid, diphtheria toxoid	Antibody response
Conjugate vaccines	*Haemophilus influenzae, Streptococcus pneumoniae* (pneumococcus)	Helper T cell-dependent antibody response to polysaccharide antigens
mRNA vaccines	SARS-CoV-2	T cell and antibody responses
Hybrid viral vaccines	SARS-CoV-2, Ebola	T cell and antibody responses
DNA vaccines	Clinical trials ongoing for several infections	T cell and antibody responses

Fig. 8.14 Vaccination strategies. A summary of different types of vaccines in use or tried, as well as the nature of the protective immune responses induced by these vaccines. *BCG,* Bacille Calmette-Guérin; *HIV,* human immunodeficiency virus.

are produced. These **hybrid viral vaccines** are now in use for SARS-CoV-2 and have been approved for Ebola virus.

SUMMARY

- Humoral immunity is the type of adaptive immunity that is mediated by antibodies. Antibodies prevent infections by blocking the ability of microbes to infect host cells, and they eliminate microbes by activating several effector mechanisms.
- In antibody molecules, the antigen-binding (Fab) regions are spatially separate from the effector (Fc) regions. The ability of antibodies to neutralize microbes and toxins is entirely a function of the antigen-binding regions. Even Fc-dependent effector functions are activated only after antibodies bind antigens.
- IgG antibodies remain in the circulation and tissues longer than most other proteins because FcRn expressed in several cell types protects endocytosed IgG from lysosomal degradation and shuttles it back into the blood or tissue fluids.
- Antibodies are produced in lymphoid tissues and bone marrow, from which they enter the circulation and are able to reach any site of infection. Heavy-chain class switching and affinity maturation enhance the protective functions of antibodies.

- Antibodies neutralize the infectivity of microbes and the pathogenicity of microbial toxins by binding to and interfering with the ability of these microbes and toxins to attach to host cells.
- Antibodies coat (opsonize) microbes and promote their phagocytosis by binding to Fc receptors on phagocytes. The binding of antibody Fc regions to Fc receptors also stimulates the microbicidal activities of phagocytes.
- The complement system is a collection of circulating and cell surface proteins that play important roles in host defense. The complement system may be activated on microbial surfaces without antibodies (alternative and lectin pathways, which are mechanisms of innate immunity) and after the binding of antibodies to antigens (classical pathway, a mechanism of adaptive humoral immunity).
- Complement proteins are sequentially cleaved, and active components, in particular C4b and C3b, become covalently attached to the surfaces on which complement is activated. The late steps of complement activation lead to the formation of the cytolytic MAC.
- Different products of complement activation promote phagocytosis of microbes, induce cell lysis, and stimulate inflammation. Mammals express cell surface and circulating regulatory proteins that prevent inappropriate complement activation on host cells.
- IgA antibody is produced in the lamina propria of mucosal organs and is actively transported by a special Fc receptor across the epithelium into the lumen, where it blocks the ability of microbes to invade the epithelium.
- Neonates acquire IgG antibodies from their mothers through the placenta, using the FcRn to capture and transport the maternal antibodies. Infants also acquire IgA antibodies from the mother's colostrum and milk by ingestion.
- Microbes have developed strategies to resist or evade humoral immunity, such as varying their antigens and becoming resistant to complement and phagocytosis.
- Most vaccines in current use work by stimulating the production of neutralizing antibodies.

REVIEW QUESTIONS

1. What regions of antibody molecules are involved in the functions of antibodies?
2. How do heavy-chain class (isotype) switching and affinity maturation improve the ability of antibodies to combat infectious pathogens?
3. In what situations does the ability of antibodies to neutralize microbes protect the host from infections?
4. How do antibodies assist in the elimination of microbes by phagocytes?
5. How is the complement system activated?
6. Why is the complement system effective against microbes but does not react against host cells and tissues?
7. What are the functions of the complement system, and what components of complement mediate these functions?
8. How do antibodies prevent infections by ingested and inhaled microbes?
9. How are neonates protected from infection before their immune system has reached maturity?

Answers to and discussion of the Review Questions may be found on p. 322.

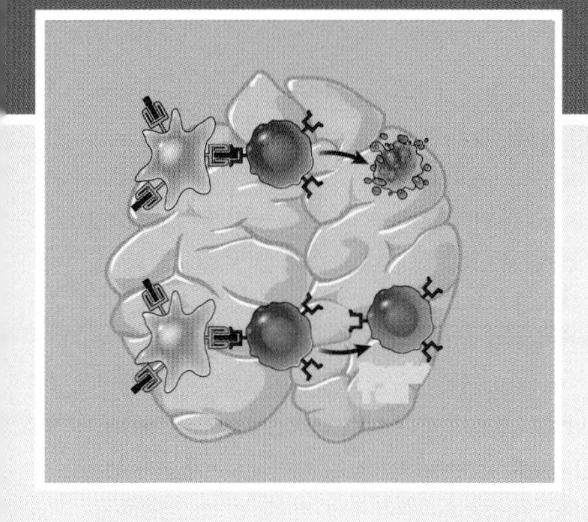

9

Immunologic Tolerance and Autoimmunity

Self–Nonself Discrimination in the Immune System and Its Failure

CHAPTER OUTLINE

One of the remarkable properties of the normal immune system is that it can react to an enormous variety of microbes but does not react against the individual's own (self) antigens. This unresponsiveness to self antigens, also called **immunologic tolerance**, is maintained despite the fact that the molecular mechanisms by which lymphocyte receptor specificities are generated cannot exclude receptors specific for self antigens. In other words, lymphocytes with the ability to recognize self antigens are constantly being produced during the normal process of lymphocyte development. Furthermore, many self antigens have ready access to the immune system, so unresponsiveness to these antigens cannot be maintained simply by concealing them from lymphocytes. The process by which antigen-presenting cells (APCs) display antigens to T cells does not distinguish between foreign and self proteins, so self antigens are normally presented by APCs and seen by T lymphocytes. It follows that there must exist mechanisms that prevent immune responses to self antigens. These mechanisms are responsible for one of the cardinal features of the adaptive immune system—namely, its ability to discriminate between self and nonself (usually microbial) antigens. If these mechanisms fail, the immune system may attack the individual's own cells and tissues. Such reactions are called **autoimmunity**, and the diseases they cause are called autoimmune diseases. In addition to tolerating the presence of self antigens, the immune system has to coexist with many commensal microbes that live immediately outside the epithelial barriers of their human hosts, often in a state of symbiosis, and it must tolerate innumerable intrinsically harmless nonmicrobial environmental antigens, including those ingested and inhaled. Furthermore, the immune system of a pregnant female has to accept the presence of a fetus that expresses antigens derived from the father, which are foreign to the mother. Unresponsiveness to commensal microbes, the fetus, and harmless environmental antigens is maintained by many of the same mechanisms involved in unresponsiveness to self.

In this chapter, we address the following questions:

- How does the immune system maintain unresponsiveness to self antigens?
- What are the factors that may contribute to the loss of self-tolerance and the development of autoimmunity?
- How does the immune system maintain unresponsiveness to commensal microbes and the fetus?
- How might self-tolerance fail, resulting in autoimmunity?

This chapter begins with a discussion of the important principles and features of self-tolerance.

IMMUNOLOGIC TOLERANCE: GENERAL PRINICIPLES AND SIGNIFICANCE

Immunologic tolerance is a lack of response to antigens that is induced by exposure of lymphocytes to these antigens. When lymphocytes with receptors for a particular antigen encounter this antigen, two outcomes are possible. The lymphocytes may be activated to proliferate and to differentiate into effector and memory cells, leading to a productive immune response; antigens that elicit such a response are said to be **immunogenic**. Alternatively, the lymphocytes may be functionally inactivated or killed, resulting in tolerance; antigens that induce tolerance are said to be **tolerogenic**. Normally, microbes are immunogenic and self antigens are tolerogenic.

The choice between lymphocyte activation and tolerance is determined largely by the nature of the antigen and the additional signals present when the antigen is displayed to the immune system. In fact, the same antigen may be administered in different ways to induce an immune response or tolerance. This experimental observation has been exploited to analyze which factors determine whether activation or tolerance develops as a consequence of encounter with an antigen.

The phenomenon of immunologic tolerance is important for several reasons. First, as we stated at the outset, self antigens normally induce tolerance, and failure of self-tolerance is the underlying cause of autoimmune diseases. Second, if we learn how to induce tolerance in lymphocytes specific for a particular antigen, we may be able to use this knowledge to prevent or control unwanted immune reactions. Strategies for inducing tolerance are being tested to treat allergic and autoimmune diseases and to prevent the rejection of organ transplants. The same strategies may be valuable in gene therapy to prevent immune responses against the products of newly expressed genes or vectors and even for stem cell transplantation if the stem cell donor is genetically different from the recipient.

Immunologic tolerance to different self antigens may be induced when developing lymphocytes encounter these antigens in the generative (central) lymphoid organs, a process called central tolerance, or when mature lymphocytes encounter self antigens in secondary (peripheral) lymphoid organs or peripheral tissues, called peripheral tolerance (Fig. 9.1). Central tolerance is a mechanism of tolerance only to self antigens that are present in the generative lymphoid organs—namely, the bone marrow and thymus—and it eliminates potentially responding lymphocytes before they have completed their maturation. Tolerance to self antigens that are not present in these organs must be induced and maintained by peripheral mechanisms, which inactivate or eliminate lymphocytes that have matured and entered peripheral tissues.

With this brief background, we proceed to a discussion of the mechanisms of immunologic tolerance and how the failure of each mechanism may result in autoimmunity. We discuss the mechanisms of central and peripheral tolerance in T and B cells separately because there are significant differences between these processes.

CENTRAL T LYMPHOCYTE TOLERANCE

A principal mechanism of central tolerance in T cells is death of immature T cells that recognize self antigens in the thymus (Fig. 9.2). T lymphocytes that

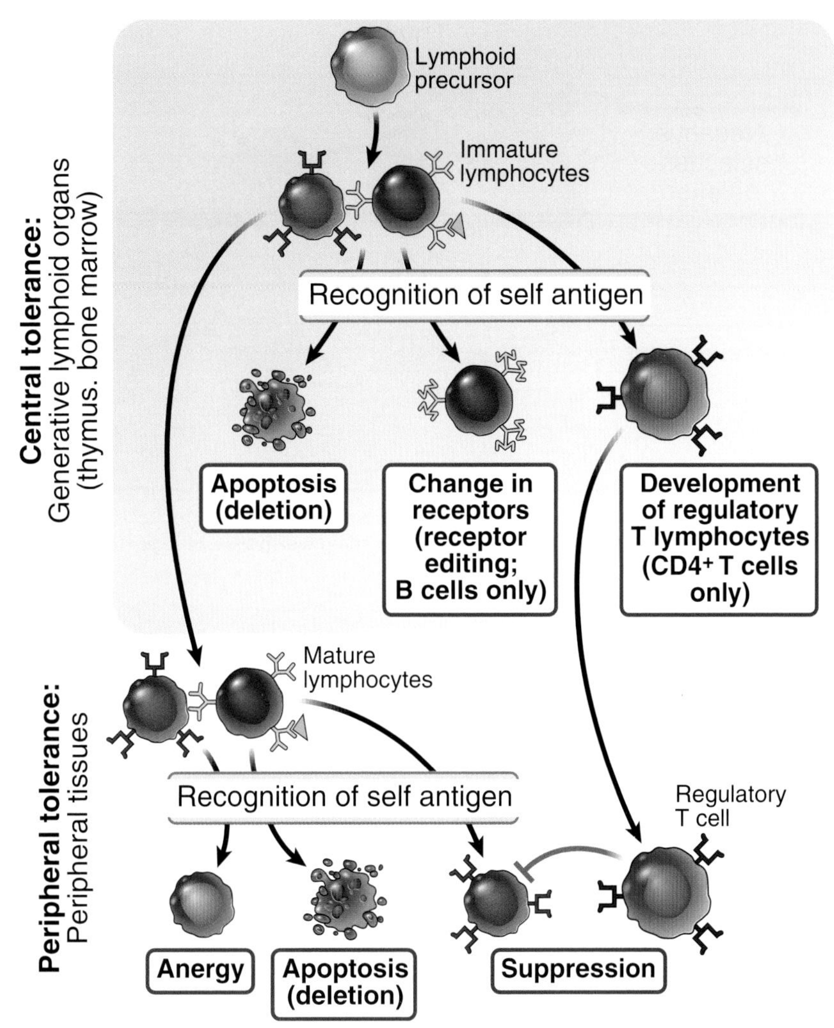

Fig. 9.1 Central and peripheral tolerance to self antigens. Central tolerance: Immature lymphocytes specific for self antigens that encounter these antigens in the generative (central) lymphoid organs may be deleted; B lymphocytes may change their specificity (receptor editing); and some T lymphocytes develop into regulatory T cells (Tregs). Some self-reactive lymphocytes may complete their maturation and enter peripheral tissues. Peripheral tolerance: Mature lymphocytes that recognize self antigens may be suppressed by Tregs, inactivated, or deleted.

develop in the thymus include cells with receptors capable of recognizing many antigens, both self and foreign. If a lymphocyte that has not completed its maturation interacts strongly with a self antigen, displayed as a peptide bound to a self major histocompatibility complex (MHC) molecule, that lymphocyte receives signals that trigger apoptosis. Thus, the self-reactive cell dies before it can mature and become functionally competent. This process, called **negative selection** (see Chapter 4), is a major mechanism of central tolerance. The process of negative selection affects self-reactive $CD4^+$ T cells and $CD8^+$ T cells, which recognize self peptides displayed by class II MHC and class I MHC molecules, respectively. Why immature lymphocytes die upon receiving strong T cell receptor (TCR) signals in the thymus, whereas mature lymphocytes that get strong TCR signals in the periphery are activated, is not fully understood.

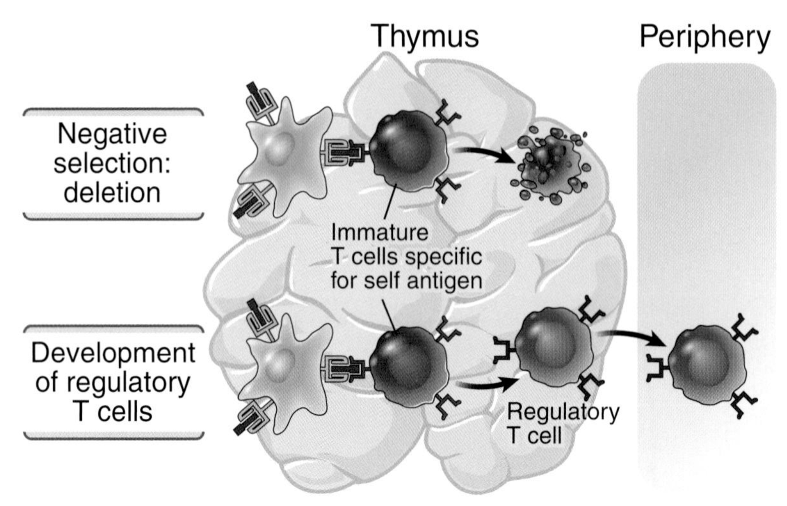

Fig. 9.2 Central T cell tolerance. Strong recognition of self antigens by immature T cells in the thymus may lead to death of the cells (negative selection, or deletion) or the development of regulatory T cells that enter peripheral tissues.

Immature lymphocytes may interact strongly with an antigen if the antigen is present at high concentrations in the thymus and if the lymphocytes express receptors that recognize the antigen with high affinity. Antigens that induce negative selection may include proteins that are abundant throughout the body, such as plasma proteins and common cellular proteins.

Surprisingly, many self proteins that are normally present only in certain peripheral tissues, called tissue-restricted antigens, are also expressed in some specialized cells in the thymus called medullary thymic epithelial cells (MTECs). MTECs transcribe and express genes encoding antigens that are otherwise only expressed by the cells in one or another peripheral tissue type (e.g., lung, liver, muscle, intestine). A protein called **AIRE** (autoimmune regulator) promotes expression of antigens characteristic of the various tissue cell types. Mutations in the *AIRE* gene are the cause of a rare disorder called autoimmune polyglandular syndrome. In this disorder, several tissue antigens are not expressed in the MTECs because of a lack of functional AIRE protein, but the antigens are expressed normally in peripheral tissues. The immature T cells specific for these antigens are not eliminated and do not develop into regulatory cells, but instead mature into functionally competent T cells that enter the periphery, where they encounter the antigens, attack the tissues, and cause disease. Although endocrine organs are the most frequent targets of this autoimmune attack, other tissues such as the skin are also affected. In addition, autoreactive helper T cells in these patients promote the production of autoantibodies by B cells that may recognize secreted proteins such as cytokines. This rare syndrome illustrates the importance of negative selection in the thymus for maintaining self-tolerance, but it is not known if defects in negative selection contribute to common autoimmune diseases.

Central tolerance by deletion is imperfect, and some self-reactive lymphocytes mature and are present in healthy individuals. As discussed next, peripheral mechanisms prevent the activation of these lymphocytes. Some of the immature $CD4^+$ T cells that recognize self antigens in the thymus do not die but rather develop into regulatory T cells (Tregs), enter secondary lymphoid organs and peripheral tissues, and inhibit the activation of mature self-reactive T cells in these tissues. The functions of Tregs are described later. What determines whether a thymic $CD4^+$ T cell that recognizes a self antigen will die or become a Treg is not established.

PERIPHERAL T LYMPHOCYTE TOLERANCE

The principal mechanism of peripheral tolerance is suppression by Tregs. Peripheral tolerance is important for preventing autoimmunity in situations in which

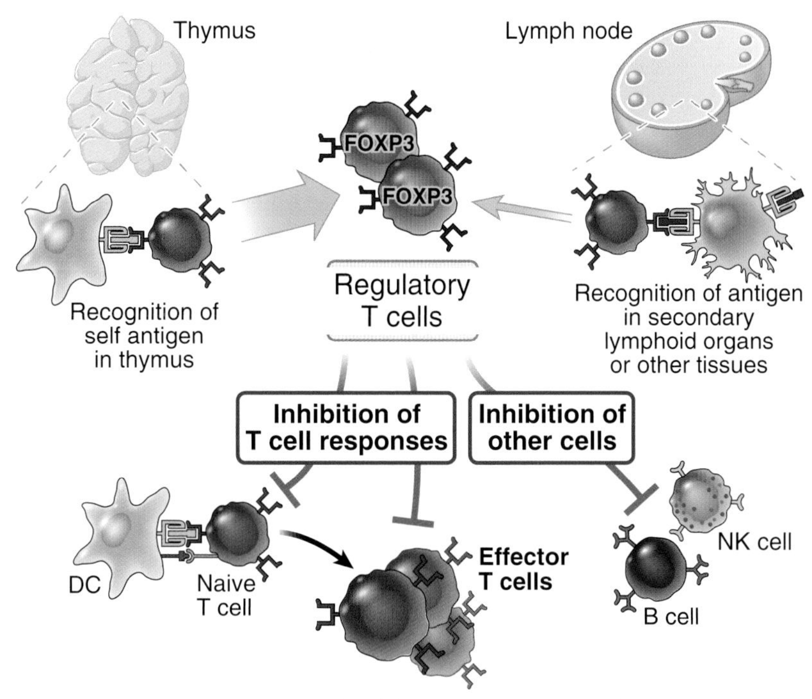

Fig. 9.3 Development and function of Tregs. CD4$^+$ T cells that recognize self antigens may differentiate into regulatory cells in the thymus or peripheral tissues, in a process that is dependent on the transcription factor FOXP3. (The *larger arrow* from the thymus, compared with the one from peripheral tissues, indicates that most of these cells probably arise in the thymus.) These regulatory cells inhibit the activation of naive T cells and their differentiation into effector T cells by contact-dependent mechanisms or by secreting cytokines that inhibit T cell responses. The generation and maintenance of Tregs also require interleukin-2 (not shown). *DC*, Dendritic cell; *NK*, natural killer.

deletion of T cells specific for antigens that are expressed in the thymus is incomplete, as well as for antigens that are not present in the thymus.

Role of Regulatory T Cells in Peripheral Tolerance

Properties and Development of Regulatory T Cells

Tregs are a unique population of CD4$^+$ T cells whose function is to inhibit the activation of other lymphocytes, primarily other T cells (Fig. 9.3). The majority of self-reactive Tregs develop in the thymus, but many also develop in peripheral tissues, especially in the intestinal tract and the placenta. Most Tregs are CD4$^+$ and express high levels of CD25, the α chain of the interleukin-2 (IL-2) receptor, as well as the inhibitory receptor cytotoxic T-lymphocyte antigen 4 (CTLA-4). They also express a transcription factor called FOXP3, which is required for the development and function of these cells. Mutations of the gene encoding FOXP3 in humans or in mice cause a systemic, multiorgan autoimmune disease, demonstrating the importance of FOXP3$^+$ Tregs for the maintenance of self-tolerance. The human disease is known by the acronym IPEX, for immune dysregulation, polyendocrinopathy, enteropathy, and X-linked syndrome.

The survival and function of Tregs are dependent on the cytokine IL-2. This role of IL-2 accounts for the autoimmune disease that develops in mice in which IL-2 or IL-2 receptor genes are deleted and in humans with homozygous mutations in the α or β chain of the IL-2 receptor. Recall that we introduced IL-2 in Chapter 5 as a cytokine made by antigen-activated T cells that stimulates proliferation of these cells. Mature Tregs do not produce IL-2 themselves and rely on IL-2 made by other T cells (Fig. 9.4). Thus, IL-2 is an example of a cytokine that serves two opposite roles: it promotes immune responses by stimulating T cell proliferation, and it inhibits immune responses by maintaining

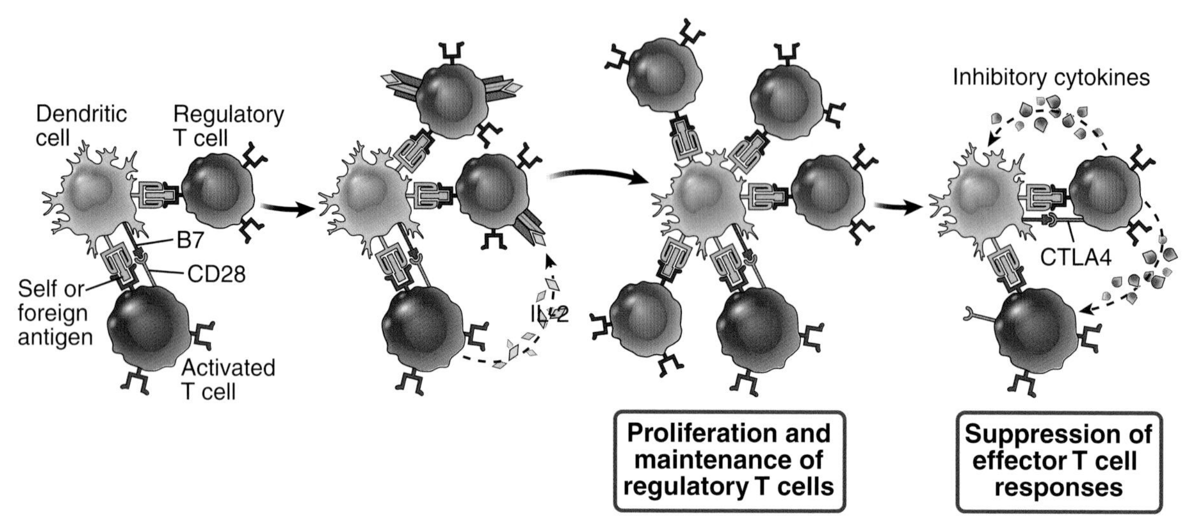

Fig. 9.4 Role of IL-2 in Treg function. If Tregs and conventional (responding, or activated) T cells recognize an antigen on an APC, the activated T cells produce IL-2. IL-2 increases the proliferation and functions of the Tregs. CTLA-4 expressed by Tregs blocks or removes B7 on the APC, and Tregs produce cytokines that inhibit the functions of APCs and responses of T cells. Thus, the activation of conventional T cells sets up a negative feedback loop in which Tregs terminate the response. Tregs also express high levels of IL-2 receptors and outcompete responding cells for this essential growth factor (not shown).

functional Tregs. Numerous clinical trials are testing the ability of IL-2 to promote regulation and control harmful immune reactions, such as inflammation in autoimmune diseases and graft rejection.

The cytokine transforming growth factor β (TGF-β) also plays a role in the generation of Tregs, in part by stimulating expression of the FOXP3 transcription factor. Many cell types can produce TGF-β, but the source of TGF-β for inducing Tregs in the thymus or peripheral tissues is not defined.

Mechanisms of Action of Regulatory T Cells

Tregs suppress immune responses by several mechanisms, including the following:

- Tregs express the inhibitory receptor (coinhibitor) CTLA-4, which blocks B7 costimulators and removes them from APCs and thus prevents T cell activation (see Fig. 5.14). As noted in previous chapters, naive T lymphocytes need at least two signals to induce their proliferation and differentiation into effector and memory cells: Signal 1 is always antigen, and signal 2 is provided by costimulators that are expressed on APCs, typically as part of the innate immune response to microbes (or to damaged host cells) (see Fig. 5.6). We described the costimulatory receptor CD28 and the inhibitory receptor CTLA-4 in Chapter 5. To summarize the key point relevant to Tregs, CD28 is the principal activating receptor for B7 costimulators and CTLA-4 is a high-affinity receptor that binds and removes B7 from the surface of APCs. As a result, when CTLA-4 is expressed, B7 is reduced on the surface of APCs, CD28 cannot be engaged effectively, responding T cells do not receive adequate signal 2, and they cannot respond well to antigens. Long before the discovery of Tregs, it was demonstrated that antigen recognition without costimulation leads to unresponsiveness in T cells, a phenomenon called **anergy.** It is likely that Tregs promote T cell anergy by reducing costimulation by APCs. CTLA-4 also may be expressed by activated T cells other than Tregs and by competing with CD28 it may terminate the responses of those activated cells.

  The essential role of CTLA-4 in maintaining self-tolerance is demonstrated by the findings that rare inherited mutations in *CTLA4* cause systemic autoimmune disease. Patients who are treated with antibodies that block CTLA-4 to enhance immune responses against tumors, a strategy called checkpoint blockade (see Chapter 10), also often develop autoimmunity.
- Some Tregs produce cytokines (e.g., IL-10, TGF-β) that inhibit the activation of lymphocytes, dendritic

cells, and macrophages. IL-10 may be especially important for controlling immune responses to self antigens and commensal microbes in intestinal tissues, as evidenced by the severe colitis that develops in infants who inherit mutations of the IL-10 receptor. IL-10 functions mainly by suppressing expression of B7 and production of cytokines by dendritic cells and macrophages. TGF-β inhibits responses of many immune cells, including lymphocytes and myeloid cells. Both IL-10 and TGF-β are produced by numerous cell types in addition to Tregs.

- Tregs, by virtue of the high level of expression of the IL-2 receptor, may bind and consume this essential T cell growth factor, thus reducing its availability for responding T cells.

The great interest in Tregs has been driven, to a large extent, by the hypothesis that the underlying abnormality in some autoimmune diseases in humans is defective Treg function or the resistance of pathogenic T cells to regulation by Tregs. There is also growing interest in cellular therapy with Tregs to treat graft-versus-host disease, graft rejection, and autoimmune disorders.

Anergy and Exhaustion

Anergy was discovered as T cell unresponsiveness resulting from antigen recognition without costimulation. Tregs may induce anergy by blocking costimulation, but the same type of unresponsiveness may be seen in other situations that are not dependent on Tregs. Several mechanisms may contribute to the development of anergy.

- Dendritic cells and other APCs in normal uninfected tissues and secondary lymphoid organs are normally in a resting (or immature) state, in which they express little or no costimulators (see Chapter 5). The low level of B7 expression is likely an intrinsic property of immature dendritic cells (DCs) and is reinforced by the ability of Tregs to block and remove B7 from the surface of APCs. These dendritic cells constantly process and display the self antigens that are present in these tissues. T lymphocytes with receptors for these self antigens are able to recognize the antigens and thus receive signals from their antigen receptors (signal 1), but the T cells do not receive strong costimulation. In infections and in response to vaccination with adjuvants, APCs are activated and they increase the expression of costimulators, leading to effective immune responses. Thus, the presence or absence of costimulation is a major factor determining whether T cells are activated or tolerized.
- Several cell-intrinsic signals promote T cell unresponsiveness. The inhibitory receptor PD-1 is expressed on T cells in response to antigen recognition. When PD-1 engages its ligands, which are expressed on APCs and other cell types, its cytoplasmic tail binds and activates a tyrosine phosphatase that inhibits tyrosine kinase–dependent signals from the TCR complex and CD28 (see Chapter 5). Thus, PD-1 limits activation by persistent stimuli, such as some microbes, tumors, and self antigens. This phenomenon has been called **exhaustion** because the T cells make an effective response that is terminated (see Chapter 6). Blockade of PD-1 for cancer immunotherapy, in order to prevent exhaustion of tumor-specific T cells, often induces autoimmune reactions, supporting a role for PD-1 in self-tolerance.
- The activation of ubiquitin ligases such as CBL-B in T cells responding to antigens limits and terminates activation. CBL-B targets TCR- and CD28-associated signaling molecules for degradation, and thus inhibits T cell activation. Genome-wide association studies have revealed variants of *CBLB* in patients with the autoimmune diseases multiple sclerosis and type I diabetes.

Deletion: Apoptosis of Mature Lymphocytes

Recognition of self antigens may trigger pathways of apoptosis that result in elimination (deletion) of the self-reactive lymphocytes (Fig. 9.5). There are two likely mechanisms of death of mature T lymphocytes that recognize self antigens:

- Antigen recognition induces the production in T cells of proapoptotic proteins of the BCL-2 family that cause mitochondrial molecules, such as cytochrome c, to leak out and activate cytosolic enzymes called caspases that induce apoptosis. In normal immune responses, the activity of these proapoptotic molecules is counteracted by other proteins of the BCL-2 family that have antiapoptotic functions and are induced by costimulation and by growth factors produced during the responses. However, self antigens, which are recognized without strong costimulation, may not stimulate production of antiapoptotic proteins, and the relative deficiency of survival signals induces death of the cells that recognize these antigens.

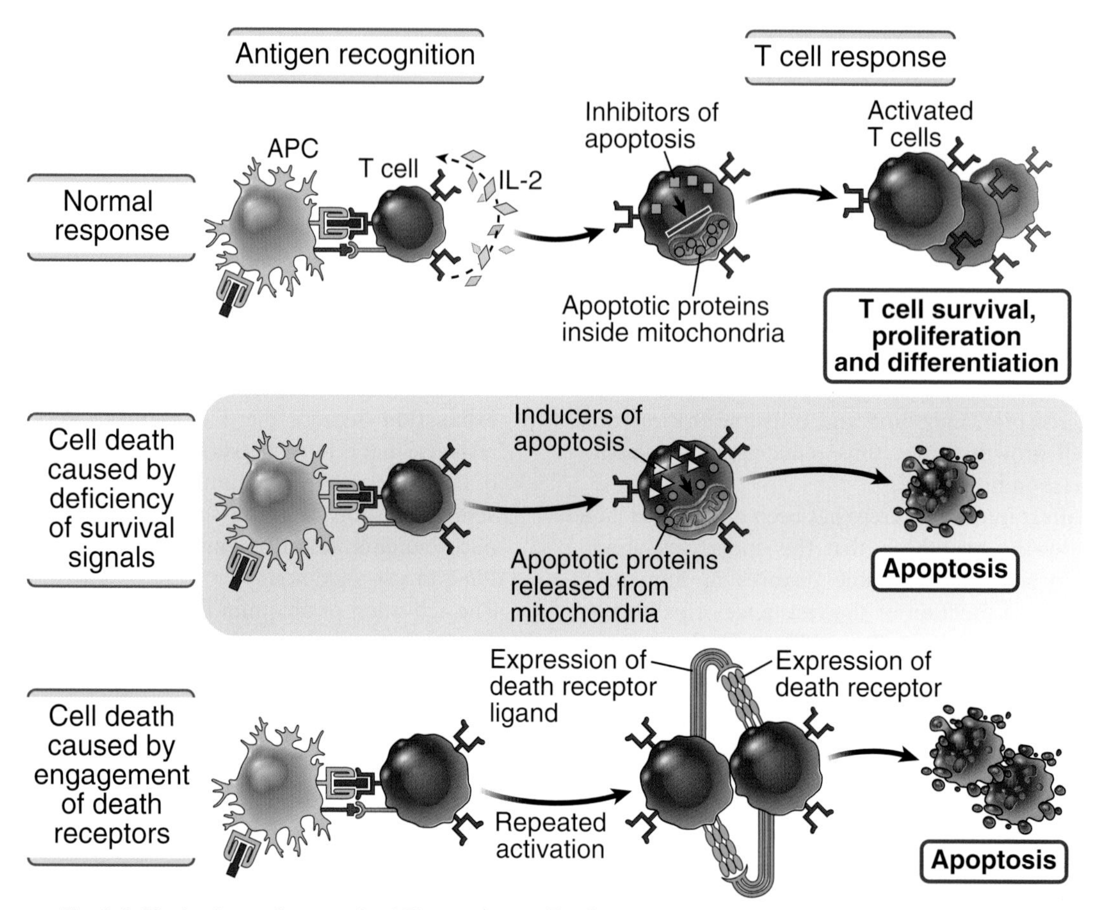

Fig. 9.5 Mechanisms of apoptosis of T lymphocytes. T cells respond to antigen presented by normal antigen-presenting cells *(APCs)* by secreting interleukin-2 (IL-2), expressing antiapoptotic (prosurvival) proteins, and undergoing proliferation and differentiation. The antiapoptotic proteins prevent the release of mediators of apoptosis from mitochondria. Self antigen recognition by T cells without costimulation may lead to relative deficiency of intracellular antiapoptotic proteins, and the excess of proapoptotic proteins causes cell death by inducing release of mediators of apoptosis from mitochondria (death by the mitochondrial [intrinsic] pathway of apoptosis). Alternatively, self antigen recognition may lead to expression of death receptors and their ligands, such as Fas and Fas ligand (FasL), on lymphocytes, and engagement of the death receptor leads to apoptosis of the cells by the death receptor (extrinsic) pathway.

- Recognition of self antigens may lead to the coexpression of death receptors and their ligands. This ligand-receptor interaction generates signals through the death receptor that culminate in the activation of caspases and apoptosis. The best-defined death receptor—ligand pair involved in self-tolerance is a protein called FAS (CD95), which is expressed on many cell types, and FAS ligand (FAS-L), which is expressed mainly on activated T cells. In the absence of FAS or FAS-L, T cells activated by antigens fail to be culled and this dysfunction may drive autoimmunity.

Evidence from genetic studies supports the role of apoptosis in self-tolerance. Eliminating the mitochondrial pathway of apoptosis in mice results in a failure of deletion of self-reactive T cells in the thymus and also in peripheral tissues, leading to autoimmunity. Mice with mutations in the *fas* and *fasl* genes and children with mutations in *FAS* all develop autoimmune diseases with lymphocyte accumulation. Children with mutations in the genes encoding caspase-8 or -10, which are downstream of FAS signaling, also have similar autoimmune diseases. These human diseases, collectively called the **autoimmune lymphoproliferative syndrome**

(ALPS), are rare and are the only known examples of an autoimmune disorder caused by defects in apoptosis. Mutations in *FAS* also have been shown to allow the accumulation of B cells in germinal centers, and this may contribute to a break in peripheral B cell tolerance (discussed later).

From this discussion of the mechanisms of T cell tolerance, it should be clear that self antigens differ from foreign microbial antigens in several ways, which contribute to the choice between tolerance induced by the former and activation by the latter.

- Self antigens are present in the thymus, where they induce deletion of immature cells that recognize self antigens and generate Tregs; by contrast, most microbial antigens tend to be excluded from the thymus because they are typically captured from their sites of entry and transported into secondary lymphoid organs (see Chapter 3).
- Tregs specific for self antigens reduce the expression of costimulators on APCs displaying these antigens and thus prevent the activation of self-reactive T lymphocytes. By contrast, microbes elicit innate immune reactions, leading to the increased expression of costimulators and cytokines that promote T cell proliferation and differentiation into effector cells.
- Self antigens are present throughout life and may therefore cause prolonged or repeated TCR engagement, which may promote the development of Tregs and other mechanisms of self-tolerance (e.g., PD-1 engagement, FAS-mediated apoptosis).

B LYMPHOCYTE TOLERANCE

Tolerance of B cells that recognize multivalent self polysaccharides, lipids, nucleic acids and membrane proteins can be induced in a T-independent manner. Self proteins may not elicit autoantibody responses because of tolerance in both helper T cells and B cells. It is suspected that diseases associated with autoantibody production, such as systemic lupus erythematosus (SLE), are caused by defective tolerance in both B lymphocytes and helper T cells.

Central B Cell Tolerance

When immature B lymphocytes interact strongly with self antigens in the bone marrow, the B cells either change their receptor specificity (receptor editing) or are killed (deletion) (Fig. 9.6).

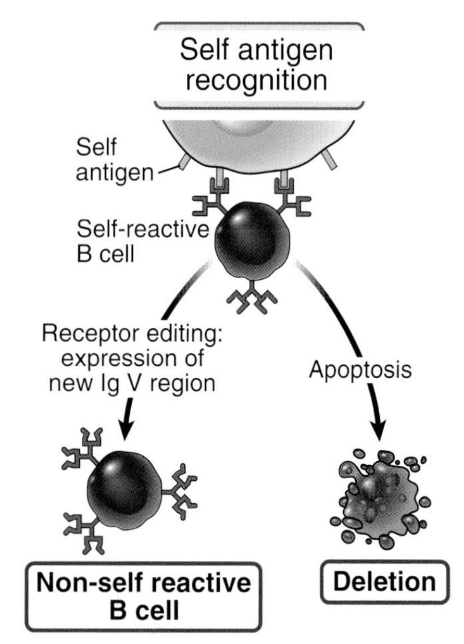

Fig. 9.6 Central tolerance in immature B lymphocytes. An immature B cell that recognizes self antigen in the bone marrow changes its antigen receptor (receptor editing) or dies by apoptosis (negative selection, or deletion). *Ig*, Immunoglobulin.

- **Receptor editing.** Immature B cells are at a stage of maturation in the bone marrow when they have rearranged their immunoglobulin (Ig) genes, express IgM, and have shut off the *RAG* genes that encode the VDJ recombinase. If these B cells recognize self antigens in the bone marrow, they may re-express *RAG* genes, resume light-chain gene recombination, and express a new Ig light chain (see Chapter 4). The previously rearranged heavy chain gene cannot recombine again because during the two steps of VDJ recombination for Ig heavy chain genes, all the other D segment genes upstream and downstream of the D segment that was used have been deleted. The new light chain associates with the previously expressed Ig heavy chain to produce a new antigen receptor that may no longer recognize the self antigen. This process of changing receptor specificity, called receptor editing, reduces the chance that potentially harmful self-reactive B cells will leave the marrow. It is estimated that 25% to 50% of mature B cells in a normal individual may have undergone receptor editing during their maturation. (There is no evidence that developing T cells can undergo receptor editing.)

- **Deletion.** If editing fails, immature B cells that strongly recognize self antigens receive death signals and die by apoptosis. This process of deletion is similar to negative selection of immature T lymphocytes. As in the T cell compartment, negative selection of B cells eliminates lymphocytes with high-affinity receptors for abundant, and usually widely expressed, cell membrane or soluble self antigens.

Peripheral B Cell Tolerance

Mature B lymphocytes that encounter self antigens in peripheral lymphoid tissues become incapable of responding to that antigen (Fig. 9.7). Some self antigens, such as soluble proteins, may be recognized with low avidity. B cells specific for these antigens survive, but antigen receptor expression is reduced and the cells become functionally unresponsive (anergic). According to one hypothesis, if B cells recognize a protein antigen but do not receive T cell help (because helper T cells have been eliminated or are tolerant), the B cells become anergic because of a block in signaling from the antigen receptor. Anergic B cells may leave lymphoid follicles and are subsequently excluded from the follicles. B cells that have recently emerged from the bone marrow, called transitional B cells, are eliminated by apoptosis when their antigen receptors are triggered by self antigens in the periphery. It has also been proposed that some B cells produce cytokines such as IL-10 and TGF-β and thus inhibit activation of other immune cells, but the significance of these putative regulatory B cells in maintaining self-tolerance and preventing autoimmunity is not established.

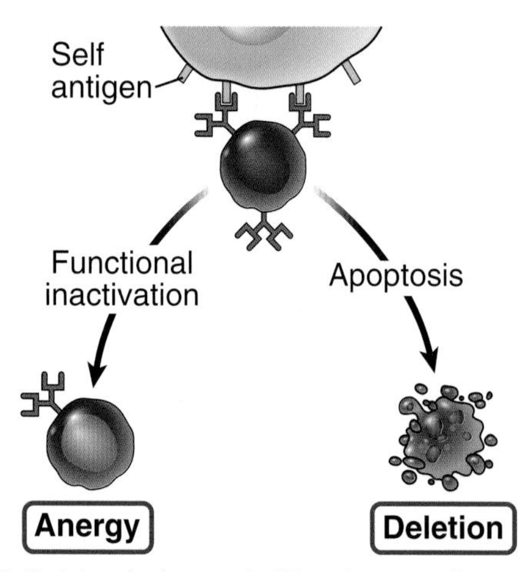

Fig. 9.7 Peripheral tolerance in B lymphocytes. A mature B cell that recognizes a self-antigen without T cell help is functionally inactivated and becomes incapable of responding to that antigen (anergy), or it dies by apoptosis (deletion), or its activation is suppressed by engagement of inhibitory receptors.

TOLERANCE TO COMMENSAL MICROBES AND FETAL ANTIGENS

Before concluding our discussion of the mechanisms of immunologic tolerance, it is useful to consider two other types of antigens that are not self but are produced by cells or tissues that have to be tolerated by the immune system. These are products of commensal microbes that live in symbiosis with humans and paternally derived antigens in the fetus. Coexistence with these antigens is dependent on many of the same mechanisms that are used to maintain peripheral tolerance to self antigens.

Tolerance to Commensal Microbes in the Intestines and Skin

The microbiome of healthy humans consists of approximately 10^{14} bacteria and viruses (which is estimated to be almost 10 times the number of nucleated human cells, prompting microbiologists to point out that we are only 10% human and 90% microbial!). These microbes reside in the intestinal and respiratory tracts and on the skin, where they serve many essential functions. For instance, in the gut, commensal bacteria aid in digestion and absorption of foods and prevent overgrowth of potentially harmful microbes. Mature lymphocytes in these tissues are capable of recognizing the commensal organisms but do not react against them, so the microbes are not eliminated and harmful inflammation is not triggered. In the gut, several mechanisms account for the inability of the healthy immune system to react against commensal microbes. These mechanisms include an abundance of IL-10–producing Tregs, and an unusual property of intestinal dendritic cells such that signaling from some Toll-like receptors leads to inhibition rather than activation. There is some evidence that commensal bacteria in the skin also induce Tregs. In addition, many commensal microbes are physically separated from the intestinal and cutaneous immune systems by the epithelium and products of the epithelial cells such as mucus in the gut and keratin on the skin.

Tolerance to Fetal Antigens

The evolution of placentation in eutherian mammals allowed the fetus to mature before birth but created the problem that paternal antigens expressed in the fetus, which are foreign to the mother, have to be tolerated by the immune system of the pregnant mother. One mechanism of this tolerance is the generation of peripheral FOXP3$^+$ Tregs specific for these paternal antigens. In fact, evolution of mammalian placentation coincides with the ability to generate stable peripheral Tregs. Other mechanisms of fetal tolerance include exclusion of inflammatory cells from the pregnant uterus, poor antigen presentation in the placenta, the low expression of conventional MHC class I molecules on fetal trophoblast cells, the expression of ligands for natural killer (NK) cell inhibition on the fetal trophoblast, an inability to generate harmful T helper 1 (Th1) responses in the healthy pregnant uterus, and chemical modifications of antigens in the fetus that make them tolerogenic.

Now that we have described the principal mechanisms of immunologic tolerance, we consider the consequences of the failure of self-tolerance—namely, the development of autoimmunity.

AUTOIMMUNITY

Autoimmunity is defined as an immune response against self (autologous) antigens. It is an important cause of disease, estimated to affect 3% to 5% of the population in developed countries, and the prevalence of several autoimmune diseases is increasing. Different autoimmune diseases may be organ-specific, affecting only one or a few organs, or systemic, with widespread tissue injury and clinical manifestations. Tissue injury in autoimmune diseases may be caused by antibodies or by T cells specfic for self antigens (see Chapter 11).

Pathogenesis

The principal factors in the development of autoimmunity are the inheritance of susceptibility genes and environmental triggers, such as infections (Fig. 9.8). It is postulated that susceptibility genes interfere with pathways of self-tolerance, leading to the persistence of self-reactive T and B lymphocytes. Environmental stimuli may cause cell and tissue injury and inflammation and activate these self-reactive lymphocytes, resulting in the generation of effector T cells and autoantibodies that are responsible for the autoimmune disease.

Despite our growing knowledge of the immunologic abnormalities that may result in autoimmunity, we still do not know the etiology of common human autoimmune diseases. This lack of understanding results from several factors: autoimmune diseases in humans usually are heterogeneous and multifactorial; the self antigens that are the inducers and targets of the autoimmune reactions are often unknown; and the diseases may be detected long after the autoimmune reactions have been initiated.

Genetic Factors

Inherited risk for most autoimmune diseases is attributable to multiple gene loci. If an autoimmune disease develops in one of two twins, the same disease is more likely to develop in the other twin than in an unrelated member of the general population. Furthermore, this increased incidence is greater among monozygotic (identical) twins than among dizygotic twins who have grown up together, excluding the role of environmental factors. These findings prove the importance of genetics in the susceptibility to autoimmunity. Genome-wide association studies have revealed some of the variations (polymorphisms) of genes that may contribute to different autoimmune diseases. Emerging results suggest that different polymorphisms are more frequent (predisposing) or less frequent (protective) in patients than in healthy controls. The importance of these polymorphisms is reinforced by the finding that many of them affect genes involved in immune responses, and the same genetic polymorphism may be associated with more than one autoimmune disease. However, these polymorphisms are frequently present in healthy individuals, and the individual contribution of each of these genes to the development of autoimmunity is very small, so many risk alleles together are needed to cause the disease. Many of these polymorphisms are in the regulatory regions of the genes (promoters and enhancers) and not in the coding sequences, suggesting that they influence expression of the encoded proteins.

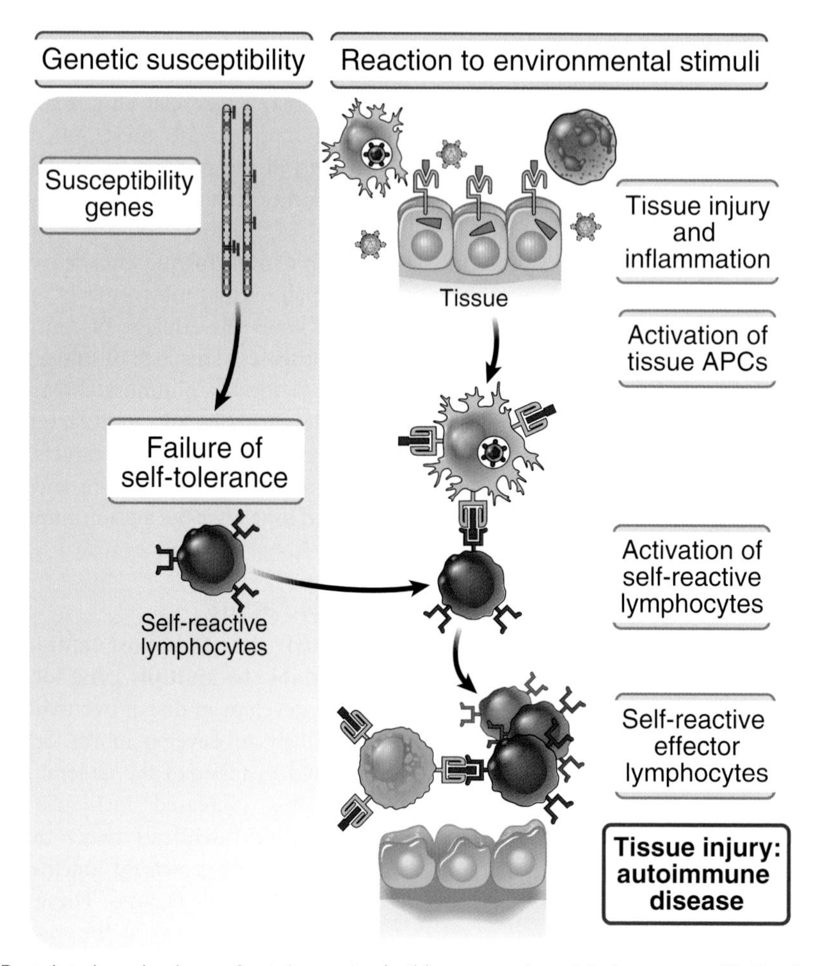

Fig. 9.8 Postulated mechanisms of autoimmunity. In this proposed model of organ-specific T cell–mediated autoimmunity, various genetic loci may confer susceptibility to autoimmunity, probably by influencing the maintenance of self-tolerance. Environmental triggers, such as infections and other inflammatory stimuli, promote the influx of lymphocytes into tissues and the activation of antigen-presenting cells *(APCs)* and subsequently of self-reactive T cells, resulting in tissue injury.

Many autoimmune diseases in humans and inbred animals are linked to particular MHC alleles (Fig. 9.9). The association between human leukocyte antigen (HLA) alleles and autoimmune diseases in humans was recognized many years ago and was one of the first indications that T cells played an important role in these disorders (because the only known function of MHC molecules is to present peptide antigens to T cells). The incidence of numerous autoimmune diseases is greater among individuals who inherit particular HLA allele(s) than in the general population. The likelihood of a particular autoimmune disease in people with versus without a particular polymorphism (such as an HLA allele) is expressed as the relative risk (or odds ratio). Most of these disease associations are with class II HLA alleles (most often HLA-DR and HLA-DQ), perhaps because class II MHC molecules control the development and activation of $CD4^+$ T cells, which are involved in both cell-mediated and humoral immune responses to proteins as well in regulating immune responses. It is important to point out that, although an HLA allele may increase the risk for developing a particular autoimmune disease, the HLA allele is not, by itself, the cause of the disease. In fact, the disease never develops in the vast majority of people who inherit an HLA allele that does confer increased risk for the

Disease	MHC allele	Relative risk
Ankylosing spondylitis	HLA-B27	90
Rheumatoid arthritis	HLA-DRB1*0401/0404	4-12
Type 1 diabetes mellitus	HLA-DRB1*0301/0401	35
Pemphigus vulgaris	HLA-DR4	14

Fig. 9.9 Association of autoimmune diseases with alleles of the major histocompatibility complex (MHC) locus. Family and linkage studies show a greater likelihood of developing certain autoimmune diseases in persons who inherit particular human leukocyte antigen (HLA) alleles than in persons who lack these alleles (odds ratio or relative risk). Selected examples of HLA disease associations are listed. For instance, in people who have the HLA-B27 allele, the risk of development of ankylosing spondylitis, an autoimmune disease of the spine, is much higher than in B27-negative people; other diseases show various degrees of association with other HLA alleles. The *asterisks* indicate HLA alleles identified by molecular (DNA-based) typing instead of the older serologic (antibody-based) methods.

disease. Despite the clear association of MHC alleles with several autoimmune diseases, how these alleles contribute to the development of the diseases remains unknown. Some hypotheses are that particular MHC alleles may be especially effective at presenting pathogenic self peptides to autoreactive T cells or that they are inefficient at displaying certain self antigens in the thymus, leading to defective negative selection of T cells.

Polymorphisms in non-HLA genes are associated with various autoimmune diseases and may contribute to failure of self-tolerance or abnormal activation of lymphocytes (Fig. 9.10A). Many such disease-associated genetic variants have been described:

- Polymorphisms in the gene encoding the tyrosine phosphatase PTPN22 (protein tyrosine phosphatase N22) may lead to uncontrolled activation of both B and T cells and are associated with numerous autoimmune diseases, including rheumatoid arthritis, SLE, and type 1 diabetes.
- Variants of the innate immune cytoplasmic microbial sensor NOD-2 that reduce resistance to intestinal microbes are associated with Crohn's disease, an inflammatory bowel disease, in some ethnic populations.
- Other polymorphisms associated with multiple autoimmune diseases include genes encoding the IL-2 receptor α chain (CD25), thought to influence the balance of effector and regulatory T cells; the receptor for the cytokine IL-23, which promotes the development of proinflammatory T helper 17 (Th17) cells; and CTLA-4, a key inhibitory receptor in T cells discussed earlier and in Chapter 5.

Some rare autoimmune disorders are caused by mutations in single genes that have high penetrance and lead to autoimmunity in most individuals who inherit these mutations, although the pattern of inheritance varies. These genes, alluded to earlier, include *AIRE*, *FOXP3*, *FAS*, and *CTLA4* (Fig. 9.10B). Mutations in these genes have been valuable for identifying key molecules and pathways involved in self-tolerance. However, these Mendelian forms of autoimmunity are rare, and common autoimmune diseases are not caused by mutations in any of these known genes.

Role of Infections and Other Environmental Influences

Infections may activate self-reactive lymphocytes, thereby triggering the development of autoimmune diseases. Clinicians have recognized for many years that the clinical manifestations of autoimmunity sometimes are preceded by infectious prodromes. This association between infections and autoimmune tissue injury has been established in some animal models.

Infections may contribute to autoimmunity in several ways (Fig. 9.11):

- An infection in a tissue may induce a local innate immune response, which may lead to increased

A

Genes that may contribute to genetically complex autoimmune diseases		
Gene(s)	Disease association	Mechanism
PTPN22	RA, several others	Abnormal tyrosine phosphatase regulation of T cell selection and activation?
NOD2	Crohn's disease	Defective resistance or abnormal responses to intestinal microbes?
IL23R	IBD, PS, AS	Component of IL-23 receptor; role in generation and maintenance of Th17 cells
CTLA4	T1D, RA	Inhibitory receptor of T cells, effector molecule of regulatory T cells
CD25 (IL-2Rα)	MS, type 1 diabetes, others	Abnormalities in effector and/or regulatory T cells?
C2, C4 (Complement proteins)	SLE	Defects in clearance of immune complexes or in B cell tolerance?
FCGRIIB (FCγRIIb)	SLE	Defective feedback inhibition of B cells

B

Single-gene defects that cause autoimmunity (Mendelian diseases)		
Gene(s)	Disease association	Mechanism
AIRE	Autoimmune polyglandular syndrome (APS-1)	Reduced expression of peripheral tissue antigens in the thymus, leading to defective elimination of self-reactive T cells
CTLA4	CTLA4 haploinsufficiency with autoimmune infiltration	Impaired regulatory T cell function leading to loss of B and T cell homeostasis
FOXP3	Immune dysregulation, polyendocrinopathy, enteropathy, X-linked (IPEX)	Deficiency of regulatory T cells
FAS	Autoimmune lymphoproliferative syndrome (ALPS)	Defective apoptosis of self-reactive T and B cells in the periphery
CD25	IPEX-like syndrome	Deficiency of regulatory T cells; defective IL-10 production
IL-10/IL-10R	Infantile enterocolitis	Defective suppression of immune responses to commensal bacteria?

Fig. 9.10 Roles of non-MHC genes in autoimmunity. A, Selected examples of variants (polymorphisms) of genes that confer susceptibility to autoimmune diseases but individually have small or no effects. **B,** Examples of single genes whose mutations result in autoimmunity. These are rare examples of autoimmune diseases with Mendelian inheritance. The pattern of inheritance varies in the different diseases. APS-1 is autosomal recessive, and both alleles of the gene *(AIRE)* have to be abnormal to cause the disease. IPEX is X-linked, so mutation in one allele of the gene *(FOXP3)* is sufficient to cause a defect in boys. ALPS is autosomal dominant because FAS and FASL are trimeric proteins and mutations in one of the alleles of either gene result in reduced expression of intact trimers. *CTLA4* mutations cause disease in an autosomal manner. *AS,* Ankylosing spondylitis; *IBD,* inflammatory bowel disease; *IL,* interleukin; *MS,* multiple sclerosis; *PS,* psoriasis; *RA,* rheumatoid arthritis; *SLE,* systemic lupus erythematosus; *T1D,* type 1 diabetes; *Th17,* T helper 17 cell.

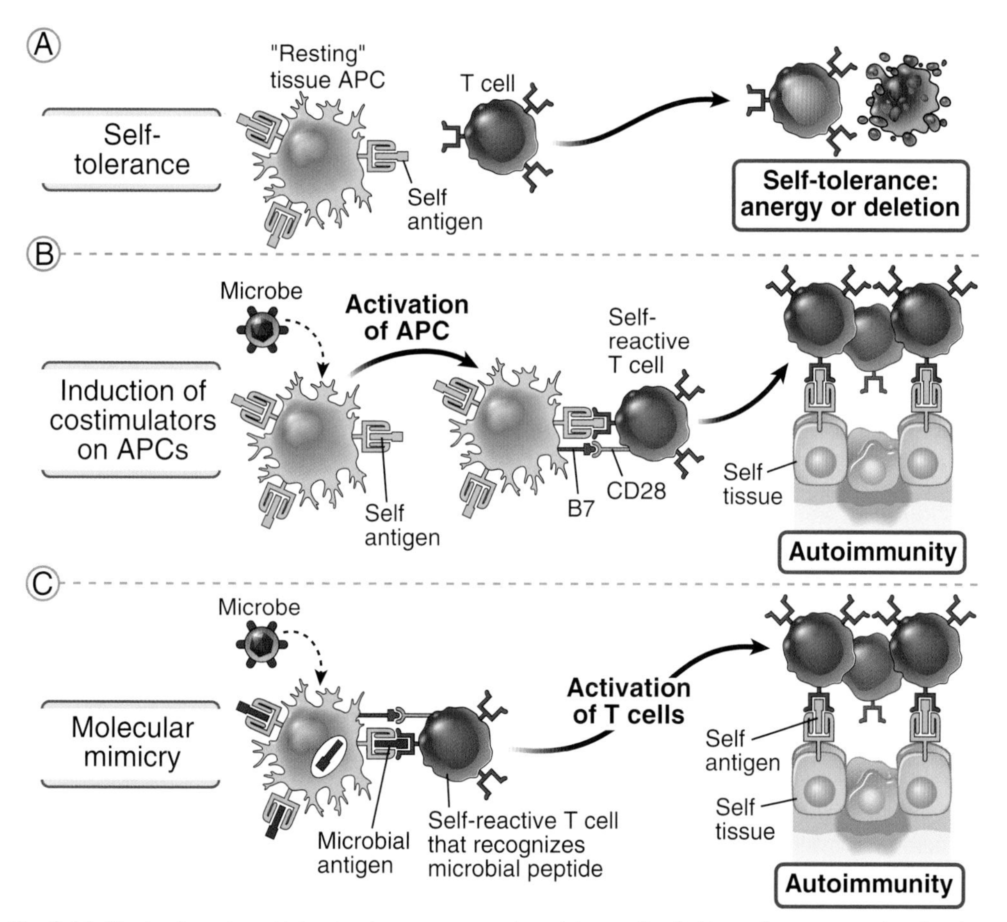

Fig. 9.11 Mechanisms by which microbes may promote autoimmunity. A, Normally, an encounter of mature T cells with self antigens presented by resting tissue antigen-presenting cells (APCs) results in peripheral tolerance. **B,** Microbes may activate the APCs to express costimulators, and when these APCs present self antigens, the specific T cells are activated, rather than being rendered tolerant. **C,** Some microbial antigens may cross-react with self antigens (mimicry). Therefore, immune responses initiated by the microbes may become directed at self cells and self tissues. This figure illustrates concepts as they apply to T cells; molecular mimicry also may apply to self-reactive B lymphocytes.

production of costimulators and cytokines by tissue APCs. These activated tissue APCs may be able to stimulate self-reactive T cells that encounter self antigens in the tissue. This may lead to disease if it occurs in people who are already genetically at risk for developing autoimmunity. Nevertheless, autoimmune disease does not develop in most infections, probably because the mechanisms of tolerance are adequate to limit autoimmune reactions.

One set of cytokines produced in innate immune responses to viruses are type I interferons (IFNs). Excessive production of type I IFNs has been associated with the development of several autoimmune diseases, notably SLE. These cytokines may activate APCs or lymphocytes, but what stimulates their production and how they contribute to autoimmunity is not well understood.

- Some infectious microbes may produce peptide antigens that are similar to, and cross-react with, self antigens. Immune responses to these microbial peptides may result in an immune attack against self antigens. Such cross-reactions between microbial and self antigens are termed **molecular mimicry.** Although the contribution of molecular mimicry to autoimmunity has fascinated immunologists, its actual significance in the development of most autoimmune diseases remains unknown. In some disorders, antibodies produced against a microbial protein bind to self proteins. For example, in rheumatic fever, a fairly common disease before the widespread use of antibiotics, antibodies against streptococci cross-react with a myocardial antigen and cause heart disease.
- The innate response to infections may alter the chemical structure of self antigens. For example, some periodontal bacterial infections are associated with rheumatoid arthritis. It is postulated that the inflammatory responses to these bacteria lead to enzymatic conversion of arginines to citrullines in self proteins, and the citrullinated proteins are recognized as nonself and elicit adaptive immune responses.
- Infections may injure tissues and release antigens that are normally sequestered from the immune system. For example, some sequestered antigens (e.g., in testis and eye) normally may not be seen by the immune system. Release of these antigens (e.g., by trauma or infection) may initiate an autoimmune reaction against the tissue.
- The abundance and composition of normal commensal microbes in the gut, skin, and other sites (the microbiome) may influence the health of the immune system and the maintenance of self-tolerance. This possibility has generated a great deal of interest, but normal variations in the microbiome of humans related to environmental exposure and diet make it difficult to define the relationship between particular microbes and the development of autoimmune diseases.

Paradoxically, some infections appear to confer protection from certain autoimmune diseases. This conclusion is based on epidemiologic data and limited experimental studies. The basis of this protective effect of infections is unknown.

Several other environmental and host factors may contribute to autoimmunity. Many autoimmune diseases are more common in women than in men, but how sex might affect immunologic tolerance or lymphocyte activation remains unknown. Exposure to sunlight is a trigger for the development of the autoimmune disease SLE, in which autoantibodies are produced against self nucleic acids and self nucleoproteins. It is postulated that these nuclear antigens may be released from cells that die by apoptosis as a consequence of exposure to ultraviolet radiation in sunlight.

SUMMARY

- Immunologic tolerance is specific unresponsiveness to an antigen induced by exposure of lymphocytes to that antigen. All individuals are tolerant of (unresponsive to) all or most of their own (self) antigens. Tolerance against antigens may be induced by administering that antigen in particular ways, and this strategy may be useful for treating immunologic diseases and for preventing the rejection of transplants.
- Central tolerance is induced in immature lymphocytes that encounter antigens in the generative lymphoid organs. Peripheral tolerance results from the recognition of antigens by mature lymphocytes in peripheral tissues.
- Central tolerance of T cells is the result of strong recognition of self antigens in the thymus by developing T cells. Some of these self-reactive T cells die (negative selection), thus eliminating the potentially

most dangerous T cells, which express high-affinity receptors for self antigens. Other T cells of the CD4 lineage develop into regulatory T cells (Tregs) that suppress self-reactivity in the periphery.

- Regulatory T cells (Tregs) generated in the thymus or periphery function to suppress immune responses by multiple mechanisms, including CTLA-4–mediated removal and blocking of B7 molecules of APCs, suppression of immune cells by inhibitory cytokines, and consumption of IL-2. Tregs express the transcription factor FOXP3, CTLA—4, and the IL-2 receptor α chain CD25, and mutations in any of these impair Treg function and result in autoimmune diseases.
- Peripheral tolerance in T cells is induced by multiple other mechanisms. Anergy (functional inactivation) results from the recognition of antigens without costimulators (second signals). Deletion (death by apoptosis) may occur when T cells encounter self antigens.
- In B lymphocytes, central tolerance occurs when immature cells recognize self antigens in the bone marrow. Some of the cells change their receptors (receptor editing), and others die by apoptosis (negative selection, or deletion). Peripheral tolerance is induced when mature B cells recognize self antigens without T cell help, which results in anergy and/or death of the B cells.
- Autoimmune diseases result from a failure of self-tolerance. Multiple factors contribute to autoimmunity, including the inheritance of susceptibility genes and environmental triggers such as infections.
- Many genes contribute to the development of autoimmunity. The strongest associations are between human leukocyte antigen (HLA) genes and various T cell—dependent autoimmune diseases.
- Infections predispose to autoimmunity by causing inflammation and stimulating the expression of costimulators or because of cross-reactions between microbial and self antigens.

REVIEW QUESTIONS

1. What is immunologic tolerance? Why is it important?
2. How is central tolerance induced in T lymphocytes and B lymphocytes?
3. Where do regulatory T cells develop, and how do they protect against autoimmunity?
4. How are T cell dysfunctional states of anergy and exhaustion induced in T cells and how may they contribute to peripheral tolerance
5. What are the mechanisms that prevent immune responses against commensal microbes and fetuses?
6. What are some of the genes that contribute to autoimmunity? How may MHC genes play a role in the development of autoimmune diseases?
7. What are some possible mechanisms by which infections promote the development of autoimmunity?

Answers to and discussion of the Review Questions may be found on p. 322.

10

Tumor and Transplantation Immunology

Immune Responses to Cancer Cells and Normal Foreign Cells

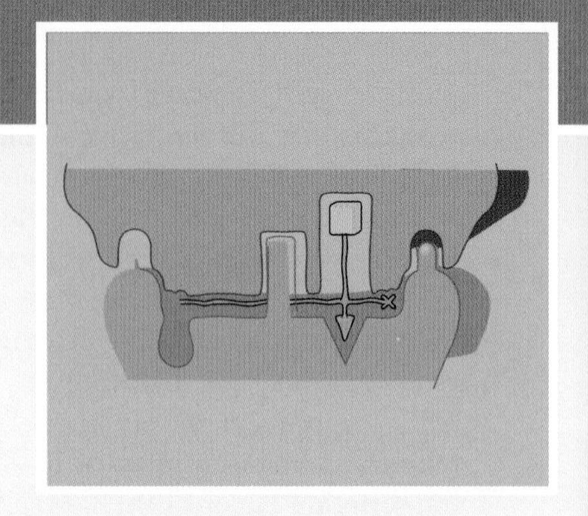

CHAPTER OUTLINE

Cancer and organ transplantation are two situations in which the immune response to human cells that are genetically distinct from normal self has important clinical consequences. In order for cancers to grow, they have to evade host immunity, and effective methods of enhancing patients' immune responses against tumors, called cancer immunotherapy, have had enormous impact on clinical oncology. In organ transplantation, the situation is the reverse: immune responses against grafted tissues from other people are a major barrier to successful transplantation, and suppressing these responses is a central focus of transplantation medicine. Because of the importance of the immune system in host responses to tumors and transplants, tumor immunology and transplantation immunology have become subspecialties in which researchers and clinicians come together to address both fundamental and clinical questions.

Immune responses against tumors and transplants share several characteristics. These are situations in which the immune system is not responding to microbes, as it usually does, but to noninfectious cells that are perceived as foreign. The antigens that mark tumors and transplants as foreign may be expressed in virtually any cell type that is, respectively, malignantly transformed or grafted from one individual to another. Therefore, immune responses against tumors and transplants may be directed against diverse cell types.

In this chapter we focus on the following questions:

- What are the antigens in tumors and tissue transplants that are recognized as foreign by the immune system?

- How does the immune system recognize and react to tumors and transplants?
- How can immune responses to tumors and grafts be manipulated to enhance tumor rejection and inhibit graft rejection?

We discuss tumor immunity first and then transplantation, and we point out the principles common to both.

IMMUNE RESPONSES AGAINST TUMORS

For over a century, scientists have proposed that a physiologic function of the adaptive immune system is to prevent the outgrowth of transformed cells and to destroy these cells before they become harmful tumors. Control and elimination of malignant cells by the immune system is called tumor **immune surveillance**. Several lines of evidence support the idea that immune surveillance against tumors is important for preventing tumor growth (Fig. 10.1). However, the fact that common malignant tumors develop in immunocompetent individuals indicates that tumor immunity is often incapable of preventing tumor growth or is easily overwhelmed by rapidly growing tumors. This has led to the growing realization that the immune response to tumors is often dominated by tolerance or regulation, not by effective immunity. Cancer biologists now consider the ability to evade immune destruction as a fundamental feature ("hallmark") of cancers. The field of tumor immunology has focused on defining the types of tumor antigens against which the immune system reacts, understanding the nature of the immune responses to tumors and mechanisms by which tumors evade them, and developing strategies for maximally enhancing antitumor immunity.

Tumor Antigens

Malignant tumors express various types of molecules that may be recognized by the immune system as foreign antigens (Fig. 10.2). Protein antigens that elicit cytotoxic T lymphocyte (CTL) responses are the most relevant for protective antitumor immunity because CTLs are the prinicipal mechanism for killing tumor cells. These tumor antigens are present in the cytosol of tumor cells and are processed and displayed by class I **major histocompatibility complex** (MHC) molecules

Evidence	Conclusion
Lymphocytic infiltrates around some tumors and enlargement of draining lymph nodes correlate with better prognosis	Immune responses against tumors inhibit tumor growth
Transplants of tumors between syngeneic animals are rejected, and more rapidly if the animals have been previously exposed to that tumor; immunity to tumor transplants can be transferred by lymphocytes from a tumor bearing animal	Tumor rejection shows features of adaptive immunity (specificity, memory) and is mediated by lymphocytes
Immunodeficient individuals have an increased incidence of some types of tumors	The immune system protects against the growth of tumors
Therapeutic blockade of T cell inhibitory receptors such as PD-1 and CTLA-4 leads to tumor remission	Tumors evade immune surveillance in part by inhibiting T cells

Fig. 10.1 Evidence supporting the concept that the immune system reacts against tumors Several lines of clinical and experimental evidence indicate that defense against tumors is mediated by reactions of the adaptive immune system. *CTLA-4,* Cytotoxic T-lymphocyte antigen 4; *PD-1,* programmed cell death protein 1.

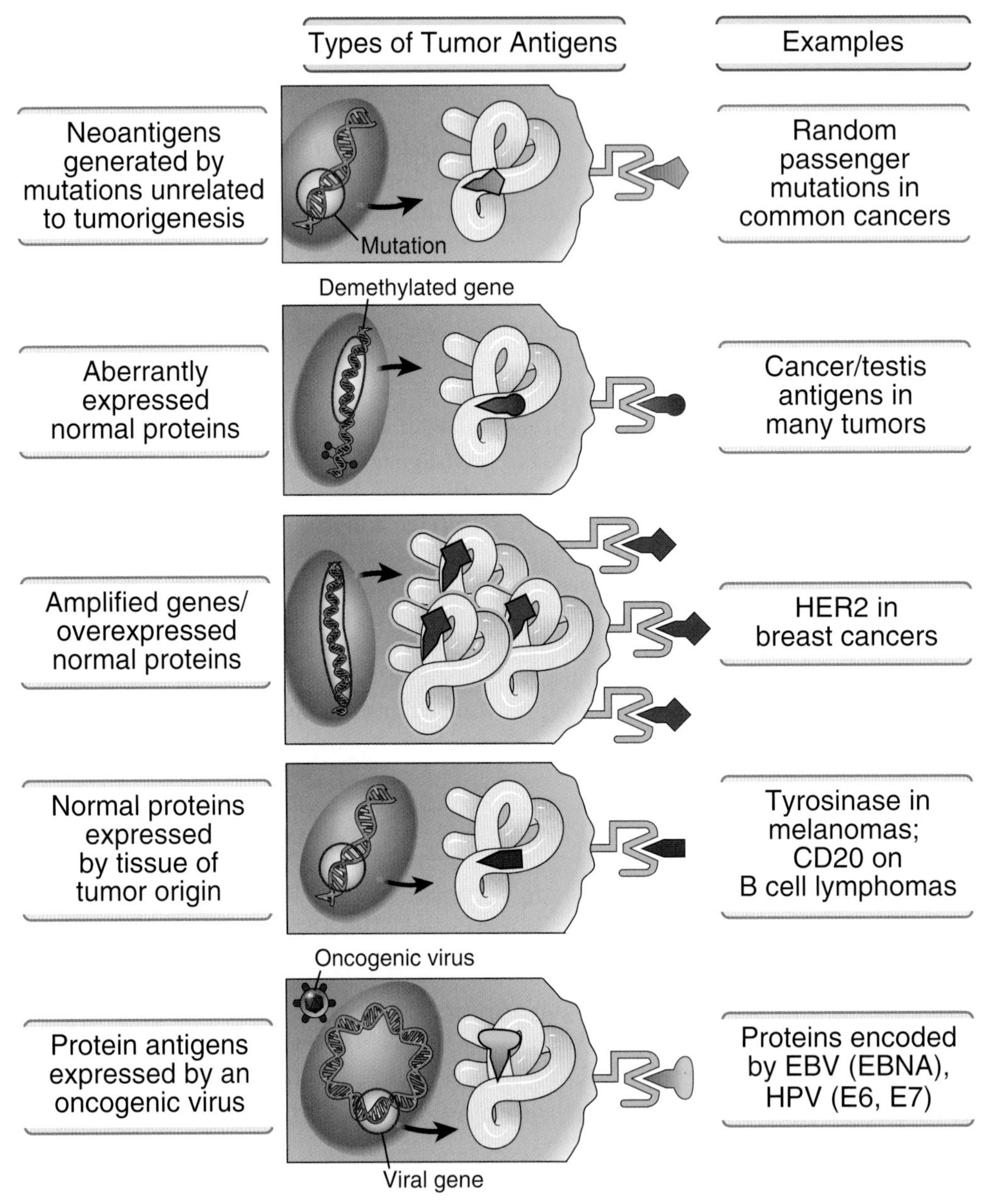

Fig. 10.2 Types of tumor antigens recognized by T cells Tumor antigens that are recognized by tumor-specific $CD8^+$ T cells may be mutated forms of various self proteins that do not contribute to malignant behavior of the tumor; products of oncogenes or tumor suppressor genes; self proteins whose expression is increased in tumor cells; and products of oncogenic viruses. Cancer/testis antigens are proteins that are normally expressed in the testis and are also expressed in some tumors. HER2 is a receptor for epidermal growth factor whose expression is increased in tumors, usually because of gene amplification. In some cases, unmutated cellular proteins that are normally expressed in low amounts (e.g., tyrosinase) or at particular stages of development (e.g., cancer/testis antigens) may be immunogenic when expressed in tumors at increased levels. Tumor antigens also may be recognized by $CD4^+$ T cells, but less is known about the role that $CD4^+$ T cells play in tumor immunity. *EBNA*, Epstein-Barr nuclear antigen; *EBV*, Epstein-Barr virus; *HPV*, human papillomavirus.

for recognition by $CD8^+$ CTLs. The tumor antigens that elicit immune responses can be classified into several groups:

- **Neoantigens encoded by randomly mutated genes.** Sequencing of tumor genomes has revealed that common human tumors harbor a large number of mutations in diverse genes, reflecting the genetic instability of malignant cells. These mutations usually play no role in tumorigenesis and are called passenger mutations. Many of these mutations result in expression of mutated proteins, called neoantigens because they are newly expressed in the tumor cells but are not present in the normal cells of origin of the tumor. Because T cells only recognize peptides bound to MHC molecules, mutated tumor proteins can be recognized by T cells if peptides carrying the mutated amino acid sequences can bind to and be presented by the patients' MHC alleles. Tumor neoantigens may not induce tolerance because they are not present in normal cells and are the most common targets of tumor-specific adaptive immune responses. In fact, the number of these mutations in human cancers correlates with the strength of the antitumor immune responses patients mount and the effectiveness of immunotherapies that enhance those responses.
- **Products of oncogenes or mutated tumor suppressor genes.** Some tumor antigens are products of mutations, called driver mutations, in genes that are involved in the process of malignant transformation. The driver mutations that encode tumor antigens may be amino acid substitutions, deletions, or new sequences generated by gene translocations, all of which can be seen as foreign.
- **Aberrantly expressed or overexpressed structurally normal proteins.** In several human tumors, antigens that elicit immune responses are normal proteins (products of unmutated genes) whose expression is dysregulated in the tumors. Sometimes, this is a consequence of epigenetic changes such as demethylation of the promoters in genes encoding these proteins, as in the case of cancer/testis antigens, which are normally expressed only in germ cells but are seen in many types of cancers. Overexpression may be a result of gene amplification such as the HER2 protein in some breast cancers. These structurally normal self antigens would not be expected to elicit immune responses, but their aberrant expression may be enough to make them immunogenic. In other cases, self proteins that are expressed only in embryonic tissues may not induce tolerance in adults, and the same proteins expressed in tumors may be recognized as foreign by the immune system.
- **Viral antigens.** In tumors caused by oncogenic viruses, such as Epstein-Barr virus (EBV) and human papillomavirus (HPV), the tumor antigens may be encoded by the viruses.

Immune Mechanisms of Tumor Rejection

The principal immune mechanism of tumor eradication is killing of tumor cells by CTLs specific for tumor antigens. The role of CTLs in tumor rejection has been established in animal models: tumors can be destroyed by transferring tumor-reactive $CD8^+$ T cells into the tumor-bearing animals. Studies of many human tumors indicate that abundant CTL infiltration predicts a more favorable clinical course compared with tumors with sparse CTLs.

CTL responses against tumors are initiated by recognition of tumor antigens on host antigen-presenting cells (APCs). The APCs ingest tumor cells or their antigens and present the antigens to naive $CD8^+$ T cells in draining lymph nodes (Fig. 10.3). Tumors may arise from virtually any nucleated cell type in any tissue. We know, however, that the activation of naive $CD8^+$ T cells to proliferate and differentiate into active CTLs requires recognition of antigen (class I MHC–associated peptide) on dendritic cells in secondary lymphoid organs and also costimulation and/or help from class II MHC–restricted $CD4^+$ T cells (see Chapter 5). How, then, can tumors of different cell types stimulate CTL responses? The likely answer is that apoptotic tumor cells or proteins released possibly from necrotic tumor cells are ingested by the host's dendritic cells and transported to lymph nodes draining the site of the tumor. The protein antigens of the tumor cells are processed in proteasomes and displayed by class I MHC molecules on the host dendritic cells. This process, called **cross-presentation** or cross-priming, was introduced in Chapter 3 (see Fig. 3.16). Dendritic cells can also present peptides derived from ingested tumor

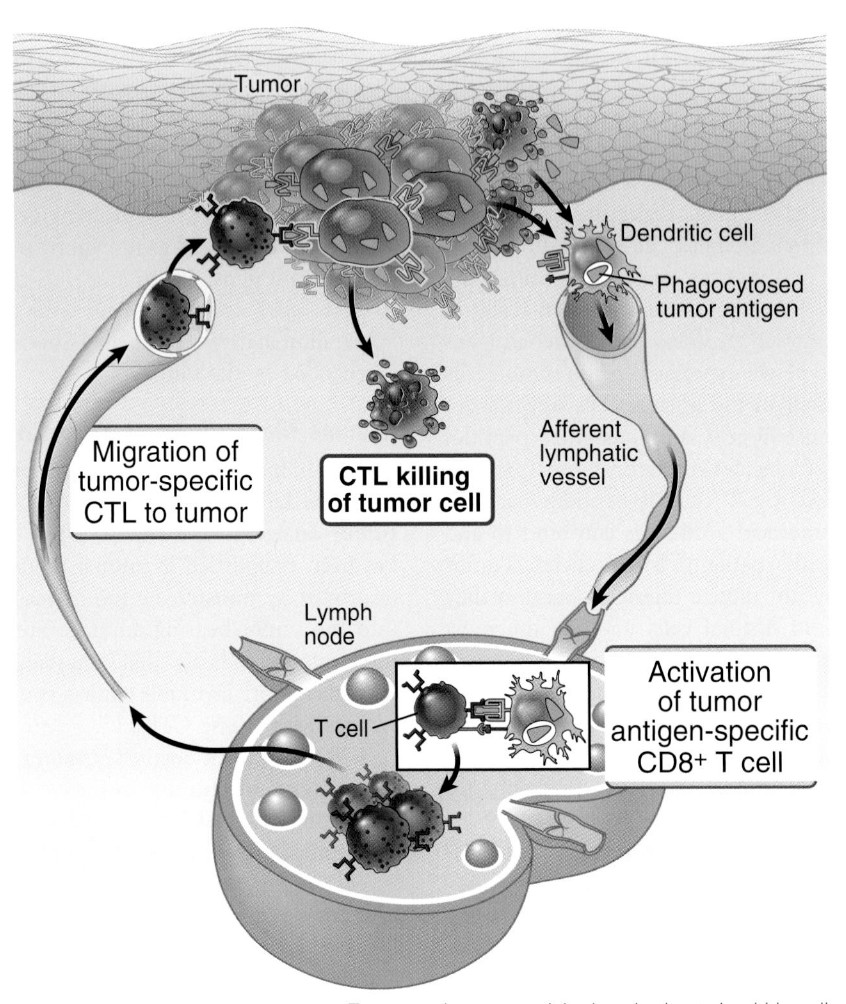

Fig. 10.3 Immune response against tumors Tumor antigens are picked up by host dendritic cells, and responses are initiated in secondary (peripheral) lymphoid organs. Tumor-specific cytotoxic T lymphocytes (CTLs) migrate back to the tumor and kill tumor cells. Other mechanisms of tumor immunity are not shown.

antigens on class II MHC molecules. Thus, tumor antigens may be recognized by $CD8^+$ T cells and by $CD4^+$ T cells.

At the same time that dendritic cells are presenting tumor antigens, they may express costimulators that provide signals for the activation of the T cells. It is not known how tumors induce the expression of costimulators on APCs because, as discussed in Chapter 5, the physiologic stimuli for the induction of costimulators are usually microbes, and tumors are generally sterile. A likely possibility is that tumor cells die if their growth outstrips their blood and nutrient supply, and adjacent normal tissue cells may be injured and die due to the invasive tumor. These dying cells release products (damage-associated molecular patterns; see Chapter 2) that stimulate innate responses. The activation of APCs to express costimulators is part of these responses.

Once naive $CD8^+$ T cells have differentiated into effector CTLs, they are able to migrate back to any site where the tumor is growing and kill tumor cells

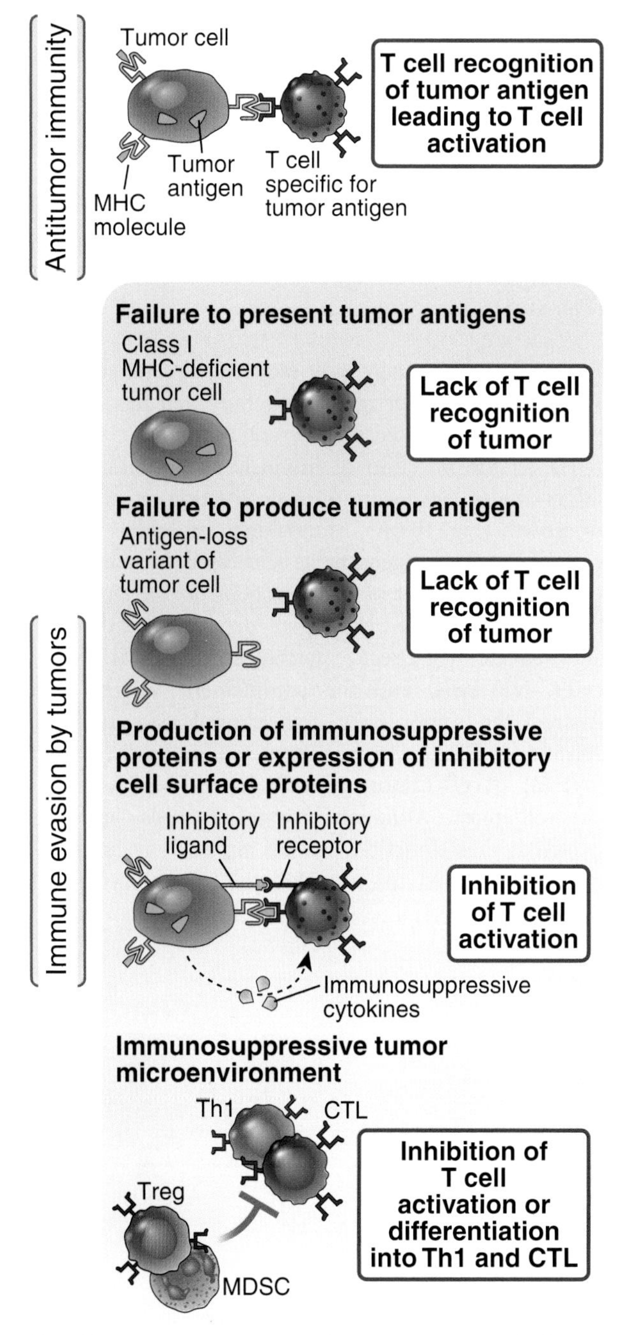

Fig. 10.4 How tumors evade immune responses Antitumor immunity develops when T cells recognize tumor antigens and are activated. Tumor cells may evade immune responses by losing expression of antigens or major histocompatibility complex (MHC) molecules or by producing immunosuppressive cytokines or ligands such as PD-L1 for inhibitory receptors such as PD-1 on T cells. Tumors may also create an immunosuppressive microenvironment with regulatory T cells and antiinflammatory myeloid cells. *CTL,* Cytotoxic T lymphocyte; *MDSC,* myeloid derived suppressor cell; *PD-L1,* programmed cell death protein ligand 1; *Th1,* T helper 1; *Treg,* regulatory T cell.

expressing the relevant antigens without a requirement for costimulation or T cell help.

Immune mechanisms in addition to CTLs may play a role in tumor rejection. Antitumor $CD4^+$ T cell responses have been detected in patients, and increased numbers of $CD4^+$ effector T cells, especially T helper 1 (Th1) and follicular helper T cells, in tumor infiltrates are associated with good prognosis. Antitumor antibodies are also detectable in some cancer patients, but whether these antibodies protect individuals against tumor growth has not been established. Experimental studies have shown that activated macrophages and natural killer (NK) cells are capable of killing tumor cells, and Th1 responses work largely by activating macrophages, but the protective role of these effector mechanisms in tumor-bearing patients is not clearly established.

Evasion of Immune Responses by Tumors

Immune responses often fail to check tumor growth because cancers evade immune recognition or resist immune effector mechanisms. Not surprisingly, tumor cells that evade the host immune response are selected to survive and grow. Tumors use several mechanisms to avoid destruction by the immune system (Fig. 10.4):

- Some tumors stop expressing class I MHC molecules or molecules involved in antigen processing or MHC assembly, so they cannot display antigens to $CD8^+$ T cells. The most frequent mutations that cause loss of class I MHC expression are those affecting β2-microglobulin, which is an essential component of the class I MHC molecule expressed on cell surfaces (see Fig. 3.7).
- Tumors induce mechanisms that inhibit T cell activation. For example, many tumors overexpress PD-L1, a ligand for the T cell inhibitory receptor programmed cell death protein 1 (PD-1), or induce the expression of PD-L1 on APCs in the tumor environment. Furthermore, tumors, being persistent, cause repeated stimulation of T cells specific for tumor antigens, which stimulates expression of PD-1. The result is that tumor-specific $CD8^+$ T cells develop an exhausted state, mediated by expression of PD-1 and

other inhibitory molecules, and become unresponsive to antigen.
- Factors in the tumor microenvironment may impair the ability of dendritic cells to induce strong antitumor immune responses. Some tumors may induce regulatory T cells, which suppress antitumor immune responses. Myeloid-derived suppressor cells, which have phenotypic features of neutrophils and monocytes but mainly antiinflammatory functions, are abundant in tumors, and are thought to contribute to immunosuppression.
- Some tumors may secrete immunosuppressive cytokines, such as transforming growth factor β (TGF-β).

Cancer Immunotherapy

The main strategies for cancer immunotherapy currently in practice and under development include the use of specific antitumor antibodies, the introduction of autologous T cells that recognize tumor antigens, enhancing preexisting host antitumor T cell immune responses by administering antibodies that block inhibitory molecules, and vaccination with tumor antigens. Until recently, most treatment protocols for disseminated cancers, which cannot be cured surgically, relied on chemotherapy and irradiation, both of which damage normal nontumor tissues and are hence associated with serious toxicities. Because the immune response is highly specific, it has long been hoped that tumor-specific immunity may be used to selectively eradicate tumors without injuring the patient. Only recently has the promise of cancer immunotherapy been realized in patients. The history of cancer immunotherapy illustrates how the initial, often empirical, approaches have been largely supplanted by rational strategies based on our improved understanding of immune responses (Fig. 10.5).

Passive Immunotherapy with Monoclonal Antibodies

A strategy for tumor immunotherapy that has been in practice for a limited number of tumors for decades relies on the injection of monoclonal antibodies that target cancer cells for immune destruction or inhibition of growth (Fig. 10.6A). Monoclonal antibodies against various tumor antigens have been used in many cancers. The antibodies bind to antigens on the surface of the tumors (not the neoantigens produced inside cells) and activate host effector mechanisms, such as phagocytes, NK cells, and the complement system, that destroy the tumor cells. For example, an antibody specific for CD20, which is expressed on B cells, is used to treat B cell tumors, usually in combination with chemotherapy. Although normal B cells are also depleted, their function can be replaced by administration of pooled immunoglobulin (Ig) from normal

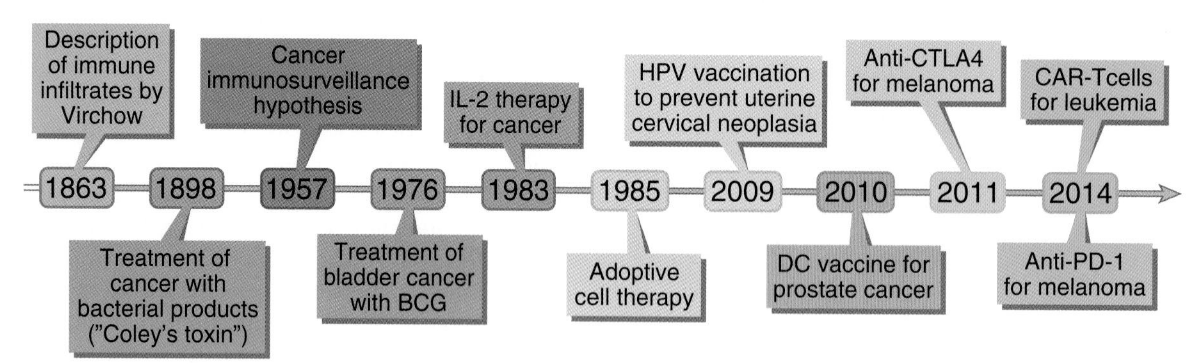

Fig. 10.5 History of cancer immunotherapy Some of the important discoveries in the field of cancer immunotherapy are summarized. *BCG,* Bacillus Calmette-Guérin; *CAR,* chimeric antigen receptor; *CTLA-4,* cytotoxic T lymphocyte antigen 4; *DC,* dendritic cell; *FDA,* Federal Drug Administration; *HPV,* human papillomavirus; *IL-2,* interleukin-2; *PD-1,* programmed cell death protein 1. (Modified from Lesterhuis WJ, Haanen JB, Punt CJ. Cancer immunotherapy—revisited, *Nat Rev Drug Discov.* 10:591–600, 2011.)

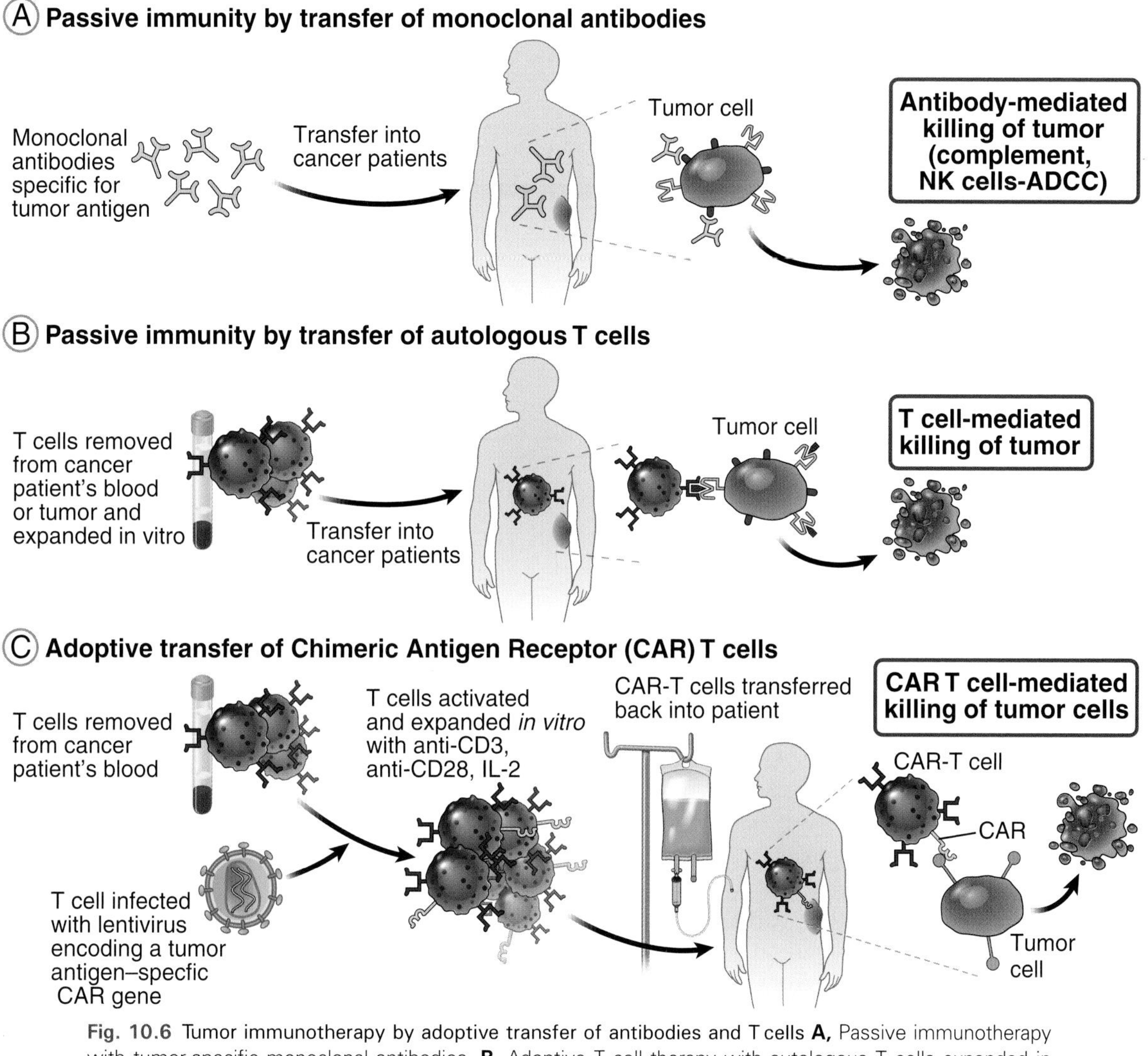

Fig. 10.6 Tumor immunotherapy by adoptive transfer of antibodies and T cells **A,** Passive immunotherapy with tumor-specific monoclonal antibodies. **B,** Adoptive T cell therapy with autologous T cells expanded in culture and transferred back into the patient. **C,** Adoptive T cell therapy with chimeric antigen receptor (CAR)-T cells: T cells isolated from the blood of a patient are expanded in culture, genetically modified to express recombinant CARs, and transferred back into the patient (see Fig. 10.7). *ADCC,* Antibody-dependent cellular cytotoxicity; *IL-2,* interleukin-2; *NK,* natural killer (cell).

donors, if necessary. CD20 is not expressed on plasma cells, so these cells survive and continue to produce antibodies. Because CD20 is also not expressed by hematopoietic stem cells, normal B cells are replenished after the antibody treatment is stopped. Other monoclonal antibodies that are used in cancer therapy may work by blocking growth factor signaling (e.g., anti-HER2 for breast cancer and anti–epidermal growth factor [EGF] receptor antibody for various tumors) or by inhibiting angiogenesis (e.g., antibody against vascular endothelial growth factor [VEGF] for colon cancer and other tumors).

Adoptive T Cell Therapy

Tumor immunologists have attempted to enhance antitumor immunity by removing T cells from cancer patients, activating the cells ex vivo to increase their numbers and differentiate them into potent effector cells, and transferring the activated cells back into the patient. Many variations of this approach, called adoptive T cell therapy, have been tried.

- **Adoptive therapy with autologous tumor-specific T cells.** T cells specific for tumor antigens can be detected in the circulation and tumor infiltrates in cancer patients. T cells can be isolated from the blood or tumor biopsy samples of a patient, expanded by culture with growth factors, and injected back into the same patient (Fig. 10.6B). Presumably, this expanded T cell population contains activated tumor-specific CTLs, which migrate into the tumor and destroy it. This approach, which has been combined with administration of T cell—stimulating cytokines such as interleukin-2 (IL-2) and traditional chemotherapy, has shown inconsistent results among different patients and tumors. One likely reason is that the frequency of tumor-specific T cells is too low to be effective in these lymphocyte populations. Attempts to overcome this limitation include isolating TCRs from the patient's tumor-specific T cell clones and introducing the TCRs into autologous T cells before transfer.
- **Chimeric antigen receptor (CAR)—expressing T cells.** In a more recent modification of adoptive T cell therapy, blood T cells from cancer patients are transduced with viral vectors that encode a chimeric antigen receptor (CAR), which is expressed on the T cell surface, recognizes a tumor antigen, and provides potent signals to activate the T cells (Fig. 10.6C). The CARs currently in use have a single-chain antibody forming the extracellular portion with both heavy- and light-chain variable domains, which together form the binding site for a tumor antigen (Fig. 10.7). The specificity of the endogenous T cell receptors (TCRs) of the transduced T cells is irrelevant to the effectiveness of this approach. The use of this antibody-based antigen recognition structure avoids the limitations of MHC restriction of TCRs and thus permits the use of the same CAR in many different patients, regardless of the human leukocyte antigen (HLA) alleles they express. Furthermore, tumors cannot evade CAR-T cells by downregulating MHC expression. In order to work in T cells, the CARs are constructed to have intracellular signaling domains of TCR complex proteins (e.g., the immunoreceptor tyrosine-based activation motifs [ITAMs] of the TCR complex ζ protein) and of costimulatory receptors such as CD28 and CD137. Therefore, these receptors provide both antigen recognition (via the extracellular Ig domain) and activating signals (via the introduced cytoplasmic domains). CAR-expressing T cells are expanded ex vivo and transferred back into the patient, where they recognize the antigen on the tumor cells and become activated to kill the cells. CAR-T cell therapy targeting the B cell proteins CD19 and CD20 has shown remarkable efficacy in treating and

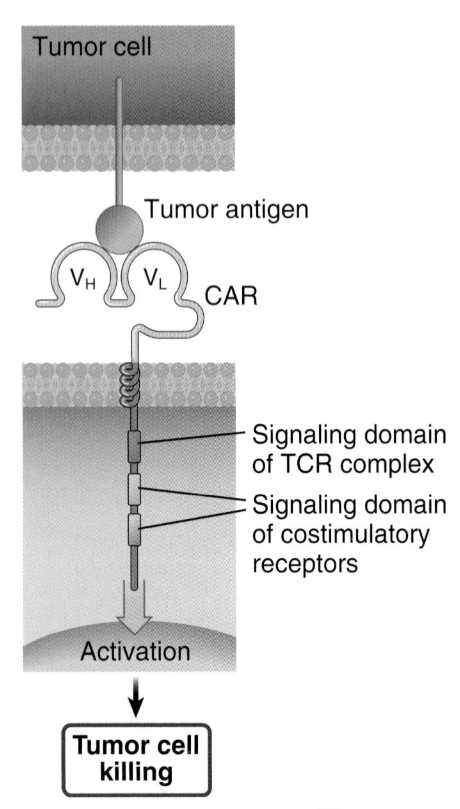

Fig. 10.7 Chimeric antigen receptor The receptor that is expressed in T cells consists of an extracellular immunoglobulin part that recognizes a surface antigen on tumor cells and intracellular signaling domains from the T cell receptor *(TCR)* complex and costimulatory receptors that provide the signals that activate the killing function of the T cells. *CAR,* Chimeric antigen receptor.

even curing B cell–derived leukemias and lymphomas that are refractory to other therapies. CARs with other specificities for plasma cell tumors are now approved and other CARs are in development and clinical trials.

The most serious toxicity associated with CAR-T cell therapy is a cytokine release syndrome, mediated by massive amounts of inflammatory cytokines, including IL-6, interferon-γ, and others, that are released because all of the injected T cells recognize and are activated by the patients' tumor cells. These cytokines cause high fever, hypotension, tissue edema, neurologic derangements, and multiorgan failure. The severity of the syndrome can be mitigated by treatment with antiinflammatory drugs and some anticytokine antibodies. CAR-T cell therapy also may be complicated by on-target, off-tumor toxicities, if the CAR-T cells are specific for an antigen that is present on normal cells as well as on the tumors. CD19- or CD20-specific CAR-T cells deplete normal B cells, which express these proteins, sometimes requiring antibody replacement therapy to prevent immunodeficiency. Such replacement may not be feasible for other tissues that are destroyed because of the reactivity of the CAR. Although CAR-T cell therapy is effective against leukemias and other tumors in the blood (to which the injected T cells have ready access), it has so far not been successful in solid tumors because of the challenge of selecting optimal tumor antigens to target without injuring normal tissues and difficulties in getting T cells into the tumor sites.

Immune Checkpoint Blockade

Blocking inhibitory receptors on T cells or their ligands stimulates antitumor immune responses. The realization that tumors evade immune attack by engaging regulatory mechanisms that suppress immune responses has led to a novel and remarkably effective new strategy for tumor immunotherapy. The principle of this strategy is to boost host immune responses against tumors by blocking normal inhibitory signals for T cells, thus removing the brakes (checkpoints) on the immune response (Fig. 10.8). This has been accomplished with blocking monoclonal antibodies specific for the T cell inhibitory molecules CTLA-4 and PD-1, first approved for treating metastatic melanoma in 2011 and 2014, respectively. Since then, the use of anti–PD-1 or anti–PD-L1 antibodies has expanded to many different cancer types. Blockade of the PD-1 pathway works by reducing the generation of exhausted T cells, promoting differentiation of memory T cells into effector cells, and thus reversing the exhaustion phenotype. The most remarkable feature of these therapies is that they have dramatically improved the chances of survival of many patients with advanced, widely metastatic tumors, which previously were almost 100% lethal within months to a few years, with some patients having gone into long-term remission. There are several novel features of immune checkpoint blockade and limitations that still need to be overcome to maximize their usefulness.

- Although the efficacy of checkpoint blockade therapies for many advanced tumors is superior to any previous form of therapy, only a subset of patients (15% to 40% for different tumors) respond to this treatment, and among patients with many tumor types (e.g., glioblastoma, pancreatic cancer, breast cancer), few or even none respond. The reasons for this poor response are not well understood. Nonresponding tumors may induce T cell expression of checkpoint molecules other than the ones being targeted therapeutically, or they may rely on evasion mechanisms other than engaging these inhibitory receptors. Oncologists and immunologists are currently investigating which biomarkers will predict responsiveness to different checkpoint blockade approaches.
- One of the most reliable indictors that a tumor will respond to checkpoint blockade therapy is if it carries a high number of mutations, which correlates with high numbers of neoantigens and host T cells that can respond to those antigens. In fact, tumors that have deficiencies in mismatch repair enzymes, which normally correct errors in DNA replication that lead to point mutations, have the highest mutation burdens of all cancers, and these cancers are the most likely to respond to checkpoint blockade therapy. Anti–PD-1 therapy is now approved for any recurrent or metastatic tumor with mismatch repair deficiencies, regardless of the cell of origin or histologic type of tumor. This is a paradigm shift in how cancer treatments are chosen, based on genetic features only regardless of histology or tissue of origin.

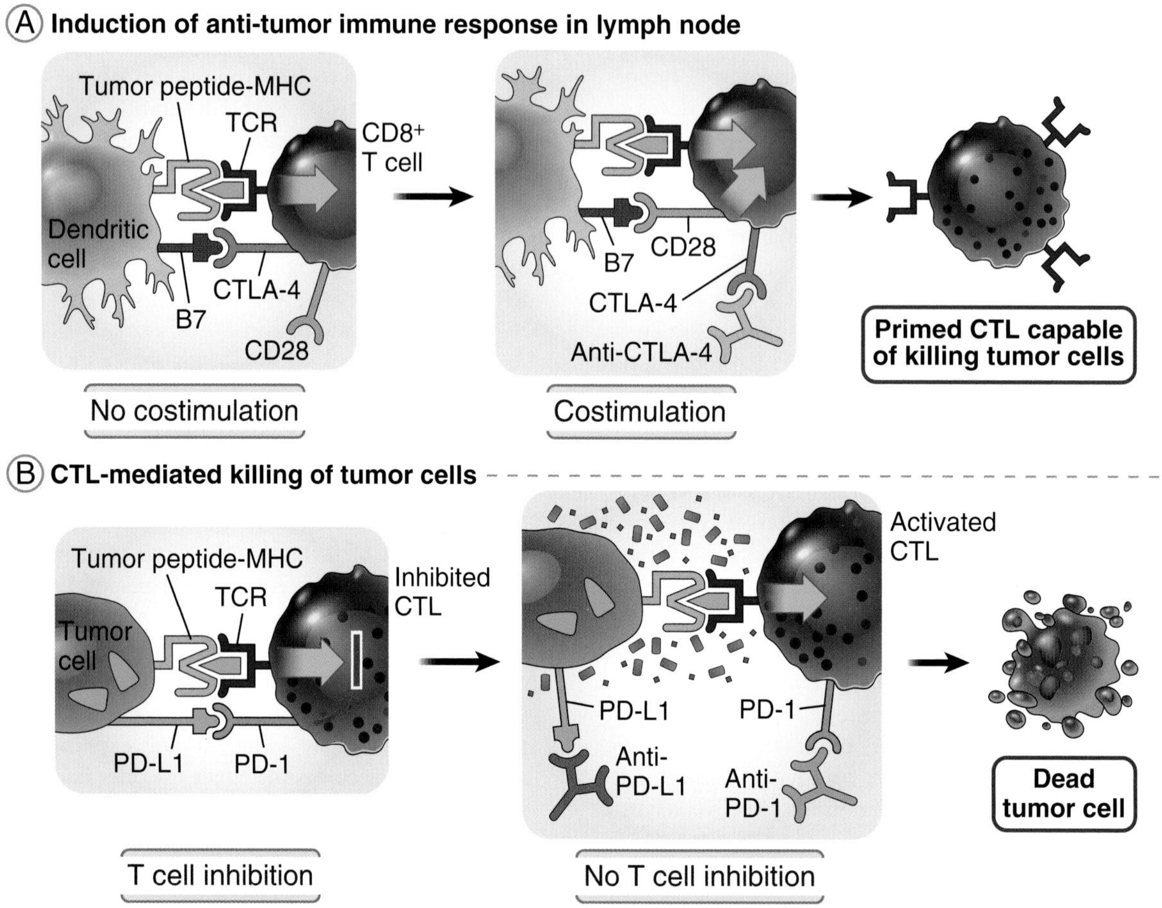

Fig. 10.8 Tumor immunotherapy by immune checkpoint blockade Tumor patients often mount ineffective T cell responses to their tumors because of the upregulation of inhibitory receptors such as CTLA-4 and PD-1 on the tumor-specific T cells, and expression of the ligand PD-L1 on the tumor cells and APCs. Blocking anti–CTLA-4 antibodies **(A)** or anti–PD-1 or anti–PD-L1 antibodies **(B)** are effective in treating several types of advanced tumors by releasing the inhibition of tumor-specific T cells by these molecules. Anti–CTLA-4 may work by blocking CTLA-4 on responding T cells (shown) or on regulatory T cells. *CTL,* Cytotoxic T lymphocyte; *CTLA-4,* cytotoxic T-lymphocyte antigen 4; *MHC,* major histocompatibility complex; *PD-1,* programmed cell death protein 1; *PD-L1,* PD-ligand 1; *TCR,* T cell receptors.

- The combined use of different checkpoint inhibitors, or one inhibitor with other modes of therapy, will likely be necessary to achieve higher rates of therapeutic success. The first approved example of this is the combined use of anti–CTLA-4 and anti–PD-1 to treat melanomas, which was shown to be more effective than anti–CTLA-4 alone. This reflects the fact that the mechanisms by which CTLA-4 and PD-1 inhibit T cell activation are different (see Fig. 10.8 and Fig. 5.14). There are numerous ongoing clinical trials blocking other checkpoint molecules, usually in combination with PD-1 blockade, or combining checkpoint blockade with other treatments, such as chemotherapy and radiation, small molecule kinase inhibitors, oncolytic viral infection of tumors, angiogenesis inhibitors, and immune stimulants (e.g., cytokines such as IL-2).
- The most common toxicities associated with checkpoint blockade are immune damage to organs. This is predictable, because the physiologic function of the inhibitory receptors is to maintain tolerance to self and other antigens, so blocking these checkpoints may unleash autoimmunity and possibly reactions to commensal organisms in the gut (see

Chapter 9). A wide range of organs may be affected, including colon, lungs, endocrine organs, heart, and skin, each requiring different clinical interventions, sometimes including cessation of the life-saving tumor immunotherapy.

Stimulation of Host Antitumor Immune Responses by Vaccination with Tumor Antigens

One strategy under investigation for stimulating active immunity against tumors is to vaccinate patients with tumor antigens. Unlike standard antimicrobial vaccines, which are prophylactic in that they prevent infections, tumor antigen vaccines are meant to be therapeutic, in that they are intended to stimulate immune responses against cancers that have already developed. An important reason for defining tumor antigens is to produce and use these antigens to vaccinate individuals against their own tumors. Most tumor vaccines tried to date have used differentiation antigens that are present on neoplastic cells and normal cells of the tissue type from which the tumor developed. There has been little success with this strategy, perhaps because there is strong tolerance to these self antigens that cannot be overcome by the vaccines.

More recently, there has been work on developing personalized cancer vaccines targeting neoantigens generated by random passenger mutations, which are unique to each patient's tumor. This approach relies on DNA sequencing to determine all the mutations in tumor DNA that would be predicted to be present in peptides that are most likely to bind to the HLA alleles of the patient. The personalized tumor vaccines are created using several of the neoantigen-containing peptides. This approach is promising, but it also has significant challenges, including the customization required for each patient and possible outgrowth of tumor clones that lose expression of MHC or the neoantigens.

Tumor-specific vaccines may be administered as a mixture of the antigen with adjuvants, just like antimicrobial vaccines. In another approach, a tumor patient's dendritic cells are expanded in vitro from blood precursors, the dendritic cells are exposed to tumor cells or tumor antigens, and these tumor-antigen—pulsed dendritic cells are used as vaccines. The dendritic cells bearing tumor antigens will theoretically mimic the normal pathway of cross-presentation and will generate CTLs against the tumor cells. The success of checkpoint blockade therapies, described previously, has raised hopes that vaccination used in combination with therapies to block immune regulation will have added benefits.

Tumors caused by oncogenic viruses can be prevented by vaccinating against these viruses. Two such vaccines that are proving to be remarkably effective are against human papillomavirus (the cause of cervical cancer and some types of oropharyngeal cancer) and hepatitis B virus (a cause of a form of liver cancer and liver failure due to fibrosis). These are prophylactic antiviral (not antitumor) vaccines given to individuals before they are infected and thus prevent infections by the tumor-causing viruses.

IMMUNE RESPONSES AGAINST TRANSPLANTS

Some of the earliest attempts to replace damaged tissues by transplantation were made during World War II as a way of treating pilots who had received severe skin burns in airplane crashes. It was soon realized that individuals reject tissue grafts from other individuals. Rejection results from inflammatory reactions that damage the transplanted tissues. Studies since the 1940s and 1950s established that graft rejection is mediated by the adaptive immune system because it shows specificity and memory and it is dependent on lymphocytes (Fig. 10.9). Much of the knowledge about the immunology of transplantation came from experiments with inbred strains of rodents, particularly mice. All members of an inbred strain are genetically identical to one another and different from the members of other strains. The experimental studies showed that grafts among members of one inbred strain are accepted and grafts from one strain to another are rejected, firmly establishing rejection as a process controlled by the animals' genes.

As mentioned in Chapter 3, the genes that contribute the most to the rejection of grafts exchanged between mice of different inbred strains are called MHC genes. The language of transplantation immunology evolved from the experimental studies. The individual who provides the graft is called the **donor**, and the individual in whom the graft is placed is the **recipient** or **host**. Animals that are identical to one another (and grafts exchanged among these animals) are said to be

Evidence	Conclusion
Prior exposure to donor MHC molecules leads to accelerated graft rejection	Graft rejection shows memory and specificity, two cardinal features of adaptive immunity
The ability to reject a graft rapidly can be transferred to a naive individual by lymphocytes from a sensitized individual	Graft rejection is mediated by lymphocytes
Depletion or inactivation of T lymphocytes by drugs or antibodies results in reduced graft rejection	Graft rejection requires T lymphocytes

Fig. 10.9 Evidence indicating that the rejection of tissue transplants is an immune reaction Clinical and experimental evidence indicates that rejection of grafts is a reaction of the adaptive immune system. *MHC,* Major histocompatibility complex.

syngeneic; animals (and grafts) of one species that differ from other animals of the same species are said to be **allogeneic**; and animals (and grafts) of different species are **xenogeneic**. Allogeneic and xenogeneic grafts, also called **allografts** and **xenografts** respectively, are always rejected by a recipient with a normal immune system. The antigens that serve as the targets of rejection are called alloantigens and xenoantigens, and the antibodies and T cells that react to these antigens are alloreactive and xenoreactive, respectively. In clinical practice, transplants are usually allografts, exchanged between two people who are members of an outbred species who differ genetically from one another unless they are identical twins). Most of the following discussion focuses on immune responses to allografts.

Transplantation Antigens

The antigens of allografts that serve as the principal targets of rejection are proteins encoded in the MHC. Homologous MHC genes and molecules are present in all mammals; the human MHC is called the **human leukocyte antigen (HLA)** complex. It took more than 20 years after the discovery of the MHC to show that the physiologic function of MHC molecules is to display peptide antigens for recognition by T lymphocytes (see Chapter 3). Recall that every person expresses six class I HLA alleles (one allele of *HLA-A*, *-B*, and *-C* from each parent) and usually more than six class II HLA alleles (one allele of *HLA-DQ* and *HLA-DP* and one or two of *HLA-DR* from each parent). MHC genes are highly polymorphic, with thousands of alleles of *HLA-A*, *HLA-B*, *HLA-C*, and *HLA-DRβ* genes and hundreds of alleles of *HLA-DQβ* and *HLA-DPβ* genes in the population. Because of this tremendous polymorphism, two unrelated individuals are very likely to express several HLA proteins that are different from, and therefore appear foreign to, each other. Because the genes in the HLA locus are tightly linked, all the HLA genes from each parent are inherited together, as a haplotype, in a Mendelian pattern, and therefore the chance that two siblings will have the same MHC alleles is 1 in 4.

The reaction to allogeneic MHC antigens on another individual's cells is one of the strongest immune responses known. TCRs for antigens have evolved to recognize MHC molecules, which is essential for surveillance of cells harboring infectious microbes. As a result of positive selection of developing T cells in the thymus, mature T cells that have some affinity for self MHC molecules survive, and many of these will, by chance, have high affinity for self MHC displaying foreign peptides (as in normal immune responses). Allogeneic MHC molecules containing peptides derived from the allogeneic cells may look like self MHC molecules plus bound foreign peptides (Fig. 10.10). Therefore, recognition of allogeneic MHC molecules in allografts is an example of an immunologic cross-reaction.

There are several reasons why recognition of allogeneic MHC molecules results in strong T cell reactions. Many clones of T cells, including memory T cells generated from prior infections, that are specific

Ⓐ Normal

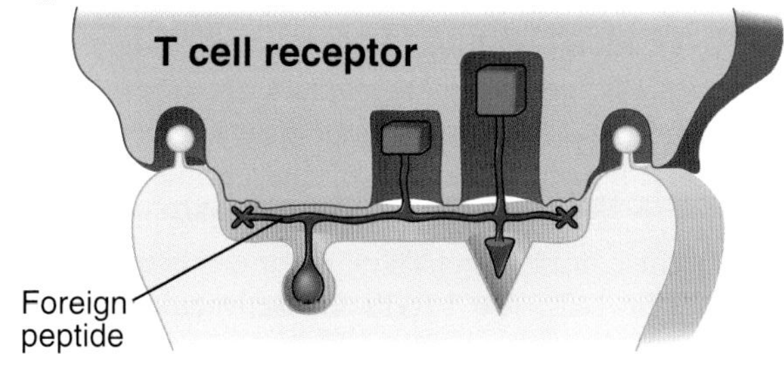

Self MHC molecule presents foreign peptide to T cell selected to recognize self MHC weakly, but may recognize self MHC-foreign peptide complexes well

Ⓑ Allorecognition

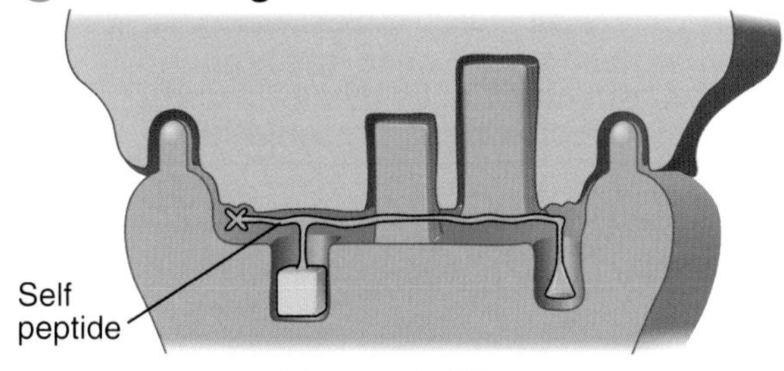

The self MHC-restricted T cell recognizes the allogeneic MHC molecule whose structure resembles a self MHC-foreign peptide complex

Ⓒ Allorecognition

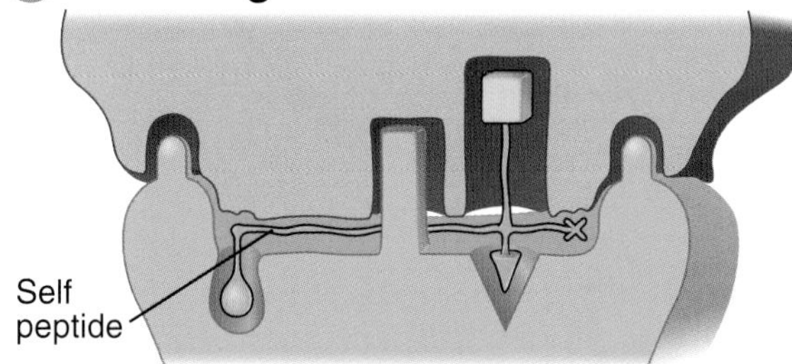

The self MHC-restricted T cell recognizes a structure formed by both the allogeneic MHC molecule and the bound peptide

for different foreign peptides bound to the same self MHC molecule may cross-react with any one allogeneic MHC molecule, regardless of the bound peptide, as long as the allogeneic MHC molecule resembles complexes of self MHC plus foreign peptides. As a result, many self MHC–restricted T cells specific for different peptide antigens may recognize any one allogeneic MHC molecule. In addition, the process of negative selection in the thymus eliminates cells that strongly recognize self MHC, but there is no mechanism for selectively eliminating T cells whose TCRs have a high affinity for allogeneic MHC molecules because these are never present in the thymus. Furthermore, a single allogeneic graft cell will express thousands of MHC molecules, every one of which may be recognized as foreign by a graft recipient's T cells. By contrast, in the case of an infected cell, only a small fraction of the self MHC molecules on the cell surface will carry a foreign microbial peptide recognized by the host's T cells. The net result of these features of allorecognition is that the frequency of alloreactive T cells in any individual is estimated to be at least 1000-fold greater than the frequency of T cells that recognize any one microbial antigen.

Although MHC proteins are the major antigens that stimulate graft rejection, other polymorphic proteins may also play a role. Non-MHC antigens that induce graft rejection are called minor histocompatibility antigens, and most are normal cellular proteins that differ in sequence between donor and recipient. These polymorphic proteins yield peptides that are presented by the recipient's MHC molecules and trigger a T cell response. The rejection reactions that minor histocompatibility antigens elicit are not as strong as reactions against foreign MHC proteins.

Fig. 10.10 Recognition of allogeneic major histocompatibility complex (MHC) molecules by T lymphocytes Recognition of allogeneic MHC molecules may be thought of as a cross-reaction in which a T cell specific for a self MHC molecule–foreign peptide complex **(A)** also recognizes an allogeneic MHC molecule whose structure resembles that of the self MHC molecule–foreign peptide complex (**B** and **C**). Peptides derived from the graft or recipient (labeled self peptide) may not contribute to allorecognition **(B)**, or they may form part of the complex that the T cell recognizes **(C)**. The type of T cell recognition depicted in **B** and **C** is called direct allorecognition.

Induction of Immune Responses Against Transplants

In order to elicit antigraft immune responses, alloantigens from the graft are transported by dendritic cells to draining lymph nodes, where they are recognized by alloreactive T cells (Fig. 10.11). The dendritic cells that present alloantigens also provide costimulators and can stimulate helper T cells as well as alloreactive CTLs. The effector T cells that are generated circulate back to the transplant and mediate rejection.

T cells in allograft recipients may recognize unprocessed donor MHC molecules on the surface of graft cells, or they may recognize peptides derived from donor MHC molecules bound to recipient MHC molecules on the surface of recipient APCs (Fig. 10.12). These two pathways of presentation of graft antigens have different features and names.

- **Direct allorecognition.** Most tissues contain dendritic cells, and when the tissues are transplanted, the dendritic cells in the graft may migrate to secondary lymphoid organs of the recipient. When naive T cells in the recipient recognize the donor allogeneic MHC molecules on these graft-derived dendritic cells, the T cells are activated; this process is called **direct recognition** (or direct presentation) of alloantigens. Direct recognition stimulates the development of alloreactive T cells (e.g., CTLs) that can then recognize the allogeneic MHC molecules on cells of the graft and destroy the graft.

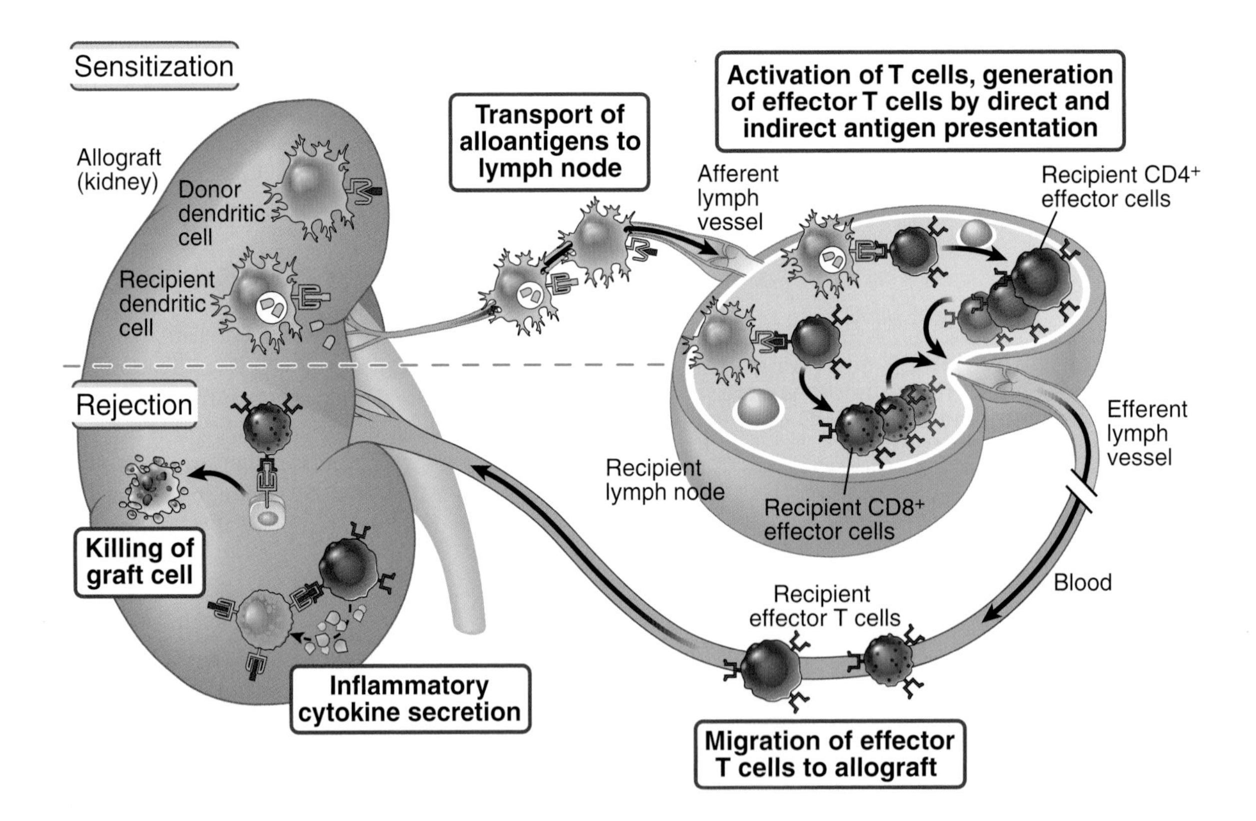

Fig. 10.11 **Immune response against transplants** Graft antigens that are expressed on donor dendritic cells or captured by recipient dendritic cells are transported to peripheral lymphoid organs where alloantigen-specific T cells are activated (the sensitization step). The T cells migrate back into the graft and destroy graft cells (rejection). Antibodies are also produced against graft antigens and can contribute to rejection (not shown). The example shown is that of a kidney graft, but the same general principles apply to all organ grafts.

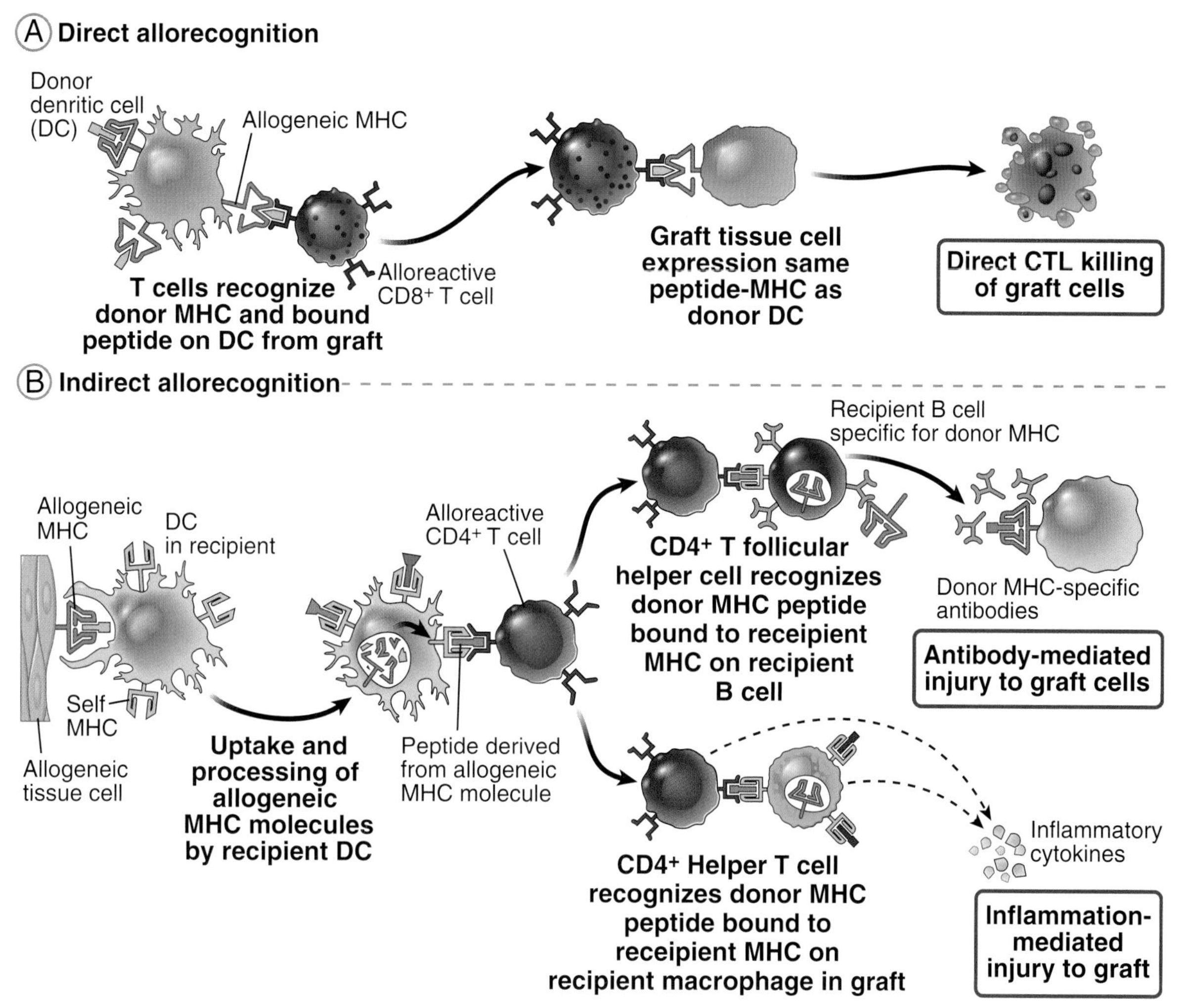

Fig. 10.12 Direct and indirect recognition of alloantigens **A,** Direct alloantigen recognition occurs when T cells bind directly to intact allogeneic major histocompatibility complex (MHC) molecules on antigen-presenting cells (APCs) in a graft, as illustrated in Fig. 10.8. $CD4^+$ T cells also may be activated upon recognition of allogeneic class II MHC molecules on graft APCs (not shown). **B,** Indirect alloantigen recognition occurs when allogeneic MHC molecules from graft cells are taken up and processed by recipient APCs, and peptide fragments of the allogeneic MHC molecules are presented by recipient (self) MHC molecules. Recipient APCs also may process and present graft proteins other than allogeneic MHC molecules.

- **Indirect allorecognition.** Graft cells or alloantigens may be ingested by recipient dendritic cells and transported to draining lymph nodes. Here, donor alloantigens (usually donor MHC molecules) are processed and presented by self MHC molecules on the recipient APCs. This process is called **indirect recognition** (or indirect presentation) and is similar to the cross-presentation of viral or tumor antigens to $CD8^+$ T cells, discussed in Chapter 3 and earlier in this chapter. If alloreactive CTLs are induced by the indirect pathway, these CTLs are specific for donor alloantigens displayed by the recipient's self MHC molecules on the recipient's APCs, so they cannot recognize and kill cells in the graft (which, of course, express donor MHC molecules). When graft alloantigens are recognized by the indirect pathway, the subsequent rejection of the graft likely is mediated mainly by alloreactive $CD4^+$ T cells.

These T cells may enter the graft together with host APCs, recognize graft antigens that are picked up and displayed by these APCs, and secrete cytokines that injure the graft by an inflammatory reaction. Indirect allorecognition by host $CD4^+$ T cells also contributes to production of host antibodies that bind to graft MHC molecules, as discussed later.

We do not know the relative importance of the direct and indirect pathways of allorecognition in T cell–mediated rejection of allografts. The direct pathway may be most important for CTL-mediated acute rejection, and the indirect pathway may play a greater role in chronic rejection, as described later.

T cell responses to allografts require costimulation, but which stimuli in grafts enhance the expression of costimulators on APCs is unclear. As with tumors, graft cells may undergo necrosis, perhaps in the period of ischemia between removal of the organ from the donor and placement in the recipient, and substances released from the injured and dead cells activate APCs by innate immune mechanisms. As we discuss later, blocking costimulation is one therapeutic strategy for promoting graft survival.

The **mixed lymphocyte reaction** (MLR) is an in vitro model of T cell recognition of alloantigens. In this model, T cells from one individual are cultured with leukocytes of another individual, and the responses of the T cells are assayed. The magnitude of this response is proportional to the extent of the MHC differences between these individuals and is a rough predictor of the outcomes of grafts exchanged between these individuals.

Although much of the emphasis on allograft rejection has been on the role of T cells, alloantibodies also contribute to rejection. Most of these antibodies are helper T cell–dependent high-affinity antibodies. In order to produce alloantibodies, recipient B cells recognize donor alloantigens and then process and present peptides derived from these antigens to helper T cells (that may have been previously activated by recipient dendritic cells presenting the same donor alloantigen), thus initiating the process of antibody production (see Fig. 10.12B). This is a good example of indirect presentation of alloantigens, in this case by B lymphocytes.

Immune Mechanisms of Graft Rejection

Graft rejection is classified into hyperacute, acute, and chronic, on the basis of clinical and pathologic features (Fig. 10.13). This historical classification was devised by clinicians for the diagnosis of kidney allograft rejection, and it has proved useful for understanding the mechanisms of graft rejection and for predicting outcomes and designing therapies. It has become apparent that each type of rejection is mediated by a particular type of immune response.

- **Hyperacute rejection** occurs within minutes of transplantation and is characterized by thrombosis of graft vessels and ischemic necrosis of the graft. Hyperacute rejection is mediated by circulating antibodies that are specific for antigens on graft endothelial cells and that are present before transplantation. These preformed antibodies may be natural IgM antibodies specific for blood group antigens (discussed later in this chapter), or they may be antibodies specific for allogeneic MHC molecules that were induced by previous exposure to allogeneic cells due to blood transfusions, pregnancy, or prior organ transplantation. Almost immediately after transplantation, the antibodies bind to antigens on the graft vascular endothelium and activate the complement and clotting systems, leading to injury to the endothelium and thrombus formation. Hyperacute rejection is not a common problem in clinical transplantation because donors and recipients are matched for blood type and potential recipients are tested for antibodies against the cells of the prospective donor. (The test for antibodies is called a cross-match.) However, hyperacute rejection is a major barrier to xenotransplantation, as discussed later.
- **Acute rejection** occurs usually within days or weeks after transplantation but may occur months later, and is the principal cause of early graft failure. Acute rejection is mediated by T cells and antibodies specific for alloantigens in the graft. The T cells may be $CD8^+$ CTLs that directly destroy graft cells or $CD4^+$ cells that secrete cytokines and induce inflammation, which destroys the graft. T cells may also react against cells in graft vessels, leading to vascular damage. Antibodies contribute especially to the vascular component of acute rejection. Antibody-mediated injury to graft vessels is caused mainly by complement activation by the classical pathway. Current immunosuppressive therapy is designed to prevent and reduce acute rejection by blocking the activation of alloreactive T cells.
- **Chronic rejection** is an indolent form of graft damage that develops over months or years, leading to

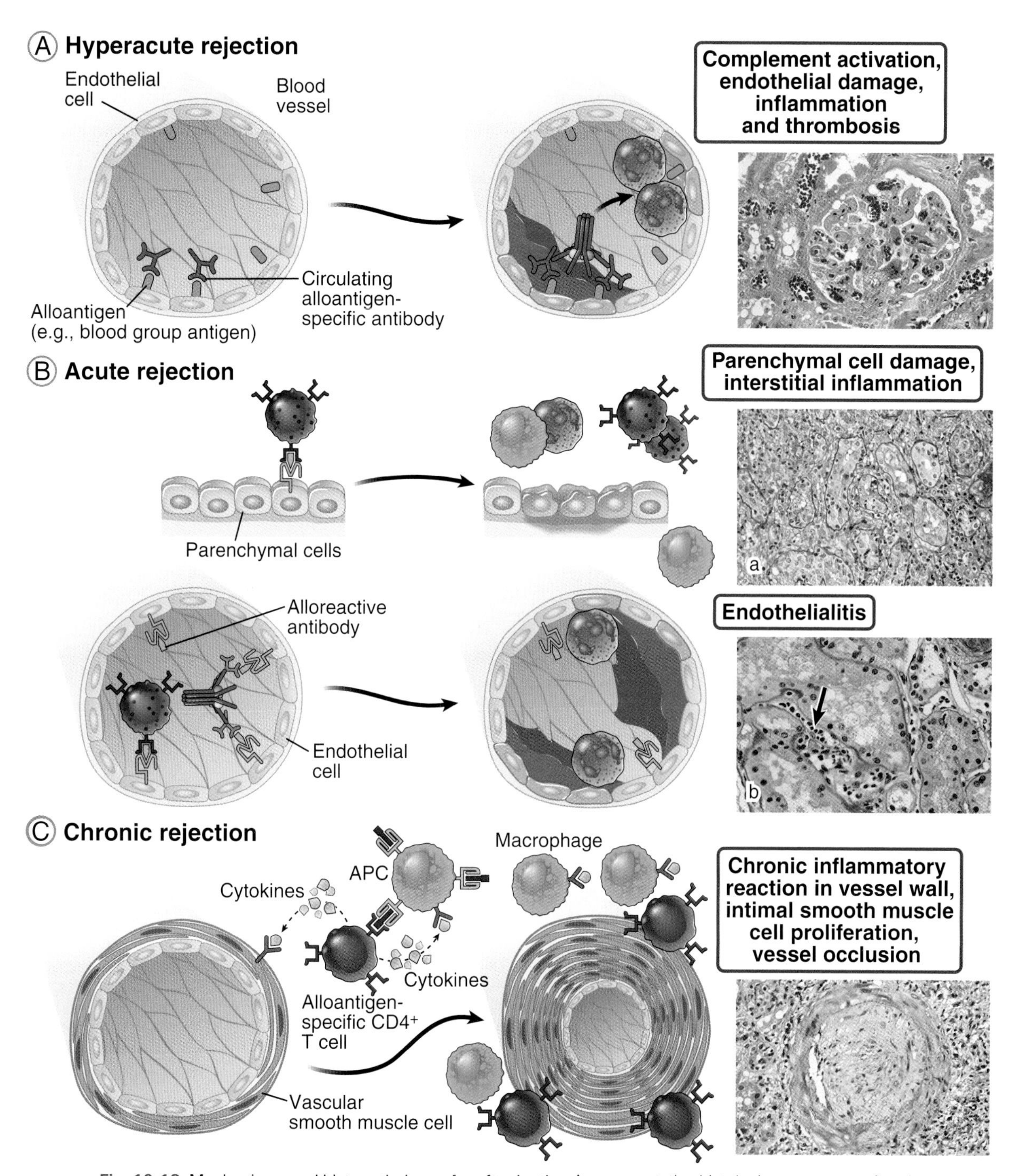

Fig. 10.13 Mechanisms and histopathology of graft rejection A representative histologic appearance of each type of rejection is shown on the right. **A,** In hyperacute rejection, preformed antibodies react with alloantigens on the vascular endothelium of the graft, activate complement, and trigger rapid intravascular thrombosis and ischemic necrosis of the graft. **B,** In acute rejection, $CD8^+$ T lymphocytes reactive with alloantigens on graft endothelial cells and parenchymal cells or antibodies reactive with endothelial cells cause damage to these cell types. Inflammation of the endothelium is called endothelialitis. The histology shows acute cellular rejection in **a** and humoral (antibody-mediated) rejection in **b**. **C,** In chronic rejection with graft arteriosclerosis, T cells reactive with graft alloantigens may produce cytokines that induce inflammation and proliferation of intimal smooth muscle cells, leading over years to luminal occlusion. *APC,* Antigen-presenting cell.

progressive loss of graft function. Chronic rejection may be manifested by gradual narrowing of graft blood vessels, called graft arteriosclerosis, and fibrosis of the graft. In both lesions, the culprits are thought to be T cells that react against graft alloantigens and secrete cytokines, which stimulate the proliferation and activities of fibroblasts and vascular smooth muscle cells in the graft. Alloantibodies may also contribute to chronic rejection. Although treatments to prevent or suppress acute rejection have steadily improved, leading to better 1-year survival of transplants, chronic rejection is refractory to most of these therapies and is becoming the principal cause of graft failure.

Prevention and Treatment of Graft Rejection

The mainstay of preventing and treating the rejection of organ transplants is immunosuppression, using drugs that deplete T cells or inhibit T cell activation and effector functions (Fig. 10.14). The development of immunosuppressive drugs launched the modern era of organ transplantation because these drugs made it feasible to transplant organs from donors that were not HLA-matched with recipients, especially in situations in which such matching was impractical, such as transplantation of heart, lung, and liver.

One of the first and still most useful classes of immunosuppressive drugs used in clinical transplantation are the calcineurin inhibitors, including cyclosporine and FK506 (tacrolimus), which function by blocking the protein phosphatase calcineurin. This enzyme is required to allow the translocation of the transcription factor NFAT (nuclear factor of activated T cells) to the nucleus, and blocking its activity inhibits the transcription of cytokine genes in the T cells. Another widely used drug is rapamycin (sirolimus), which inhibits a kinase called molecular target of rapamycin (mTOR) required for T cell activation. Many other immunosuppressive agents are now used as adjuncts to or instead of calcineurin and mTOR inhibitors (see Fig. 10.14).

All of these immunosuppressive drugs carry the problem of nonspecific immunosuppression (i.e., the drugs inhibit responses to more than the graft). Therefore, patients receiving these drugs as part of their post-transplantation treatment regimen become susceptible to infections, particularly by intracellular microbes, and the patients have an increased risk for developing cancers, especially skin cancers and others caused by oncogenic viruses.

The matching of donor and recipient HLA alleles by tissue typing had an important role in minimizing graft rejection before cyclosporine became available for clinical use. Although MHC matching is critical for the success of transplantation of some types of tissues (e.g., hematopoietic stem cell transplants) and improves survival of other types of organ grafts (e.g., renal allografts), modern immunosuppression is so effective that HLA matching is not considered necessary for many types of organ transplants (e.g., heart and liver), mainly because the number of donors is limited and the recipients often are too sick to wait for well-matched organs to become available.

The long-term goal of transplant immunologists is to induce immunologic tolerance specifically for the graft alloantigens. If this is achieved, it will allow graft acceptance without shutting off other immune responses in the host. However, many years of experimental and clinical attempts to induce graft-specific tolerance have not yet resulted in clinically practical methods.

A major problem in transplantation is the shortage of suitable donor organs. **Xenotransplantation** has been considered a possible solution for this problem, and there has been extensive research on the use of pigs as a source of grafts because their organs are similar in size to human organs. Experimental studies show that hyperacute rejection is a frequent cause of pig and other mammalian xenotransplant loss. The reasons for the high incidence of hyperacute rejection of xenografts are that individuals often have antibodies that cross-react with carbohydrate antigens on cells from other species and the xenograft cells lack regulatory proteins that can inhibit human complement activation. These antibodies, similar to antibodies against blood group antigens, are called natural antibodies because their production does not require prior exposure to the xenoantigens. It is thought that these antibodies are produced against bacteria that normally inhabit the gut and that the antibodies cross-react with cells of other

Drug	Mechanism of action
Cyclosporine, FK506 (tacrolimus)	Blocks T cell cytokine production by inhibiting the phosphatase calcineurin and thus blocking activation of the NFAT transcription factor
Mycophenolate mofetil	Blocks lymphocyte proliferation by inhibiting guanine nucleotide synthesis in lymphocytes
Rapamycin	Blocks lymphocyte proliferation by inhibiting mTOR and IL-2 signaling
Corticosteroids	Reduce inflammation by effects on multiple cell types
Anti-thymocyte globulin	Binds to and depletes T cells by promoting phagocytosis or complement-mediated lysis (Used to treat acute rejection)
Anti-IL-2 receptor (CD25) antibody	Inhibits T cell proliferation by blocking IL-2 binding; may also opsonize and help eliminate activated IL-2R-expressing T cells
CTLA4-Ig (belatacept)	Inhibits T cell activation by blocking B7 costimulator binding to T cell CD28
Anti-CD52 (alemtuzumab)	Depletes lymphocytes by complement-mediated lysis

Fig. 10.14 Treatments for graft rejection. Agents used to treat rejection of organ grafts and their mechanisms of action. Like cyclosporine, tacrolimus (FK506) is a calcineurin inhibitor. *CTLA4-Ig,* Cytotoxic T lymphocyte–associated protein 4–immunoglobulin (fusion protein), not widely used; *IL,* interleukin; *mTOR,* molecular target of rapamycin; *NFAT,* nuclear factor of activated T cells.

species. Xenografts are also subject to acute rejection, much like allografts but often more severe than rejection of allografts. One solution to xenograft rejection has been to genetically modify the donor animals so that they do not produce the relevant xenoantigens and/or express human complement regulatory proteins. The only patient so far to receive a heart graft from a genetically modified pig, in early 2022, survived for 2 months.

Transplantation of Blood Cells and Hematopoietic Stem Cells

Transfer of blood cells between humans, called transfusion, is the oldest form of transplantation in clinical medicine. The major barrier to transfusion is the presence of different blood group antigens in different individuals, the prototypes of which are the ABO antigens, and natural antibodies produced against these (Fig. 10.15). The antigens are expressed on red blood cells, endothelial cells, and many other cell types. ABO antigens are carbohydrates on membrane glycoproteins or glycosphingolipids; they contain a core glycan that may be enzymatically modified by addition of either of two types of terminal sugar residues. There are three alleles of the gene encoding the enzyme that adds these sugars: one encodes an enzyme that adds N-acetylgalactosamine, one that adds galactose, and one that is inactive and cannot add either. Therefore, depending on the alleles inherited, an individual may be one of four different ABO blood groups: Blood group A individuals have N-acetylgalactosamine added to the core glycan; blood group B individuals have a terminal galactose; blood group AB individuals express both terminal sugars on different glycolipid or glycoprotein molecules; and individuals with blood group O express the core glycan without either of the terminal sugars.

Individuals are tolerant of the blood group antigens they express but make antibodies specific for the antigens they do not express. Thus, type A individuals make anti-B antibodies, type B individuals make anti-A antibodies, O group individuals make both anti-A and anti-B, and type AB individuals do not make anti-A or anti-B antibodies. These antibodies are called natural antibodies because they are made in the absence of overt exposure to the blood group antigens they recognize. They are likely produced by B cells in response to structurally similar antigens of intestinal microbes, and the antibodies cross-react with ABO blood group antigens. Because the blood group antigens are sugars, they do not elicit T cell responses that drive isotype switching, and the antibodies specific for A or B antigens are largely IgM. The preformed antibodies react against transfused blood cells expressing

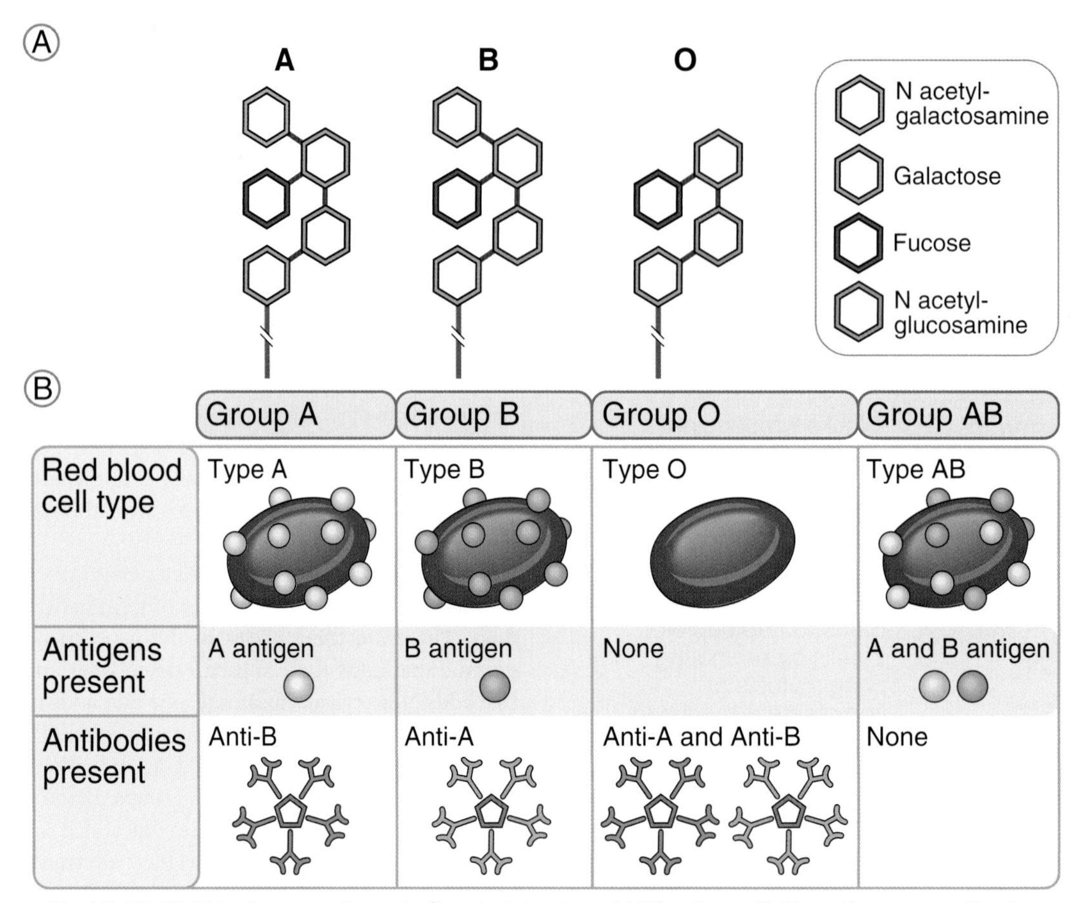

Fig. 10.15 ABO blood group antigens **A,** Chemical structure of ABO antigens. **B,** The antigens and antibodies present in people with the major ABO blood groups.

the target antigens and activate complement, which lyses the red cells; the result may be a severe **transfusion reaction**, characterized by a systemic inflammatory response, intravascular thrombosis, and kidney damage. This problem is avoided by matching blood donors and recipients so there are no antigens on the donor cells that can be recognized by preformed antibodies in the recipient, a standard practice in medicine.

Blood group antigens other than the ABO antigens are also involved in transfusion reactions, and these usually are less severe. One important example is the RhD antigen, which is a red cell membrane protein expressed by about 90% of people. Pregnant women who are RhD-negative can be immunized by exposure to RhD-expressing red cells from the baby during childbirth if the baby inherited the RhD gene from the father. The mother will produce anti-RhD antibodies that can cross the placenta during subsequent pregnancies and attack Rh-positive fetal cells, causing hemolytic disease of the fetus and newborn. This problem is prevented by treating the mother with anti-RhD antibody at the time of the first delivery; the injected antibody clears Rh-positive fetal cells from the maternal circulation and blocks the Rh antigen, thus reducing its immunogenicity.

Hematopoietic stem cell (HSC) transplantation is being used increasingly to treat blood cancers (e.g., leukemias, myeloma) and correct hematopoietic defects. Either bone marrow cells or, more often, HSCs mobilized in a donor's blood are injected into the circulation of a recipient, and the cells home to the marrow. The transplantation of HSCs poses many special problems. Before transplantation, some of the bone marrow of the recipient has to be destroyed to create space to receive the transplanted stem cells, and this depletion of the recipient's marrow inevitably causes deficiency of blood cells, including immune cells, resulting in potentially serious immune deficiencies before the transplanted stem cells generate enough replacement blood cells. The immune system reacts strongly against allogeneic HSCs, so successful transplantation requires careful HLA matching of donor and recipient. If mature allogeneic T cells are transplanted with the stem cells, these mature T cells can attack the recipient's tissues, resulting in **graft-versus-host disease**. When the donor is an HLA-identical sibling, this reaction is directed against minor histocompatibility antigens. The same reaction is exploited to kill leukemia cells (so-called graft-versus-leukemia effect), so depleting T cells from the donor HSCs to prevent graft-vs-host disease also decreases the antileukemia effect.

Despite these problems, hematopoietic stem cell transplantation is a successful therapy for a wide variety of diseases affecting the hematopoietic and lymphoid systems.

SUMMARY

- The adaptive immune system is able to eradicate or prevent the growth of tumors.
- Tumors may induce antibody, $CD4^+$ T cell, and $CD8^+$ T cell responses, but $CD8^+$ cytotoxic T lymphocyte (CTL) killing of tumor cells appears to be the most important antitumor effector mechanism.
- Most cancer antigens that induce T cell responses are neoantigens encoded by randomly mutated genes (passenger mutations), which do not contribute to the malignant phenotype of the cancer cells. Other tumor antigens include products of oncogenes and tumor suppressor genes, overexpressed or aberrantly expressed structurally normal molecules, and products of oncogenic viruses.
- CTLs recognize mutant peptides derived from tumor antigens displayed by class I major histocompatibility complex (MHC) molecules. The induction of CTL responses against tumor antigens involves ingestion of tumor cells or their antigens by dendritic cells, cross-presentation of the antigens to naive $CD8^+$ T cells, activation of the T cells and differentiation into CTLs, CTL migration from the blood into tumors, CTL recognition of the tumor antigens on the tumor cells, and killing of the tumor cells.
- Tumors may evade immune responses by losing expression of their antigens, shutting off expression of MHC molecules or molecules involved in antigen processing, expressing ligands for T cell inhibitory receptors, and inducing regulatory T cells or secreting cytokines that suppress immune responses.
- CAR-T cell immunotherapy is a breakthrough approach now in clinical practice. CAR-T cells are generated in vitro by transducing a cancer patient's T cells to express a recombinant receptor with an antibody-like binding site for a tumor antigen and a cytoplasmic tail with potent T cell signaling functions. Adoptive transfer of CAR-T cells back into patients has been successful in treating B cell–derived leukemias, myelomas, and lymphomas.
- Immune checkpoint blockade is the major cancer immunotherapy strategy in current practice. Monoclonal antibodies that block the function of T cell inhibitory molecules, such as cytotoxic T-lymphocyte antigen 4 (CTLA-4) and programmed cell death protein 1 (PD-1), are injected into the patient, which enhances the activation of tumor-specific T cells by tumor antigens. This approach has been highly successful in treating patients with many kinds of advanced cancers, but only 15% to 40% of patients with different cancers respond, many tumor types are unresponsive, and many patients develop autoimmune side effects.
- Personalized neoantigen vaccines are now in clinical trials. The creation of these vaccines relies on cancer genome sequencing to identify neoantigen peptides unique to an individual patient's tumor, which bind to that patient's MHC molecules.
- Organ and tissue transplantation from one individual to another is widely used to treat many diseases, but a major barrier to successful transplantation of foreign tissues is rejection by adaptive immune responses, including $CD8^+$ CTLs, $CD4^+$ helper T cells, and antibodies.
- The most important antigens that stimulate graft rejection are allogeneic MHC molecules, which resemble peptide-loaded self MHC molecules that the graft recipient's T cells can recognize. Allogeneic MHC molecules are either presented by graft antigen-presenting cells without processing to recipient T cells (direct presentation), or are processed

and presented as peptides bound to self MHC by host APCs (indirect presentation).

- Grafts may be rejected by different mechanisms. Hyperacute rejection is mediated by preformed antibodies to blood group antigens or human leukocyte antigen molecules, which cause endothelial injury and thrombosis of blood vessels in the graft. Acute rejection is mediated by T cells, which injure graft cells and endothelium, and by antibodies that bind to the endothelium. Chronic rejection is caused by T cells that produce cytokines that stimulate growth of vascular smooth muscle cells and tissue fibroblasts.
- Treatment for graft rejection is designed to suppress T cell responses and inflammation. The mainstay of treatment has been immunosuppressive drugs, including calcineurin inhibitors, molecular target of rapamycin (mTOR) inhibitors, and many others.
- Blood cell transfusion is the oldest and most widely used form of transplantation and requires ABO blood group compatibility of donor and recipient. ABO blood group antigens are sugars expressed on the surfaces of red blood cells, endothelial cells, and other cells, and humans produce natural antibodies specific for the ABO antigens they do not express, which can destroy transfused blood cells from incompatible donors.
- Hematopoietic stem cell transplants are widely used to treat cancers of blood cells and to replace defective components of the immune or hematopoietic system. These cell transplants elicit strong rejection reactions, carry the risk for graft-versus-host disease, and often lead to temporary immunodeficiency in recipients.

REVIEW QUESTIONS

1. What are the main types of tumor antigens that the immune system reacts against?
2. What is the evidence that tumor rejection is an immunologic phenomenon?
3. How do naive $CD8^+$ T cells recognize tumor antigens, and how are these cells activated to differentiate into effector CTLs?
4. What are some of the mechanisms by which tumors may evade the immune response?
5. What are some strategies for enhancing host immune responses to tumor antigens?
6. Why do normal T cells, which recognize foreign peptide antigens bound to self MHC molecules, react strongly against the allogeneic MHC molecules of a graft?
7. What are the principal mechanisms of rejection of allografts?
8. How is the likelihood of graft rejection reduced in clinical transplantation?
9. What are some of the problems associated with the transplantation of hematopoietic stem cells?

Answers to and discussion of the Review Questions may be found on p. 322.

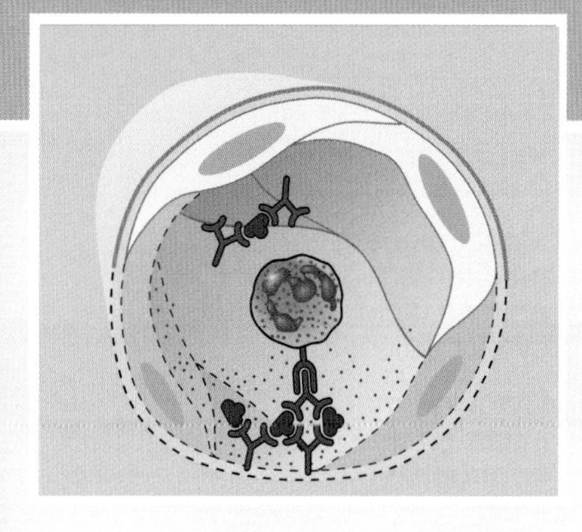

11

Hypersensitivity
Disorders Caused by Immune Responses

CHAPTER OUTLINE

The concept that the immune system is required for defending the host against infections has been emphasized throughout this book. However, immune responses are themselves capable of causing tissue injury and disease. Injurious, or pathologic, immune reactions are called **hypersensitivity reactions** because they reflect excessive or aberrantly directed immune responses. Hypersensitivity reactions may occur in two situations. First, responses to foreign antigens (microbes and noninfectious environmental antigens) may cause tissue injury, especially if the reactions are repetitive or poorly controlled. Second, the immune responses may be directed against self (autologous) antigens, as a result of the failure of self-tolerance (see Chapter 9). Responses against self antigens are termed **autoimmunity**, and disorders caused by such responses are called **autoimmune diseases**.

This chapter describes the important features of hypersensitivity reactions and the resulting diseases, focusing on their pathogenesis. Their clinicopathologic features are described only briefly and can be found in other medical textbooks. The following questions are addressed:

- What are the mechanisms of different types of hypersensitivity reactions?
- What are the major clinical and pathologic features of diseases caused by these reactions?
- What principles underlie treatment of such diseases?

TYPES OF HYPERSENSITIVITY REACTIONS

Hypersensitivity reactions are classified on the basis of the principal immunologic mechanism that is responsible for tissue injury and disease (Fig. 11.1).

Type of hypersensitivity	Pathologic immune mechanisms	Mechanisms of tissue injury and disease
Immediate hypersensitivity (Type I)	Th2 cells, IgE antibody, mast cells, eosinophils 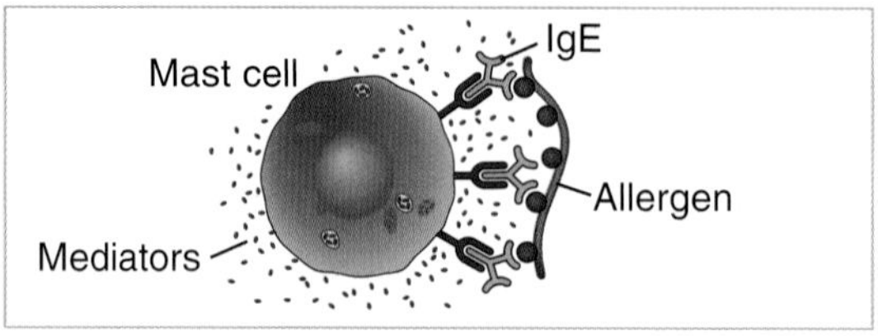	Mast cell-derived mediators (vasoactive amines, lipid mediators, cytokines) Cytokine-mediated inflammation (eosinophils, neutrophils)
Antibody-mediated diseases (Type II)	IgM, IgG antibodies against cell surface or extracellular matrix antigens 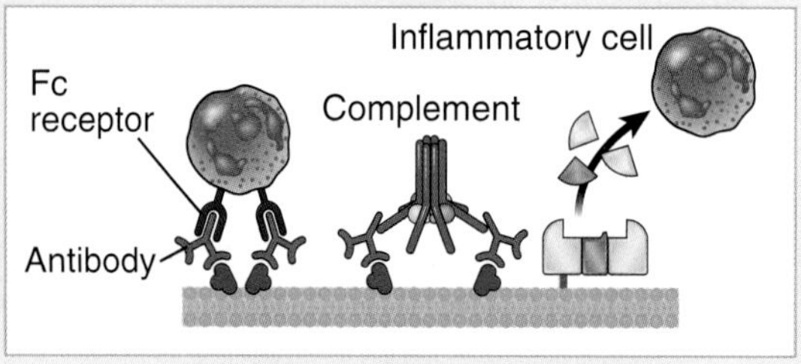	Complement- and Fc receptor-mediated recruitment and activation of leukocytes (neutrophils, macrophages) Opsonization and phagocytosis of cells Abnormalities in cellular function, e.g. hormone receptor signaling
Immune complex-mediated diseases (Type III)	Immune complexes of circulating antigens and IgM or IgG antibodies deposited in vascular basement membrane 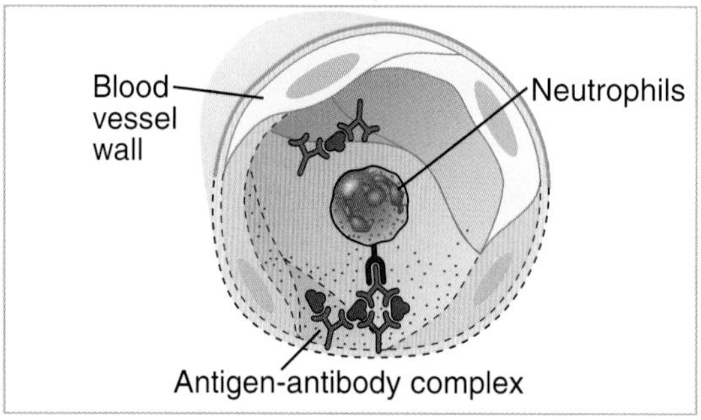	Complement- and Fc receptor-mediated recruitment and activation of leukocytes
T cell-mediated diseases (Type IV)	1. Cytokine-mediated inflammation ($CD4^+$ T cells) 2. T cell-mediated killing ($CD8^+$ CTLs) 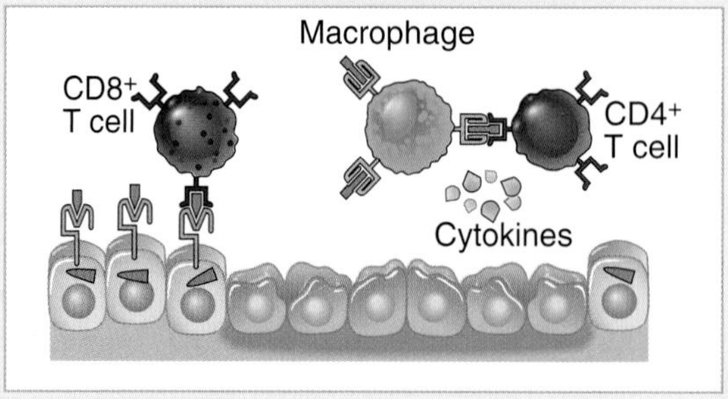	1. Macrophage activation, cytokine-mediated inflammation 2. Direct target cell lysis, cytokine-mediated inflammation

Fig. 11.1 **Types of hypersensitivity reactions.** In the four major types of hypersensitivity reactions, different immune effector mechanisms cause tissue injury and disease. *CTLs,* Cytotoxic T lymphocytes; *Ig,* immunoglobulin; *Th2,* T helper 2 cell.

We will use the informative descriptive classifications throughout this chapter, but we will also indicate the numerical designations for each type because they are widely used.

- Immediate hypersensitivity, or type I hypersensitivity, is caused by the release of mediators from mast cells. This reaction most often depends on the production of immunoglobulin E (IgE) antibody against environmental antigens and the binding of IgE to mast cells in various tissues.
- Antibodies (typically IgG) that are directed against cell or tissue antigens can damage these cells or tissues or can impair their function. These diseases are said to be antibody-mediated or type II hypersensitivity.
- Antibodies (also usually IgG) against soluble antigens in the blood may form complexes with the antigens, and the immune complexes may deposit in blood vessels in various tissues, causing inflammation and tissue injury. Such disorders are called immune complex diseases or type III hypersensitivity.
- Some diseases result from the reactions of T lymphocytes specific for self antigens or microbes in tissues. These are T cell–mediated diseases or type IV hypersensitivity.

This classification scheme is useful because it distinguishes the mechanisms of immune-mediated tissue injury. In many human immunologic diseases, however, the damage may result from a combination of antibody-mediated and T cell–mediated reactions, so it is often difficult to classify these diseases neatly into one type of hypersensitivity.

IMMEDIATE HYPERSENSITIVITY

Immediate hypersensitivity is an IgE antibody– and mast cell–mediated reaction to certain antigens that causes rapid vascular leakage, excessive mucosal secretions, and contraction of bronchial and intestinal smooth muscle, often followed by inflammation. A disorder in which IgE-mediated immediate hypersensitivity is prominent is also called an **allergy**, or atopy, and individuals with a propensity to develop these reactions are said to be atopic. The term immediate hypersensitivity arose from the rapid reaction to skin challenge by allergens in previously sensitized individuals, but many allergies are associated with chronic inflammation and are not only immediate reactions. Common types of allergic disorders include hay fever, food allergies, asthma, atopic dermatitis, and anaphylaxis. Allergies are the most frequent disorders of the immune system, estimated to affect 10% to 20% of people, and the incidence of allergic diseases has been increasing, especially in industrialized societies.

The sequence of events in the development of allergic reactions includes activation of T helper 2 (Th2) cells and interleukin-4 (IL-4)– and IL-13–secreting follicular helper T (Tfh) cells, which stimulate the production of IgE antibodies in response to an antigen; binding of the IgE to IgE-specific Fc receptors of mast cells; and on subsequent exposure to the antigen, cross-linking of the bound IgE by the antigen, leading to activation of the mast cells and release of various mediators (Fig. 11.2). Some of these reactions consist of two phases (Fig. 11.3), depending on the kinetics of production of different mediators. Mast cell mediators cause a rapid increase in vascular permeability and smooth muscle contraction, resulting in many of the symptoms of these reactions. This vascular and smooth muscle reaction may occur within minutes of reintroduction of antigen into a previously sensitized individual, hence the designation immediate. Other mast cell mediators are cytokines that recruit neutrophils and eosinophils to the site of the reaction over several hours. This inflammatory component is called the late-phase reaction, and it is mainly responsible for the tissue injury that results from repeated bouts of immediate hypersensitivity.

With this background, we proceed to a discussion of the steps in immediate hypersensitivity reactions.

Activation of Th2 Cells and Production of IgE Antibody

In individuals who are prone to allergies, exposure to some antigens results in the activation of Th2 cells and IL-4– and IL-13–secreting Tfh cells, and the production of IgE antibody (see Fig. 11.2A). Most individuals do not mount strong Th2 responses to environmental antigens. For unknown reasons, when some individuals encounter certain antigens, such as proteins in pollen, certain foods, insect venoms, or animal dander, or if they are treated with certain drugs, such as penicillin, there is a strong response characterized by production of IL-4, IL-5, and IL-13 by Th2 cells. Group 2 innate lymphoid cells (ILC2s) may produce IL-5 and IL-13 early, but their role in allergic

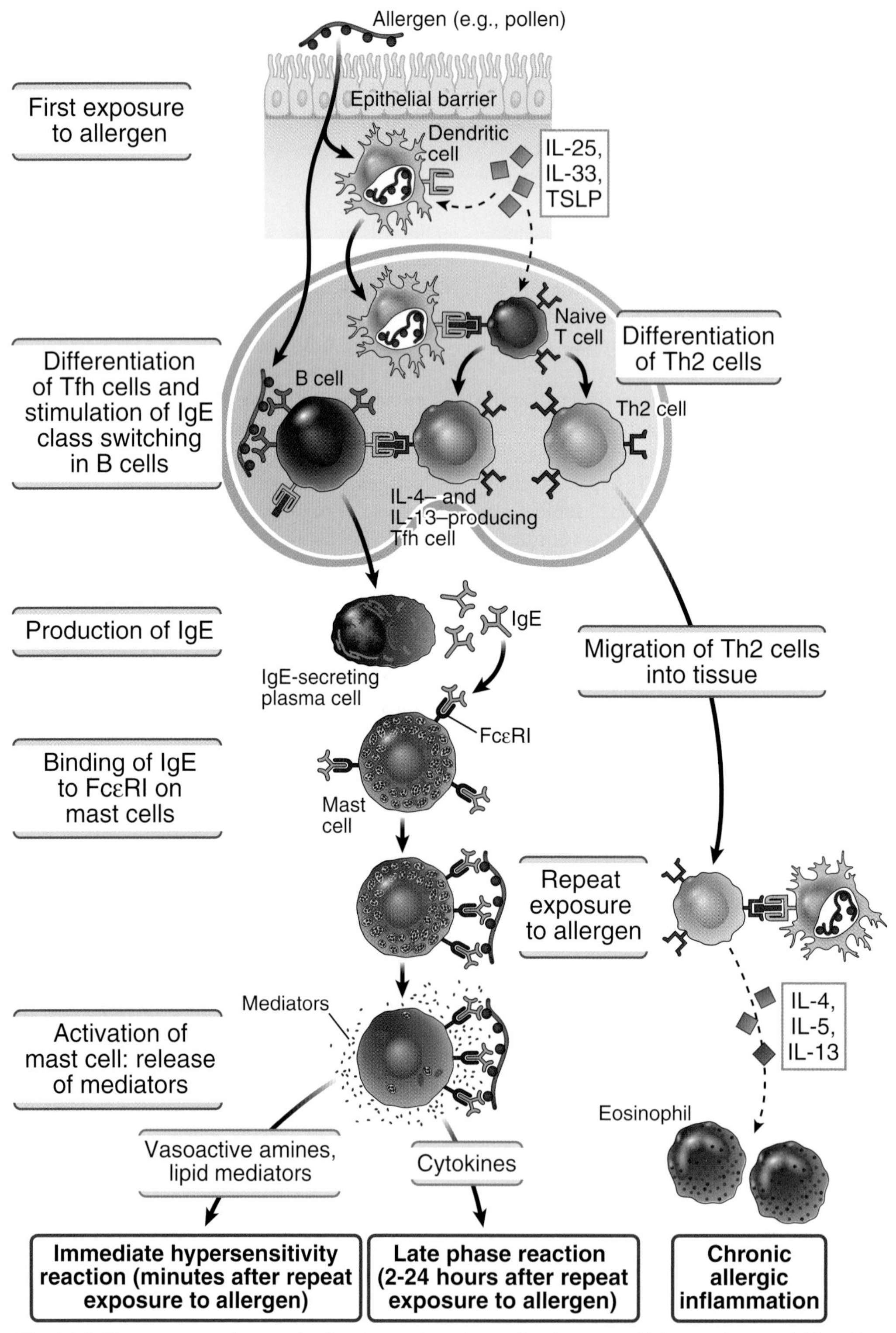

Fig. 11.2 The sequence of events in allergic reactions. Immediate hypersensitivity reactions are initiated by the introduction of an allergen, which, together with cytokines from epithelial cells, stimulates Th2 and IL-4/IL-13–producing Tfh cells. Immunoglobulin E (IgE) produced in response to antigen and Tfh cells binds to Fc receptors (FcεRI) on mast cells, and subsequent exposure to the allergen activates the mast cells to secrete the mediators that are responsible for the pathologic reactions of immediate hypersensitivity. Mast cells, Th2 cells, and other cells produce cytokines that elicit inflammation in the late-phase reaction and in chronic allergic diseases, often with abundant eosinophils. *APC,* Antigen-presenting cell; *TSLP,* thymic stromal lymphopoietin.

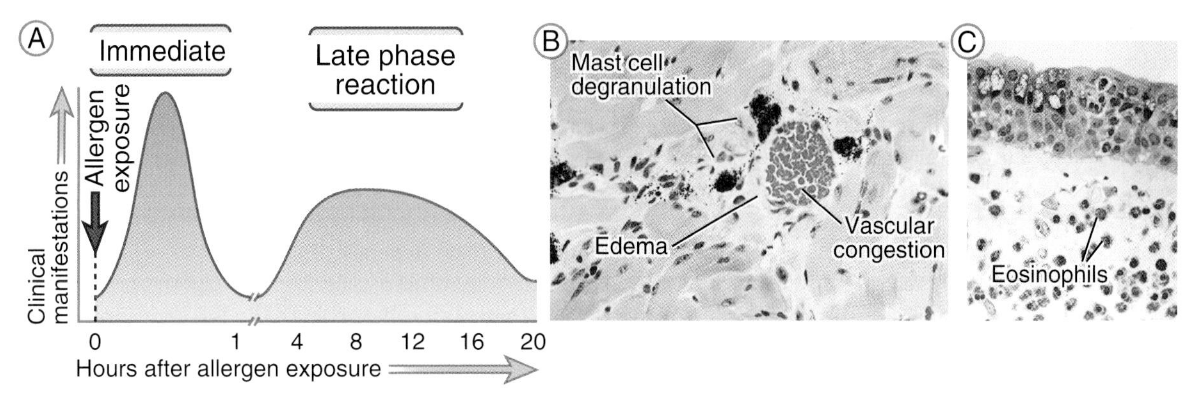

Fig. 11.3 Phases of allergic reactions. **A,** Kinetics of the immediate and late-phase reactions. The immediate vascular and smooth muscle reaction to allergen develops within minutes after challenge (allergen exposure in a previously sensitized individual), and the late-phase reaction develops 2 to 24 hours later. **B,** Morphology of the immediate reaction is characterized by vasodilation, congestion, and edema. **C,** The late-phase reaction is characterized by an inflammatory infiltrate rich in eosinophils, neutrophils, and T cells. (Micrographs courtesy the late Dr. Daniel Friend, Department of Pathology, Brigham and Women's Hospital, Boston.)

disease is not established. The similar reaction of Th2 cells and ILC2s is collectively called a type 2 response. Immediate hypersensitivity develops as a consequence of the activation of Th2 and IL-4— and IL-13—secreting Tfh cells in response to protein antigens or chemicals that bind to proteins. Antigens that elicit allergic reactions are called allergens. Any atopic individual may be allergic to one or more of these antigens. It is not understood why only a small subset of common environmental antigens elicit Th2-mediated reactions and IgE production, or what characteristics of these antigens are responsible for their behavior as allergens.

In some chronic allergic diseases, the initiating event may be epithelial barrier injury, which induces epithelial cells to secrete IL-25, IL-33, and thymic stromal lymphopoietin (TSLP), cytokines that induce type 2 immune responses. Epithelial injury in the skin is sometimes related to mutations in filaggrin, a keratinocyte protein needed to maintain normal barrier function. In the bronchial tree of the lung, viral infections are considered a potential cause of the initial injury. IL-25, IL-33, and TSLP activate ILC2s to produce IL-5 and IL-13, and dendritic cells (DCs) exposed to these cytokines drive differentiation of naive T cells in the lymph nodes toward Th2 and IL-4— or IL-13—producing Tfh cells.

In secondary lymphoid organs, IL-4 and IL-13 secreted by Tfh cells stimulate B lymphocytes to switch to IgE-producing cells that differentiate into plasma cells. As a result, atopic individuals produce large amounts of IgE antibody in response to antigens that do not elicit IgE responses in other people. IL-4 and IL-13 secreted by Th2 cells induce some of the responses of tissues in allergic reactions, such as increased intestinal motility and excess mucus secretions. Th2 cells also secrete IL-5, which promotes eosinophilic inflammation that is characteristic of tissues affected by allergic diseases.

The propensity for differentiation of antigen-activated T cells to IL-4— and IL-5—producing effector T cells, and resulting atopic diseases such as asthma, has a strong genetic basis. A major known risk for developing allergies is a family history of atopic disease, and gene association studies indicate that many different genes play contributory roles. Some of these genes encode cytokines or receptors known to be involved in T and B lymphocyte responses, including IL-4, IL-5, and IL-13, and the IL-4 receptor; how these gene variants alter lymphocyte function or expression and function of cytokines that contribute to atopic diseases is not known.

Various environmental factors have a profound influence on the tendency to develop allergies. In industrialized societies, air pollutants may contain many potential allergens. As mentioned earlier, viral infections of the respiratory tract promote the development of allergic responses. By contrast, some infections and allergen exposures in early childhood may reduce the development of subsequent allergic disease. This has led to the hygiene hypothesis, which proposes that higher levels of exposure to microbes and allergens in early life results in less allergy (and autoimmunity) later. The mechanism responsible for this effect, and even its significance, are unclear. Environmental pollution and reduced early life exposure to microbes and allergens may account for the increasing incidence of allergic diseases in industrialized countries. An example of the protective effect of early exposure to allergens is seen in infants and young children given small amounts of peanut-containing foods; this practice reduces the incidence of peanut allergy later in life.

Activation of Mast Cells and Secretion of Mediators

IgE antibody produced in response to an allergen binds to high-affinity Fc receptors, specific for the ε heavy chain, that are expressed on mast cells (see Fig. 11.2B). In an atopic individual, mast cells are coated with IgE antibody specific for the antigen(s) to which the individual is allergic. This process of coating mast cells with IgE is called sensitization, because it makes the mast cells sensitive to activation by subsequent encounter with that antigen. In nonallergic individuals, by contrast, mast cells may carry IgE molecules of many different specificities because many antigens may elicit small IgE responses, and the amount of IgE specific for any one antigen is not enough to cause immediate hypersensitivity reactions upon exposure to that antigen.

Mast cells are present in all connective tissues, especially under epithelia, and they are usually located adjacent to blood vessels. Which of the body's mast cells are activated by binding of an allergen often depends on the route of entry of the allergen. For example, inhaled allergens activate mast cells in the submucosal tissues of the bronchus, whereas ingested allergens activate mast cells in the wall of the intestine. Allergens that enter the blood via absorption from the intestine or by direct injection may be delivered to all tissues, resulting in systemic mast cell activation.

The high-affinity receptor for IgE, called FcεRI, consists of three polypeptide chains, one of which binds the Fc portion of the ε heavy chain very strongly, with a K_d of approximately 10^{-11} M. (The concentration of IgE in the plasma is approximately 10^{-9} M, which explains why even in nonallergic individuals, mast cells are always coated with IgE bound to FcεRI.) The other two chains of the receptor are signaling proteins. The same FcεRI is also present on basophils, which are circulating cells with many of the features of mast cells, but normally the number of basophils in the blood is very low and they are not present in tissues, so their role in immediate hypersensitivity is not as well established as the role of mast cells.

When mast cells sensitized by IgE are exposed to the allergen, they are activated to secrete inflammatory mediators (Fig. 11.4). Mast cell activation results from binding of the allergen to two or more IgE antibodies on the cell. When this happens, the FcεRI molecules that are carrying the IgE are cross-linked, triggering biochemical signals from the signal-transducing chains of FcεRI. The signals lead to the release of inflammatory mediators.

The most important mediators produced by mast cells are vasoactive amines and proteases stored in and released from granules, products of arachidonic acid metabolism, and cytokines (see Fig. 11.4). These mediators have different actions. The major amine, histamine, causes increased vascular permeability and vasodilation, leading to the leakage of fluid and plasma proteins into tissues, and stimulates the transient contraction of bronchial and intestinal smooth muscle. Proteases may damage local tissues. Arachidonic acid metabolites include prostaglandins, which cause vascular dilation, and leukotrienes, which stimulate prolonged bronchial smooth muscle contraction. Cytokines induce local inflammation (the late-phase reaction, described next). Thus, mast cell mediators are responsible for acute vascular and smooth muscle reactions and more prolonged inflammation, the hallmarks of allergy.

Cytokines produced by mast cells stimulate the recruitment of leukocytes, which cause the late-phase reaction. The principal leukocytes involved in this reaction are eosinophils, neutrophils, and Th2 cells. Mast cell—derived tumor necrosis factor (TNF) and IL-4 promote neutrophil- and eosinophil-rich inflammation. Chemokines produced by mast cells and by epithelial cells in the tissues also contribute to leukocyte recruitment. Eosinophils and neutrophils liberate proteases, which cause tissue damage, and Th2 cells may exacerbate the reaction by producing more cytokines. Eosinophils are prominent in many allergic reactions and are an important cause of tissue injury in these reactions. These

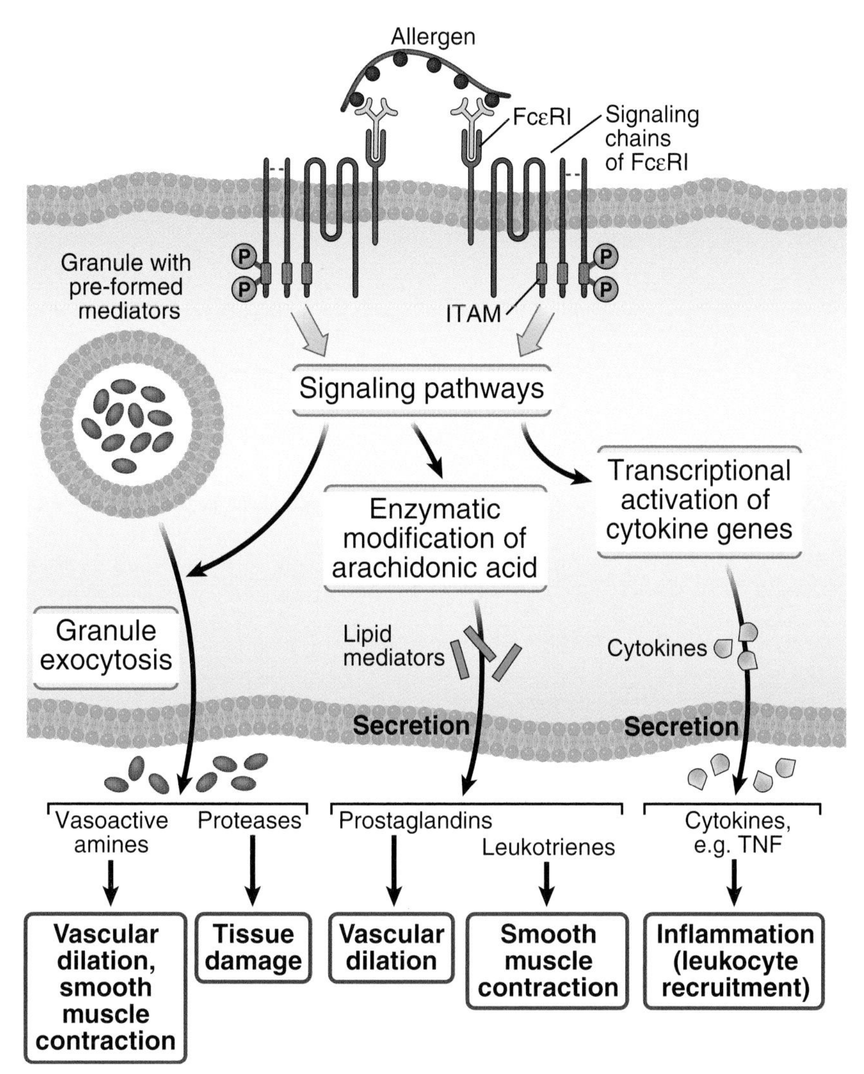

Fig. 11.4 Production and actions of mast cell mediators. Cross-linking of IgE on a mast cell by an allergen stimulates phosphorylation of immunoreceptor tyrosine-based activation motifs *(ITAMs)* in the signaling chains of the IgE Fc receptor *(FcεRI)*, which then initiates multiple signaling pathways. These signaling pathways stimulate the release of mast cell granule contents (amines, proteases), the synthesis of arachidonic acid metabolites (prostaglandins, leukotrienes), and the synthesis of various cytokines. *TNF,* Tumor necrosis factor.

cells are activated by the cytokine IL-5, which is produced by Th2 cells and ILC2s.

Allergic Diseases and Therapy

Immediate hypersensitivity reactions have diverse clinical and pathologic features, all of which are attributable to mediators produced by mast cells in different amounts and in different tissues (Fig. 11.5).

- Some manifestations of allergy, such as rhinitis and sinusitis, which are components of **hay fever**, are reactions to inhaled allergens, such as proteins in the pollen of many plants. Mast cells in the nasal mucosa produce histamine, and Th2 cells produce IL-13; these two mediators cause increased production of mucus. Late-phase reactions may lead to more prolonged inflammation.
- In **food allergies**, ingested allergens trigger mast cell degranulation, and the released histamine and other mediators causes increased peristalsis, resulting in vomiting and diarrhea.

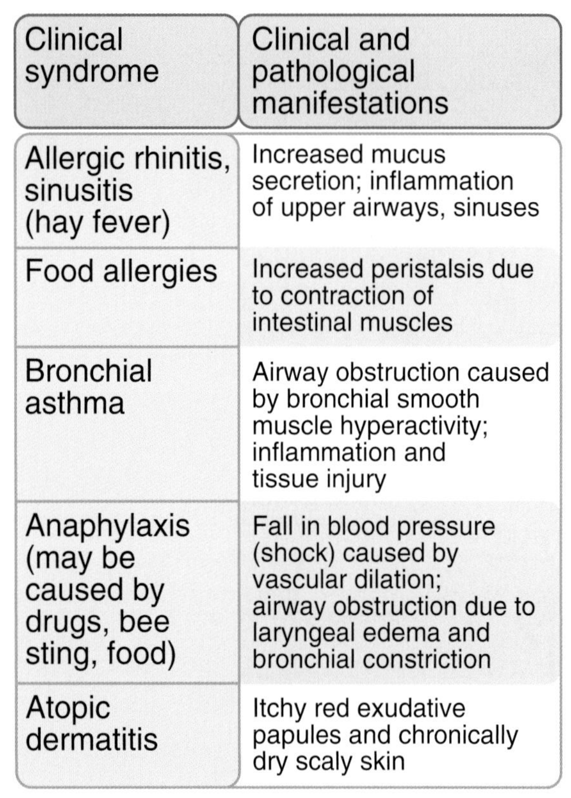

Clinical syndrome	Clinical and pathological manifestations
Allergic rhinitis, sinusitis (hay fever)	Increased mucus secretion; inflammation of upper airways, sinuses
Food allergies	Increased peristalsis due to contraction of intestinal muscles
Bronchial asthma	Airway obstruction caused by bronchial smooth muscle hyperactivity; inflammation and tissue injury
Anaphylaxis (may be caused by drugs, bee sting, food)	Fall in blood pressure (shock) caused by vascular dilation; airway obstruction due to laryngeal edema and bronchial constriction
Atopic dermatitis	Itchy red exudative papules and chronically dry scaly skin

Fig. 11.5 Clinical manifestations of immediate hypersensitivity reactions. Immediate hypersensitivity may be manifested in many other ways, as in development of skin lesions (e.g., urticaria, eczema).

- **Asthma** is a clinical syndrome characterized by difficulty in breathing, cough, and wheezing, due to intermittent obstruction of expiratory airflow. The most common cause of asthma is respiratory allergy in which inhaled allergens stimulate bronchial mast cells to release mediators, including leukotrienes, which cause repeated bouts of bronchial constriction and airway obstruction. In chronic asthma, large numbers of eosinophils accumulate in the bronchial mucosa, excessive secretion of mucus occurs in the airways, and the bronchi are constricted because bronchial smooth muscle becomes hyperreactive to various stimuli and hypertrophied over time. Some cases of asthma are not associated with IgE production and may be triggered by cold or exercise, which cause mast cell degranulation in some people by poorly understood mechanisms.
- **Atopic dermatitis** (commonly called eczema) is characterized by flares of itchy red exudative papules and chronically dry scaly skin. Defective skin barrier function (e.g., associated with filaggrin mutations) results in frequent colonization by *Staphylococcus aureus* and the activation of keratinocytes to secrete cytokines that promote type 2 immune responses. Patients with eczema often go on to develop chronic late-phase reactions in the skin, and children with eczema are at increased risk for developing food allergies and asthma. The occurrence of these three allergic reactions—atopic dermatitis, food allergy, and asthma—in the same patient is sometimes called the atopic triad, and their sequential development, starting from childhood, is known as the atopic march.
- The most severe form of immediate hypersensitivity is **anaphylaxis**, a systemic reaction characterized by edema in many tissues, including the larynx, accompanied by a fall in blood pressure (anaphylactic shock) and hypoxemia due to airway obstruction by laryngeal edema, mucus plugging, and bronchospasm. Some of the most frequent inducers of anaphylaxis include bee stings, injected or ingested penicillin-family antibiotics, and ingested nuts or shellfish. The reaction is caused by widespread mast cell degranulation in response to the systemic distribution of the antigen, and it is life threatening because of the sudden fall in blood pressure and airway obstruction.

The therapy for immediate hypersensitivity diseases is aimed at inhibiting mast cell degranulation, antagonizing the effects of mast cell mediators, and reducing inflammation (Fig. 11.6). Common drugs include antihistamines for hay fever; inhaled β-adrenergic agonists, leukotriene and prostaglandin antagonists, and corticosteroids that relax bronchial smooth muscles and reduce airway inflammation in asthma; and epinephrine in anaphylaxis. Many patients benefit from repeated administration of small doses of allergens, called desensitization or allergen-specific immunotherapy. This treatment may work by changing the T cell response away from Th2 dominance or the antibody response away from IgE, by inducing tolerance in allergen-specific T cells, or by stimulating regulatory T cells (Tregs). Monoclonal antibody drugs that block IgE binding to their receptors on mast cells or block various cytokines or their receptors, including IL-4, IL-5, TSLP, and IL-33, are now approved for the treatment of some forms of asthma and atopic dermatitis.

Syndrome	Therapy	Mechanism of action
Anaphylaxis	Epinephrine	Causes vascular smooth muscle contraction, increases cardiac output (to counter shock), and inhibits bronchial smooth muscle cell contraction
Asthma	Corticosteroids	Reduce inflammation
	Leukotriene antagonists	Relax bronchial smooth muscle and reduce inflammation
	Beta adrenergic receptor agonists	Relax bronchial smooth muscles
Various allergic diseases	Desensitization (repeated administration of low doses of allergens)	Unknown; may inhibit IgE production and increase production of other Ig isotypes; may induce T cell tolerance
	Anti-IgE antibody	Neutralize and eliminate IgE
	Antihistamines	Block actions of histamine on vessels and smooth muscles
	Cromolyn	Inhibits mast cell degranulation
	Antibodies that block cytokines and their receptors: TSLP, IL-33, IL-4, IL-5R (asthma); IL-4R (atopic dermatitis)	Block cytokine-driven inflammation

Fig. 11.6 Treatment of immediate hypersensitivity reactions. The table summarizes the principal mechanisms of action of currently used therapies for allergic disorders. *Ig,* Immunoglobulin; *IL,* interleukin; *TSLP,* thymic stromal lymphopoietin.

Before concluding the discussion of immediate hypersensitivity, it is important to address the question of why evolution has preserved an IgE antibody– and mast cell–mediated immune response whose major effects are pathologic. There is no definitive answer to this puzzle, but immediate hypersensitivity reactions likely evolved to protect against pathogens or toxins. It is known that eosinophils are important mediators of defense against helminthic infections, and mast cells play a role in innate immunity against some bacteria and in destroying venomous toxins produced by arachnids and snakes.

DISEASES CAUSED BY ANTIBODIES SPECIFIC FOR CELL AND TISSUE ANTIGENS

Antibodies, typically of the IgG class, may cause disease, called type II hypersensitivity disorders, by binding to their target antigens in different tissues. Antibodies against cells or extracellular matrix components may deposit in any tissue that expresses the relevant target antigen; thus, diseases caused by such antibodies are usually specific for a particular tissue. The antibodies that cause disease are most often autoantibodies against self antigens. The production of autoantibodies results from a failure of self-tolerance. In Chapter 9, we discussed the mechanisms by which self-tolerance may fail, but why this happens in most human autoimmune diseases is still not understood.

Mechanisms of Antibody-Mediated Tissue Injury and Disease

Antibodies specific for cell and tissue antigens may deposit in tissues and cause injury by inducing local inflammation, by inducing phagocytosis and destruction of antibody-coated cells, or by interfering with normal cellular functions (Fig. 11.7).

- **Inflammation.** Antibodies against tissue antigens induce inflammation by attracting and activating leukocytes. When IgG antibodies of the IgG1 and IgG3 subclasses are cross-linked by antigens, they bind to neutrophil and macrophage Fc receptors and activate these leukocytes, resulting in inflammation (see Chapter 8). The same antibodies as well as IgM, when bound to antigens, activate the complement system by the classical pathway, resulting in the production of complement by-products that recruit leukocytes and induce inflammation. When leukocytes are activated at sites of antibody deposition, these cells release reactive oxygen species and lysosomal enzymes that damage the adjacent tissues.
- **Opsonization and phagocytosis.** If antibodies bind to antigens on cells, such as erythrocytes and platelets, the cells are opsonized and may be ingested and destroyed by host phagocytes.
- **Abnormal cellular responses.** Some antibodies may cause disease without directly inducing tissue injury. For example, in pernicious anemia, autoantibodies specific for intrinsic factor, a protein required for absorption of vitamin B_{12}, decrease the intestinal uptake of this vitamin, leading to a multisystem disease in which anemia is a major component. Other antibodies may directly activate receptors, mimicking their physiologic ligands. The only known example is a form of hyperthyroidism called Graves disease, in which antibodies against the receptor for thyroid-stimulating hormone activate thyroid cells even in the absence of the hormone.

Examples and Treatment of Diseases Caused by Cell-Specific or Tissue-Specific Antibodies

Antibodies specific for cell and tissue antigens are the cause of many human diseases involving blood cells, heart, kidney, lung, and skin (Fig. 11.8). Examples of antitissue antibodies are those that react with the glomerular basement membrane and induce inflammation, a form of glomerulonephritis. Antibodies against cells include those that opsonize blood cells and target them for phagocytosis, as in autoimmune hemolytic anemia (red cell destruction) and autoimmune thrombocytopenia (destruction of platelets). Antibodies that interfere with hormones or their receptors were mentioned earlier. In most of these cases, the antibodies are autoantibodies, but less commonly, antibodies produced against a microbe may cross-react with an antigen in the tissues. For instance, in rare instances, streptococcal infection stimulates the production of antibacterial antibodies that cross-react with antigens in the heart, producing the cardiac inflammation that is characteristic of rheumatic fever.

Therapy for antibody-mediated diseases is intended mainly to limit inflammation and its injurious consequences with drugs such as corticosteroids. In severe cases, plasmapheresis is used to reduce levels of circulating antibodies. In hemolytic anemia and thrombocytopenia, splenectomy is of clinical benefit because the spleen is the major organ in which opsonized blood cells are phagocytosed. Some of these diseases respond to treatment with intravenous IgG (IVIG) pooled from healthy donors. How IVIG works is not known; it may bind to the inhibitory Fc receptor on myeloid cells and B cells and thus block activation of these cells (see Fig. 7.15), or it may reduce the half-life of pathogenic antibodies by competing for binding to the neonatal Fc receptor in endothelial cells and macrophages (see Fig. 8.2). A monoclonal antibody that blocks the complement protein C5 is used to treat myasthenia gravis. Treatment of patients with an antibody specific for CD20, a surface protein of mature B cells, results in depletion of the B cells. Anti-CD20 was first used for treating B cell—derived lymphomas and leukemias but is also now used for some autoimmune diseases. A synthetic Fc portion of IgG that binds to FcRn and thereby increases clearance of IgG, including autoantibodies (see Chapter 8), is approved for the treatment of generalized myasthenia gravis. Other approaches in development for

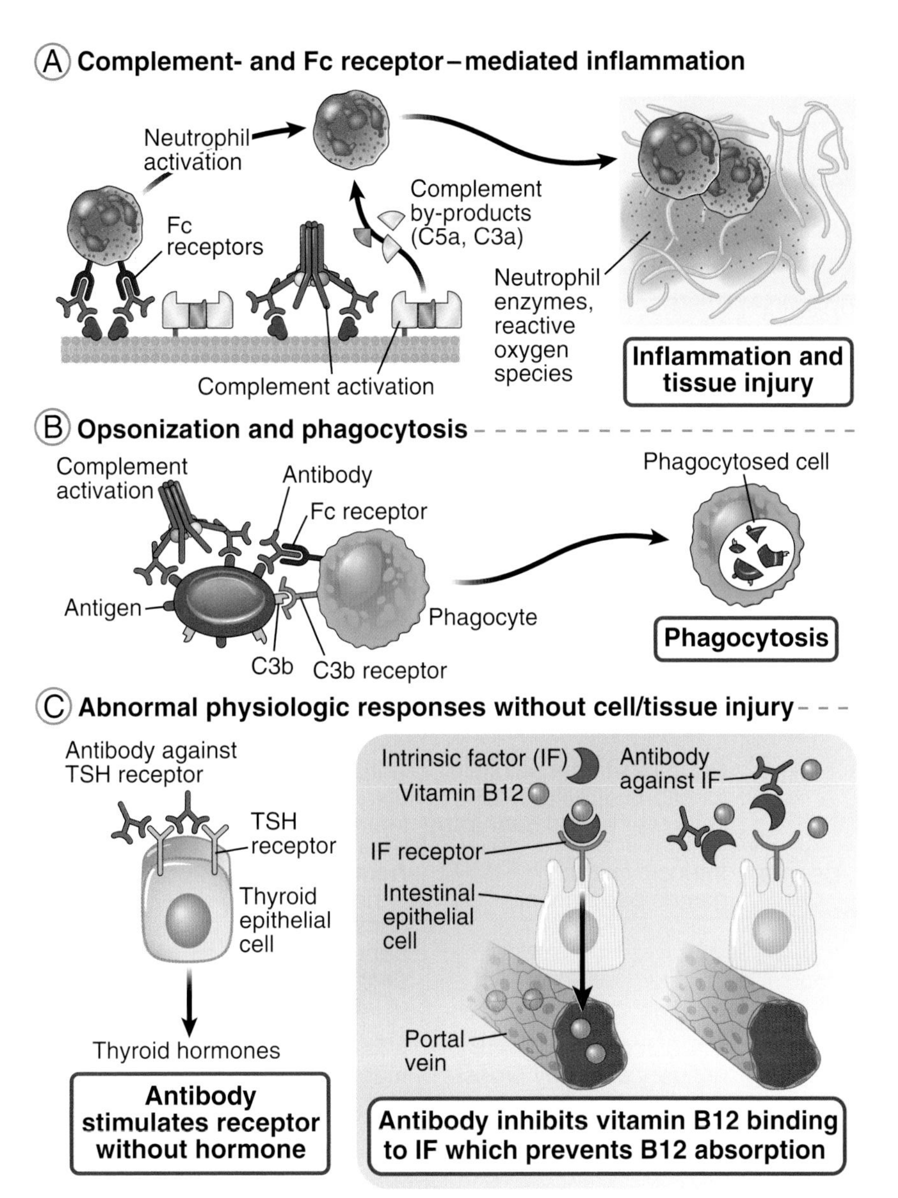

Fig. 11.7 Pathogenesis of diseases caused by antibodies specific for tissue or cell antigens. Antibodies can cause disease by **(A)** inducing inflammation at the site of deposition, **(B)** opsonizing cells (such as red cells) for phagocytosis, and **(C)** interfering with normal cellular functions, such as hormone receptor signaling or absorption of a necessary dietary factor. *TSH,* Thyroid-stimulating hormone.

inhibiting the production of autoantibodies include treating patients with antibodies that block CD40 or its ligand and thus inhibit helper T cell–dependent B cell activation, as well as antibodies to block cytokines that promote the survival of B cells and plasma cells. There is also interest in inducing tolerance in cases in which the autoantigens are known.

DISEASES CAUSED BY ANTIGEN-ANTIBODY COMPLEXES

Antibodies (typically IgG) may cause disease by forming immune complexes that deposit in blood vessels (Fig. 11.9). Many acute and chronic hypersensitivity disorders are caused by, or are associated with, immune

Antibody-mediated disease	Target antigen	Mechanisms of disease	Clinicopathologic manifestations
Autoimmune hemolytic anemia	Erythrocyte membrane proteins (Rh blood group antigens, I antigen)	Opsonization and phagocytosis of erythrocytes	Hemolysis, anemia
Autoimmune (idiopathic) thrombocytopenic purpura	Platelet membrane proteins (gpIIb/IIIa integrin)	Opsonization and phagocytosis of platelets	Bleeding
Goodpasture's syndrome	Protein in basement membranes of kidney glomeruli and lung alveoli	Complement and Fc receptor-mediated inflammation	Nephritis, lung hemorrhage
Graves' disease (hyperthyroidism)	Thyroid stimulating hormone (TSH) receptor	Antibody-mediated stimulation of TSH receptors	Hyperthyroidism
Myasthenia gravis	Acetylcholine receptor	Antibody inhibits acetycholine binding, complement-mediated damage to neuro-muscular junction	Muscle weakness, paralysis
Pemphigus vulgaris	Proteins in intercellular junctions of epidermal cells (epidermal cadherin)	Antibody-mediated disruption of intercellular adhesions	Skin vesicles (bullae)
Pernicious anemia	Intrinsic factor, gastric parietal cells	Neutralization of intrinsic factor and damage to gastric parietal cells causing decreased absorption of vitamin B_{12}	Anemia due to abnormal erythropoiesis, nerve damage
Rheumatic fever	Streptococcal cell wall antigen; antibody cross-reacts with myocardial antigen	Inflammation, macrophage activation	Myocarditis, arthritis

Fig. 11.8 Human antibody-mediated diseases (type II hypersensitivity). The figure lists examples of human diseases caused by antibodies. In most of these diseases, the role of antibodies is inferred from the detection of antibodies in the blood or the lesions, and in some cases by similarities with experimental models in which the involvement of antibodies can be formally established by transfer studies. In some of these diseases, concurrent T cell–mediated tissue damage also occurs. *TSH,* Thyroid-stimulating hormone.

complexes (Fig. 11.10); these are called type III hypersensitivity disorders.

Mechanisms and Examples of Immune Complex–Mediated Diseases

Antigen-antibody complexes, which are produced during normal immune responses, cause disease only when they are formed in excessive amounts, are not efficiently removed by phagocytes, and become deposited in tissues. Complexes containing positively charged antigens are particularly pathogenic because they bind avidly to negatively charged components of the basement membranes of blood vessels and kidney glomeruli. Immune complexes usually deposit in blood vessels, especially vessels through which plasma is filtered at high

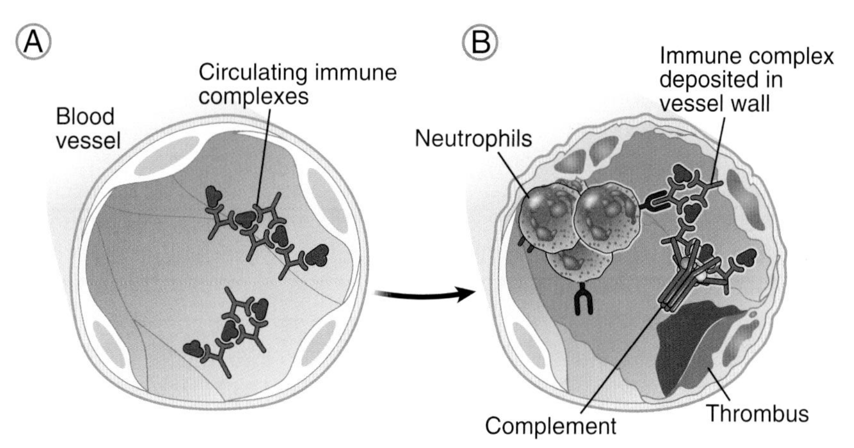

Fig. 11.9 Pathogenesis of immune complex–mediated diseases. Immune complexes are formed in the circulation and deposit in blood vessels, where they elicit complement- and Fc receptor–mediated inflammation.

Immune complex disease	Antibody specificity	Clinicopathologic manifestations
Systemic lupus erythematosus	DNA, nucleoproteins, others	Nephritis, arthritis, vasculitis
Polyarteritis nodosa	In some cases, microbial antigens (e.g., Hepatitis B virus surface antigen); most cases unknown	Vasculitis
Post-streptococcal glomerulonephritis	Streptococcal cell wall antigen(s)	Nephritis
Serum sickness (clinical and experimental)	Various protein antigens	Systemic vasculitis, nephritis, arthritis
Arthus reaction (experimental)	Various protein antigens	Cutaneous vasculitis

Fig. 11.10 Immune complex diseases (type III hypersensitivity). Examples of human diseases caused by the deposition of immune complexes, as well as two experimental models. In the diseases, immune complexes are detected in the blood or in the tissues that are the sites of injury. In all the disorders, injury is caused by complement-mediated and Fc receptor–mediated inflammation.

pressure (e.g., in renal glomeruli and joint synovium), but vessels in any organ may be affected. Therefore, in contrast to diseases caused by antibodies specific for particular cell or tissue antigens, immune complex diseases tend to be systemic and often manifest as widespread vasculitis involving sites that are particularly susceptible to immune complex deposition, such as kidneys and joints. Once deposited in the vessel walls, the Fc regions of the antibodies activate complement and bind Fc receptors on neutrophils, activating the cells to release damaging proteases and reactive oxygen species. This inflammatory response within the vessel wall, called vasculitis, may cause local hemorrhage or thrombosis leading to ischemic tissue injury. In the kidney glomerulus, glomerulonephritis can impair normal filtration function, leading to renal failure.

The first immune complex disease studied was **serum sickness**, seen in subjects who received antitoxin-containing serum from immunized animals for the treatment of infections. Some of these treated individuals subsequently developed a systemic inflammatory disease with fever, rashes, and arthritis. The disease was caused by antibodies made against the injected animal proteins and formation of immune complexes with these foreign proteins, followed by deposition of the complexes in various tissues. This illness could be recreated in experimental animals by systemic administration of a protein antigen, which elicits an antibody response and leads to the formation of circulating immune complexes. This can occur as a complication of any therapy involving injection of foreign proteins, such as antibodies against microbial toxins, snake venoms, or T cells, that are usually made in goats or rabbits, and even some humanized monoclonal antibodies that are used to treat different diseases and may differ only slightly from normal human Ig.

A localized immune complex reaction called the **Arthus reaction** was first studied in experimental animals. It is induced by subcutaneous administration of a protein antigen to a previously immunized animal; it results in the formation of immune complexes at the site of antigen injection and a local vasculitis.

In human immune complex diseases, the antibodies may be specific for self antigens or microbial antigens. In several systemic autoimmune diseases, many of the clinical manifestations are caused by vascular injury when complexes of the antibodies and self antigens deposit in vessels in different organs. For example, in systemic lupus erythematosus, immune complexes of self DNA and anti-DNA antibodies can deposit in the blood vessels of almost any organ, causing vasculitis and impaired blood flow, leading to a multitude of different organ pathologies and symptoms. Several immune complex diseases are initiated by infections. For example, in response to some streptococcal infections, individuals make antistreptococcal antibodies that form complexes with the bacterial antigens. These complexes deposit in kidney glomeruli, or the antibodies form complexes with bacterial antigens planted in the glomeruli, causing an inflammatory process called poststreptococcal glomerulonephritis that can lead to renal failure. Other immune complex diseases caused by complexes of antimicrobial antibodies and microbial antigens lead to vasculitis. This may occur in patients with chronic infections with certain viruses (e.g., the hepatitis virus) or parasites (e.g., malaria).

The current mainstay of therapy for immune complex diseases are antiinflammatory and immunosuppressive drugs, mainly steroids.

DISEASES CAUSED BY T LYMPHOCYTES

T cells play a central role in chronic immunologic diseases in which inflammation is a prominent component. Many of the newly developed therapies that have shown efficacy in such diseases are drugs that inhibit the recruitment and activities of T cells.

Etiology of T Cell–Mediated Diseases

The major causes of T cell–mediated hypersensitivity reactions are autoimmunity and exaggerated or persistent responses to microbial or other environmental antigens. The autoimmune reactions usually are directed against cellular antigens with restricted tissue distribution. Therefore, T cell–mediated autoimmune diseases tend to be limited to a few organs and are not systemic. Examples of T cell–mediated hypersensitivity reactions against environmental antigens include contact sensitivity to chemicals (e.g., upon skin contact with some therapeutic drugs, substances found in plants such as poison ivy, and metals such as nickel in jewelry). In these cases, the drugs or chemicals bind to and modify self proteins in the skin, creating neoantigens that can be recognized by T cells. Tissue injury may also accompany T cell responses to microbes. For example, in tuberculosis, a T cell–mediated immune response develops against protein antigens of *Mycobacterium tuberculosis*, and the response becomes chronic because the infection is difficult to eradicate. The resultant granulomatous inflammation causes injury to tissues at the site of infection.

Excessive polyclonal T cell activation by certain microbial toxins produced by some bacteria and viruses can lead to production of large amounts of inflammatory cytokines, causing a syndrome similar to septic shock. These toxins are called **superantigens** because they stimulate large numbers of T cells. Superantigens bind to nonpolymorphic regions of MHC molecules on antigen-presenting cells and simultaneously to invariant parts of T cell receptors on many different clones of T cells, regardless of antigen specificity, thereby activating these cells.

Mechanisms of T Cell–Mediated Tissue Injury

In T cell–mediated diseases, tissue injury is caused most often by inflammation induced by cytokines that are produced mainly by CD4$^+$ T cells, and in some cases by killing of host cells by CD8$^+$ cytotoxic T lymphocytes (CTLs) (Fig. 11.11). These mechanisms of tissue injury are the same as the mechanisms used by T cells to eliminate cell-associated microbes.

CD4$^+$ T cells may react against cell or tissue antigens and secrete cytokines that induce local inflammation and activate macrophages. Different diseases may be associated with activation of Th1 and Th17 cells. Th1 cells are the source of interferon-γ (IFN-γ), the principal macrophage-activating cytokine, and Th17 cells are responsible for the recruitment of leukocytes, including neutrophils. The actual tissue injury in these diseases is caused mainly by the macrophages and neutrophils.

The typical reaction mediated by T cell cytokines is **delayed-type hypersensitivity** (DTH), so called because it occurs 24 to 48 hours after an individual previously exposed to a protein antigen is challenged with the antigen (i.e., the reaction is delayed). The delay occurs because it takes a day or two for circulating effector T lymphocytes to home to the site of antigen challenge, respond to the antigen at this site, and secrete cytokines that induce a detectable reaction. DTH reactions are manifested by infiltrates of T cells and monocytes in the tissues (Fig. 11.12), edema and fibrin deposition caused by increased vascular permeability in response to cytokines produced by CD4$^+$ T cells, and tissue damage induced by leukocyte products, mainly from macrophages that are activated by the T cells. DTH reactions are often used to determine if people have been previously exposed to and have responded to an antigen. For example, a DTH reaction to a mycobacterial antigen, purified protein derivative (PPD), applied to the skin, is an indicator of past or active mycobacterial infection.

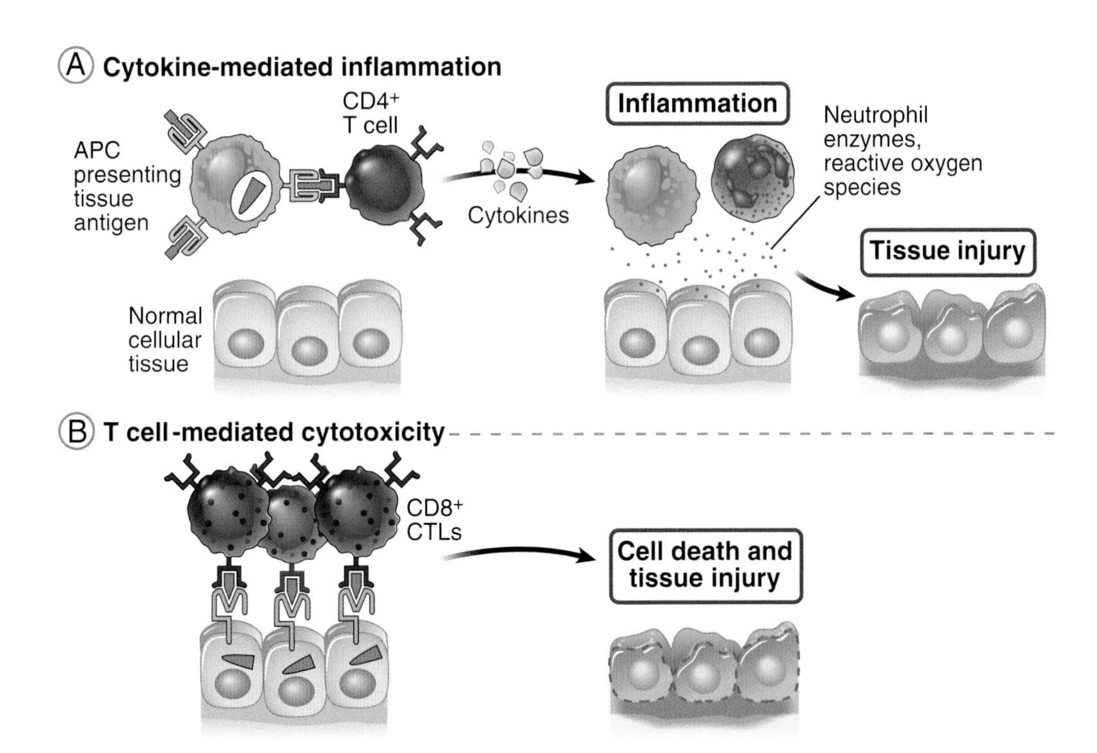

Fig. 11.11 Mechanisms of T cell–mediated tissue injury (type IV hypersensitivity). T cells may cause tissue injury and disease by two mechanisms. **A,** Inflammation triggered by cytokines produced mainly by CD4$^+$ T cells in which tissue injury is caused by activated macrophages and neutrophils. **B,** Direct killing of target cells is mediated by CD8$^+$ cytotoxic T lymphocytes (CTLs). *APC,* Antigen-presenting cell.

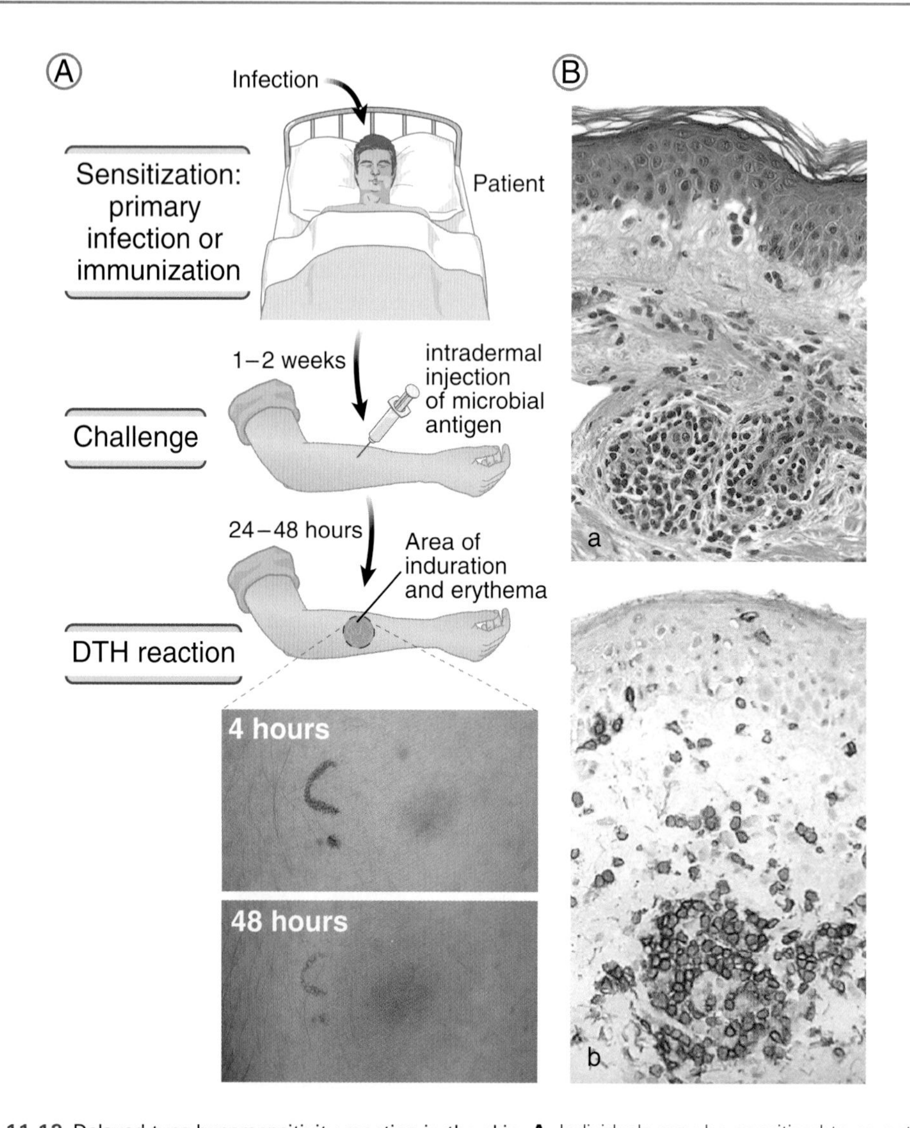

Fig. 11.12 **Delayed-type hypersensitivity reaction in the skin. A,** Individuals may be sensitized to an antigen (e.g., mycobacterial protein) by infection or vaccination. Subsequent cutaneous challenge with the antigen elicits a visible reaction (erythema, swelling) within 48 hours. **B,** A biopsy sample of the reaction shows perivascular accumulation (cuffing) of mononuclear inflammatory cells (lymphocytes and macrophages), with associated dermal edema and fibrin deposition ***(a).*** Immunoperoxidase staining reveals a predominantly perivascular cellular infiltrate that marks positively with anti-CD4 antibodies ***(b).*** *DTH,* Delayed-type hypersensitivity. (B, Courtesy Dr. Louis Picker, Department of Pathology, Oregon Health Sciences University, Portland, OR.)

$CD8^+$ T cells specific for antigens in host cells may directly kill these cells (e.g., in type 1 diabetes and autoimmune myocarditis). $CD8^+$ T cells also produce cytokines, including IFN-γ that may induce inflammation in some hypersensitivity diseases (e.g., contact sensitivity). In many T cell–mediated autoimmune diseases, both $CD4^+$ T cells and $CD8^+$ T cells specific for self antigens are present, and both contribute to tissue injury.

Examples and Therapy of T Cell–Mediated Diseases

Many organ-specific autoimmune diseases in humans are thought to be caused by T cells, based on the identification of these cells in lesions and similarities with animal models in which the diseases are known to be T cell mediated (Fig. 11.13). These disorders are typically chronic and progressive, in part because long-lived memory T cells are generated, and the inciting antigens, such as tissue antigens or proteins expressed by persistent microbes, are often not cleared. Also, tissue injury causes release and alteration of self proteins, which may result in reactions against these newly encountered proteins. This phenomenon has been called epitope spreading to indicate that the initial immune response against one or a few self antigen epitopes may spread to include responses against more self antigens.

The therapy for T cell–mediated hypersensitivity disorders is designed to reduce inflammation and to inhibit T cell activation. The mainstay of treatment of such diseases has been antiinflammatory steroids, but these drugs have significant side effects. The development of more targeted therapies based on understanding of the fundamental mechanisms of these diseases has been one of the most impressive accomplishments of immunology. Antagonists of inflammatory cytokines have proved to be very effective in patients with various inflammatory and autoimmune diseases. For example, monoclonal antibodies that block TNF or the IL-6 receptor are now used to treat rheumatoid arthritis; antibodies specific for IL-17, IL-23, and TNF are used to treat psoriasis; and an antibody that blocks the Th1- and Th17-inducing cytokines IL-12

Disease	Specificity of pathogenic T cells	Clinicopathologic manifestations
Rheumatoid arthritis	Unknown antigens in joint	Inflammation of synovium and erosion of cartilage and bone in joints
Type 1 diabetes	Pancreatic islet antigens	Impaired glucose metabolism, vascular disease
Crohn's disease	Unknown, ? role of intestinal microbes	Inflammation of the bowel wall; abdominal pain, diarrhea, hemorrhage
Psoriasis	Unknown	Chronic skin inflammation
Multiple sclerosis	Myelin proteins	Demyelination in the central nervous system, sensory and motor dysfunction
Contact sensitivity (e.g. poison ivy, drug reaction)	Modified skin proteins	DTH reaction in skin, rash
Chronic infections (e.g., tuberculosis)	Microbial proteins	Chronic (e.g., granulomatous) inflammation

Fig. 11.13 Human T cell–mediated diseases. Diseases in which T cells play a dominant role in causing tissue injury; antibodies and immune complexes may also contribute. Note that rheumatoid arthritis and type 1 diabetes are autoimmune disorders. Crohn's disease, an inflammatory bowel disease, is likely caused by reactions against microbes in the intestine and may have a component of autoimmunity. The other diseases are caused by reactions against foreign (microbial or environmental) antigens. In most of these diseases, the role of T cells is inferred from the detection and isolation of T cells reactive with various antigens from the blood or lesions, and from the similarity with experimental models in which the involvement of T cells has been established by a variety of approaches. The specificity of pathogenic T cells has been defined in animal models and in some of the human diseases. Although multiple sclerosis (MS) has long been considered a T cell–mediated disease, the most successful therapy for MS is depletion of B cells. Viral hepatitis and toxic shock syndrome are disorders in which T cells play an important pathogenic role, but these are not considered examples of hypersensitivity. *DTH,* Delayed-type hypersensitivity.

and IL-23 is used for inflammatory bowel diseases. Small molecule inhibitors of the inflammatory cytokine signaling molecules Janus kinases (JAKs) are also approved for all three diseases. Other agents developed to inhibit T cell responses include drugs that block costimulators such as B7. In some diseases that are considered to be T cell–mediated, including multiple sclerosis and rheumatoid arthritis, one effective therapy is the depletion of B cells using anti-CD20. Whether the role of B cells in these diseases is as a source of pathogenic antibodies or as APCs for pathogenic T cells, especially memory cells, has not been established. Clinical trials are underway to test the efficacy of transferring in vitro expanded Tregs and administering IL-2 to expand endogenous Tregs for the treatment of autoimmune diseases such as type 1 diabetes and lupus. Research is ongoing on methods for inducing tolerance in pathogenic T cells.

SUMMARY

- Immune responses that cause tissue injury are called hypersensitivity reactions, and the diseases caused by these reactions are called hypersensitivity diseases.
- Hypersensitivity reactions may arise from uncontrolled or abnormal responses to foreign antigens or autoimmune responses against self antigens.
- Hypersensitivity reactions are classified according to the mechanism of tissue injury.
- Immediate hypersensitivity (type I, commonly called allergy) is caused by the activation of T helper 2 (Th2) cells and interleukin-4 (IL-4)— and IL-13—producing follicular helper (Tfh) cells and production of immunoglobulin E (IgE) antibody against environmental antigens or drugs (allergens), sensitization of mast cells by the IgE, and degranulation of these mast cells on subsequent encounter with the allergen.
- Clinicopathologic manifestations of immediate hypersensitivity result from the actions of mediators secreted by the mast cells: amines increase vascular permeability of and dilate blood vessels, arachidonic acid metabolites cause bronchial smooth muscle contraction, and cytokines induce inflammation, the hallmark of the late-phase reaction. Treatment of allergies is designed to inhibit the production of mediators, antagonize their actions, and counteract their effects on end organs.
- Antibodies against cell and tissue antigens may cause tissue injury and disease (type II hypersensitivity). IgM and IgG antibodies activate complement, which promotes phagocytosis of cells to which they bind, induces inflammation, and causes cell lysis. IgG also promotes Fc receptor–mediated phagocytosis of cells and leukocyte recruitment. Antibodies may interfere with the functions of cells by binding to essential molecules and receptors.
- In immune complex diseases (type III hypersensitivity), antibodies may bind to circulating antigens to form immune complexes, which deposit in vessels, leading to inflammation in the vessel wall (vasculitis), which secondarily causes tissue injury due to impaired blood flow.
- T cell–mediated diseases (type IV hypersensitivity) result from inflammation caused by cytokines produced by $CD4^+$ Th1 and Th17 cells, or killing of host cells by $CD8^+$ cytotoxic T lymphocytes.

REVIEW QUESTIONS

1. What are the major types of hypersensitivity reactions?
2. What types of antigens may induce immune responses that cause hypersensitivity reactions?
3. What is the sequence of events in a typical immediate hypersensitivity reaction? What is the late-phase reaction, and how is it caused?
4. What are some examples of immediate hypersensitivity disorders, what is their pathogenesis, and how are they treated?
5. How do antibodies cause tissue injury and disease?
6. What are some examples of diseases caused by antibodies specific for cell surface or tissue matrix antigens?
7. How do immune complexes cause disease, and how are the clinical manifestations different from most diseases caused by antibodies specific for cell surface or tissue matrix proteins?
8. What are some examples of diseases caused by T cells, what is their pathogenesis, and what are their principal clinical and pathologic manifestations?

Answers to and discussion of the Review Questions may be found on p. 322.

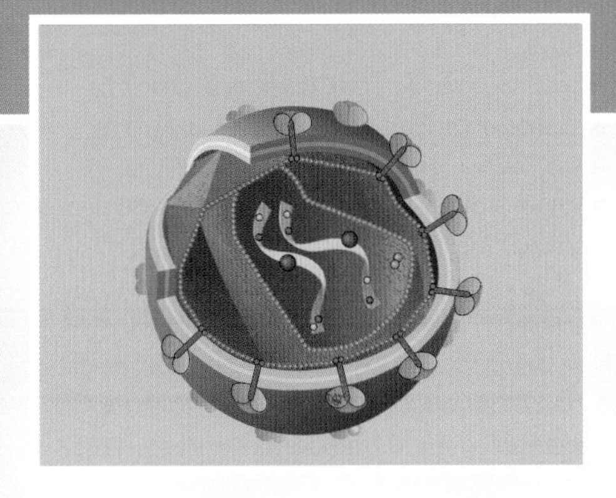

12

Immunodeficiency Diseases
Disorders Caused by Defective Immunity

CHAPTER OUTLINE

Defects in the development and functions of the immune system result in increased susceptibility to infections. The infections may be newly acquired or the reactivation of latent infections such as cytomegalovirus, Epstein-Barr virus (EBV), and tuberculosis, in which the normal immune response keeps the infection in check but does not eradicate it. These consequences of defective immunity are predictable because, as emphasized throughout this book, the normal function of the immune system is to defend individuals against infections. Disorders caused by defective immunity are called **immunodeficiency diseases**. Some of these diseases are also associated with increased incidence of certain cancers and autoimmunity. Several immunodeficiencies result from mutations in single genes that encode components of the immune system; these are called **primary** (or **congenital**) **immunodeficiencies**. Other defects in immunity may result from infections, nutritional abnormalities, or medical treatments that cause loss or inadequate function of various components of the immune system; these are called **acquired** (or **secondary**) **immunodeficiencies**.

In this chapter we describe the causes and pathogenesis of immunodeficiencies. Among the acquired diseases, we emphasize acquired immunodeficiency syndrome (AIDS), which results from infection by human immunodeficiency virus (HIV) and is one of the most devastating health problems worldwide. We address the following questions:

- What are the mechanisms by which immunity is compromised in the most common primary immunodeficiency diseases?
- How does HIV cause the clinical and pathologic abnormalities of AIDS?

- What approaches are being used to treat immunodeficiency diseases?

Information about the clinical features of these disorders can be found in textbooks of pediatrics and medicine.

PRIMARY (CONGENITAL) IMMUNODEFICIENCIES

Primary immunodeficiencies are diseases caused by defects usually in only one or two genes in each disease that lead to impaired maturation or function of different components of the immune system. It is estimated that as many as 1 in 500 individuals in the United States and Europe suffer from congenital immune deficiencies of varying severity. These immunodeficiencies share several features, the most common being increased susceptibility to infections (Fig. 12.1). Primary immunodeficiency diseases may, however, differ considerably in clinical and pathologic manifestations. Some of these disorders result in greatly increased incidence of infections that may manifest early after birth and may be fatal unless the immunologic defects are corrected. Other primary immunodeficiencies lead to mild infections and may first be detected in adult life.

Mutations in over 450 different genes have been identified as causes of over 400 primary immunodeficiency disorders. Although X-linked recessive immunodeficiency diseases were the first primary immunodeficiency disorders in which the genetic defects were identified, the majority of primary immunodeficiencies exhibit an autosomal recessive inheritance. Autosomal recessive alleles are often detected in consanguineous families when the same mutation is inherited from both parents. In other cases, especially in offspring of nonconsanguineous couples, one defective allele of a specific gene is inherited from one parent and a different mutation in the same gene is inherited from the other parent; individuals with this kind of autosomal recessive inheritance pattern are referred to as compound heterozygotes. Occasionally, the causative mutation arises de novo in the patient and is not present in either parent. The expression and clinical presentation of diseases caused by the same mutation may be variable. Multiple factors may contribute to the phenotypic variability, including the

Type of immunodeficiency	Histopathology and laboratory abnormalities	Common infectious consequences
B cell deficiencies	Often absent or reduced follicles and germinal centers in lymphoid organs Reduced serum Ig levels	Pyogenic bacterial infections, enteric bacterial and viral infections
T cell deficiencies	May be reduced T cell zones in lymphoid organs Reduced DTH reactions to common antigens Defective T cell proliferative responses to mitogens *in vitro*	Viral and other intracellular microbial infections (e.g., *Pneumocystis jiroveci*, other fungi, non-tuberculous mycobacteria) Some cancers (e.g., EBV-associated lymphomas, skin cancers)
Innate immune deficiencies	Variable, depending on which component of innate immunity is defective	Variable; pyogenic bacterial and viral infections

Fig. 12.1 Features of immunodeficiency diseases. Summary of the important diagnostic features and clinical manifestations of immunodeficiencies affecting different components of the immune system. Within each group, different diseases, and even different patients with the same disease, may show considerable variation. Reduced numbers of circulating B or T cells are often detected in some of these diseases. *DTH,* Delayed-type hypersensitivity; *EBV,* Epstein-Barr virus; *Ig,* immunoglobulin.

coinheritance of modifier genes, environmental factors, and epigenetic modifications of genes that vary from one individual to another. However, in most cases, different phenotypic manifestations of the same mutation remain unexplained.

The following discussion summarizes the pathogenesis of select immunodeficiencies, several of which are linked to genes and proteins mentioned in earlier chapters. These examples illustrate the physiologic importance of various components of the immune system.

Defects in Innate Immunity

Abnormalities in two components of innate immunity, phagocytes and the complement system, are important causes of immunodeficiency (Fig. 12.2).

- **Chronic granulomatous disease** (CGD) is caused by mutations in genes encoding subunits of the enzyme phagocyte NADPH oxidase, which catalyzes the production of microbicidal reactive oxygen species in lysosomes (see Chapter 2). Affected neutrophils and macrophages are unable to kill the microbes they phagocytose. The most common infections in CGD patients are bacteria such as *Staphylococcus* and fungi such as *Aspergillus* and *Candida.* Many of the organisms that are particularly troublesome in patients with CGD produce catalase, which destroys the microbicidal hydrogen peroxide that may be produced by host cells from residual reactive oxygen radicals. The immune system tries to compensate for this defective microbial killing by calling in more macrophages and by activating T cells, which stimulate recruitment and activation of phagocytes. Therefore, collections of macrophages accumulate around foci of infections to try to control the infections. These collections resemble granulomas, giving

Disease	Functional Deficiencies	Genetic Defect
Chronic granulomatous disease	Defective production of reactive oxygen species by phagocytes; recurrent intracellular bacterial and fungal infections	Mutations in genes encoding phagocyte oxidase complex; phox-91 (cytochrome b_{558} α subunit) is mutated in X-linked form
Leukocyte adhesion deficiency type 1	Defective leukocyte adhesion to endothelial cells and migration into tissues linked to decreased or absent expression of β_2 integrins; recurrent bacterial and fungal infections	Mutations in gene encoding the β chain (CD18) of β2 integrins
Leukocyte adhesion deficiency type 2	Defective leukocyte rolling on endothelium and migration into tissues because of decreased or absent expression of leukocyte ligands for endothelial E- and P-selectins; recurrent bacterial and fungal infections	Mutations in gene encoding GDP-fucose transporter-1, required for transport of fucose into the Golgi and its incorporation into sialyl-Lewis X
Chediak-Higashi Syndrome	Defective vesicle fusion and lysosomal function in neutrophils, macrophages, dendritic cells, NK cells, cytotoxic T cells, and many other cell types; recurrent infections by pyogenic bacteria	Mutations in gene encoding LYST, a protein involved in fusion of vesicles (including lysosomes)
Toll-like receptor signaling defects	Recurrent infections caused by defects in TLR signaling	Mutations in genes encoding TLR3 and MyD88 compromise NF-κB activation and type I interferon production in response to microbes

Fig. 12.2 Primary immunodeficiencies caused by defects in innate immunity. Immunodeficiency diseases caused by defects in various components of the innate immune system. *NF-κB,* Nuclear factor κB; *NK,* natural killer, *TLR,* Toll-like receptors.

rise to the name of this disease. The most common form of CGD is X-linked, caused by mutations in a subunit of the phagocyte oxidase that is encoded by the *PHOX91* gene on the X chromosome.

- **Leukocyte adhesion deficiency** is caused by mutations in genes encoding an essential integrin chain, a Golgi transporter required for the expression of the ligand for selectins, or signaling molecules activated by chemokine receptors that are required to activate integrins. Integrins and selectins are involved in the adhesion of leukocytes to other cells. As a result of these mutations, blood leukocytes do not bind firmly to vascular endothelium and are not recruited normally to sites of infection (see Chapter 2).
- Deficiencies of almost every complement protein, and many complement regulatory proteins, have been described (see Chapter 8). C3 deficiency results in severe infections and may be fatal. Deficiencies of C2 and C4, two components of the classical pathway of complement activation, occasionally result in increased bacterial or viral infection but more often in increased incidence of systemic lupus erythematosus, in part because of defective clearance of immune complexes. Deficiencies of complement regulatory proteins lead to various syndromes associated with excessive complement activation.
- The **Chédiak-Higashi syndrome** is an immunodeficiency disease in which lysosomal trafficking and the transport of granules are defective. The defect compromises many immune cells, including phagocytes, which normally kill ingested microbes in their lysosomes, and natural killer (NK) cells and cytotoxic T cells, which normally use proteins in specialized secretory lysosomes to kill other infected host cells. In all cases, the result is increased susceptibility to bacterial infection.
- Rare patients have been described with mutations affecting Toll-like receptors (TLRs) or signaling pathways downstream of TLRs, including molecules required for activation of the nuclear factor κB (NF-κB) transcription factor. Several of these mutations make patients susceptible to only a limited set of infections. For example, mutations affecting MyD88, an adaptor protein required for signaling by most TLRs, are associated with severe bacterial (most often pneumococcal) pneumonias, and mutations affecting TLR3 are associated with recurrent herpesvirus encephalitis and severe influenza.

Defects in Lymphocyte Maturation

Many primary immunodeficiencies are the result of genetic abnormalities that cause blocks in the maturation of B lymphocytes, T lymphocytes, or both (Figs. 12.3 and 12.4).

Severe Combined Immunodeficiency (SCID)

Disorders manifesting as defects in both the B cell and T cell arms of the adaptive immune system are classified as SCID. The underlying cause of most cases of SCID is a defect in T cell development or function; defective humoral immunity is largely a consequence of the lack of T cell helper function. Several different genetic abnormalities may cause SCID.

- **X-linked SCID**, affecting only male children, accounts for about half of the cases of SCID. More than 99% of these cases are caused by mutations in the common γ (γc) chain signaling subunit of the receptors for several cytokines, including IL-2, IL-4, IL-7, IL-9, IL-15, and IL-21. (Because the γc chain was first identified as one of the three chains of the IL-2 receptor, it is also called the IL-2Rγ chain.) When the γc chain is not functional, immature lymphocytes, especially pro-T cells, cannot proliferate in response to IL-7, which is the major growth factor for these cells. Defective responses to IL-7 result in reduced survival and maturation of lymphocyte precursors. In humans, the defect affects T cell maturation, but B cell development proceeds normally. The consequence of this developmental block is a profound decrease in the numbers of mature T cells, deficient cell-mediated immunity, and defective humoral immunity because of absence of T cell help (even though B cells may mature almost normally). NK cells are also deficient, because the γc chain is part of the receptor for IL-15, the major cytokine involved in NK cell proliferation and maturation.
An autosomal recessive form of SCID is caused by mutations in the gene encoding a kinase called Janus kinase 3 (JAK3) that is involved in signaling by the γc cytokine receptor chain. Such mutations result in the same abnormalities as those in X-linked SCID caused by γc mutations.
- About half the cases of **autosomal recessive SCID** are caused by mutations in adenosine deaminase (ADA),

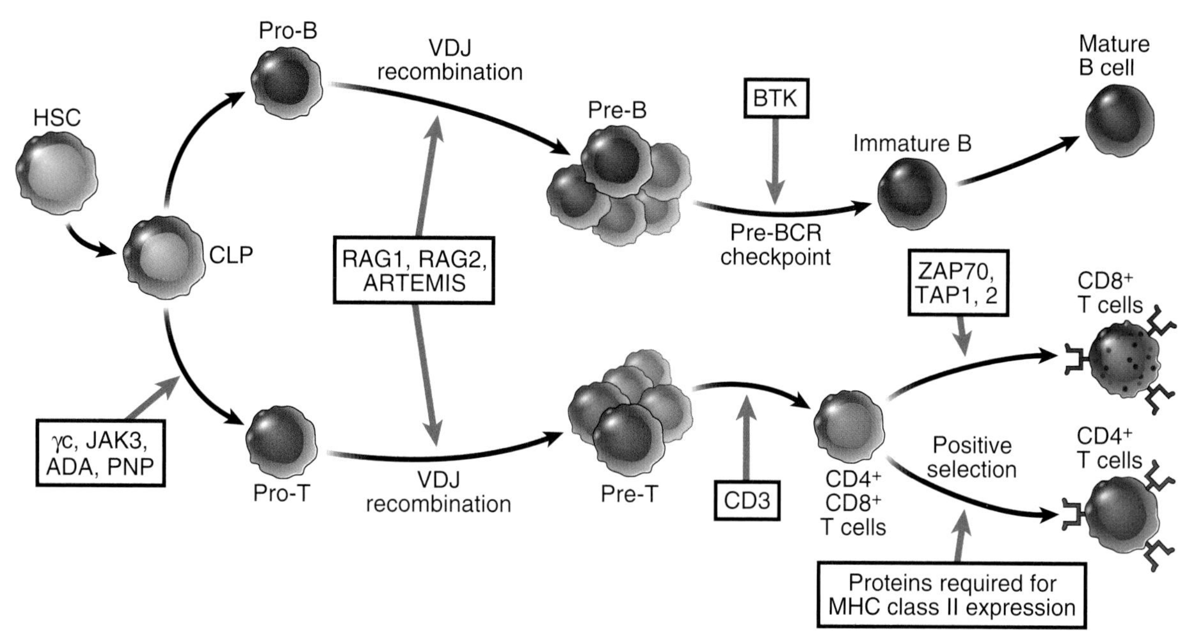

Fig. 12.3 **Primary immunodeficiencies caused by genetic defects in lymphocyte maturation.** Lymphocyte maturation pathways are described in Chapter 4. The proteins for which expression or functions are impaired by genetic mutations are listed in the boxes. The functions of the proteins are discussed in the text. *ADA,* Adenosine deaminase; *BCR,* B cell receptor; *CLP,* common lymphoid progenitor; *HSC,* hematopoietic stem cell; *PNP,* purine nucleoside phosphorylase; *RAG,* recombination-activating gene; *TCR,* T cell receptor.

an enzyme involved in the breakdown of adenosine. Deficiency of ADA leads to the accumulation of toxic purine metabolites in cells that are actively synthesizing DNA—namely, proliferating cells. Lymphocytes are particularly susceptible to injury by purine metabolites because these cells undergo tremendous proliferation during their maturation. ADA deficiency results in a block in T cell maturation more than in B cell maturation. A similar phenotype is seen in individuals who have a deficiency in purine nucleoside phosphorylase (PNP).

- Other causes of autosomal recessive SCID include mutations in the *RAG1* or *RAG2* gene, which encode the recombinase that is required for immunoglobulin (Ig) and T cell receptor (TCR) gene recombination and lymphocyte maturation. In the absence of RAG1 or RAG2, B and T cells fail to develop (see Chapter 4). Mutations in the *ARTEMIS* gene, which encodes an endonuclease involved in VDJ recombination, also result in failure of B and T cell development.
- **DiGeorge syndrome** (also known as 22q11 deletion syndrome) is characterized in part by a defect in T cell maturation. It results from a deletion on chromosome 22, which interferes with the development of the thymus (and parathyroid glands). The condition tends to improve with age, probably because the small amount of thymic tissue that does develop is able to support some T cell maturation.

With the increasing application of newborn screening to identify primary immunodeficiencies, many other rare causes of SCID have been discovered.

Selective B Cell Deficiencies

The most common clinical syndrome caused by a block in B cell maturation is **X-linked agammaglobulinemia** (initially called Bruton agammaglobulinemia). In this disorder, pre-B cells in the bone marrow fail to survive, resulting in a marked decrease or absence of mature B lymphocytes and serum Igs. The disease is caused by mutations in the gene encoding Bruton tyrosine kinase (BTK), resulting in defective production or function of the enzyme. BTK is activated by the pre-B cell receptor expressed in pre-B cells, and it delivers signals that promote the survival, proliferation, and maturation of these cells. The

Defects in T and B cell development:		
Disease	**Functional deficiencies**	**Mechanism of defect**
Severe Combined Immunodeficiency Disease (SCID)		
X-linked SCID	Markedly decreased T cells; normal or increased B cells; reduced serum Ig	Cytokine receptor common γ chain gene mutations, defective T cell maturation due to lack of IL-7 signals
Autosomal recessive SCID due to ADA, PNP deficiency	Progressive decrease in T and B cells (mostly T); reduced serum Ig in ADA deficiency, normal B cells and serum Ig in PNP deficiency	ADA or PNP deficiency leads to accumulation of toxic metabolites in lymphocytes
Autosomal recessive SCID due to defective VDJ recombination	Markedly decreased T and B cells; reduced serum Ig	Mutations in *RAG* genes and other genes involved in VDJ recombination
Defective class II MHC expression: The bare lymphocyte syndrome	Impaired CD4+ T cell development and activation; defective cell-mediated and humoral immunity	Mutations in genes encoding transcription factors required for class II MHC gene expression
DiGeorge syndrome (22q11 deletion syndrome)	Decreased T cells; normal B cells; normal or decreased serum Ig	Anomalous development of 3rd and 4th branchial pouches, leading to thymic hypoplasia
Impaired B cell development		
X-linked agammaglobulinemia	Decrease in all serum Ig isotypes; reduced B cell numbers	Block in maturation beyond pre-B cells, because of mutation in Bruton tyrosine kinase (BTK)

Fig. 12.4 Features of primary immunodeficiencies caused by defects in lymphocyte maturation. The figure summarizes the principal features of the most common primary immunodeficiencies in which the genetic blocks are known. *ADA,* Adenosine deaminase; *Ig,* immunoglobulin; *IL-7R,* interleukin-7 receptor; *MHC,* major histocompatibility complex; *PNP,* purine nucleoside phosphorylase; *RAG,* recombination-activating gene.

BTK gene is located on the X chromosome. Therefore, females who carry a mutant *BTK* allele on one of their X chromosomes are carriers of the disease, but male offspring who inherit the abnormal X chromosome suffer from the B cell deficiency. In about a fourth of patients with X-linked agammaglobulinemia, autoimmune diseases, notably arthritis, develop as well. A link between an immunodeficiency and autoimmunity seems paradoxical. One possible explanation for this association is that BTK contributes to B cell receptor signaling, which is required for B cell tolerance, so defective BTK may result in the accumulation of autoreactive B cells.

Defects in Lymphocyte Activation and Function

Numerous immunodeficiency diseases are caused by mutations affecting molecules involved in lymphocyte activation (Fig. 12.5).

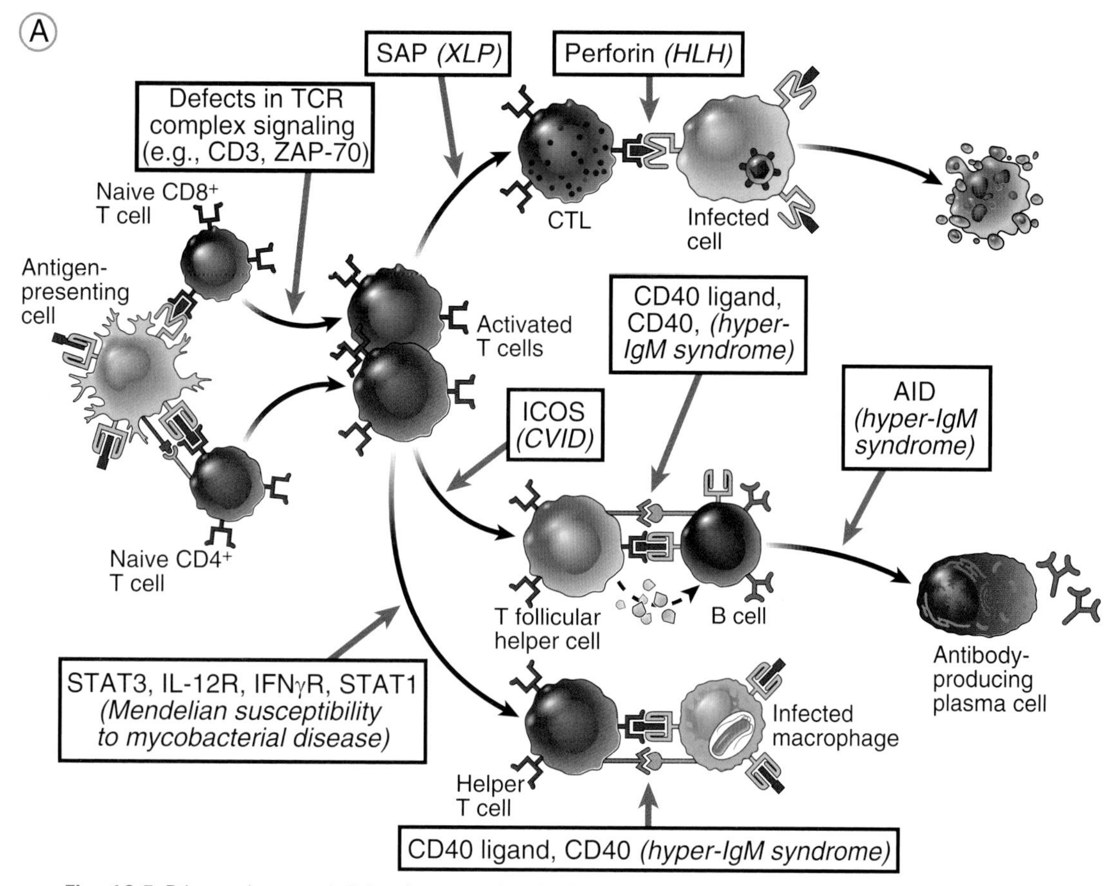

Fig. 12.5 Primary immunodeficiencies associated with defects in lymphocyte activation and effector functions. These immunodeficiencies may be caused by genetic defects in the expression of molecules listed in the boxes, required for antigen presentation to T cells, T or B lymphocyte antigen receptor signaling, helper T cell activation of B cells and macrophages, and differentiation of antibody-producing B cells. **A,** Examples showing the sites at which immune responses may be blocked. *AID,* Activation-induced deaminase; *CVID,* common variable immunodeficiency; *ICOS,* inducible costimulatory; *IFNγR,* IFN-γ receptor; *Ig,* immunoglobulin; *IL-12R,* IL-12 receptor; *MHC,* major histocompatibility complex; *SAP,* SLAM-associated protein; *ZAP-70,* ζ chain–associated protein of 70 kD.

Defects in B Cell Responses

Defective antibody production may result from abnormalities in B cells or in helper T cells.

- **X-linked hyper-IgM syndrome**. This disease is characterized by defective B cell heavy-chain class (isotype) switching, so IgM is the major serum antibody, and by deficient cell-mediated immunity against intracellular microbes. The disease is caused by mutations in the X chromosome gene encoding CD40 ligand (CD40L), the helper T cell protein that binds to CD40 on B cells, dendritic cells, and macrophages and thus mediates T cell–dependent activation of these cells (see Chapters 6 and 7). Failure to express functional CD40L leads to defective germinal center reactions in T cell–dependent B cell responses, so there is poor humoral immunity with reduced Ig class switching and no affinity maturation. In addition, the defective T cell–dependent macrophage and dendritic cell activation results in reduced cell-mediated immunity. Boys with this disease are especially susceptible to infection by *Pneumocystis jiroveci*, a fungus that survives within phagocytes in the absence of T cell help. An autosomal recessive form of hyper-IgM syndrome with a similar phenotype to that seen in the X-linked

Defects in T and B cell activation and function		
Disease	**Functional deficiencies**	**Mechanism of defect**
X-linked hyper-IgM syndrome	Defects in helper T cell-dependent B cell and macrophage activation	Mutations in CD40 ligand
Selective Ig deficiency	Reduced or no production of selective Ig isotypes; susceptibility to infections or no clinical problem	Mutations in Ig genes or unknown mutations
Common variable immunodeficiency	Reduced immunoglobulins; susceptibility to bacterial infections	Mutations in receptors for B cell growth factors, costimulators
Hemophagocytic lymphohistiocytosis	Impaired CTL and NK cell killing function; uncontrolled compensatory macrophage activation	Mutations in perforin gene and other genes required for CTL and NK cell granule exocytosis
Mendelian susceptibility to mycobacterial disease	Decreased Th1-mediated macrophage activation; susceptibility to infection by atypical mycobacteria and other intracellular pathogens	Mutations in genes encoding IL-12, the receptors for IL-12 or interferon-γ, STAT1
Defects in T cell receptor complex expression or signaling	Decreased T cells or abnormal ratios of $CD4^+$ and $CD8^+$ subsets; decreased cell-mediated immunity	Mutations or deletions in genes encoding CD3 proteins, ZAP-70
Mucocutaneous candidiasis, bacterial skin abcesses	Decreased Th17-mediated inflammatory responses	Mutations in genes encoding STAT3, IL-17, IL-17R
X-linked lymphoproliferative syndrome	Uncontrolled EBV-induced B cell proliferation and CTL activation; defective NK cell and CTL function and antibody responses	Mutations in gene encoding SAP (involved in signaling in lymphocytes)

Fig. 12.5, cont'd B, Summary of the features of immunodeficiency disorders whose genetic basis is shown in part **A.** Note that abnormalities in class II MHC expression and TCR complex signaling can cause defective T cell maturation (see Fig. 12.2), as well as defective activation of the cells that do mature, as shown here. *CTL,* Cytotoxic T lymphocyte; *EBV,* Epstein-Barr virus; *NK,* natural killer.

disease is observed in individuals with mutations in CD40. Another autosomal recessive form of hyper-IgM syndrome in which there are humoral abnormalities but no defect in cellular immunity is seen in individuals with mutations affecting the enzyme activation-induced deaminase (AID), which is induced by CD40 signaling and is involved in B cell class switching and affinity maturation (see Chapter 7).

- **Selective Ig class deficiencies.** Genetic deficiencies in the production of selected Ig classes are quite common. **IgA deficiency** is thought to affect as many as 1 in 700 people but causes no clinical problems in most patients and sinus, lung, and intestinal infections in a minority. The defect causing these deficiencies is not known in a majority of cases; rarely, the deficiencies may be caused by mutations of Ig heavy-chain constant (C) region genes.

- **Common variable immunodeficiency (CVID).** CVID is a heterogeneous group of disorders that are characterized by poor antibody responses to infections and reduced serum levels of IgG, IgA, and sometimes IgM. The underlying causes of CVID include mutations in genes encoding signaling molecules and transcription factors involved in B cell activation or encoding receptors that play a role in T cell–B cell interactions. Patients have recurrent infections, autoimmune disease, and lymphomas.

Defective Activation of T Lymphocytes

A variety of inherited abnormalities may interfere with T cell activation.

- **Bare lymphocyte syndrome**. This disease is caused by a failure to express class II major histocompatibility complex (MHC) molecules, as a result of mutations in the transcription factors that normally induce class II MHC expression. Recall that class II MHC molecules display peptide antigens for recognition by $CD4^+$ T cells, and this recognition is critical for maturation and activation of the T cells. The disease is manifested by a profound decrease in $CD4^+$ T cells because of defective maturation of these cells in the thymus, resulting in a form of SCID, and poor activation of $CD4^+$ T cells in secondary lymphoid organs.
- **Hemophagocytic lymphohistiocytosis** (HLH) syndromes are characterized by systemic, sometimes life-threatening, activation of immune cells, principally macrophages, usually in response to infections. Many cases of HLH occur as a manifestation of genetic disorders in which cytotoxic $CD8^+$ T cells and NK cells are unable to kill virus-infected target cells. These include patients with mutations in the gene encoding perforin and mutations in genes that encode proteins involved in granule exocytosis. These mutations result in persistent infections, usually viral, and excessive production of IFN-γ by T cells and NK cells, which in turn causes excessive macrophage activation. Some of these highly activated macrophages in the bone marrow ingest red blood cells, giving the syndrome its name.
- **Mendelian susceptibility to mycobacterial disease.** Mutations in the genes encoding components of interleukin-12 (IL-12), the IL-12 receptor, the interferon-γ (IFN-γ) receptor, or associated signaling molecules result in deficient cell-mediated immunity because of defects in Th1 development or Th1 cell-mediated macrophage activation (see Chapter 6). Patients present with increased susceptibility to weakly virulent environmental *Mycobacterium* species (often called atypical mycobacteria) as well as other intracellular pathogens, including *Salmonella* and various other bacterial, fungal, and viral species.
- Rare cases of selective T cell deficiency are caused by mutations affecting various signaling pathways or cytokines and receptors involved in differentiation of naive T cells into effector cells. Depending on the mutation and the extent of the defect, affected patients show severe T cell deficiency or deficiency in particular arms of T cell–mediated immunity, such as in T helper 1 (Th1) responses (associated with nontuberculous mycobacterial infections, discussed earlier) and Th17 responses (associated with fungal and bacterial infections). These defects have revealed the importance of various pathways of T cell activation, but these are rare disorders.

Lymphocyte Abnormalities Associated with Other Diseases

Some systemic diseases that involve multiple organ systems, and whose major manifestations are not immunologic, may have a component of immunodeficiency.

- **Wiskott-Aldrich syndrome** is characterized by eczema, reduced blood platelets, and immunodeficiency. This X-linked disease is caused by a mutation in a gene that encodes a regulator of the actin cytoskeleton required for signal transduction, immune synapse formation, and cytoskeletal reorganization. Because of the absence of this protein, platelets and leukocytes do not develop normally, are small, and fail to migrate normally.
- **Ataxia-telangiectasia** is characterized by gait abnormalities (ataxia), vascular malformations (telangiectasia), and immunodeficiency. The disease is caused by mutations in a gene that encodes a protein involved in DNA repair. Defects in this protein lead to abnormal DNA repair (e.g., during recombination

of antigen receptor gene segments), resulting in defective lymphocyte maturation.

- Some patients with autoimmune diseases, such as the disease caused by *AIRE* mutations (see Chapter 9), develop autoantibodies against their own cytokines and manifestations of immunodeficiency because of the resulting cytokine depletion. Autoantibodies against type I IFN (see Chapter 2) are seen even in the absence of overt autoimmunity and have been associated with severe cases of COVID-19.

Therapy of Primary Immunodeficiencies

Treatment of primary immunodeficiencies varies with the disease. SCID is fatal in early life unless the patient's immune system is reconstituted. The most widely used treatment is hematopoietic stem cell transplantation, with careful matching of donor and recipient to avoid potentially serious graft-versus-host disease. For selective B cell defects (e.g., X-linked agammaglobulinemia), patients may be given intravenous injections of pooled immunoglobulin (IVIG) from healthy donors to provide passive immunity. Although the ideal treatment for all congenital immunodeficiencies is to replace the defective gene, this remains a distant goal for most diseases. Successful gene therapy has been achieved in some patients with X-linked SCID; a normal γc gene is introduced into their hematopoietic stem cells, which are then transplanted back into the patients. Patients with ADA deficiency have been treated with enzyme replacement and gene therapy. In all patients with these diseases, infections are treated with antibiotics as needed.

ACQUIRED (SECONDARY) IMMUNODEFICIENCIES

Deficiencies of the immune system often develop because of abnormalities that are not genetic but are acquired during life (Fig. 12.6). The most serious of these abnormalities worldwide is HIV infection, described later. The most frequent causes of secondary immunodeficiencies in developed countries are cancers involving the bone marrow (leukemias) and immunosuppressive therapies. Cancer treatment with chemotherapeutic drugs and irradiation may damage proliferating cells, including precursors of leukocytes in the bone marrow and mature lymphocytes, resulting in immunodeficiency. Immunosuppressive drugs used to prevent graft rejection or treat

Cause	Mechanism
Human immunodeficiency virus infection	Depletion of CD4+ helper T cells
Irradiation and chemotherapy treatments for cancer	Decreased bone marrow precursors for all leukocytes
Immunosuppression for graft rejection and inflammatory diseases	Depletion or functional impairment of lymphocytes
Involvement of bone marrow by cancers (metastases, leukemias)	Reduced site of leukocyte development
Protein-calorie malnutrition	Metabolic derangements inhibit lymphocyte maturation and function
Loss of spleen (surgically to treat trauma or by infarction)	Decreased phagocytosis of microbes

Fig. 12.6 Acquired (secondary) immunodeficiencies. The most common causes of acquired immunodeficiency diseases and how they lead to defects in immune responses.

inflammatory diseases, including some of the newer therapies (e.g., cytokine antagonists, leukocyte adhesion molecule blockers), are designed to blunt immune responses. Therefore, immunodeficiency is a complication of such therapies. Protein-calorie malnutrition results in deficiencies of virtually all components of the immune system and is a common cause of immunodeficiency in countries with widespread poverty or famines.

ACQUIRED IMMUNODEFICIENCY SYNDROME (AIDS)

Since AIDS was first recognized as a distinct entity in the 1980s, it has become one of the most devastating afflictions in history. AIDS is caused by infection with HIV. Of the estimated 38 million HIV-infected people worldwide, about 70% are in Africa and 20% in Asia. More than 35 million deaths are attributable to HIV/AIDS, with almost 1 million deaths annually. Effective antiretroviral drugs have been developed, but the infection continues to spread in parts of the world where these therapies are not widely available, and in some African countries, more than 30% of the population has HIV infection. This section describes the important features of HIV, how it infects humans, and the disease it causes, ending with a brief discussion of the current status of therapy and vaccine development.

Human Immunodeficiency Virus (HIV)

HIV is a retrovirus that infects cells of the immune system, mainly CD4$^+$ T lymphocytes, and causes progressive destruction of these cells. An infectious HIV particle consists of two RNA strands within a protein core, surrounded by a lipid envelope derived from infected host cells but containing viral proteins (Fig. 12.7). The viral RNA encodes structural proteins, various enzymes, and proteins that regulate transcription of viral genes and the viral life cycle.

The life cycle of HIV consists of the following sequential steps: infection of cells, production of a DNA copy of viral RNA and integration into the host genome, expression of viral genes, and production of viral particles (Fig. 12.8). HIV infects cells by virtue of its major envelope glycoprotein, called gp120 (for 120-kD glycoprotein), which binds to CD4 and to particular chemokine receptors on human cells (mainly CXCR4 and CCR5). The major cell types that express these surface molecules and thus may be infected by HIV are CD4$^+$ T lymphocytes; additionally, some HIV strains can infect macrophages, and macrophages and dendritic cells may acquire the virus by phagocytosis. After binding to cellular receptors, the viral membrane fuses with the host cell membrane, and the virus enters the cell's cytoplasm. Here the virus is uncoated and its RNA is released. A DNA copy of the viral RNA is synthesized by the viral reverse transcriptase enzyme (a process characteristic of all retroviruses); this is converted to double-stranded DNA, which then is integrated into the host cell's DNA by the action of the viral integrase enzyme. The integrated viral DNA is called a provirus. If the infected T cell is activated by some extrinsic stimulus, such as another infectious microbe or cytokines, the cell responds by turning on the transcription of many of its own genes and often by producing cytokines itself. A negative consequence of this normal protective response is that the cytokines, and the process of cellular activation, may induce the transcription of proviral genes, leading to production of viral RNAs. The RNAs are translated into a precursor protein, which is processed into mature forms by the viral and cellular proteases. RNA replication generates copies of the viral RNA that are coated with structural proteins to form a core structure, which migrates to the cell membrane, acquires a lipid envelope from the host, and buds out as an infectious viral particle, ready to infect another cell. The integrated HIV DNA provirus may remain latent within infected cells for months or years, hidden from the patient's immune system (and even from antiviral therapies, discussed later).

Most cases of AIDS are caused by HIV-1 (i.e., HIV type 1). A closely related virus, HIV-2, causes some cases of the disease.

Pathogenesis of AIDS

AIDS develops over many years as latent HIV becomes activated and destroys cells of the immune system. Virus production leads to death of infected cells, as well as to death of uninfected lymphocytes, subsequent immune deficiencies, and clinical AIDS (Fig. 12.9). HIV infection is acquired by sexual intercourse, sharing of contaminated needles among intravenous drug users, transplacental transfer, or, in rare cases, transfusion of infected blood or blood products. After infection there may be a brief acute viremia, when

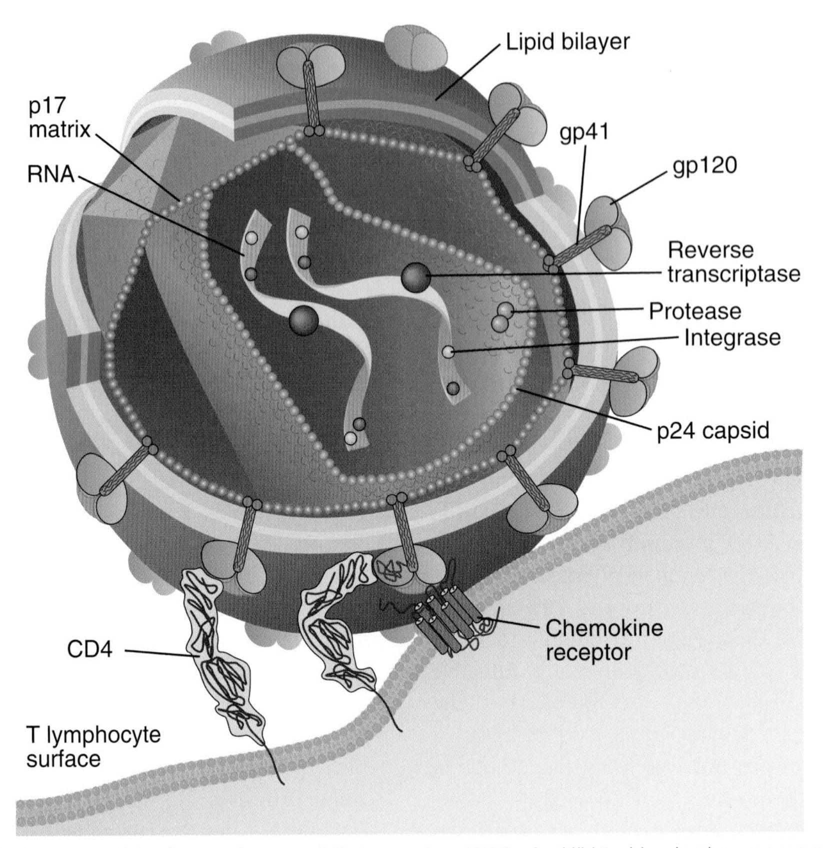

Fig. 12.7 Structure of the human immunodeficiency virus (HIV). An HIV-1 virion is shown next to a T cell surface. HIV-1 consists of two identical strands of RNA (the viral genome) and associated enzymes, including reverse transcriptase, integrase, and protease, packaged in a cone-shaped core composed of the p24 capsid protein with a surrounding p17 protein matrix, all surrounded by a phospholipid membrane envelope derived from the host cell. Virally encoded envelope proteins (gp41 and gp120) bind to CD4 and chemokine receptors on the host cell surface. *MHC,* Major histocompatibility complex. (Adapted from © 1996 Terese Winslow. Reproduced with permission.)

the virus is detected in the blood. The virus primarily infects $CD4^+$ T cells at sites of entry through mucosal epithelia, where there may be considerable destruction of infected T cells. Because a large fraction of the body's lymphocytes, and especially memory T cells, reside in mucosal tissues, the result of the local destruction may be a significant functional deficit that is not initially reflected in the presence of infected cells in the blood or the depletion of circulating T cells. If the integrated provirus is activated in infected cells, as described previously, the result is increased production of viral particles and spread of the infection. During the course of HIV infection, the major source of infectious viral particles are activated $CD4^+$ T cells. Follicular helper T cells and macrophages may become reservoirs of infection, wherein the virus may lie dormant for months or years.

The depletion of $CD4^+$ T cells after HIV infection is caused mainly by a cytopathic effect of the virus resulting from production of viral particles in infected cells. Active viral gene expression and protein production may interfere with the synthetic machinery of the infected T cells. Therefore, T cells in which the virus is replicating are killed during this process. The number of T cells lost during the progression to AIDS appears to be greater than the number of infected cells. The mechanism of this loss of uninfected T cells remains poorly defined.

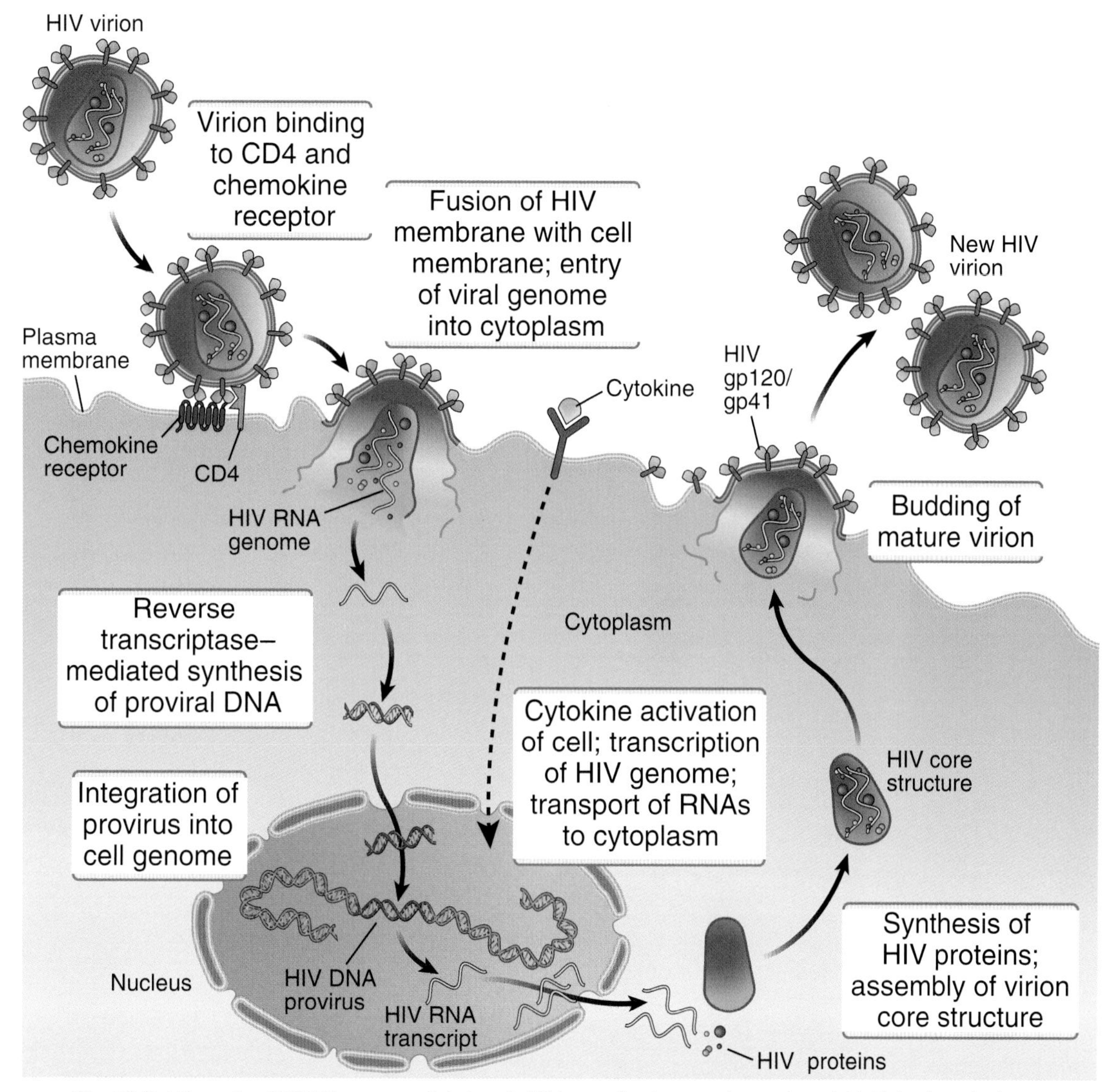

Fig. 12.8 Life cycle of HIV. The sequential steps in HIV reproduction are shown, from initial infection of a host cell to release of new virus particles (virions).

Other cells harboring the virus, such as dendritic cells and macrophages, may also die, resulting in destruction of the architecture of lymphoid organs. Many studies have suggested that immune deficiency results not only from depletion of T cells but also from various functional abnormalities in T lymphocytes and other immune cells. The significance of these functional defects has not been established, however, and loss of T cells (followed by a fall in the blood $CD4^+$ T cell count) remains the most reliable indicator of disease progression.

Clinical Features of HIV Infection and AIDS

The clinical course of HIV infection is characterized by several phases, culminating in immune deficiency (Fig. 12.10A).

- **Acute HIV syndrome.** Early after HIV infection, patients may experience a mild acute illness with fever

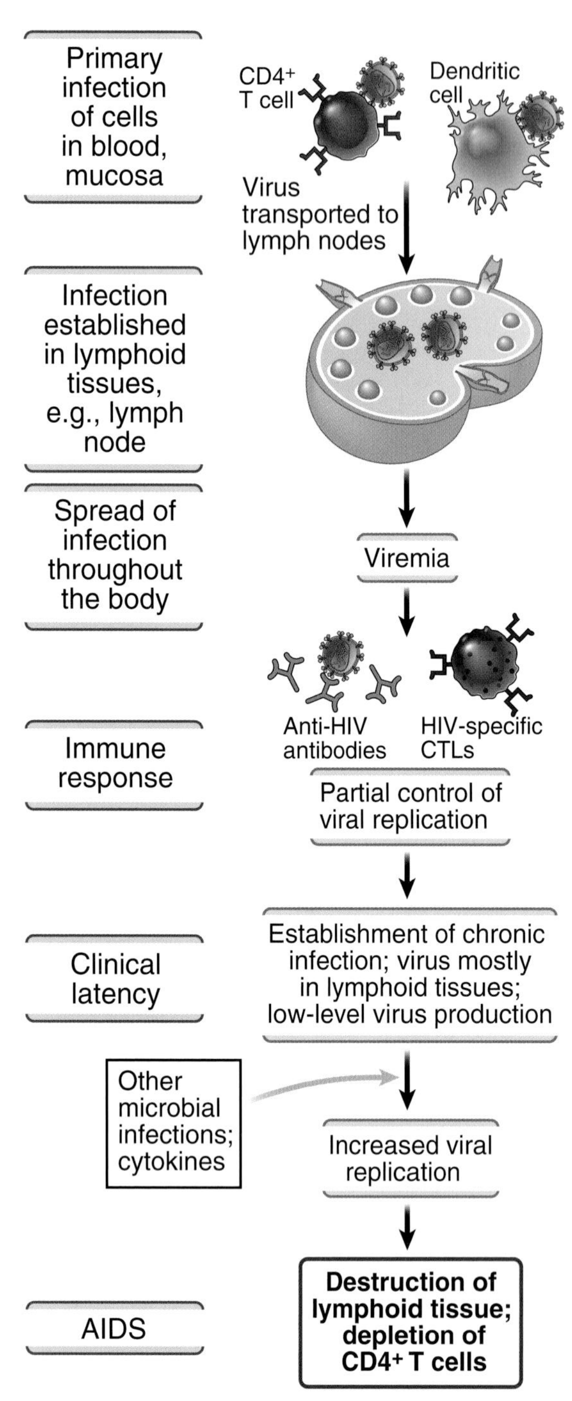

Fig. 12.9 Pathogenesis of disease caused by HIV. The development of HIV disease is associated with the spread of HIV from the initial site of infection to lymphoid tissues throughout the body. The immune response of the host temporarily controls acute infection but does not prevent establishment of chronic infection of cells in lymphoid tissues. Cytokines produced in response to HIV and other microbes serve to enhance HIV production and progression to acquired immunodeficiency syndrome (AIDS). *CTLs*, Cytotoxic T lymphocytes.

and malaise, correlating with the initial viremia. This illness subsides within a few days, and the disease enters a period of clinical latency.

- **Latency.** During latency, there may be few clinical problems, but there usually is a progressive loss of $CD4^+$ T cells in lymphoid tissues and destruction of the architecture of these tissues. Eventually, the blood $CD4^+$ T cell count begins to decline, and when the count falls below 200 cells/mm^3 (normal ~ 1500 cells/mm^3), patients become susceptible to infections and are diagnosed as having AIDS.
- **Clinical AIDS. AIDS ultimately causes increased susceptibility to infections and some cancers as a consequence of immune deficiency.** Patients not given antiretroviral drugs are often infected by intracellular microbes, such as viruses, the fungal pathogen *P. jirovecii*, and nontuberculous mycobacteria, all of which are normally eradicated by T cell–mediated immunity. Many of these microbes are present in the environment, but they do not infect healthy persons with intact immune systems. Because these infections are seen in immunodeficient persons, in whom the microbes have an opportunity to establish infection, these types of infections are said to be opportunistic. Latent viruses, such as cytomegalovirus and EBV, which are normally kept in check by cytotoxic T lymphocyte (CTL) responses, may be reactivated in patients with AIDS because of CTL defects, resulting in severe disease. Even though HIV does not infect $CD8^+$ T cells, the CTL responses are defective probably because $CD4^+$ helper T cells (the main targets of HIV) are required for full $CD8^+$ CTL responses against many viruses (see Chapters 5 and 6). AIDS patients are at increased risk for infections by extracellular bacteria, probably because of impaired helper T cell–dependent antibody responses to bacterial antigens. Patients also become susceptible to cancers caused by oncogenic viruses. The two most common types of cancers are B cell lymphomas, caused by EBV, and a tumor of small blood vessels called Kaposi sarcoma, caused by a herpesvirus. Patients with advanced AIDS often

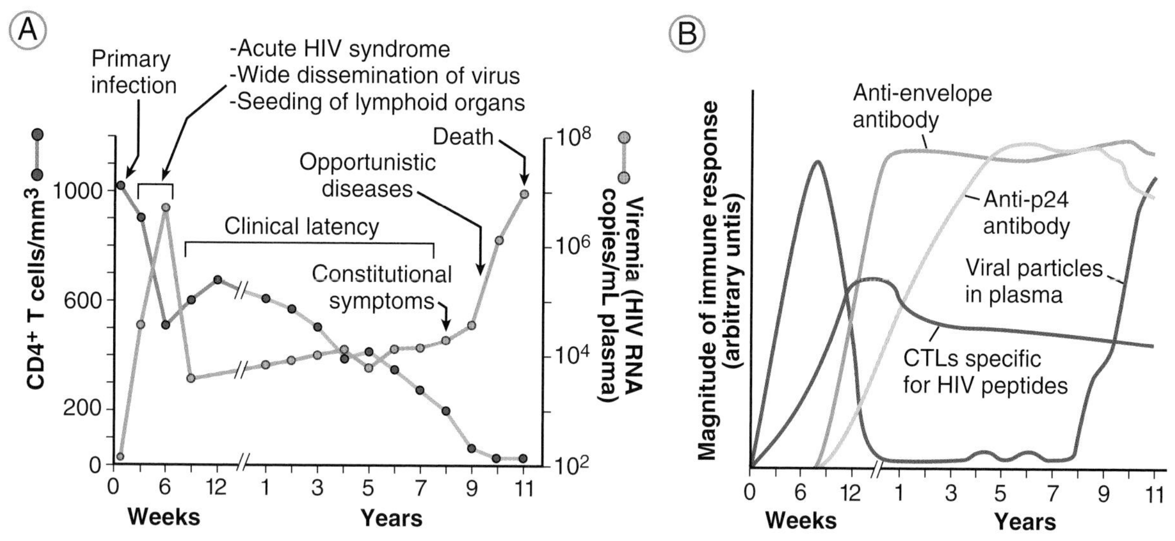

Fig. 12.10 Clinical course of *HIV* disease. **A,** Blood-borne virus (plasma viremia) is detected early after infection and may be accompanied by systemic symptoms typical of acute HIV syndrome. The virus spreads to lymphoid organs, but plasma viremia falls to very low levels (detectable only by sensitive reverse transcriptase–polymerase chain reaction assays) and stays this way for many years. $CD4^+$ T cell counts steadily decline during this clinical latency period because of active viral replication and T cell for of infection and other clinical components of acquired immunodeficiency syndrome. **B,** Magnitude and kinetics of immune responses, shown in arbitrary relative units. *CTLs,* Cytotoxic T lymphocytes. (Reproduced with permission from Pantaleo G, Graziosi C, Fauci AS: The immunopathogenesis of human immunodeficiency virus infection, *N Engl J Med* 328:327–335, 1993.)

have a wasting syndrome with significant loss of body mass, caused by altered metabolism and reduced calorie intake. The neurocognitive disorder that develops in some patients with AIDS may be caused by infection of macrophages (microglial cells) in the brain or the effects of cytokines or shed viral particles.

The clinical course of HIV infection has been dramatically changed by effective antiretroviral drug therapy. With appropriate treatment, patients exhibit much slower progression of the disease, fewer opportunistic infections, and greatly reduced incidence of cancers and neurocognitive disorder.

The immune response to HIV is ineffective in controlling spread of the virus and its pathologic effects. Infected patients produce antibodies and CTLs against viral antigens, and the responses help limit the early, acute HIV syndrome (Fig. 12.10B). But these immune responses usually do not prevent progression of the disease. Antibodies against envelope glycoproteins, such as gp120, may be ineffective because the virus rapidly mutates the region of gp120 that is the target of most antibodies. CTLs are often ineffective in killing infected cells because the virus inhibits the expression of class I MHC molecules by the infected cells. Immune responses to HIV may paradoxically promote spread of the infection. Antibody-coated viral particles may bind to Fc receptors on macrophages and follicular dendritic cells in lymphoid organs, thus increasing virus entry into these cells and creating additional reservoirs of infection. If CTLs are able to kill infected cells, the dead cells may be cleared by macrophages, which can migrate to other tissues and spread the infection. By infecting and thus interfering with the function of immune cells, the virus is able to prevent its own eradication.

A small fraction of patients control HIV infection without therapy; these individuals are often referred to as elite controllers or long-term nonprogressors. There has been great interest in defining the mechanisms that may protect these individuals, because elucidation of these mechanisms may suggest therapeutic approaches.

The presence of certain human leukocyte antigen (HLA) alleles, such as HLA-B57 and HLA-B27, are protective, and it has been shown that these HLA molecules are particularly efficient at presenting certain HIV peptides (which the virus cannot afford to mutate without losing viability) to $CD8^+$ T cells. In addition, rare individuals homozygous for a 32bp deletion in the gene encoding the CCR5 chemokine receptor (this variant is relatively frequent in Northern Europeans) lack functional CCR5, rendering them resistant to HIV infection.

Therapy and Vaccination Strategies

The current treatment for AIDS is aimed at controlling replication of HIV and the infectious complications of the disease. Combinations of drugs that block the activity of the viral reverse transcriptase, protease, and integrase enzymes are now being administered early in the course of the infection, and inhibitors of viral entry and fusion also have been developed. This therapeutic approach is called highly active or combination antiretroviral therapy (ART). In societies with widely available ART, opportunistic infections (e.g., by *P. jirovecii*) and some tumors (e.g., Kaposi sarcoma, EBV-induced lymphoma), which were devastating complications in the past, are now rarely seen in HIV-infected patients. In fact, treated patients are living long life spans and are dying of cardiovascular and other diseases that also afflict individuals who age without HIV (although they may be accelerated as a consequence of HIV infection, for unknown reasons). Even these highly effective antiretroviral drugs do not completely eradicate HIV infection. The virus is capable of mutating its genes, which may render it resistant to the drugs used, and reservoirs of latent virus (e.g., in lymphoid tissues) may be inaccessible to these drugs.

The development of effective vaccines will be necessary for control of HIV infection worldwide. A successful vaccine probably needs to induce high titers of broadly neutralizing antibodies that can recognize a wide range of virus isolates and a strong T cell response, as well as mucosal immunity. It has proved difficult to achieve all these goals with current vaccination strategies. The tremendous mutability of the virus allows it to mutate away from most neutralizing antibodies. The goal of creating vaccines that can elicit broadly neutralizing antibodies has not yet been met, and so far, vaccine trials for HIV have proved disappointing.

SUMMARY

- Immunodeficiency diseases are caused by defects in various components of the immune system that result in increased susceptibility to infections and some cancers. Primary (congenital) immunodeficiency diseases are caused by genetic abnormalities. Acquired (secondary) immunodeficiencies are the result of infections, cancers, malnutrition, or treatments for other conditions that adversely affect the cells of the immune system.
- SCID results from blocks in lymphocyte maturation. It may be caused by mutations in the cytokine receptor γc chain that reduce the IL-7–driven proliferation of immature lymphocytes, by mutations in enzymes involved in purine metabolism, or by other defects in lymphocyte maturation.
- Selective B cell maturation defects are seen in X-linked agammaglobulinemia, caused by abnormalities in an enzyme involved in B cell maturation (BTK), and selective T cell maturation defects are seen in the DiGeorge syndrome, in which the thymus does not develop normally.
- Some immunodeficiency diseases are caused by defects in lymphocyte activation. The X-linked hyper-IgM syndrome is caused by mutations in the gene encoding CD40 ligand, resulting in defective helper T cell–dependent B cell responses (e.g., Ig heavy-chain class switching) and T cell–dependent macrophage activation. The bare lymphocyte syndrome is caused by reduced expression of class II MHC proteins, resulting in impaired maturation and activation of $CD4^+$ T cells.
- AIDS is caused by the retrovirus HIV, which infects $CD4^+$ T cells, macrophages, and dendritic cells by using an envelope protein (gp120) to bind to CD4 and chemokine receptors. The viral RNA is reverse transcribed, and the resulting DNA integrates into the host genome, where it may be activated to produce infectious virus. Infected cells die during this

process of virus replication, and death of cells of the immune system is the principal mechanism by which the virus causes immune deficiency.

- The clinical course of HIV infection typically consists of acute viremia, clinical latency with progressive destruction of $CD4^+$ T cells and dissolution of lymphoid tissues, and ultimately AIDS, with severe immunodeficiency resulting in opportunistic infections, some cancers, weight loss, and a neurocognitive disorder. Treatment of HIV infection is designed to interfere with the life cycle of the virus. Vaccine development is ongoing.

REVIEW QUESTIONS

1. What are the most common clinicopathologic manifestations of immunodeficiency diseases?
2. What are some of the proteins affected by mutations that may block the maturation of T and B lymphocytes in human immunodeficiency diseases?
3. What are some of the mutations that may block activation or effector functions of both mature $CD4^+$ T cells and B cells, and what are the clinicopathologic consequences of these mutations?
4. How does HIV infect cells and replicate inside infected cells?
5. What are the principal clinical manifestations of advanced HIV infection, and what is the pathogenesis of these manifestations?

Answers to and discussion of the Review Questions may be found on p. 322.

GLOSSARY

A

$\alpha\beta$ T cell receptor ($\alpha\beta$ TCR) The most common form of TCR, expressed on both $CD4^+$ and $CD8^+$ T cells. The $\alpha\beta$ TCR recognizes peptide antigen bound to an MHC molecule. Both α and β chains contain highly variable (V) regions that together form the antigen-binding site as well as constant (C) regions. TCR V and C regions are structurally homologous to the V and C regions of immunoglobulin molecules.

ABO blood group antigens Carbohydrate antigens attached mainly to cell surface proteins or lipids that are present on many cell types, including red blood cells and endothelial cells. These antigens differ among individuals, depending on inherited alleles encoding the enzymes required for synthesis of the carbohydrate. The ABO antigens act as alloantigens that are responsible for blood transfusion reactions and hyperacute rejection of allografts.

Acquired immunodeficiency A deficiency in the immune system that is acquired after birth, because of infection (e.g., acquired immunodeficiency syndrome [AIDS]), malnutrition, aging, and other conditions, and that is not related to a genetic defect. Synonymous with **secondary immunodeficiency**.

Acquired immunodeficiency syndrome (AIDS) A disease caused by human immunodeficiency virus (HIV) infection that is characterized by depletion of $CD4^+$ T cells, leading to a profound defect in immunity. Clinically, AIDS includes opportunistic infections, malignant tumors, wasting, and neurocognitive disorder.

Activation-induced cell death (AICD) Apoptosis of activated lymphocytes, generally used for T cells.

Activation-induced (cytidine) deaminase (AID) An enzyme expressed in B cells that catalyzes the conversion of cytosine into uracil in DNA, which is a step required for somatic hypermutation and affinity maturation of antibodies and for immunoglobulin class switching.

Activator protein 1 (AP-1) A family of DNA-binding transcription factors composed of dimers of two proteins that bind to one another through a shared structural motif called a leucine zipper. The best-characterized AP-1 factor is composed of the proteins FOS and JUN. AP-1 is involved in transcriptional regulation of many different genes that are important in the immune system, such as cytokine genes.

Active immunity The form of adaptive immunity that is induced by exposure to a foreign antigen and activation of lymphocytes and in which the immunized individual plays an active role in responding to the antigen. This type contrasts with passive immunity, in which an individual receives antibodies or lymphocytes from another individual who was previously actively immunized.

Acute-phase proteins Proteins, mostly synthesized in the liver in response to inflammatory cytokines such as IL-1, IL-6, and TNF, whose plasma concentrations increase shortly after infection as part of the acute phase response. Examples include C-reactive protein, complement proteins, fibrinogen, and serum amyloid A protein. The acute-phase proteins play various roles in the innate immune response to microbes. Also called **acute phase reactants.**

Acute-phase response The increase in plasma concentrations of several proteins, called acute-phase proteins (reactants), that usually occurs as part of the early innate immune response to infections.

Acute rejection A form of graft rejection involving vascular and parenchymal injury mediated by T cells, macrophages, and antibodies that usually occurs days or weeks after transplantation but may occur later if pharmacologic immunosuppression becomes inadequate.

Adaptive immunity The form of immunity that is mediated by lymphocytes and stimulated by exposure to foreign antigens. In contrast to innate immunity, adaptive immunity is characterized by exquisite specificity for distinct antigens and by long-term and specific memory, manifest as more rapid and vigorous responses upon repeated exposure to the same microbe. Adaptive immunity is also called specific immunity or acquired immunity.

Adaptor protein A protein involved in intracellular signal transduction pathways that serves as a bridge molecule or scaffold for the recruitment of other signaling molecules. During lymphocyte antigen receptor or cytokine receptor signaling, adaptor proteins may be phosphorylated on tyrosine residues to enable them to bind other proteins containing SRC homology 2 (SH2) domains. Adaptor proteins involved in T cell activation include LAT, SLP76, and GRB2.

Addressin An adhesion molecule expressed on endothelial cells in different anatomic sites that directs organ-specific lymphocyte homing. Mucosal addressin cell adhesion molecule 1 (MAdCAM-1) is an example of an addressin expressed in Peyer patches in the intestinal wall that binds to the integrin $\alpha 4\beta 7$ on gut-homing T cells.

Adhesion molecule A cell surface molecule whose function is to promote adhesive interactions with other cells or the extracellular matrix. Leukocytes and endothelial cells express various types of adhesion molecules, such as selectins, integrins, and members of the immunoglobulin superfamily, and these molecules play crucial roles in cell migration and cellular activation in innate and adaptive immune responses.

Adjuvant A substance, distinct from antigen, that enhances T and B cell activation, mainly by promoting innate immune responses, which enhance the accumulation and activation of antigen-presenting cells (APCs) at the site of antigen exposure. Adjuvants, which are routinely used in clinical vaccines and experimental animal immunizations, stimulate expression of T cell—activating costimulators and cytokines by APCs and may also prolong the expression of peptide-MHC complexes on the surface of APCs.

Adoptive transfer The process of transferring cells from one individual into another or back into the same individual after in vitro expansion and activation. Adoptive transfer is used in research to define the role of a particular cell population (e.g., T cells) in an immune response. Clinically, adoptive transfer of tumor-specific T lymphocytes is used in cancer therapy.

Affinity The strength of the binding between a single binding site of a molecule (e.g., an antibody) and a ligand (e.g., an antigen). The affinity of a molecule X for a ligand Y is represented by the dissociation constant (K_d), which is the concentration of Y that is required to occupy the combining sites of half the X molecules present in a solution. A smaller K_d indicates a stronger or higher affinity interaction, because a lower concentration of ligand is needed to occupy the sites.

Affinity maturation The process that leads to increased affinity of antibodies for a particular antigen as a T cell–dependent antibody response progresses. Affinity maturation takes place in germinal centers of lymphoid tissues and is the result of somatic mutation of immunoglobulin genes, followed by selective survival of the B cells producing the highest affinity antibodies.

Alarmin A general term for molecules released from stressed or damaged cells that activate inflammatory responses. Alarmin is sometimes used synonymoulsy with danger associated molecular patterns (DAMPs), which are self molecules that bind to innate patten recognition receptors, but some cytokines (e.g., IL-33, IL-25) and some antimicrobial peptides (e.g., defensins, cathelicidin) are also often called alarmins.

Allele One of different variations of the same gene in different individuals present at a particular chromosomal locus, each differing in nucleotide sequence, and often by amino acid sequence of the encoded protein. An individual who is heterozygous at a locus has two different alleles, each on a different member of a pair of chromosomes, one inherited from the mother and one from the father. If a particular gene in a population has different alleles, the gene or locus is said to be polymorphic. MHC genes have many alleles (i.e., they are highly polymorphic).

Allelic exclusion The exclusive expression of only one of two inherited alleles encoding Ig heavy and light chains and TCR-β chains. Allelic exclusion occurs when the protein product of one productively recombined antigen receptor locus on one chromosome blocks rearrangement and expression of the corresponding locus on the other chromosome. This ensures that most lymphocytes will express a single antigen receptor and that all antigen receptors expressed by one clone of lymphocytes will have the identical specificity. Because the TCR-α chain locus does not show allelic exclusion, some T cells express two different TCRs with different α chains.

Allergen An antigen that elicits an immediate hypersensitivity (allergic) reaction. Allergens are proteins or chemicals bound to proteins that induce IgE antibody responses in atopic individuals.

Allergy A disorder characterized by immediate hypersensitivity reactions, often named according to the type of antigen (allergen) that elicits the disease, such as food allergy, bee sting allergy, and penicillin allergy. All of these conditions are the result of IgE production stimulated by IL-4– and IL-13–producing helper T cells, followed by allergen and IgE-dependent mast cell activation.

Alloantibody An antibody specific for an alloantigen (i.e., specific for an antigen present in some individuals of a species but not in others).

Alloantigen A cell or tissue antigen that is present in some individuals of a species but not in others and that is recognized as foreign on an allograft. Alloantigens are usually products of polymorphic genes.

Alloantiserum The alloantibody-containing serum of an individual who has previously been exposed to one or more alloantigens.

Allograft An organ or tissue graft from a donor who is of the same species but genetically nonidentical to the recipient (also called an allogeneic graft).

Alloreactive Reactive to alloantigens; describes T cells or antibodies from one individual that will recognize antigens on cells or tissues of another genetically nonidentical individual.

Alternative macrophage activation Macrophage activation by IL-4 and IL-13 leading to an antiinflammatory and tissue-reparative phenotype, in contrast to classical macrophage activation induced by interferon-γ and TLR ligands.

Alternative pathway of complement activation An antibody-independent pathway of activation of the complement system that occurs when the spontaneously generated C3b fragment of the C3 protein binds to microbial cell surfaces. The alternative pathway is a component of the innate immune system and mediates inflammatory responses to infection as well as direct lysis of microbes. The alternative pathway, as well as the classical and lectin pathways, terminate with formation of the membrane attack complex.

Anaphylatoxins The C5a, C4a, and C3a complement fragments that are generated during complement activation. The anaphylatoxins bind specific cell surface receptors and promote acute inflammation by activating mast cells (similar to the more severe reactions seen in anaphylaxis) and stimulating neutrophil chemotaxis.

Anaphylaxis A severe form of immediate hypersensitivity in which there is systemic mast cell or basophil activation, and the released mediators cause bronchial constriction, edema in multiple tissues, and cardiovascular collapse.

Anchor residues The amino acid residues of a peptide whose side chains fit into pockets in the floor of the peptide-binding cleft of an MHC molecule. The side chains bind to complementary amino acids in the MHC molecule and therefore serve to anchor the peptide in the cleft of the MHC molecule.

Anergy A state of unresponsiveness to antigenic stimulation. Lymphocyte anergy (also called clonal anergy) is the failure of clones of T or B cells to react to an antigen and is a mechanism of maintaining immunologic tolerance to self. Clinically, anergy refers to the lack of T cell–dependent cutaneous delayed-type hypersensitivity reactions to common (usually microbial) antigens.

Angiogenesis New blood vessel formation regulated by a variety of protein factors elaborated by cells of the innate and adaptive immune systems and often accompanying chronic inflammation and tumor growth.

Antibody A type of glycoprotein molecule, also called **immunoglobulin** (Ig), produced only by B lymphocytes and plasma cells derived from B cells, which binds antigens, often with a high degree of specificity and affinity. Membrane-bound antibodies serve as antigen receptors that initiate B cell activation. Secreted antibodies perform various effector functions, including neutralizing antigens, activating complement, and promoting leukocyte-dependent destruction of microbes. The basic structural unit of an antibody is composed of two identical heavy chains and two identical light chains. The N-terminal variable regions of the heavy and light chains form the antigen-binding sites, whereas the C-terminal constant regions of the heavy chains of secreted antibodies interact with other molecules in the immune system. Every individual has millions of different antibodies, each with a unique antigen-binding site.

Antibody-dependent cell-mediated cytotoxicity (ADCC) A process by which NK cells are targeted to cells coated with IgG specific for cell surface antigens, resulting in lysis of the antibody-coated cells. A specific receptor for the constant region of IgG, called FcγRIII (CD16), is expressed on the NK cell membrane and mediates binding of the cells to the IgG attached to other cells.

Antibody feedback The downregulation of antibody production by secreted IgG antibodies that occurs when antigen-antibody complexes simultaneously engage B cell membrane Ig and one type of Fcγ receptor (FcγRIIB). Under these conditions, the cytoplasmic tail of FcγRIIB transduces inhibitory signals inside the B cell and terminates the B cell response.

Antibody repertoire The collection of different antibody specificities expressed in an individual.

Antibody-secreting cell A B lymphocyte that has undergone differentiation and produces the secreted form of Ig. Antibody-secreting cells are generated from naive B cells in response to antigen and other stimuli and reside in the spleen and lymph nodes, as well as in the bone marrow. Often used synonymously with **plasma cells.**

Antigen A molecule that binds to an antibody or a TCR. Antigens that bind to antibodies include all classes of molecules. Most TCRs bind only peptide fragments of proteins complexed with MHC molecules.

Antigen presentation The display of peptides bound by molecules on the surface of an antigen-presenting cell that permits specific recognition by TCRs and activation of T cells.

Antigen-presenting cell (APC) A cell that displays peptide fragments of protein antigens, in association with MHC molecules, on its surface and activates antigen-specific T cells. In addition to displaying peptide-MHC complexes, APCs also express costimulatory molecules to optimally activate T lymphocytes.

Antigen processing The intracellular conversion of protein antigens derived from the extracellular space or the cytosol into peptides and loading of these peptides onto MHC molecules for display to T lymphocytes.

Antigenic variation The process by which antigens expressed by microbes may change by various genetic mechanisms and therefore allow the microbe to evade immune responses. Examples of antigenic variation include the change in influenza virus surface proteins hemagglutinin and neuraminidase, which necessitates the use of new vaccines each year, and the emergence of new SARS-CoV-2 variants, which often evade immunity induced by earlier viral strains.

Antimicrobial peptide One of a group of peptides with both positively charged and hydrophobic amino acid residues, which insert into and disrupt the integrity of the outer membranes of bacteria and some viruses. Antimicrobial peptides include cathelicidins, defensins, and RegIII peptides. They are produced by epithelial barrier cells and leukocytes and serve as innate immune effector molecules.

Antiretroviral therapy (ART) Combination chemotherapy for HIV infection, usually consisting of two nucleoside reverse transcriptase inhibitors and either one viral protease inhibitor or one non-nucleoside reverse transcriptase inhibitor. ART can reduce plasma virus titers to below detectable levels for more than 1 year and slow the progression of HIV disease. Also called **highly active antiretroviral therapy (HAART).**

Antiserum Serum from an individual previously immunized with an antigen that contains antibody specific for that antigen.

Apoptosis A process of cell death characterized by activation of intracellular caspases, DNA cleavage, nuclear condensation and fragmentation, and plasma membrane blebbing that leads to phagocytosis of cell fragments without inducing an inflammatory response. This type of cell death is important in the return to homeostasis after an immune response to an infection, maintenance of tolerance to self antigens, and killing of infected cells by cytotoxic T lymphocytes and natural killer cells.

Arthus reaction A localized form of experimental immune complex–mediated cutaneous vasculitis induced by injection of an antigen subcutaneously into a previously immunized animal or into an animal that has been given intravenous antibody specific for the antigen. Circulating antibodies bind to the injected antigen and form immune complexes that are deposited in the walls of small arteries at the injection site, giving rise to a local cutaneous vasculitis and tissue necrosis.

Asthma A pulmonary disease usually caused by repeated immediate hypersensitivity reactions in the bronchi that leads to intermittent and reversible airway obstruction, chronic bronchial inflammation with eosinophils, and bronchial smooth muscle cell hypertrophy and hyperreactivity.

Atopy The propensity of an individual to produce IgE antibodies in response to various environmental antigens and to develop strong immediate hypersensitivity (allergic) responses. People who have allergies to environmental antigens, such as pollen or house dust, are said to be atopic.

Autoantibody An antibody produced in an individual that is specific for a self antigen. Autoantibodies can damage cells and tissues and are produced in excess in autoimmune diseases, such as systemic lupus erythematosus and myasthenia gravis.

Autocrine factor A molecule that acts on the same cell that produces the factor. For example, IL-2 is an autocrine T cell growth factor that stimulates mitotic activity of the T cell that produces it.

Autoimmune disease A disease caused by a breakdown of self-tolerance such that the adaptive immune system responds to self antigens and causes cell and tissue damage. Autoimmune diseases can be caused by immune attack against one organ or tissue (e.g., multiple sclerosis, thyroiditis, or type 1 diabetes) or against multiple and systemically distributed antigens (e.g., systemic lupus erythematosus).

Autoimmune lymphoproliferative syndrome (ALPS) A syndrome caused by genetic mutations that impair regulation of lymphocyte apoptosis, characterized by lymphadenopathy, hepatomegaly, splenomegaly, high risk of lymphoma, and autoimmunity affecting hematopoietic cells. Most cases of ALPS are caused by mutations in the gene encoding FAS, but some cases are due to mutations in the genes encoding FAS-ligand or CASPASE-10

Autoimmune polyglandular syndrome type 1 (APS-1) Also known as autoimmune polyendocrinopathy—candidiasis—ectodermal dystrophy/dysplasia (APECED). A rare autoimmune disease caused by genetic deficiency of the autoimmune regulator protein AIRE required for central T cell tolerance to many different tissue antigens. APS-1 patients suffer from immune injury to multiple endocrine organs and skin.

Autoimmune regulator (AIRE) A protein that functions to induce expression of peripheral tissue protein antigens in medullary thymic epithelial cells, which is required for tolerance to those antigens. Mutations in the gene encoding AIRE in humans and mice impair expression of tissue antigens in the thymus and cause autoimmune disease (autoimmune polyglandular syndrome type 1) affecting multiple organs because of a failure to delete T cells or to generate regulatory T cells specific for these antigens.

Autoimmunity Responses of the adaptive immune system to self antigens that occurs when mechanisms of self-tolerance fail.

Autologous graft A tissue or organ graft in which the donor and recipient are the same individual. Autologous bone marrow and skin grafts are performed in clinical medicine.

Autophagy The normal process by which a cell degrades its own components by lysosomal catabolism. Autophagy plays a role in innate immune defense against infections, and polymorphisms of genes that regulate autophagy are linked to risk for some autoimmune diseases.

Avidity The overall strength of interaction between two molecules, such as an antibody and antigen. Avidity depends on both the affinity and the valency of interactions. Therefore, the avidity of a pentameric IgM antibody, with 10 antigen-binding sites, for a multivalent antigen is much greater than the affinity of a single-antibody combining site specific for the same antigen. Avidity can be used to describe the strength of cell-cell interactions, which are mediated by many binding interactions between cell surface molecules.

B

B lymphocyte The only cell type (including plasma cells derived from B lymphocytes) capable of producing antibody molecules and therefore the mediator of humoral immune responses. B lymphocytes, or B cells, develop in the bone marrow, and mature B cells are found mainly in lymphoid follicles in secondary lymphoid tissues, in the bone marrow, and in the circulation.

B-1 lymphocytes A subset of B lymphocytes that develop earlier during ontogeny than do conventional (follicular) B cells, express a limited repertoire of V genes with little junctional diversity, and secrete IgM antibodies that bind T-independent antigens. Many B-1 cells express the CD5 molecule.

Bare lymphocyte syndrome An immunodeficiency disease characterized by a lack of class II MHC molecule expression that leads to defects in maturation and activation of $CD4^+$ T cells and cell-mediated immunity. The disease is caused by mutations in genes encoding factors that regulate class II MHC gene transcription.

Basophil A type of bone marrow—derived circulating granulocyte with structural and functional similarities to mast cells that has granules containing many of the same inflammatory mediators as mast cells and expresses a high-affinity Fc receptor for IgE. Basophils that are recruited into tissue sites where antigen is present may contribute to immediate hypersensitivity reactions.

BCL-2 family proteins A family of structurally homologous cytoplasmic and mitochondrial membrane proteins that regulate apoptosis by influencing mitochondrial outer membrane permeability. Members of this family can be pro-apoptotic (such as BAX, BAD, and BAK) or antiapoptotic (e.g., BCL-2 and BCL-X_L).

BCL-6 A transcriptional repressor that is required for germinal center B cell development and for follicular T—helper cell development.

B cell receptor (BCR) The cell surface antigen receptor on B lymphocytes, which is a membrane-bound immunoglobulin molecule.

B cell receptor complex (BCR complex) A multiprotein complex expressed on the surface of B lymphocytes that recognizes antigen and transduces activating signals into the cell. The BCR complex includes membrane Ig, which is responsible for binding antigen, and Igα and Igβ proteins, which initiate signaling events.

BLIMP-1 A transcriptional repressor that is required for plasma cell generation.

Bi-specific T cell engagers (BiTEs) Recombinant antibody-derived molecules consisting of two fused Ig single-chain variable fragments (scFvs), one specific for the CD3 molecule on T cells and the other for a tumor antigen. They bring together T cells with tumor cells and activate the T cell to kill the tumor cell. BiTEs are used clinically to treat B cell tumors.

Bone marrow The tissue within the central cavity of bone that is the site of generation of all circulating blood cells after birth. B lymphocytes undergo most of the steps of their development in the bone marrow, while T cell progenitors arsing in the bone migrate to the thymus, where they develop into mature T cells.

Bone marrow transplantation See hematopoietic stem cell transplantation.

Bruton tyrosine kinase (BTK) A TEC family tyrosine kinase that is essential for B cell maturation. Mutations in the gene encoding BTK cause X-linked agammaglobulinemia, a disease characterized by failure of B cells to mature beyond the pre-B cell stage.

Burkitt lymphoma A malignant B cell tumor that is diagnosed by histologic features but almost always carries a reciprocal chromosomal translocation involving *Ig* gene loci and the cellular *MYC* gene on chromosome 8. Many cases of Burkitt lymphoma in Africa are associated with Epstein-Barr virus infection.

C

C (constant region) gene segments The DNA sequences in the *Ig* and *TCR* gene loci that encode the invariant portions of Ig heavy and light chains and TCRs α, β, γ, and δ chains.

C1 A plasma complement system protein composed of several polypeptide chains that initiates the classical pathway of complement activation by attaching to the Fc portions of immunoglobulin G (IgG) or IgM antibody that has bound antigen.

C1-inhibitor (C1-INH) A plasma protease inhibitor that blocks esterase functions of the C1r and C1s components of complement C1, which initiate the classical pathway of complement activation. C1-INH is a serine protease inhibitor (serpin), which inhibits other mediators of inflammation as well by blocking proteases in fibrinolytic, coagulation, and kinin pathways. A genetic deficiency in C1-INH causes the disease hereditary angioedema, characterized mainly by deregulated bradykinin activity. A disease with similar manifestations but occurring later in life called acquired angioedema is usually caused by autoantibodies specific for C1-INH.

C2 A classical complement pathway protein that is proteolytically cleaved by activated C1 to generate C2a, which forms part of the classical pathway C3 convertase.

C3 The central and most abundant complement system protein; it is involved in the classical, alternative and lectin complement pathways. C3 is proteolytically cleaved during complement activation to generate a C3b fragment, which covalently attaches to cell or microbial surfaces, and a C3a fragment, which is released and has various proinflammatory activities.

C3 convertase A multiprotein enzyme complex generated by the early steps of the classical, alternative and lectin pathways of complement activation. C3 convertase cleaves C3, which gives rise to two proteolytic products called C3a and C3b. C3b binds covalently to microbial surfaces where it acts as an opsonin and initiates the late steps of complement activation, and released C3a, sometimes called an anaphylatoxin, has various proinflammatory activities.

C4 A classical complement pathway protein that is proteolytically cleaved by activated C1 to generate C4b, which forms part of the classical pathway C3 convertase, and a released C4a fragment, sometimes called an anaphylatoxin, which has various proinflammatory activities.

C5 A protein that is cleaved by C5 convertases in all complement pathways, generating a C5b fragment, which initiates formation of the membrane attack complex, and a released C5a fragment, sometimes called an anaphylatoxin, which has various proinflammatory activities.

C5 convertase A multiprotein enzyme complex generated by C3b binding to C3 convertase. C5 convertase cleaves C5 and initiates the late steps of complement activation leading to formation of the membrane attack complex and lysis of cells.

Calcineurin A cytoplasmic serine/threonine phosphatase that dephosphorylates the transcription factor NFAT, thereby allowing NFAT to enter the nucleus. Calcineurin is activated by calcium signals generated through TCR signaling in response to antigen recognition, and the immunosuppressive drugs cyclosporine and FK506 (tacrolimus) work by blocking calcineurin activity.

Carcinoembryonic antigen (CEA, CD66) A highly glycosylated membrane protein; increased expression of CEA in many carcinomas of the colon, pancreas, stomach, and breast results in a rise in blood levels. The level of blood CEA is used to monitor the persistence or recurrence of metastatic carcinoma after treatment.

Caspases Intracellular proteases with cysteines in their active sites that cleave substrates at the C-terminal sides of aspartic acid residues. Most are components of enzymatic cascades that cause apoptotic death of cells, but caspase-1, which is part of the inflammasome, drives inflammation by processing inactive precursor forms of the cytokines IL-1 and IL-18 into their active forms.

Cathelicidins Polypeptides produced by neutrophils and various barrier epithelia that serve various functions in innate immunity, including direct toxicity to microorganisms, activation of leukocytes, and neutralization of lipopolysaccharide. Along with defensins, cathelicidins are often called **antimicrobial peptides (AMPs).**

Cathepsins Thiol and aspartyl proteases with broad substrate specificities, which are abundant in the endosomes in APCs, and play an important role in generating peptide fragments from ingested exogenous protein antigens that bind to class II MHC molecules.

CD molecules Cell surface molecules expressed on various cell types in the immune system that are designated by the "cluster of differentiation" or CD number. See Appendix I for a list of CD molecules.

Cell-mediated immunity The form of adaptive immunity that is mediated by T lymphocytes and serves as the defense mechanism against various types of microbes that are taken up by phagocytes or infect nonphagocytic cells. Cell-mediated immune responses include $CD4^+$ T cell–mediated activation of phagocytes and $CD8^+$ CTL–mediated killing of infected cells.

Central tolerance A form of self-tolerance induced in generative (central) lymphoid organs as a consequence of immature self-reactive lymphocytes recognizing self antigens and subsequently leading to their death or inactivation. Central tolerance prevents the emergence of lymphocytes with high-affinity receptors for the self antigens that are expressed in the bone marrow or thymus.

Centroblasts Rapidly proliferating B cells in the dark zone of germinal centers of secondary lymphoid tissues, which give rise to thousands of progeny, express activation-induced deaminase, and undergo somatic mutation of their *V* genes.

Centrocytes B cells in the light zone of germinal centers of secondary lymphoid organs, which are the progeny of proliferating centroblasts of the dark zone but do not themselves proliferate much. Centrocytes that express high-affinity Ig are selected to survive, resulting in affinity maturation of the B cell response, and undergo isotype switching and further differentiation into long-lived plasma cells and memory B cells.

Checkpoint blockade A form of cancer immunotherapy in which blocking antibodies specific for T cell inhibitory molecules, including PD-1, PD-L1, and CTLA-4, are administered to cancer patients to boost antitumor T cell responses; also called immune checkpoint blockade. This approach has been successful in effectively treating several kinds of metastatic cancers that are unresponsive to other therapies.

Chédiak-Higashi syndrome A rare autosomal recessive immunodeficiency disease caused by a defect in the trafficking of cytoplasmic granules of various cell types that affects the lysosomes of neutrophils and macrophages as well as the granules of CTLs and NK cells. Patients show reduced resistance to infection with pyogenic bacteria.

Chemokine receptors Cell surface receptors for chemokines that transduce signals stimulating the migration of leukocytes. There are at least 19 different mammalian chemokine receptors, each of which binds a different set of chemokines; all are members of the seven-transmembrane α-helical, G protein—coupled receptor family.

Chemokines A large family of structurally homologous low-molecular-weight cytokines that stimulate leukocyte chemotaxis, regulate the migration of leukocytes from the blood to tissues by activating leukocyte integrins, and maintain the spatial organization of different subsets of lymphocytes and antigen-presenting cells within lymphoid organs.

Chemotaxis Movement of a cell directed by a chemical concentration gradient. Leukocyte chemotaxis within various tissues is often directed by gradients of chemokines, leukotrienes, and the bacterial peptide N-formylmethionyl-leucyl-phenylalanine.

Chimeric antigen receptor (CAR) Genetically engineered receptors with tumor antigen—specific binding sites encoded by recombinant Ig-variable genes and cytoplasmic tails containing signaling domains of the TCR complex and T cell costimulatory receptors. When T cells are engineered to express chimeric antigen receptors, these cells can recognize and kill cells that the extracellular domain recognizes. Adoptive transfer of CAR-expressing T cells has been used successfully for the treatment of some types of hematologic cancers.

Chromosomal translocation A chromosomal abnormality in which a segment of one chromosome is transferred to another. Many malignant diseases of lymphocytes are associated with chromosomal translocations involving an Ig or TCR locus and a chromosomal segment containing a cellular oncogene.

Chronic granulomatous disease (CGD) A rare inherited immunodeficiency disease caused by mutations in genes encoding components of the phagocyte oxidase enzyme complex that is needed for microbial killing by polymorphonuclear leukocytes. The disease is characterized by recurrent intracellular bacterial and fungal infections, often accompanied by chronic cell-mediated immune responses and the formation of granulomas.

Chronic rejection A form of allograft rejection characterized by fibrosis with loss of normal organ structures occurring during a prolonged period. In many cases, the major pathologic lesion in chronic rejection is graft arterial occlusion caused by proliferation of intimal smooth muscle cells, which is called graft arteriosclerosis.

c-KIT ligand (stem cell factor) A protein required for hematopoiesis, early steps in T cell development in the thymus, and mast cell development. c-KIT ligand is produced in membrane-bound and secreted forms by stromal cells in the bone marrow and thymus, and it binds to the c-KIT tyrosine kinase membrane receptor on pluripotent stem cells.

Class I major histocompatibility complex (MHC) molecule One of two forms of polymorphic heterodimeric membrane proteins that bind and display peptide fragments of protein antigens on the surface of APCs for recognition by T lymphocytes. Class I MHC molecules usually display peptides derived from proteins that are proteolytically processed by proteasomes in the cytosol of the cell, for recognition by $CD8^+$ T cells.

Class II-associated invariant chain peptide (CLIP) A peptide remnant of the invariant chain that sits in the class II MHC peptide-binding cleft and is removed by action of the HLA-DM molecule before the cleft becomes accessible to peptides produced from extracellular protein antigens that are internalized into vesicles.

Class II major histocompatibility complex (MHC) molecule One of two major classes of polymorphic heterodimeric membrane proteins that bind and display peptide fragments of protein antigens on the surface of APCs for recognition by T lymphocytes. Class II MHC molecules usually display peptides derived from extracellular proteins that are internalized into phagocytic or endocytic vesicles, for recognition by $CD4^+$ T cells.

Classical macrophage activation Macrophage activation by interferon-γ, Th1 cells, and TLR ligands, leading to a proinflammatory and microbicidal phenotype. Classically activated macrophages are also called M1 macrophages.

Classical pathway of complement activation The complement pathway that is an effector arm of humoral immunity, generating inflammatory mediators, opsonins for phagocytosis of antigens, and lytic complexes that destroy cells. The classical pathway is initiated by binding of the C1 molecule to the Fc portions of IgG or IgM antibodies in antigen-antibody complexes, leading to proteolytic cleavage of C4 and C2 proteins to generate the classical pathway C3 convertase. The classical pathway, as well as the alternative and lectin pathways, terminate with formation of the membrane attack complex.

Clonal anergy A state of antigen unresponsiveness of a clone of T lymphocytes experimentally induced by recognition of antigen in the absence of additional signals (costimulatory signals) required for functional activation. Clonal anergy is considered a model for one mechanism of tolerance to self antigens and may be applicable to B lymphocytes as well.

Clonal deletion A mechanism of lymphocyte tolerance in which an immature T cell in the thymus or an immature B cell in the bone marrow undergoes apoptotic death as a consequence of recognizing a self antigen.

Clonal expansion The approximately 1000- to 100,000-fold increase in the number of lymphocytes specific for an antigen that results from antigen stimulation and proliferation of naive T and B cells. Clonal expansion occurs in lymphoid tissues and is required to generate enough antigen-specific effector T lymphocytes and plasma cells from rare naive precursors to eradicate infections.

Clonal ignorance A form of lymphocyte unresponsiveness in which self antigens are ignored by the immune system even though lymphocytes specific for those antigens remain viable and functional.

Clonal selection A fundamental tenet of the immune system meaning that every individual possesses numerous clonally derived lymphocytes, each clone having arisen from a single precursor, expresses one antigen receptor, and is capable of recognizing and responding to a distinct antigenic determinant. When an antigen enters, it selects a specific preexisting clone and activates it.

Clone A group of cells, all derived from a single common precursor, that maintain many of the genotypic and phenotypic features shared by the cell of origin. In adaptive immunity, all members of a clone of lymphocytes share the same clonally unique recombined *Ig* or *TCR* genes. The rearranged *Ig V* genes of different cells within a clone of B cells may change in sequence as a result of somatic hypermutation that occurs after activation of mature B cells.

Coinhibitory pathway A physiologic mechanism of regulation of T cell activation that involves the binding of molecules on an APC to receptors on a T cell, resulting in inhibition of T cell activation by antigen plus costimulators. Examples include PD-L1 on APCs binding to PD-1 on T cells, which generates inhibitory signals that block TCR and costimulatory signals in the T cell, or CTLA-4 on a T cell binding to B7-1 and B7-2 on an APC, preventing costimulation by B7-1 and B7-2 binding to CD28 on the T cell. Blockade of coinhibitory pathways is widely used as a therapeutic strategy to enhance antitumor immunity (checkpoint blockade).

Collectins A family of proteins, including mannose-binding lectin, that is characterized by a collagen-like domain and a lectin (i.e., carbohydrate-binding) domain. Collectins play a role in the innate immune system by acting as microbial pattern recognition receptors, and they may activate the complement system by binding to C1q.

Colony-stimulating factors (CSFs) Cytokines that promote the expansion and differentiation of bone marrow progenitor cells. CSFs are essential for the maturation of red blood cells, granulocytes, monocytes, and lymphocytes. Examples of CSFs include granulocyte-monocyte colony-stimulating factor (GM-CSF), granulocyte colony-stimulating factor (G-CSF), and IL-3.

Combinatorial diversity The diversity of Ig and TCR specificities generated by the use of many different combinations of different variable, diversity, and joining segments during somatic recombination of DNA in the *Ig* and *TCR* loci in developing B and T cells. Combinatorial diversity is one mechanism, which works together with junctional diversity, for the generation of large numbers of different antigen receptor genes from a limited number of DNA gene segments.

Common variable immunodeficiency disease (CVID) One of a group of heterogeneous disorders characterized by reduced circulating antibody, impaired antibody responses to infection and vaccines, increased incidence of infections, typically with *H. influenzae* and *S. pneumoniae,* various autoimmune manifestations, and a high incidence of lymphomas. Most cases are sporadic, but up to 25% of patients have a family history, some of which are result of monogenic mutations, such as mutations of the *CTLA4* gene.

Complement A system of plasma and cell surface proteins that interact with one another and with other molecules of the immune system to generate important effectors of innate and adaptive immune responses. The classical, alternative, and lectin pathways of the complement system are activated by antigen-antibody complexes, microbial surfaces, and plasma lectins binding to microbes, respectively, and consist of a cascade of proteolytic enzymes that generate inflammatory mediators and opsonins. All three pathways lead to the formation of a common terminal cell lytic complex that is inserted in cell membranes.

Complement receptor type 1 (CR1) A receptor for the C3b and C4b fragments of complement. Phagocytes use CR1 to mediate internalization of C3b- or C4b-coated particles. CR1 on erythrocytes serves in the clearance of immune complexes from the circulation. CR1 is also a regulator of complement activation.

Complement receptor type 2 (CR2) A receptor expressed on B cells and follicular dendritic cells that binds proteolytic fragments of the C3 complement protein, including C3d, C3dg, and iC3b. CR2 (also called CD21) functions to stimulate humoral immune responses by enhancing B cell activation by antigen and by promoting the trapping of antigen-antibody complexes in germinal centers. CR2 is also the receptor for Epstein-Barr virus.

Complement receptor type 3 (CR3) An integrin expressed mainly on neutrophils and macrophages, which binds to a peptide fragment of C3 called iC3b, which is deposited on microbes as a result of complement pathway activation. CR3 mediates the phagocytosis of iC3b coated microbes. CR3 is also called CD11b/CD18.

Complementarity-determining regions (CDRs) Short segments of Ig and TCR proteins that contain most of the sequence differences between antibodies or TCRs expressed by different clones of B cells and T cells and make contact with antigen; also called hypervariable regions. Three CDRs are present in the variable domain of each antigen receptor polypeptide chain, and six CDRs are present in an intact Ig or TCR molecule. These hypervariable segments assume loop structures that together form a surface complementary to the three-dimensional structure of the bound antigen.

Congenic mouse strains Inbred mouse strains that are identical to one another at every genetic locus except the one for which they are selected to differ; such strains are created by repetitive back cross breeding and selection for a particular trait. Congenic strains that differ from one another only at a particular MHC allele have been useful in defining the function of MHC molecules.

Congenital immunodeficiency See **primary immunodeficiency**.

Constant (C) region The portion of Ig or TCR polypeptide chains that does not vary in sequence among different clones and is not involved in antigen binding.

Contact sensitivity A state of immune responsiveness to certain chemical agents leading to T cell–mediated delayed-type hypersensitivity reactions upon skin contact. Substances that elicit contact sensitivity, including nickel ions, urushiols in poison ivy, and many therapeutic drugs, bind to and modify self proteins on the surfaces of APCs, which are then recognized by $CD4^+$ or $CD8^+$ T cells.

Coreceptor A lymphocyte surface receptor that binds to a molecule physically associated with antigen at the same time that membrane Ig or TCR binds the antigen and delivers signals required for optimal lymphocyte activation. CD4 and CD8 are T cell coreceptors that bind nonpolymorphic parts of an MHC molecule concurrently with the TCR binding to polymorphic MHC residues and the displayed peptide. CR2 is a coreceptor on B cells that binds to complement proteins attached to antigens at the same time that membrane Ig binds the antigen.

Costimulator A molecule expressed on the surface of APCs in response to innate immune stimuli, which provides a stimulus (the "second signal"), in addition to antigen (the "first signal"), required for the activation of naive T cells. The best defined costimulators are the B7 molecules (CD80 and CD86) on APCs that bind to the CD28 receptor on T cells.

COVID-19 A disease caused by the highly infectious coronavirus SARS-CoV-2, mainly affecting the respiratory system, often with pneumonia, but also involving various other organ systems. An exuberant inflammatory response to the virus can cause irreversible and often lethal lung injury in a minority of patients. The COVID-19 pandemic of 2020 to 2022 has led to the infection of hundreds of millions of people worldwide, and an estimated 6 to 15 million people have died of the disease.

CpG dinucleotides Cytidine-guanine sequences that are repeated in certain stretches of DNA and regulate gene expression dependent on methylation status. Most human CpG oligonucleotides are methylated, while most microbial CPGs are unmethylated and are recognized by Toll-like receptor 9, stimulating innate immune responses. CpG oligonucleotides are used as adjuvants in vaccines.

C-reactive protein (CRP) A member of the pentraxin family of plasma proteins involved in innate immune responses to bacterial infections. CRP binds to the capsule of pneumococcal bacteria. CRP also binds to C1q and may thereby activate complement or act as an opsonin by interacting with phagocyte C1q receptors. CRP is an acute-phase protein, and increased plasma concentration of CRP is a clinically used marker of inflammation.

Cross-matching A screening test performed to minimize the chance of adverse transfusion reactions or graft rejection, in which a patient in need of a blood transfusion or organ allograft is tested for the presence of preformed antibodies against donor cell surface antigens (usually blood group antigens or MHC molecules). The test involves mixing the recipient serum with leukocytes or red blood cells from potential donors and analyzing for agglutination or complement-dependent lysis of the cells.

Cross-presentation A mechanism by which a dendritic cell activates (or primes) a naive $CD8^+$ CTL specific for the antigens of a third cell (e.g., a virus-infected or tumor cell). Cross-presentation occurs, for example, when protein antigens from an infected cell are ingested by a dendritic cell and the microbial antigens are processed and presented in association with class I MHC molecules, unlike the general rule for phagocytosed antigens, which are presented in association with class II MHC molecules. The dendritic cell also provides costimulation for the T cells. Also called **cross-priming**.

CTLA-4 An Ig superfamily protein, expressed on the surface of activated effector T cells and Treg, that binds B7-1 and B7-2 with high affinity and plays an essential role in inhibiting T cell responses. CTLA-4 (also called CD152) is essential for Treg function and T cell tolerance to self antigens.

C-type lectin A member of a large family of calcium-dependent carbohydrate-binding proteins, many of which play important roles in innate and adaptive immunity. For example, soluble C-type lectins bind to microbial carbohydrate structures and mediate phagocytosis or complement activation (e.g., mannose-binding lectin, dectins, collectins, ficolins).

Cutaneous immune system The components of the innate and adaptive immune systems found in the skin that function together in a specialized way to detect and respond to pathogens on or in the skin and to maintain homeostasis with commensal microbes. Components of the cutaneous immune system include keratinocytes, Langerhans cells, dermal dendritic cells, intraepithelial lymphocytes, and dermal lymphocytes.

Cyclic GMP-AMP synthase A cytosolic DNA sensor of the innate immune system that generates cyclic GMP-AMP as a second messenger that interacts with the STING adaptor to induce type I interferon synthesis.

Cyclosporine A calcineurin inhibitor widely used as an immunosuppressive drug to prevent allograft rejection by blocking T cell activation. Cyclosporine (also called cyclosporin A) binds to a cytosolic protein called cyclophilin, and cyclosporine-cyclophilin complexes bind to and inhibit calcineurin, thereby inhibiting activation and nuclear translocation of the transcription factor NFAT.

Cytokines Proteins that are produced and secreted by many different cell types, and mediate inflammatory and immune reactions. Cytokines are principal mediators of communication between cells of the immune system (see Appendix II).

Cytokine storm A very strong, potentially life-threatening inflammatory reaction caused by large amounts of inflammatory cytokines rapidly released from activated T cells and macrophages in the setting of infections, auotimmune disease, or cancer immunotherapies. Clinical features include fever, chills, low white blood cell and platelet counts, edema, respiratory failure, and shock.

Cytosolic DNA sensors (CDSs) Molecules that detect microbial double-stranded DNA in the cytosol and activate signaling pathways that initiate antimicrobial responses, including type I interferon production and autophagy.

Cytotoxic (or cytolytic) T lymphocyte (CTL) A type of T lymphocyte whose major effector function is to recognize and kill host cells infected with viruses or other intracellular microbes as well as tumor cells. CTLs usually express CD8 and recognize peptides derived from cytosolic microbial and tumor antigens displayed by class I MHC molecules. CTL killing of infected and tumor cells involves delivery of the contents of cytoplasmic granules into the cytosol of the cells, leading to apoptotic death.

D

Damage-associated molecular patterns (DAMPs) Endogenous molecules that are produced by or released from damaged and dying cells that bind to pattern recognition receptors and stimulate innate immune responses. Examples include high-mobility group box 1 (HMGB1) protein, extracellular ATP, and uric acid crystals.

Death receptors Plasma membrane receptors expressed on various cell types that, upon ligand binding, transduce signals that lead to recruitment of the FAS-associated protein with death domain (FADD) adaptor protein, which activates caspase-8, leading to apoptotic cell death. All death receptors, including FAS, TRAIL, and TNFR, belong to the TNF receptor superfamily.

Dectins Pattern recognition receptors expressed on dendritic cells that recognize fungal cell wall carbohydrates and induce signaling events that promote inflammation and activate the dendritic cells.

Defensins Cysteine-rich peptides produced by epithelial barrier cells in the skin, gut, lung, and other tissues and in neutrophil granules that act as broad-spectrum antibiotics to kill a wide variety of bacteria and fungi. The synthesis of defensins is increased in response to stimulation of innate immune system receptors such as Toll-like receptors and inflammatory cytokines such as IL-1 and TNF. Defensins and cathelicidins are two types of antimicrobial peptides (AMPs).

Delayed-type hypersensitivity (DTH) An immune reaction in which T cell–dependent macrophage activation and inflammation cause tissue injury. A DTH reaction to the subcutaneous injection of antigen is often used as an assay for cell-mediated immunity (e.g., the purified protein derivative [PPD] skin test to screen for immunity to *Mycobacterium tuberculosis*).

Dendritic cells Bone marrow–derived cells found in epithelial barriers, the stroma of most organs, and lymphoid tissues, which are morphologically characterized by thin membranous projections. Many subsets of dendritic cells exist with diverse functions. Classical (conventional) dendritic cells function as innate sentinel cells and APCs for naive T lymphocytes, and they are important for initiation of adaptive immune responses to protein antigen. Plasmacytoid dendritic cells produce type I interferons in response to exposure to viruses. Monocyte-derived dendritic cells (MoDCs) are derived from blood monocytes during inflammatory reactions. Immature (resting) dendritic cells are important for induction of tolerance to self antigens.

Desensitization A method of treating immediate hypersensitivity disease (allergies) that involves repetitive administration of low doses of an antigen to which individuals are allergic. This process often prevents severe allergic reactions on subsequent environmental exposure to the allergen, but the mechanisms are not well understood.

Determinant The specific portion of a macromolecular antigen to which an antibody or TCR binds. In the case of a protein antigen recognized by a T cell, the determinant is the peptide portion that binds to an MHC molecule for recognition by the TCR. Synonymous with **epitope**.

Diacylglycerol (DAG) A signaling molecule generated by phospholipase C (PLCγ1)-mediated hydrolysis of the plasma membrane phospholipid phosphatidylinositol 4,5-bisphosphate (PIP_2) during antigen activation of lymphocytes and various other immune cells. The main function of DAG is to activate an enzyme called protein kinase C that participates in the generation of active transcription factors.

DiGeorge syndrome A selective T cell deficiency caused by a congenital malformation that results in defective development of the thymus, parathyroid glands, and other structures that arise from the third and fourth pharyngeal pouches.

Direct antigen presentation (direct allorecognition) Presentation of cell surface allogeneic MHC molecules by graft APCs to a graft recipient's T cells that leads to activation of the alloreactive T cells. In direct recognition of an allogeneic MHC molecule, a TCR that was selected to recognize a self MHC molecule plus foreign peptide cross-reacts with the intact allogeneic MHC molecule plus any bound peptide. Direct presentation is partly responsible for strong T cell responses to allografts.

Diversity The existence of a large number of lymphocytes with different antigenic specificities in any individual. Diversity is a fundamental property of the adaptive immune system and is the result of variability in the structures of the antigen-binding sites of lymphocyte receptors for antigens (antibodies and TCRs).

Diversity (D) segments Short coding sequences between the variable *(V)* and constant *(C)* gene segments in the Ig heavy chain and TCR β and γ loci that together with *J* segments are somatically recombined with V segments during lymphocyte development. The resulting recombined *VDJ* DNA codes for the carboxyl-terminal ends of the antigen receptor V regions, including the third hypervariable (CDR) regions. Random use of *D* segments contributes to the diversity of the antigen receptor repertoire.

DNA vaccine A vaccine composed of a bacterial plasmid containing a complementary DNA encoding a protein antigen. DNA vaccines presumably work because APCs are transfected in vivo by the plasmid and express immunogenic peptides that elicit specific responses. Furthermore, the plasmid DNA contains CpG nucleotides that act as adjuvants.

Double-negative thymocyte A subset of developing T cells in the thymus (thymocytes) that express neither CD4 nor CD8. Most double-negative thymocytes are at an early developmental stage and do not express antigen receptors. They will later express both CD4 and CD8 during the intermediate double-positive stage before further maturation to single-positive T cells expressing only CD4 or CD8.

Double-positive thymocyte A subset of developing T cells in the thymus (thymocytes) that express both CD4 and CD8 and are at an intermediate developmental stage. Double-positive thymocytes also express TCRs and are subject to selection processes, and they mature to single-positive T cells expressing only CD4 or CD8.

E

Ectoparasites Parasites that live on the surface of an animal, such as ticks and mites. Both the innate and adaptive immune systems may play a role in protection against ectoparasites, often by destroying the larval stages of these organisms.

Effector cells The cells that perform effector functions during an immune response, such as secreting cytokines (e.g., helper T cells), killing microbes (e.g., macrophages), killing microbe-infected host cells (e.g., CTLs), or secreting antibodies (e.g., plasma cells).

Effector phase The phase of an immune response in which a foreign antigen is destroyed or inactivated. In a humoral immune response, the effector phase may be characterized by antibody-dependent complement activation and phagocytosis of antibody- and complement-opsonized bacteria. In a cell-mediated immune response, the effector phase is characterized by activation of macrophages and other leukocytes by helper T cells and killing of infected cells by CTLs.

Endosome An intracellular membrane-bound vesicle into which extracellular proteins are internalized during antigen processing. Endosomes are formed by invagination of the plasma membrane, and they may mature into late endosomes and lysosomes, which have progressively lower pH and more hydrolytic enzymes. The proteolytic enzymes in endosomes degrade internalized proteins into peptides that bind to class II MHC molecules, and the endosomes containing these peptides fuse with Golgi-derived vesicles containing class II MHC molecules. (Endosomes are found in all cells and participate in internalization events that are not linked to antigen presentation.)

Endotoxin A component of the cell wall of gram-negative bacteria, also called **lipopolysaccharide** (LPS), that is released from dying bacteria and stimulates innate immune inflammatory responses by binding to TLR4 on many different cell types, including phagocytes, endothelial cells, dendritic cells, and barrier epithelial cells. Endotoxin contains both lipid components and carbohydrate (polysaccharide) moieties.

Enhancer A regulatory nucleotide sequence in a gene that is located either upstream or downstream of the promoter, binds transcription factors, and increases the activity of the promoter. In cells of the immune system, enhancers are responsible for integrating cell surface signals that lead to induced transcription of genes encoding many of the effector proteins of an immune response, such as cytokines.

Envelope glycoprotein (Env) A membrane glycoprotein encoded by a retrovirus that is expressed on the plasma membrane of infected cells and on the host cell–derived membrane coat of viral particles. Env proteins are often required for viral infectivity. The Env proteins of HIV include gp41 and gp120, which bind to CD4 and chemokine receptors, respectively, on human T cells and mediate fusion of the viral and T cell membranes.

Enzyme-linked immunosorbent assay (ELISA) A method of quantifying an antigen immobilized on a solid surface by use of a specific antibody with a covalently coupled enzyme. The amount of antibody that binds the antigen is proportional to the amount of antigen present and is determined by spectrophotometrically measuring the conversion of a clear substrate to a colored product by the coupled enzyme.

Eosinophil A bone marrow–derived granulocyte that is abundant in the inflammatory infiltrates of immediate hypersensitivity late-phase reactions and contributes to many of the pathologic processes in allergic diseases. They are recognizable by their bright red cytosolic granules in standard Wright-Giemsa–stained blood smears and hematoxylin and eosin–stained tissue sections. Eosinophils are important in defense against extracellular parasites, including helminths.

Epitope The specific portion of a macromolecular antigen to which an antibody or TCR binds. In the case of a protein antigen recognized by a T cell, an epitope is the peptide portion that binds to an MHC molecule for recognition by the TCR. Synonymous with **determinant**.

Epitope spreading In autoimmunity, the development of immune responses to multiple epitopes as an autoimmune disease originally targeting one epitope progresses, likely caused by further breakdown in tolerance and release of additional tissue antigens due to the inflammatory process stimulated by the initial response.

Epstein-Barr virus (EBV) A double-stranded DNA virus of the herpesvirus family that is the etiologic agent of infectious mononucleosis and is associated with some B-cell malignant tumors and nasopharyngeal carcinoma. EBV infects B lymphocytes and some epithelial cells by specifically binding to complement receptor 2 (CR2, CD21).

Experimental autoimmune encephalomyelitis (EAE) An animal model of multiple sclerosis, an autoimmune demyelinating disease of the central nervous system. EAE is induced in rodents by immunization with components of the myelin sheath of nerves (e.g., myelin basic protein), mixed with an adjuvant. The disease is mediated in large part by cytokine-secreting $CD4^+$ T cells specific for the myelin sheath proteins.

F

Fab (fragment, antigen-binding) A part of an antibody, first produced by proteolysis of IgG, that includes one complete light chain paired with one heavy chain fragment containing the variable domain and only the first constant domain. Fab fragments, which can be generated from all antibodies, retain the ability to monovalently bind an antigen but cannot interact with IgG Fc receptors on cells or with complement. Therefore, Fab preparations are used in research and therapeutic applications when antigen binding is desired without activation of effector functions. (The Fab′ fragment retains the hinge region of the heavy chain.)

F(ab′)2 A part of an Ig molecule (first produced by proteolysis of IgG) that includes two complete light chains but only the variable domain, first constant domain, and hinge region of the two heavy chains. $F(ab')_2$ fragments retain the entire bivalent antigen-binding region of an intact Ig molecule but cannot bind complement or Fc receptors. They are used in research when antigen binding is desired without antibody effector functions.

FAS (CD95) A death receptor of the TNF receptor family that is expressed on the surface of T cells and many other cell types and initiates a signaling cascade leading to apoptotic death of the cell. The death pathway is initiated when FAS ligand expressed on activated T cells binds to FAS on the same or other cells. FAS-mediated killing of lymphocytes is important for the maintenance of self-tolerance. Mutations in the *FAS* gene cause the disease autoimmune lymphoproliferative syndrome (ALPS). (See also **death receptors**.)

FAS ligand (CD95 ligand) A membrane protein that is a member of the TNF family of proteins expressed on activated T cells. FAS ligand binds to the death receptor FAS, thereby activating a signaling pathway that leads to apoptotic death of the FAS-expressing cell. Mutations in the FAS ligand gene cause systemic autoimmune disease in mice.

Fc (fragment, crystalline) A region of an antibody molecule that can be isolated by proteolysis of IgG that contains only the disulfide-linked carboxyl-terminal regions of the two heavy chains. The Fc region of Ig molecules mediates effector functions by binding to cell surface receptors or the C1q complement protein. (Fc fragments are so named because they tend to crystallize out of solution.)

Fc receptor A cell surface receptor specific for the carboxyl-terminal constant region of an Ig molecule. Fc receptors are typically multichain protein complexes that include signaling components and Ig-binding components. Several types of Fc receptors exist, including those specific for different IgG isotypes, IgE, and IgA. Fc receptors mediate many of the cell-dependent effector functions of antibodies, including phagocytosis of antibody-bound antigens, antigen-induced activation of mast cells, and targeting and activation of NK cells.

FcεRI A high-affinity receptor for the carboxyl-terminal constant region of IgE molecules that is expressed mainly on mast cells and basophils. FcεRI molecules on mast cells are usually occupied by IgE, and antigen-induced cross-linking of these IgE-FcεRI complexes activates the mast cell and initiates immediate hypersensitivity reactions.

Fcγ receptor (FcγR) A cell surface receptor specific for the carboxyl-terminal constant region of IgG molecules. There are several different types of Fcγ receptors, including a high-affinity FcγRI that mediates phagocytosis by macrophages and neutrophils, a low-affinity FcγRIIB that transduces inhibitory signals in B cells and myeloid cells, and a low-affinity FcγRIIIA that mediates recognition of opsonized cells by and activation of NK cells.

Fibroblastic reticular cell (FRC) Mesenchymally derived cells that drive formation of secondary lymphoid organs during embryonic development and contribute in multiple ways to the structure and functions of these organs.

Ficolins Hexameric innate immune system plasma proteins, containing collagen-like domains and fibrinogen-like carbohydrate-recognizing domains, which bind to cell wall components of gram-positive bacteria, opsonizing them and activating complement.

First-set rejection Allograft rejection in an individual who has not previously received a graft or otherwise been exposed to tissue alloantigens from the same donor. First-set rejection usually takes approximately 7 to 14 days.

FK506 See **Tacrolimus**.

Flow cytometry A method of analysis of the phenotype of cell populations requiring a specialized instrument (flow cytometer) that can detect fluorescence on individual cells in a suspension and thereby determine the number of cells expressing the molecule to which a fluorescent probe binds, as well as the relative amount of the molecule expressed. Suspensions of cells are incubated with fluorescently labeled antibodies or other probes, and the amount of probe bound by each cell in the population is measured by passing the cells one at a time through a fluorimeter with a laser-generated incident beam.

Fluorescence-activated cell sorter (FACS) An adaptation of the flow cytometer that is used for the purification of cells from a mixed population according to which and how much fluorescent probe the cells bind. Cells are first stained with a fluorescently labeled probe, such as an antibody specific for a surface antigen of a cell population. The cells are then passed one at a time through a fluorimeter with a laser-generated incident beam and are deflected into different collection tubes by electromagnetic fields whose strength and direction are varied according to the measured intensity of the fluorescence signal.

Follicle See **lymphoid follicle**.

Follicular dendritic cells (FDCs) Cells in lymphoid follicles of secondary lymphoid organs that express complement receptors and Fc receptors and have long cytoplasmic processes that form a meshwork integral to the architecture of the follicles. Follicular dendritic cells display antigens on their surface for B cell recognition and are involved in the activation and selection of B cells expressing high-affinity membrane Ig during the process of affinity maturation. They are nonhematopoietic cells (not of bone marrow origin).

Follicular helper T cell (Tfh cell) See **T follicular helper (Tfh) cells**.

***N*-Formylmethionine** An amino acid that initiates all bacterial proteins and no mammalian proteins (except those synthesized within mitochondria) and serves as a signal to the innate immune system of infection. Specific receptors for *N*-formylmethionine–containing peptides are expressed on neutrophils and mediate activation and chemotaxis of the neutrophils.

FOXP3 A forkhead family transcription factor expressed by and required for the development and function of $CD4^+$ regulatory T cells. Mutations in the gene ending FOXP3 in mice and humans result in an absence of $CD25^+$ regulatory T cells and a multisystem autoimmune disease called IPEX.

G

γδ T cell receptor (γδ TCR) A form of TCR that is distinct from the more common αβ TCR and is expressed on a subset of T cells found mostly in epithelial barrier tissues. Although the γδ TCR is structurally similar to the αβ TCR, the forms of antigen recognized by γδ TCRs are poorly understood; they do not recognize peptide complexes bound to polymorphic MHC molecules.

G protein–coupled receptor family A diverse family of receptors for hormones, lipid inflammatory mediators, and chemokines that use associated trimeric G proteins for intracellular signaling.

G proteins Proteins that bind guanyl nucleotides and act as exchange molecules by catalyzing the replacement of bound guanosine diphosphate (GDP) by guanosine triphosphate (GTP). G proteins with bound GTP can activate a variety of cellular enzymes in different signaling cascades. Trimeric GTP-binding proteins are associated with the cytoplasmic portions of many cell surface receptors, such as chemokine receptors. Other small soluble G proteins, such as RAS and RAC, are recruited into signaling pathways by adaptor proteins.

GATA-3 A transcription factor that plays an essential role in the differentiation of T helper 2 cells from naive T cells and group 2 ILCs.

Generative lymphoid organ An organ in which lymphocytes develop from immature precursors. The bone marrow and thymus are the major generative lymphoid organs in which B cells and T cells develop, respectively. Generative lymphoid organs are also called **primary lymphoid organs** or **central lymphoid organs**.

Germinal centers Specialized structures in secondary (peripheral) lymphoid organs generated during T-dependent humoral immune responses, where extensive B cell proliferation, isotype switching, somatic mutation, affinity maturation, memory B cell generation, and induction of long-lived plasma cells take place. Germinal centers appear as lightly staining regions within lymphoid follicles in spleen, lymph node, and mucosal lymphoid tissue.

Germline organization The inherited arrangement of variable, diversity, joining, and constant region gene segments of the antigen receptor loci in nonlymphoid cells or in immature lymphocytes. In developing B or T lymphocytes, the germline organization is modified by somatic recombination to form functional *Ig* or *TCR* genes.

Glomerulonephritis Inflammation of the renal glomeruli, often initiated by immunopathologic mechanisms such as deposition of circulating antigen-antibody complexes in the glomerular basement membrane or binding of antibodies to antigens expressed in the glomerulus. The antibodies can activate complement and phagocytes, and the resulting inflammatory response can lead to renal failure.

Graft A tissue or organ that is removed from one site and placed in another site, usually in a different individual.

Graft arteriosclerosis Occlusion of graft arteries caused by proliferation of intimal smooth muscle cells. This process occurs gradually over years after transplantation and is largely responsible for chronic rejection and failure of vascularized organ grafts. The mechanism is likely to be a chronic immune response to vessel wall alloantigens. Graft arteriosclerosis is also called accelerated arteriosclerosis.

Graft rejection A specific immune response to an organ or tissue graft that leads to inflammation, graft damage, and possibly graft failure.

Graft-versus-host disease A disease occurring in hematopoietic stem cell (HSC) transplant recipients that is caused by mature T cells present in the HSC inoculum reacting with alloantigens on host cells. The disease most often affects the skin, liver, and intestines.

Granulocyte colony-stimulating factor (G-CSF) A cytokine made by activated T cells, macrophages, and endothelial cells at sites of infection that acts on progenitors in the bone marrow to increase the production of and mobilize neutrophils to replace those consumed in inflammatory reactions.

Granulocyte-monocyte colony-stimulating factor (GM-CSF) A cytokine made by activated T cells, macrophages, endothelial cells, and stromal fibroblasts that acts on bone marrow to increase the production of neutrophils and monocytes. GM-CSF is also a macrophage-activating factor and promotes the maturation of dendritic cells.

Granuloma A nodule of inflammatory tissue composed of clusters of activated macrophages and T lymphocytes, usually with associated fibrosis. Granulomatous inflammation is a form of chronic delayed-type hypersensitivity, often in response to persistent microbes, such as *Mycobacterium tuberculosis* and some fungi, or in response to particulate antigens that are not readily phagocytosed.

Granulysin A lipid-binding cationic peptide found in granules of CTLs and NK cells, which can damage cholesterol-poor membranes, typical of bacteria but not mammalian cells, and thereby can kill intracellular microbes.

Granzyme B A serine protease found in the granules of CTLs and NK cells that is released by exocytosis, enters target cells, and proteolytically cleaves and activates caspases, which in turn cleave several substrates and induce target cell apoptosis.

Gut-associated lymphoid tissue (GALT) Organized collections of lymphocytes and APCs within the mucosa of the gastrointestinal tract, such as Peyer patches in the small bowel, where adaptive immune responses to intestinal microbial flora and ingested antigens are initiated (see also **mucosa-associated lymphoid tissues**).

H

H-2 molecule A major histocompatibility complex (MHC) molecule in the mouse. The mouse MHC was originally called the H-2 locus.

Haplotype The set of tightly linked major histocompatibility complex alleles inherited together from one parent and present on one chromosome.

Hapten A small chemical that can bind to an antibody but must be attached to a macromolecule (carrier) to stimulate an adaptive immune response specific for that chemical. For example, immunization with dinitrophenol (DNP) alone will not stimulate an anti-DNP antibody response, but immunization with a protein with covalently attached DNP hapten will.

Heavy-chain class (isotype) switching The process by which a B lymphocyte changes the class, or isotype, of the antibodies that it produces, from IgM to IgG, IgE, or IgA, without changing the antigen specificity of the antibody. Heavy-chain class switching is stimulated by cytokines and CD40 ligand expressed by T follicular helper cells and involves recombination of B cell *VDJ* segments with downstream heavy-chain gene segments.

Helminth A parasitic worm. Helminthic infections often elicit Th2-dependent immune responses characterized by eosinophil-rich inflammatory infiltrates and IgE production.

Helper T (Th) cells The class of T lymphocytes whose main functions are to activate macrophages and to promote inflammation in cell-mediated immune responses and to promote B cell antibody production in humoral immune responses. These functions are mediated by secreted cytokines and by T cell CD40 ligand binding to macrophage or B cell CD40. Helper T cells express the CD4 molecule and recognize peptides displayed by class II MHC molecules.

Hematopoiesis The development of mature blood cells, including erythrocytes, leukocytes, and platelets, from pluripotent stem cells in the bone marrow and fetal liver. Hematopoiesis is regulated by several different colony-stimulating factors produced by bone marrow stromal cells, T cells, and other cell types.

Hematopoietic stem cell (HSC) Undifferentiated bone marrow cells that continuously asymmetrically divide throughout life to give rise to additional stem cells and cells that differentiate into cells of the lymphoid, myeloid, and erythrocytic lineages. Hematopoietic stem cells are also present in fetal liver.

Hematopoietic stem cell transplantation The transplantation of hematopoietic stem cells taken from the blood or bone marrow, performed clinically to treat inherited defects in and cancers of blood cells and also used in various immunologic experiments in animals. Also called bone marrow transplantation because in the past, HSCs were harvested from the bone marrow.

High endothelial venules (HEVs) Specialized venules that are the sites of lymphocyte migration from the blood into the stroma of secondary lymphoid tissues. HEVs are lined by plump endothelial cells that protrude into the vessel lumen and express unique adhesion molecules involved in binding naive (and central memory) B and T cells.

Hinge region The region of Ig heavy chains between the first two constant domains that can assume multiple conformations, thereby imparting flexibility in the orientation of the two antigen-binding sites. Because of the hinge region, an antibody molecule can simultaneously bind two epitopes that are separated by variable distances from one another.

Histamine A vasoactive amine stored in the granules of mast cells that is one of the important mediators of immediate hypersensitivity. Histamine binds to specific receptors in various tissues and causes increased vascular permeability and contraction of bronchial and intestinal smooth muscle.

HLA See **Human leukocyte antigens**.

HLA-DM A peptide exchange molecule that plays a critical role in the class II MHC pathway of antigen presentation. HLA-DM is found in the endosomes involved in class II—associated antigen presentation, where it facilitates removal of the invariant chain—derived CLIP peptide and the binding of other peptides to class II MHC molecules. HLA-DM is encoded by a gene in the MHC and is structurally similar to class II MHC molecules, but it is not polymorphic.

Homeostasis In the adaptive immune system, the maintenance of a constant number and diverse repertoire of resting lymphocytes, despite the emergence of new lymphocytes and tremendous expansion of individual clones that may occur during responses to immunogenic antigens or loss of lymphocytes (e.g., after irradiation or antibody-mediated depletion).

Homing receptor Adhesion molecules expressed on the surface of lymphocytes that are responsible for the different pathways of lymphocyte recirculation and tissue homing. Homing receptors bind to ligands (addressins) expressed on endothelial cells in particular vascular beds.

Human immunodeficiency virus (HIV) The etiologic agent of AIDS. HIV is a retrovirus that infects a variety of cell types, including $CD4^+$ helper T cells, macrophages, and dendritic cells and causes chronic progressive destruction of the immune system.

Human leukocyte antigens (HLA) MHC molecules expressed on the surface of human cells. Human MHC molecules include three types of class I MHC molecules (HLA-A, HLA-B, and HLA-C) and three types of class II MHC molecules (HLA-DP, HLA-DQ, and HLA-DR). (See also **Major histocompatibility complex [MHC] molecule**).

Humanized monoclonal antibody A monoclonal antibody encoded by a recombinant hybrid gene and composed of the antigen-binding sites from a murine monoclonal antibody and the constant region of a human antibody. Humanized antibodies are less likely than mouse monoclonal antibodies to induce an anti-antibody response in humans. They are used clinically in the treatment of inflammatory diseases, tumors, and transplant rejection. In current drug development, fully human recombinant monoclonal antibodies have largely replaced humanized mouse antibodies.

Humoral immunity The type of adaptive immune response mediated by antibodies produced by B lymphocyte-derived plasma cells. Humoral immunity is the principal adaptive immune defense mechanism against extracellular microbes and their toxins.

Hybridoma A cell line derived by fusion, or somatic cell hybridization, between a normal lymphocyte and an immortalized lymphocyte tumor line. B cell hybridomas created by fusion of normal B cells of defined antigen specificity with a myeloma cell line are used to produce monoclonal antibodies. T cell hybridomas created by fusion of a normal T cell of defined specificity with a T cell tumor line have been used in research.

Hyper-IgM syndrome Primary immune deficiency disorders caused by defective CD40-dependent functions in B cells, with impaired class switch recombination (CSR) and somatic hypermutation, leading to poor antibody-mediated immunity against extracellular pathogens and compromised defense against intracellular infections due to impaired CD40-dependent macrophage activation. The most common cause is mutations of the CD40 ligand gene on the X chromosome, but mutations in CD40 and downstream signaling molecules cause similar disorders. Mutations in the genes encoding activation-induced cytidine deaminase or uracil-DNA glycosylase result in the B cell defects seen in CD40 ligand deficiency but do not affect macrophages. Patients suffer from both pyogenic bacterial and protozoal infections.

Hyperacute rejection A form of allograft or xenograft rejection that begins within minutes to hours after transplantation and that is characterized by thrombotic occlusion of the graft vessels. Hyperacute rejection is mediated by preexisting antibodies in the host circulation that bind to donor endothelial alloantigens, such as blood group antigens or MHC molecules, and activate the complement system.

Hypersensitivity diseases Disorders caused by immune responses. Hypersensitivity diseases include autoimmune diseases, in which immune responses are directed against self antigens, and diseases that result from uncontrolled or excessive responses against foreign antigens, such as microbes and allergens. The tissue damage that occurs in hypersensitivity diseases is caused by the same effector mechanisms used by the immune system to protect against microbes.

Hypervariable region Short segments of approximately 10 amino acid residues within the variable regions of antibody or TCR proteins that form loop structures that contact antigen. Three hypervariable loops are present in each antibody heavy chain and light chain and in each TCR α and β chain. Most of the variability between different antibodies or TCRs is located within these loops (also called complementarity determining region [CDR]).

I

Idiotype The unique sequence of the antigen binding site of the antibodies or TCRs made by a single clone of B or T cells that can be recognized by a specific antibody.

Igα and Igβ Proteins that are required for surface expression and signaling functions of membrane Ig on B cells. Igα and Igβ are disulfide-linked to one another and noncovalently associated with the cytoplasmic tail of membrane Ig to form the BCR complex. The cytoplasmic domains of Igα and Igβ contain ITAMs that are involved in early signaling events during antigen-induced B cell activation.

IL-1 receptor antagonist (IL-1RA) A natural inhibitor of IL-1 produced by macrophages and other cells that is structurally homologous to IL-1 and binds to the same receptors but does not induce signaling. Recombinant IL-1RA is used as a drug to treat autoinflammatory syndromes caused by excessive IL-1 production as well as rheumatoid arthritis.

Immature B lymphocyte A membrane IgM^+, IgD^- B cell, recently derived from marrow precursors, that does not proliferate or differentiate in response to antigens but rather may undergo receptor editing, apoptotic death or become functionally unresponsive. This property is important for the negative selection of B cells that are specific for self antigens present in the bone marrow.

Immediate hypersensitivity The type of immune reaction responsible for allergic diseases, which is dependent on antigen-mediated activation of IgE-coated tissue mast cells. The mast cells release mediators that cause increased vascular permeability, vasodilation, bronchial and visceral smooth muscle contraction, and local inflammation.

Immune complex A multimolecular complex of antibody molecules and bound antigen. Because each antibody molecule has 2 to 10 antigen-binding sites and many antigens are multivalent, immune complexes can vary greatly in size. Immune complexes activate effector mechanisms of humoral immunity, such as the classical complement pathway and Fc receptor–mediated phagocytosis. Deposition of circulating immune complexes in blood vessel walls or renal glomeruli can lead to inflammation and disease (vasculitis and glomerulonephritis, respectively).

Immune complex disease An inflammatory disease caused by the deposition of antigen-antibody complexes in blood vessel walls, resulting in local complement activation and inflammation. Immune complexes may form because of overproduction of antibodies against microbial antigens or as a result of autoantibody production in the setting of an autoimmune disease such as systemic lupus erythematosus. Immune complex deposition in the specialized capillary basement membranes of renal glomeruli can cause glomerulonephritis and impair renal function. Deposition of immune complexes in joints can cause arthritis, and deposition in arterial walls can cause vasculitis with thrombosis and ischemic damage to various organs.

Immune response A collective and coordinated response to the introduction of foreign substances, mediated by the cells and molecules of the immune system.

Immune response *(Ir)* genes Originally defined as genes in inbred strains of rodents that were inherited in a dominant Mendelian manner and that controlled the ability of the animals to make antibodies against simple synthetic polypeptides. We now know that *Ir* genes are the polymorphic genes that encode class II MHC molecules, which display peptides to T lymphocytes and are therefore required for T cell activation and helper T cell–dependent B cell (antibody) responses to protein antigens.

Immune surveillance The concept that a physiologic function of the immune system is to recognize and destroy clones of transformed cells before they grow into tumors and to kill tumors after they are formed. The term *immune surveillance* is sometimes used in a general sense to describe the function of T lymphocytes to detect and destroy any cell, not necessarily a tumor cell, that is expressing foreign (e.g., microbial) antigens.

Immune system The molecules, cells, tissues, and organs that collectively function to provide immunity, or protection, against foreign pathogens and cancers.

Immunity Protection against disease, usually infectious disease, mediated by the cells and tissues that are collectively called the immune system. In a broader sense, immunity refers to the ability to respond to foreign substances, including microbes and noninfectious molecules.

Immunoblot An analytical technique in which antibodies are used to detect the presence of an antigen bound to (i.e., blotted on) a solid matrix such as filter paper (also known as a Western blot).

Immunodeficiency See **Acquired immunodeficiency** and **Primary immunodeficiency**.

Immunodominant epitope The epitope of a protein antigen that elicits most of the T cell response in an individual immunized with the native protein. Immunodominant epitopes correspond to the peptides of the protein that are proteolytically generated within APCs, bind most avidly to MHC molecules, and are most likely to stimulate T cells.

Immune dysregulation polyendocrinopathy enteropathy X-linked syndrome (IPEX) A rare autoimmune disease, caused by mutations of the FOXP3 transcription factor, resulting in a failure to produce regulatory T cells. IPEX patients suffer from immune-mediated destruction of multiple endocrine organs, as well as allergies and skin and gastrointestinal inflammation.

Immunofluorescence A technique in which a molecule is detected by use of an antibody labeled with a fluorescent probe. For example, in immunofluorescence microscopy, cells that express a particular surface antigen can be stained with a fluorescein-conjugated antibody specific for the antigen and then visualized with a fluorescent microscope.

Immunogen An antigen that induces an immune response. Not all antigens are immunogens. For example, low-molecular-weight compounds (haptens) can bind to antibodies, and are therefore antigens, but will not stimulate an immune response unless they are linked to macromolecules (carriers), and thus are not immunogens.

Immunoglobulin (Ig) Synonymous with antibody (see **Antibody**).

Immunoglobulin domain A three-dimensional globular structural motif (also called an Ig fold) found in many proteins in the immune system, including Igs, TCRs, and MHC molecules. Ig domains are approximately 110 amino acid residues in length, include an internal disulfide bond, and contain two layers of β-pleated sheets, each layer composed of three to five strands of antiparallel polypeptide chain.

Immunoglobulin heavy chain One of two types of polypeptide chains in an antibody molecule. The basic structural unit of an antibody includes two identical disulfide-linked heavy chains and two identical light chains. Each heavy chain is composed of a variable (V) Ig domain and three or four constant (C) Ig domains. The different antibody isotypes, including IgM, IgD, IgG, IgA, and IgE, are distinguished by structural differences in their heavy chain constant regions. The heavy chain constant regions mediate effector functions, such as complement activation and engagement of phagocytes.

Immunoglobulin light chain One of two types of polypeptide chains in an antibody molecule. The basic structural unit of an antibody includes two identical light chains, each disulfide linked to one of two identical heavy chains. Each light chain is composed of one variable (V) Ig domain and one constant (C) Ig domain. There are two light chain isotypes, called κ and λ, which are functionally identical. Approximately 60% of human antibodies have κ light chains, and 40% have λ light chains.

Immunoglobulin (Ig) superfamily A large family of proteins that contain a globular structural motif called an Ig domain, or Ig fold, originally described in antibodies. Many proteins of importance in the immune system, including antibodies, TCRs, MHC molecules, CD4, and CD8, are members of this superfamily.

Immunohistochemistry A technique to detect the presence of an antigen in histologic tissue sections by use of an enzyme-coupled antibody that is specific for the antigen. The enzyme converts a colorless substrate to a colored insoluble substance that precipitates at the site where the antibody and thus the antigen are localized. The position of the colored precipitate, and therefore the antigen, in the tissue section is observed by light microscopy. Immunohistochemistry is commonly used in diagnostic pathology and various fields of research.

Immunologic synapse (also called **immune synapse)** The tight juxtaposition of membranes of a T cell and an antigen presenting cell (APC). Membrane proteins of both cell types organized at the point of juxtaposition, including the TCR complex, CD4 or CD8, costimulatory receptors, and integrins on the T cell, which bind to peptide-MHC complexes, costimulators, and integrin ligands, respectively, on the antigen presenting cell. The immune synapse is required for bidirectional functional interactions between the T cell and APC, and enhances specific delivery of secreted products from the T cell to the antigen-presenting cell, such as granule contents from a CTL to its target cell.

Immunologic tolerance See **Tolerance**.

Immunologically privileged site A site in the body that is inaccessible to or actively suppresses immune responses. The anterior chamber of the eye, the testes, and the brain are examples of immunologically privileged sites.

Immunoperoxidase technique A common immunohistochemical technique in which a horseradish peroxidase–coupled antibody is used to identify the presence of an antigen in a tissue section. The peroxidase enzyme converts a colorless substrate to an insoluble brown product that is observable by light microscopy.

Immunoprecipitation A technique for the isolation of a molecule from a solution by binding it to an antibody and then rendering the antigen-antibody complex insoluble, either by precipitation with a second antibody or by coupling the first antibody to an insoluble particle or bead.

Immunoreceptor tyrosine-based activation motif (ITAM) A conserved protein motif composed of two copies of the sequence tyrosine-x-x-leucine (where x is an unspecified amino acid) found in the cytoplasmic tails of various membrane proteins in the immune system that are involved in signal transduction. ITAMs are present in the ζ and CD3 proteins of the TCR complex, in Igα and Igβ proteins in the BCR complex, and in several Ig Fc receptors. When these receptors bind their ligands, the tyrosine residues of the ITAMs become phosphorylated and form docking sites for other molecules involved in propagating cell-activating signal transduction pathways.

Immunoreceptor tyrosine-based inhibition motif (ITIM) A six-amino-acid (isoleucine-x-tyrosine-x-x-leucine) motif found in the cytoplasmic tails of various inhibitory receptors in the immune system, including FcγRIIB on B cells, killer cell Ig-like receptors (KIRs) on NK cells, and some coinhibitory receptors of T cells. When these receptors bind their ligands, the ITIMs become phosphorylated on their tyrosine residues and form a docking site for protein tyrosine phosphatases, which in turn function to inhibit other signal transduction pathways.

Immunoreceptor tyrosine-based switch motif (ITSM) A six-amino-acid (tyrosine-x-tyrosine-x-x-valine/isoleucine) motif found in the cytoplasmic tails of some receptors, which can sometimes function as an inhibitor by binding tyrosine phosphatases, as with the ITSM in the cytosolic tail of PD-1, and in other receptors (e.g., in the SLAM family), can switch from tyrosine phosphatase to tyrosine kinase binding, thereby mediating a change from an inhibitory to an activating function.

Immunosuppression Inhibition of one or more components of the adaptive or innate immune system as a result of an underlying disease or intentionally induced by drugs for the purpose of preventing or treating graft rejection or autoimmune disease. A commonly used immunosuppressive drug used to treat graft rejection is cyclosporine, which inhibits T cell cytokine production.

Immunotherapy The treatment of a disease with therapeutic agents that promote or inhibit immune responses. For example, cancer immunotherapy involves stimulating active immune responses to tumor antigens or administering antitumor antibodies or T cells to establish passive immunity.

Immunotoxins Reagents that may be used in the treatment of cancer and consist of covalent conjugates of a potent cellular toxin, such as ricin or diphtheria toxin, with antibodies specific for antigens expressed on the surface of tumor cells. It is hoped that such reagents can specifically target and kill tumor cells without damaging normal cells.

Inbred mouse strain A strain of mice created by repetitive mating of siblings that is characterized by homozygosity at every genetic locus. Every mouse of an inbred strain shares an identical set of inherited genes and is said to be syngeneic to every other mouse of the same strain.

Indirect antigen presentation (indirect allorecognition) In transplantation immunology, a pathway of presentation of donor (allogeneic) MHC molecules by recipient APCs that involves the same mechanisms used to present microbial proteins. The allogeneic MHC proteins are processed by recipient dendritic cells, and peptides derived from the allogeneic MHC molecules are presented, in association with recipient (self) MHC molecules, to host T cells. In contrast to indirect antigen presentation, direct antigen presentation involves recipient T cell recognition of unprocessed allogeneic MHC molecules on the surface of graft cells.

Inflammasomes Multiprotein complexes that assemble in the cytosol of mononuclear phagocytes, dendritic cells, and other cell types, which use caspase-1 to proteolytically generate the active form of the inflammatory cytokines IL-1β and IL-18 from inactive precursors. The formation of the inflammasome complex is stimulated by a variety of microbial products and cell damage–associated molecules, and involves assembly of multiple copies of an innate recognition protein with adaptor proteins and procaspase-1 molecules, the latter undergoing proteolysis upon inflammasome assembly to generate active caspase-1.

Inflammation A complex reaction of vascularized tissue to infection or cell injury that involves extravascular accumulation of plasma proteins and leukocytes. Acute inflammation is a common result of innate immune responses, and local adaptive immune responses can also promote inflammation. Although inflammation serves a protective function in controlling infections and promoting tissue repair, it can also cause tissue damage and disease.

Inflammatory bowel disease (IBD) A group of disorders, including ulcerative colitis and Crohn's disease, characterized by chronic inflammation in the gastrointestinal tract. Some evidence indicates that IBD is caused by inadequate regulation of T cell and innate immune responses, possibly against intestinal commensal bacteria. IBD develops in gene knockout mice lacking IL-2 or IL-10.

Innate immunity Protection against infection that relies on mechanisms that exist before infection, are capable of a rapid response to microbes, and react in essentially the same way to repeated infections. The innate immune system includes epithelial barriers, phagocytic cells (neutrophils, macrophages), ILCs and NK cells, the complement system, and cytokines, most of which are produced by dendritic cells and mononuclear phagocytes. Innate immune reactions also eliminate damaged and necrotic host tissues.

Innate lymphoid cells (ILCs) Tissue-resident cells that produce cytokines similar to those made by helper T (Th) cells but lack antigen-specific TCRs. ILCs arise from a common lymphoid progenitor in the bone marrow, have a lymphocyte morphology, and produce cytokines similar to those secreted by T cells. Three subsets of innate lymphoid cells, called ILC1, ILC2, and ILC3, produce cytokines and express different transcription factors analogous to the Th1, Th2, and Th17 subsets of $CD4^+$ effector T lymphocytes, respectively. Natural killer cells are one type of ILC with functions similar to those of cytotoxic T lymphocytes.

Integrins Heterodimeric cell surface proteins whose major functions are to mediate the adhesion of cells to other cells or to extracellular matrix. Integrins are important for T cell interactions with APCs and for migration of leukocytes from blood into tissues. Signals induced by chemokines binding to chemokine receptors increase the affinity of integrins for their ligands. Two examples of integrins important in the immune system are VLA-4 (very late antigen 4) and LFA-1 (leukocyte function-associated antigen 1).

Interferon regulatory factors (IRFs) A family of transcription factors that are activated by signals from various innate immune receptors for DNA and RNA and stimulate production of type I interferons, cytokines that inhibit viral replication.

Interferons (IFNs) Cytokines originally named for their ability to interfere with viral infections but that have other important functions in the immune system. Type I IFNs include IFN-α and IFN-β, whose main function is to prevent viral replication in cells. IFN-γ, sometimes called type II interferon, activates macrophages and various other cell types. Type III interferons, including three kinds of IFN-λ, have similar anti-vrial functions as Type 1 interferons. (see Appendix II).

Interleukins Molecularly defined cytokines that are named with a number roughly sequentially in order of discovery or molecular characterization (e.g., interleukin-1 [IL-1], IL-2). Some cytokines were originally named for their biologic activities and do not have an IL designation (see Appendix II).

Intracellular bacterium A bacterium that survives and may replicate within cells, usually in endosomes of phagocytes. The principal defense against intracellular bacteria, such as *Mycobacterium tuberculosis*, is T cell–mediated immunity.

Intraepithelial lymphocytes T lymphocytes present in the epidermis of the skin and in mucosal epithelia that typically express a limited diversity of antigen receptors. Some of these lymphocytes, called invariant NKT cells, may recognize microbial products, such as glycolipids, associated with nonpolymorphic class I MHC-like molecules. Others, called γδ T cells, recognize various nonpeptide antigens, not presented by MHC molecules. Intraepithelial T lymphocytes may be effector cells of innate immunity.

Invariant chain (I_i) A nonpolymorphic protein that binds to newly synthesized class II MHC molecules in the endoplasmic reticulum (ER). The invariant chain prevents loading of the class II MHC peptide-binding cleft with peptides present in the ER, promotes folding and assembly of class II molecules, and directs them to the endosomal compartment, where loading of peptides derived from internalized proteins takes place.

Isotype One of five types of antibodies determined by which of five different forms of heavy chain is present. Antibody isotypes (also called **classes**) include IgM, IgD, IgG, IgA, and IgE, and each isotype performs a different set of effector functions. Additional structural variations characterize four distinct subtypes of IgG and two of IgA.

J

J (joining) chain A small polypeptide that is disulfide linked to the tail pieces of IgM and IgA antibodies that joins the antibody molecules to form pentamers of IgM and dimers of IgA. The J chain also contributes to the transepithelial transport of these immunoglobulins.

JAK-STAT signaling pathway A signaling pathway initiated by cytokine binding to type I and type II cytokine receptors that sequentially involves activation of receptor-associated Janus kinase (JAK) tyrosine kinases, JAK-mediated tyrosine phosphorylation of the cytoplasmic tails of cytokine receptors, docking of signal transducers and activators of transcription (STATs) to the phosphorylated receptor chains, JAK-mediated tyrosine phosphorylation of the associated STATs, dimerization and nuclear translocation of the STATs, and STAT binding to regulatory regions of target genes leads to transcriptional activation of those genes.

Janus kinases (JAKs) A family of four related tyrosine kinases that associate with the cytoplasmic tails of several different cytokine receptors, including the receptors for IL-2, IL-3, IL-4, IL-7, IFN-γ, IL-12, and others. In response to cytokine binding and receptor dimerization, JAKs phosphorylate the cytokine receptors to permit the binding of STATs, and then the JAKs phosphorylate and thereby activate the STATs. Different JAKs associate with different cytokine receptors.

Joining (J) segments Short coding sequences between the variable *(V)* and constant *(C)* gene segments in all *Ig* and *TCR* loci, which together with *D* segments are somatically recombined with *V* segments during lymphocyte development. The resulting recombined *VDJ* DNA codes for the carboxyl-terminal ends of the antigen receptor V regions, including the third hypervariable regions (CDR3). Random use of different *J* segments contributes to the diversity of the antigen receptor repertoire.

Junctional diversity The diversity in antibody and TCR repertoires that is attributed to the addition or removal of nucleotide sequences at junctions between *V, D,* and *J* gene segments during B and T cell development.

K

Kaposi sarcoma A tumor of vascular cells that frequently arises in patients with AIDS, particularly patients not treated with antiretroviral therapy. Kaposi sarcoma in immunocompromised patients is associated with infection by the Kaposi sarcoma–associated herpesvirus (human herpesvirus 8).

Killer cell Ig-like receptors (KIRs) Ig superfamily receptors expressed by NK cells that recognize different alleles of HLA-A, HLA-B, and HLA-C molecules. Some KIRs have signaling components with ITIMs in their cytoplasmic tails, and these deliver inhibitory signals that block the activation of the NK cells. Some members of the KIR family have short cytoplasmic tails without ITIMs but associate with other ITAM-containing polypeptides and function as activating receptors.

Knockout mouse A mouse with a targeted disruption of one or more genes that is created by homologous recombination or CRISPR-Cas9 gene editing techniques. Knockout mice lacking functional genes encoding cytokines, cell surface receptors, signaling molecules, and transcription factors have provided valuable information about the roles of these molecules in the immune system.

L

Lamina propria A layer of loose connective tissue underlying the epithelium in mucosal tissues such as the intestines and airways, where dendritic cells, mast cells, lymphocytes, and macrophages mediate immune responses to invading pathogens.

Langerhans cells Immature dendritic cells found as a meshwork in the epidermal layer of the skin whose major function is to trap microbes and antigens that enter through the skin and transport the antigens to draining lymph nodes. During their migration to the lymph nodes, Langerhans cells may differentiate into mature dendritic cells, which can efficiently present antigen to naive T cells.

Large granular lymphocyte Another name for an NK cell based on the morphologic appearance of this cell type in blood smears.

Late-phase reaction A component of the immediate hypersensitivity reaction that develops 2 to 24 hours after the rapid mast cell degranulation after antigen challenge and is characterized by an inflammatory infiltrate of eosinophils, basophils, neutrophils, and lymphocytes. Repeated bouts of the late-phase reaction can cause tissue damage (as in asthma).

LCK An SRC family nonreceptor tyrosine kinase that noncovalently associates with the cytoplasmic tails of CD4 and CD8 molecules in T cells and is involved in the early signaling events of antigen-induced T cell activation. LCK mediates tyrosine phosphorylation of the cytoplasmic tails of CD3 and ζ proteins of the TCR complex.

Lectin pathway of complement activation A pathway of complement activation triggered by the binding of microbial polysaccharides to the circulating protein mannose binding lectin (MBL). MBL is structurally similar to C1q and activates the C1r-C1s enzyme complex (like C1q) or activates another serine esterase, called mannose-binding protein-associated serine esterase (MASP). The remaining steps of the lectin pathway, beginning with cleavage of C4, are the same as the classical pathway.

Leishmania An obligate intracellular protozoan parasite that infects macrophages and can cause a chronic inflammatory disease involving many tissues. *Leishmania* infection in mice has served as a model system to study the effector functions of several cytokines and the helper T cell subsets that produce them.

Lethal hit A term used to describe the events that result in irreversible damage to a target cell when a CTL binds to it. The lethal hit includes CTL granule exocytosis and perforin-dependent delivery of apoptosis-inducing granule enzymes (granzymes) into the target cell cytoplasm.

Leukemia A malignant disease of bone marrow precursors of blood cells in which large numbers of neoplastic cells usually occupy the bone marrow and often circulate in the blood. Lymphocytic leukemias are derived from B or T cell precursors, myelogenous leukemias are derived from granulocyte or monocyte precursors, and erythroid leukemias are derived from red cell precursors.

Leukocyte adhesion deficiency (LAD) One of a rare group of congenital (primary) immunodeficiency diseases with infectious complications that is caused by defective expression or function of leukocyte adhesion molecules required for tissue recruitment of phagocytes and lymphocytes. LAD-1 is due to mutations in the gene encoding the CD18 protein, which is part of β2 integrins. LAD-2 is caused by mutations in a gene that encodes a fucose transporter involved in the synthesis of leukocyte ligands for endothelial selectins. LAD-3 is due to mutations affecting proteins required for chemokine-induced activation of integrins.

Leukotrienes A class of arachidonic acid–derived lipid inflammatory mediators produced by the lipoxygenase pathway in many cell types. Mast cells make abundant leukotriene C_4 (LTC_4) and its degradation products LTD_4 and LTE_4, which bind to specific receptors on smooth muscle cells and cause prolonged bronchoconstriction. Leukotrienes contribute to pathology of asthma.

Lipopolysaccharide Synonymous with **endotoxin**.

Live virus vaccine A vaccine composed of a live but nonpathogenic (attenuated) form of a pathogenic virus. Attenuated viruses carry mutations that interfere with the viral life cycle or virulence. Because live virus vaccines actually infect the recipient cells, they can effectively stimulate immune responses that are optimal for protecting against wild-type viral infection. Commonly used live virus vaccines include those for measles, mumps, rubella, influenza, varicella, and yellow fever.

Lymph Interstitial fluid derived from the blood that has drained into lymphatic vessels and is eventually returned to the blood circulation. Lymph carries soluble antigens and dendritic cells from most tissues and organs of the body into lymph nodes for recognition by specific lymphocytes to initiate adaptive immune responses, and also carries lymphocytes into and out of lymph nodes and back into the blood circulation.

Lymph node Small nodular, encapsulated lymphocyte-rich organs situated along lymphatic channels throughout the body where adaptive immune responses to lymph-borne antigens are initiated. Lymph nodes, which are secondary or peripheral lymphoid organs, have a specialized anatomic architecture that regulates the interactions of B cells, T cells, dendritic cells, macrophages, and antigens to maximize the induction of protective immune responses. Lymph nodes also perform a filtering function, trapping microorganisms and other potentially harmful constituents in tissue fluids and preventing them from draining via the lymph into the blood.

Lymphatic system A system of vessels throughout the body that collects tissue fluid called lymph, originally derived from the blood, and returns it, through the thoracic and right lymphatic ducts, to the circulation. Lymph nodes are interspersed along these vessels and trap and retain antigens present in the lymph.

Lymphocyte homing The directed migration of subsets of circulating lymphocytes into particular tissue sites. Lymphocyte homing is regulated by the selective expression of endothelial adhesion molecules and chemokines in different tissues. For example, some lymphocytes preferentially home to the intestinal mucosa, which is regulated by the chemokine CCL25 and the endothelial adhesion molecule MAdCAM, both expressed in the gut, which bind respectively to the CCR9 chemokine receptor and the α4β7 integrin on gut-homing lymphocytes.

Lymphocyte maturation The process by which pluripotent hematopoietic stem cells develop into mature, antigen receptor–expressing naive B or T lymphocytes that populate peripheral lymphoid tissues. This process takes place in the specialized environments of the bone marrow (for B cells) and the thymus (for T cells). Synonymous with **lymphocyte development**.

Lymphocyte migration The movement of lymphocytes from the circulation into peripheral tissues.

Lymphocyte recirculation The continuous movement of naive and some memory lymphocytes from the blood to secondary lymphoid organs and back into the blood.

Lymphocyte repertoire The complete collection of antigen receptors and therefore antigen specificities expressed by all the B and T lymphocyte clones of an individual. The repertories for B and T cells are each estimated to include about 10^7 receptors.

Lymphoid follicle A B cell–rich region of a lymph node or the spleen that is the site of antigen-induced B cell proliferation and differentiation. In T cell–dependent B cell responses to protein antigens, a germinal center forms within the follicles.

Lymphoid tissue inducer cells (LTi) A type of hematopoietically derived innate lymphoid cell that stimulates the development of lymph nodes and other secondary lymphoid organs, in part through production of the cytokines lymphotoxin-α (LTα) and lymphotoxin-β (LTβ). LTis are a subset of type 3 innate lymphoid cells (ILC3).

Lymphokine An old name for a cytokine (soluble protein mediator of immune responses) produced by lymphocytes.

Lymphokine-activated killer (LAK) cells NK cells with enhanced cytotoxic activity for tumor cells as a result of exposure to high doses of IL-2. LAK cells generated in vitro have been adoptively transferred back into patients with cancer to treat their tumors.

Lymphoma A malignant tumor of B or T lymphocytes usually arising in and spreading between lymphoid tissues but that may spread to other tissues. Lymphomas often express phenotypic characteristics of the normal lymphocytes from which they were derived.

Lymphotoxin α (LTα, previously called TNF-β) A cytokine produced by T and B cells that is homologous to and binds to the same receptors as TNF. Like TNF, LT has proinflammatory effects, including endothelial and neutrophil activation. LT is also critical for the normal development of lymphoid organs.

Lysosome A membrane-bound, acidic organelle abundant in phagocytic cells that contains proteolytic enzymes that degrade proteins derived both from the extracellular environment and from within the cell. Lysosomes are involved in the class II MHC pathway of antigen processing.

M

M cells Specialized gastrointestinal mucosal epithelial cells overlying Peyer patches in the gut that play a role in delivery of antigens to Peyer patches.

M1 macrophages See **Classical macrophage activation**.

M2 macrophages See **Alternative macrophage activation**.

Macrophage A hematopoietically derived phagocytic cell that plays important roles in innate and adaptive immune responses. Macrophages are activated by microbial products such as endotoxin and by T cell cytokines such as IFN-γ. Activated macrophages phagocytose and kill microorganisms, secrete proinflammatory cytokines, and present antigens to helper T cells. Macrophages include cells derived from recently recruited blood monocytes at sites of inflammation and long-lived tissue-resident cells derived mainly from fetal hematopoietic organs. Tissue-resident macrophages are given different names and may serve special functions; these include the microglia of the central nervous system, Kupffer cells in the liver, alveolar macrophages in the lung, and osteoclasts in bone.

Major histocompatibility complex (MHC) A large genetic locus (on human chromosome 6 and mouse chromosome 17) that includes the highly polymorphic genes encoding the peptide-binding molecules recognized by T lymphocytes. The *MHC* locus also includes genes encoding cytokines, molecules involved in antigen processing, and complement proteins.

Major histocompatibility complex (MHC) molecule A heterodimeric membrane protein encoded in the *MHC* locus that displays peptides for recognition by T lymphocytes. There are two structurally distinct types of MHC molecules. Class I MHC molecules, sometimes called MHC I, are present on most nucleated cells, bind peptides derived from cytosolic proteins, and are recognized by $CD8^+$ T cells. Class II MHC molecules, sometimes called MHC II, are restricted largely to dendritic cells, macrophages, and B lymphocytes; bind peptides derived from endocytosed proteins; and are recognized by $CD4^+$ T cells.

Mannose-binding lectin (MBL) A plasma protein, also called mannose binding protein (MBP), that binds to mannose residues on microbial surfaces, thereby initiating the lectin pathway of complement activation. Macrophages express a surface receptor for C1q that can also bind MBL and mediate uptake of the MBL-opsonized organisms.

Mannose receptor A carbohydrate-binding protein (lectin) expressed by macrophages that binds mannose and fucose residues on microbial cell walls and mediates phagocytosis of the organisms.

Marginal zone A peripheral region of splenic lymphoid follicles containing macrophages that are particularly efficient at trapping polysaccharide antigens. Such antigens may persist for prolonged periods on the surfaces of marginal zone macrophages, where they are recognized by specific B cells, or they may be transported into follicles.

Marginal zone B lymphocytes A subset of B lymphocytes, found exclusively in the marginal zone of the spleen, that respond rapidly to blood-borne microbial antigens by producing IgM antibodies with limited diversity.

Mass cytometry A method of simultaneous detection and analysis of many different molecules expressed in mixed cell populations, requiring a specialized instrument based on the single cell analysis of flow cytometer coupled with a time-of-flight mass spectrometer. This technique uses antibodies labeled with different heavy metal isotopes, rather than fluorochromes used in flow cytometry.

Mast cell The major effector cell of immediate hypersensitivity (allergic) reactions. Mast cells are derived from hematopoietic stem cells, reside in most tissues adjacent to blood vessels, express a high-affinity Fc receptor for IgE (FcεRI) and contain numerous mediator-filled granules. Antigen-induced cross-linking of IgE bound to the mast cell FcεRI causes release of their granule contents as well as new synthesis and secretion of other mediators, leading to an immediate hypersensitivity reaction.

Mature B cell IgM- and IgD-expressing, functionally competent naive B cells that represent the final stage of B cell maturation in the spleen and that populate peripheral lymphoid organs.

Medullary thymic epithelial cells (MTECs) A type of stromal cell in the medulla of the thymus that plays a critical role in inducing central T cell tolerance to proteins normally expressed only in certain tissues, called tissue-restricted antigens (TRAs). MTECs consist of subpopulations that transcriptionally resemble different peripheral tissue cells and express TRAs of those tissues. The MTECs present those antigens to developing T cells, resulting in death of TRA-specific T cells or development of TRA-specific regulatory T cells. The AIRE protein promotes expression of these antigens.

Membrane attack complex (MAC) A lytic complex of the terminal components of the complement cascade, including complement proteins C5, C6, C7, C8, and multiple copies of C9, which forms in the membranes of target cells. The MAC causes lethal ionic and osmotic changes in cells.

Memory The property of the adaptive immune system to respond more rapidly, with greater magnitude, and more effectively to a repeated exposure to an antigen compared with the response to the first exposure.

Memory lymphocytes Memory B and T cells are produced by antigen stimulation of naive lymphocytes and survive in a functionally quiescent state for many years after the antigen is eliminated. Memory lymphocytes mediate rapid and enhanced (i.e., memory or recall) responses to second and subsequent exposures to antigens.

MHC restriction The characteristic of T lymphocytes that they recognize a foreign peptide antigen only when it is bound to a particular allelic form of an MHC molecule.

MHC tetramer A reagent used to identify and enumerate T cells that specifically recognize a particular MHC-peptide complex. The reagent consists of four recombinant, biotinylated MHC molecules (usually class I) loaded with a peptide and bound to a fluorochrome-labeled avidin molecule. T cells that bind the MHC tetramer can be detected by flow cytometry.

β2-Microglobulin The light chain of a class I MHC molecule, required for stability of the MHC molecule. β2-Microglobulin is encoded by a nonpolymorphic gene outside the MHC, is structurally homologous to an Ig domain, and is invariant among all class I molecules.

Mitogen-activated protein (MAP) kinase cascade One of several different intracellular signal transduction cascades initiated by ligand binding to a membrane receptor, which is characterized by successive kinase-mediated phosphorylation steps leading to the terminal activation by dual phosphorylation of one of a broad family of MAP kinases. The activated MAP kinases then phosphorylate substrates such as transcription factors that result in a functional cellular response. In T lymphocytes, antigen binding to the TCR initiates a MAP kinase cascade that involves the RAS protein and the sequential activation of three kinases, the last one being the MAP kinase called ERK.

Mixed leukocyte reaction (MLR) An in vitro reaction of alloreactive T cells from one individual against MHC antigens on blood cells from another individual. The MLR involves proliferation of and cytokine secretion by both $CD4^+$ and $CD8^+$ T cells.

Molecular mimicry A postulated mechanism of autoimmunity triggered by infection with a microbe containing antigens that are structurally homologous to and therefore cross-react with self antigens. Immune responses to the microbe are postulated to result in reactions against self antigens.

Monoclonal antibody An antibody that is specific for one antigen and is produced by a B cell hybridoma (a cell line derived by the fusion of a single normal B cell and an immortal B cell tumor line) or by phage display technology. Monoclonal antibodies are widely used in research, clinical diagnosis, and therapy.

Monocyte A type of bone marrow –derived circulating blood cell that is recruited into sites of infection or tissue injury. Once in the tissue, monocytes differentiate into macrophages that function to destroy microbes and repair injured tissues.

Mononuclear phagocytes Cells with a common bone marrow lineage whose primary function is phagocytosis. These cells function as effector cells in innate and adaptive immunity. Mononuclear phagocytes circulate in the blood in an incompletely differentiated form called monocytes, and after they settle in tissues, they mature into macrophages; they are also resident in tissues.

Mucosa-associated lymphoid tissue (MALT) Collections of lymphocytes, dendritic cells, and other cell types within the mucosa of the gastrointestinal and respiratory tracts that are sites of adaptive immune responses to antigens. MALTs are unencapsulated but organized collections of lymphocytes, with T and B cell zones similar to lymph nodes, located below mucosal epithelia, such as Peyer's patches in the gut or pharyngeal tonsils.

mRNA vaccine A vaccine comprised of messenger RNA (mRNA) encoding microbial antigens. mRNA vaccines that are currently in use are antiviral vaccines in which the viral mRNA is modified to enhance both translation and stability and to reduce ability to strongly activate innate immune responses The mRNA is encapsulated in lipid nanoparticles that protect the RNA from degradation and facilitate uptake by dendritic cells. SARS-CoV-2 mRNA vaccines have been used to immunize millions of people and have proven to offer robust protection against serious disease after infection by several variants of the virus.

Mucosal-associated invariant T (MAIT) cells A subset of T cells that express an invariant αβ TCR specific for fungal and bacterial riboflavin metabolites presented by a nonpolymorphic class I MHC-related molecule called MR1. Most MAIT cells are $CD8^+$, are activated either by microbial riboflavin derivatives or by cytokines, and may have inflammatory and cytotoxic functions. MAIT cells account for 20% to 40% of T cells in the human liver.

Mucosal immune system A part of the immune system that responds to and protects against microbes that enter the body through mucosal surfaces, such as the gastrointestinal and respiratory tracts, but also maintains tolerance to commensal organisms that live on the outside of the mucosal epithelium. The mucosal immune system is composed of organized mucosa-associated lymphoid tissues, such as Peyer patches, as well as diffusely distributed cells within the lamina propria.

Multiple myeloma A malignant tumor of antibody-producing plasma cells that often secretes intact antibodies or parts of antibody molecules. The monoclonal antibodies produced by multiple myelomas were critical for early biochemical analyses of antibody structure.

Multiple sclerosis A chronic progressive autoimmune disease of the central nervous system characterized by inflammatory damage to the myelin sheath of neurons, mediated by B cells, autoreactive $CD4^+$ T cells, and macrophages, leading to impairment of sensory and motor functions.

Multivalency See **polyvalency**.

Mycobacterium A genus of bacteria, many species of which can survive within phagocytes and cause disease. The principal host defense against mycobacteria such as *Mycobacterium tuberculosis* is cell-mediated immunity.

Myeloid cells Cells derived from the myeloid lineage of hematopoietic precursors, including granulocytes, monocytes, and dendritic cells. Myeloid cells are distinct from lymphoid cells, which include B cells, T cells, innate lymphoid cells, and natural killer cells, all derived from a common lymphoid progenitor.

Myeloid-derived suppressor cells (MDSCs) A heterogeneous group of myeloid cells derived from the same precursors that give rise to neutrophils and monocytes but have antiinflammatory and immunosuppressive properties. MDSCs are found in lymphoid tissues, blood, or tumors of cancer-bearing animals and cancer patients and are thought to suppress antitumor immune responses.

N

N nucleotides The name given to nucleotides randomly added to the junctions between *V, D,* and *J* gene segments in *Ig* or *TCR* genes during lymphocyte development. The addition of up to 20 of these nucleotides, which is mediated by the enzyme terminal deoxyribonucleotidyl transferase (TdT), contributes to the diversity of the antibody and TCR repertoires.

Naive lymphocyte A mature B or T lymphocyte that has not previously encountered antigen. When naive lymphocytes are stimulated by antigen, they differentiate into effector lymphocytes, such as antibody-secreting B cells, cytokine-producing helper T cells, and CTLs capable of killing target cells. Naive lymphocytes have surface markers and recirculation patterns that are distinct from those of previously activated lymphocytes. (*Naive* also refers to an unimmunized individual.)

Natural antibodies IgM antibodies, produced without overt antigen exposure' largely by B-1 cells specific for bacteria that are common in the environment and gastrointestinal tract. Normal individuals contain natural antibodies without any evidence of infection, and these antibodies may serve as a preformed defense mechanism against microbes that succeed in penetrating epithelial barriers. Natural antibodies specific for ABO blood group antigens are responsible for transfusion reactions.

Natural killer (NK) cells A subset of lymphoid cells, related to ILC1, that function in innate immune responses to kill microbe-infected cells by direct lytic mechanisms and by secreting IFN-γ. NK cells do not express clonally distributed antigen receptors such as Ig receptors or TCRs, and their activation is regulated by a combination of cell surface stimulatory and inhibitory receptors, the latter recognizing self class I MHC molecules.

Natural killer T cells (NKT cells) A numerically small subset of lymphocytes that express T cell receptors and some surface molecules characteristic of NK cells. Some NKT cells, called invariant NKT (iNKT), express αβ T cell antigen receptors with very little diversity and recognize lipid antigens presented by CD1 molecules. The physiologic functions of NKT cells are not well defined.

Negative selection The process by which developing lymphocytes that express self-reactive antigen receptors are eliminated, thereby contributing to the maintenance of self-tolerance. Negative selection of developing T lymphocytes (thymocytes) is best understood and involves high-avidity binding of a thymocyte to self MHC molecules with bound peptides on thymic APCs, leading to apoptotic death of the thymocyte.

Neoantigen A part of a macromolecule that is newly changed, either by chemical modification, or in the case of proteins, by mutation of the encoding gene, such that the new structure is recognized by antibodies or T cells; sometimes called **neoepitope**. Neoantigens encoded by mutated genes are the major inducers of T cell responses against many tumors.

Neonatal Fc receptor (FcRn) An IgG-specific Fc receptor that mediates the transport of maternal IgG across the placenta and the neonatal intestinal epithelium and promotes the long half-life of IgG molecules in the blood by protecting them from catabolism by phagocytes and endothelial cells.

Neonatal immunity Passive humoral immunity to infections in mammals in the first months of life, before full development of the immune system. Neonatal immunity is mediated by maternally produced antibodies transported across the placenta into the fetal circulation before birth or derived from ingested milk and transported across the gut epithelium.

Neutrophil (also called **polymorphonuclear leukocyte, PMN)** A phagocytic cell characterized by a segmented multilobed nucleus and cytoplasmic granules filled with degradative enzymes. Neutrophils are the most abundant type of circulating white blood cells and the most numerous cell type recruited into tissues as part of acute inflammatory responses to microbial infections or tissue damage.

Nitric oxide A molecule with a broad range of activities that in macrophages functions as a potent microbicidal agent to kill ingested organisms.

Nitric oxide synthase A member of a family of enzymes that synthesize the vasoactive and microbicidal compound nitric oxide from L-arginine. Macrophages express an inducible form of this enzyme on activation by various microbial or cytokine stimuli.

NOD-like receptors (NLRs) A family of cytosolic multidomain proteins that sense cytoplasmic PAMPs and DAMPs and recruit other proteins to form signaling complexes that promote inflammation.

Notch-1 A cell surface signaling receptor that is proteolytically cleaved after ligand binding, and the cleaved intracellular portion translocates to the nucleus and regulates gene expression. Notch-1 signaling is required for commitment of developing T cell precursors to the αβ T cell lineage.

Nuclear factor κB (NF-κB) A family of transcription factors composed of homodimers or heterodimers of proteins homologous to the c-REL protein. NF-κB proteins are required for the inducible transcription of many genes important in both innate and adaptive immune responses.

Nuclear factor of activated T cells (NFAT) A transcription factor required for the expression of IL-2, IL-4, and other cytokine genes. There are four different NFATs, each encoded by a separate gene; NFATp and NFATc are found in T cells. Cytoplasmic NFAT is activated by calcium/calmodulin-dependent, calcineurin-mediated dephosphorylation that permits NFAT to translocate into the nucleus and bind to consensus binding sequences in the regulatory regions of IL-2, IL-4, and other cytokine genes, usually in association with other transcription factors such as AP-1.

Nude mouse A strain of mice that lacks development of the thymus, and therefore T lymphocytes, as well as hair follicles, caused by mutation affecting the transcription factor TBX1. Nude mice have been used experimentally to define the role of T lymphocytes in immunity and disease.

O

Oncofetal antigen Proteins that are expressed at high levels on some types of cancer cells and in normal developing fetal (but not adult) tissues. Antibodies specific for these proteins are often used in histopathologic identification of tumors or to monitor the progression of tumor growth in patients. CEA (CD66) and α-fetoprotein are two oncofetal antigens commonly expressed by certain carcinomas.

Opsonin A molecule that becomes attached to the surface of a microbe and can be recognized by surface receptors of neutrophils and macrophages, thereby increasing the efficiency of phagocytosis of the microbe. Opsonins include IgG antibodies, which are recognized by the Fcγ receptor on phagocytes, and fragments of the C3 complement proteins, which are recognized by the complement receptors CR1 (CD35) and the leukocyte integrin Mac-1.

Opsonization The process of attaching opsonins, such as IgG or complement fragments, to microbial surfaces to target the microbes for phagocytosis.

Oral tolerance The suppression of systemic humoral and cell-mediated immune responses to an antigen after oral administration of that antigen, which may occur as a result of anergy of antigen-specific T cells or the production of immunosuppressive cytokines such as transforming growth factor-β. Oral tolerance is a possible mechanism for preventing immune responses to food antigens and to bacteria that normally reside as commensals in the intestinal lumen.

P

P nucleotides Short inverted repeat nucleotide sequences in the VDJ junctions of rearranged *Ig* and *TCR* genes that are generated by RAG-1– and RAG-2–mediated asymmetric cleavage of hairpin DNA intermediates during somatic recombination events. P nucleotides contribute to the junctional diversity of antigen receptors.

Paracrine factor A molecule that acts on cells in proximity to the cell that produces the factor. Most cytokines act in a paracrine fashion.

Passive immunity The form of immunity to an antigen that is established in one individual by transfer of antibodies or lymphocytes from another individual who is immune to that antigen. The recipient of such a transfer can become immune to the antigen without ever having been exposed to or having responded to the antigen. Transplacental transfer of IgG from mother to fetus is a physiologic form of passive immunity essential for protecting newborn babies from infections. An example of therapeutic passive immunity is the transfer of human sera containing antibodies specific for potentially lethal microbial toxins, viruses, or snake venom to exposed individuals.

Pathogen-associated molecular patterns (PAMPs) Structures produced by microorganisms but not mammalian (host) cells, which are recognized by and stimulate the innate immune system. Examples include bacterial lipopolysaccharide and viral double-stranded RNA.

Pathogenicity The ability of a microorganism to cause disease. Multiple mechanisms may contribute to pathogenicity, including production of toxins, stimulation of host inflammatory responses, and perturbation of host cell metabolism.

Pattern recognition receptors Signaling receptors of the innate immune system that recognize pathogen-associated molecular patterns (PAMPs) and damage-associated molecular patterns (DAMPs) and thereby activate innate immune responses. Examples include Toll-like receptors (TLRs) and NOD-like receptors (NLRs).

PD-1 An inhibitory receptor homologous to CD28 that is expressed on activated T cells and binds to PD-L1 or PD-L2, members of the B7 protein family expressed on various cell types. PD-1 is upregulated on T cells after repeated or prolonged stimulation (e.g., in the setting of chronic infection or tumors), and blockade of PD-1 with monoclonal antibodies enhances antitumor immune responses.

Pentraxins A family of plasma proteins that contain five identical globular subunits; includes the acute-phase reactant C-reactive protein.

Peptide-binding cleft The portion of an MHC molecule that binds peptides for display to T cells. The cleft is composed of paired α helices resting on a floor made up of an eight-stranded β-pleated sheet. The polymorphic residues, which are the amino acids that vary among different MHC alleles, are located in and around this cleft. Also called peptide-binding groove.

Perforin A protein present in the granules of CTLs and NK cells. When perforin is released from the granules of activated CTLs or NK cells, it inserts into the plasma membrane of the adjacent infected or tumor cells and promotes entry of granzymes into the cytosol, leading to apoptotic death of the target cell.

Periarteriolar lymphoid sheath (PALS) A cuff of lymphocytes surrounding small arterioles in the spleen, adjacent to lymphoid follicles. A PALS contains mainly T lymphocytes, approximately two-thirds of which are $CD4^+$ and one-third $CD8^+$. In humoral immune responses to protein antigens, B lymphocytes are activated at the interface between the PALS and follicles and then migrate back into the follicles to form germinal centers.

Peripheral lymphoid organs and tissues See **Secondary lymphoid organs and tissues**

Peripheral tolerance Unresponsiveness to self antigens that are present in peripheral tissues and may not be abundant in the generative lymphoid organs. Peripheral tolerance is induced by various mechanisms, including the recognition of antigens without adequate levels of the costimulators required for lymphocyte activation or by regulatory T cell–mediated suppression.

Peyer's patches Organized lymphoid tissue in the lamina propria of the small intestine in which immune responses to intestinal pathogens and other ingested antigens may be initiated. Peyer's patches are composed mostly of B cells, with smaller numbers of T lymphocytes and other cells, all arranged in follicles similar to those found in lymph nodes, often with germinal centers.

Phagocytosis The process by which certain cells of the innate immune system, including macrophages and neutrophils, engulf large particles (>0.5 μm in diameter) such as intact microbes. The cell surrounds the particle with extensions of its plasma membrane by an energy- and cytoskeleton-dependent process resulting in the formation of an intracellular vesicle called a phagosome containing the ingested particle. Phagosomes fuse with lysosomes that contains enzymes that generate free-radicals and proteases, which mediate killing of ingested microbes.

Phagosome A membrane-bound intracellular vesicle that contains microbes or particulate material ingested from the extracellular environment. Phagosomes are formed during the process of phagocytosis. They fuse with lysosomes to form phagolysosomes, leading to enzymatic degradation of the ingested material.

Phosphatase (protein phosphatase) An enzyme that removes phosphate groups from the side chains of certain amino acid residues of proteins. Protein phosphatases in lymphocytes, such as SHP-1 and SHP-2, CD45 and calcineurin, regulate the activity of various signal transduction molecules and transcription factors. Some protein phosphatases may be specific for phosphotyrosine residues and others for phosphoserine and phosphothreonine residues.

Phospholipase Cγ (PLCγ) An enzyme that catalyzes hydrolysis of the plasma membrane phospholipid PIP_2 to generate two signaling molecules, inositol 1,4,5-trisphosphate (IP_3) and diacylglycerol (DAG). PLCγ becomes activated in lymphocytes after antigen binding to the antigen receptor.

Phytohemagglutinin (PHA) A carbohydrate-binding protein, or lectin, produced by plants that cross-links T cell surface molecules, including the T cell receptor, thereby inducing polyclonal activation and agglutination of T cells. PHA was used in experimental immunology to study T cell activation. In clinical medicine, PHA is used to assess whether a patient's T cells are functional or to induce T cell mitosis for the purpose of generating karyotypic data.

Plasmablast Circulating antibody-secreting cells that are precursors of the plasma cells that reside in the bone marrow and other tissues.

Plasma cell A terminally differentiated antibody-secreting B lymphocyte with a characteristic histologic appearance, including an oval shape, eccentric nucleus, and perinuclear halo. Plasma cells are found in bone marrow, mucosal tissues, and many sites of chronic inflammation.

Platelet-activating factor (PAF) A lipid mediator derived from membrane phospholipids in several cell types, including mast cells and endothelial cells. PAF can cause bronchoconstriction and vascular dilation and leak, and it may be a mediator in asthma.

Polyclonal activators Agents that are capable of activating many clones of lymphocytes, regardless of their antigen specificities. Examples of polyclonal activators include anti-IgM antibodies for B cells and anti-CD3 antibodies, bacterial superantigens, and PHA for T cells.

Poly-Ig receptor An Fc receptor expressed by mucosal epithelial cells that mediates the transport of IgA and IgM secreted by plasma cells in the intestinal lamina propria through the intestinal epithelial cells into the lumen.

Polymerase chain reaction (PCR) A rapid method of copying and amplifying specific DNA sequences up to about 1 kb in length that is widely used as a preparative and analytical technique in all branches of molecular biology. The method relies on the use of short oligonucleotide primers complementary to the sequences at the ends of the DNA to be amplified and involves repetitive cycles of melting, annealing, and synthesis of DNA.

Polymorphism The existence of two or more alternative forms, or variants, of a gene that are present at stable frequencies in a population. Each common variant of a polymorphic gene is called an allele, and one individual may carry two different alleles of a gene, each inherited from a different parent. The MHC genes, some of which have thousands of alleles, are the most polymorphic genes in the mammalian genome.

Polyvalency The presence of multiple identical epitopes of a single antigen molecule, cell surface, or particle. Polyvalent antigens, such as bacterial capsular polysaccharides, are often capable of activating B lymphocytes independent of helper T cells. Used synonymously with **multivalency**.

Positive selection The process by which developing T cells in the thymus (thymocytes) whose TCRs bind to self MHC molecules are rescued from programmed cell death, whereas thymocytes whose receptors do not recognize self MHC molecules die by default. Positive selection ensures that mature T cells are self MHC restricted and that $CD8^+$ T cells are specific for complexes of peptides with class I MHC molecules and $CD4^+$ T cells for complexes of peptides with class II MHC molecules.

Pre-B cell A developing B cell present only in hematopoietic tissues that is at a maturational stage characterized by expression of cytoplasmic Ig μ heavy chains and surrogate light chains but not Ig light chains. Pre-B cell receptors composed of μ chains and surrogate light chains deliver signals that stimulate further maturation of pre-B cells into immature B cells.

Pre-B cell receptor A receptor expressed on developing B lymphocytes at the pre-B cell stage that is composed of Ig μ heavy chains and invariant surrogate light chains. The pre-B cell receptor associates with the Igα and Igβ signal transduction proteins to form the pre-B cell receptor complex. Pre-B cell receptors are required for stimulating the proliferation and continued maturation of the developing B cell, serving as a checkpoint that ensures productive μ heavy chain VDJ rearrangement. It is not known whether the pre-B cell receptor binds a specific ligand.

Pre-T cell A developing T lymphocyte in the thymus at a maturational stage characterized by expression of the TCR β chain but not the α chain or CD4 or CD8. In pre-T cells, the TCR β chain is found on the cell surface as part of the pre-T cell receptor.

Pre-T cell receptor A receptor expressed on the surface of pre-T cells that is composed of the TCR β chain and an invariant pre-Tα protein. This receptor associates with CD3 and ζ molecules to form the pre-T cell receptor complex. The function of this complex is similar to that of the pre-B cell receptor in B cell development, namely, the delivery of signals that stimulate further proliferation, antigen receptor gene rearrangements, and other maturational events. The pre-T cell receptor serves as a checkpoint that ensures productive TCR β chain VDJ rearrangement. It is not known whether the pre-T cell receptor binds a specific ligand.

Pre-Tα An invariant transmembrane protein with a single extracellular Ig-like domain that associates with the TCR β chain in pre-T cells to form the pre-T cell receptor.

Primary immune response An adaptive immune response that occurs after the first exposure of an individual to a foreign antigen. Primary responses are characterized by relatively slow kinetics and small magnitude compared with the secondary (memory) responses after a second or subsequent exposure.

Primary immunodeficiency A genetic defect in which an inherited deficiency in some aspect of the innate or adaptive immune system leads to an increased susceptibility to infections. Primary immunodeficiency is frequently manifested early in infancy and childhood but is sometimes clinically detected later in life. Synonymous with **Congenital immunodeficiency**.

Pro-B cell A developing B cell in the bone marrow that is the earliest cell committed to the B lymphocyte lineage. Pro-B cells do not produce Ig, but they can be distinguished from other immature cells by the expression of B lineage–restricted surface molecules such as CD19 and CD10.

Pro-T cell A developing T cell in the thymic cortex that is a recent arrival from the bone marrow and does not express TCRs, CD3, ζ chains, or CD4 or CD8 molecules. Pro-T cells are also called double-negative thymocytes.

Professional antigen-presenting cells (professional APCs) A term sometimes used to refer to APCs that activate T lymphocytes; includes mainly dendritic cells, but also mononuclear phagocytes, and B lymphocytes, all of which are capable of expressing both class I and class II MHC molecules and costimulators. The most important professional APCs for initiating primary T-cell responses are dendritic cells.

Programmed cell death See **Apoptosis**.

Promoter A DNA sequence immediately 5′ of the transcription start site of a gene where the proteins that initiate transcription bind. The term *promoter* is often used to mean the entire 5′ regulatory region of a gene, including enhancers, which are additional sequences that bind transcription factors and interact with the basal transcription complex to increase the rate of transcriptional initiation. Other enhancers may be located at a significant distance from the promoter, either 5′ of the gene, in introns, or 3′ of the gene.

Prostaglandins A class of lipid inflammatory mediators that are derived from arachidonic acid in many cell types through the cyclooxygenase pathway and that have vasodilator, bronchoconstrictor, and chemotactic activities. Prostaglandins made by mast cells are important mediators of allergic reactions. Many commonly used antiinflammatory drugs are cyclooxygenase inhibitors that block the synthesis of prostaglandins.

Proteasome A large multiprotein enzyme complex with a broad range of proteolytic activity, found in the cytoplasm of most cells, important for degrading misfolded cytosolic proteins. Proteins are targeted for proteasomal degradation by covalent linkage of ubiquitin molecules. A specialized form of the proteasome in antigen-presenting cells, called the immunoproteasome, degrades cytosolic proteins into peptides that are transported into the endoplasmic reticulum and bind to newly synthesized class I MHC molecules.

Protein kinase C (PKC) Any of several isoforms of an enzyme that mediates the phosphorylation of serine and threonine residues in many different protein substrates and thereby serves to propagate various signal transduction pathways leading to transcription factor activation. In T and B lymphocytes, PKC is activated by diacyl glycerol (DAG), which is generated in response to antigen receptor ligation.

Protein tyrosine kinases (PTKs) Enzymes that mediate the phosphorylation of tyrosine residues in proteins and thereby promote phosphotyrosine-dependent protein-protein interactions. PTKs are involved in numerous signal transduction pathways in cells of the immune system.

Protozoa Single-celled eukaryotic organisms, many of which are human parasites and cause diseases. Examples of pathogenic protozoa include *Entamoeba histolytica,* which causes amebic dysentery; *Plasmodium,* which causes malaria; and *Leishmania,* which causes leishmaniasis. Protozoa stimulate both innate and adaptive immune responses.

Provirus A DNA copy of the genome of a retrovirus that is integrated into the host cell genome and from which viral genes are transcribed and the viral genome is reproduced. HIV proviruses can remain inactive for long periods and thereby represent a latent form of HIV infection that is not accessible to immune defense.

Pyogenic bacteria Bacteria, such as gram-positive staphylococci and streptococci, that induce inflammatory responses rich in polymorphonuclear leukocytes (giving rise to pus).

Pyroptosis A form of programmed cell death of macrophages and DCs induced by canonical inflammasome activation of caspase-1 (and also noncanonical inflammasome pathways that use human caspase-4 or caspase-5), characterized by cell swelling, loss of plasma membrane integrity, and release of inflammatory mediators, such as IL-1β. In pyroptosis, the activated caspases proteolytically generates a fragment of the protein gasdermin D, which polymerizes to form pores in the plasma membrane. Pyroptosis results in the death of certain microbes that gain access to the cytosol and enhances inflammatory clearance of bacteria, but also contributes to septic shock.

R

Radioimmunoassay A highly sensitive and specific immunologic method of quantifying the concentration of an antigen in a solution that relies on a radioactively labeled antibody specific for the antigen. Usually, two antibodies specific for the antigen are used. The first antibody is unlabeled but attached to a solid support, where it binds and immobilizes the antigen whose concentration is being determined. The amount of the second, labeled antibody that binds to the immobilized antigen, as determined by radioactive decay detectors, is proportional to the concentration of antigen in the test solution. Radioimmunoassays have largely been replaced by nonradioactive solid-phase immunoassays, such as enzyme-linked immunoassays (ELISAs).

Rapamycin An immunosuppressive drug (also called sirolimus) used to treat allograft rejection. Rapamycin inhibits the activation of a protein called molecular target of rapamycin (mTOR), which is a key signaling molecule in a variety of metabolic and cell growth pathways, including the pathway required for interleukin-2–mediated T cell proliferation.

RAS A member of a family of 21-kD guanine nucleotide-binding G proteins with intrinsic GTPase activity that are involved in many different signal transduction pathways in diverse cell types. Mutated *RAS* genes are associated with neoplastic transformation. In T cell activation, RAS is recruited to the plasma membrane by tyrosine-phosphorylated adaptor proteins, where it is activated by GDP-GTP exchange factors. GTP•RAS then initiates the MAP kinase cascade, which leads to expression of the *FOS* gene and assembly of the AP-1 transcription factor.

Reactive oxygen species (ROS) Highly reactive metabolites of oxygen, including superoxide anion, hydroxyl radical, and hydrogen peroxide, that are produced by activated phagocytes, particularly neutrophils. Reactive oxygen species are used by the phagocytes to form oxyhalides that damage ingested bacteria. They also may be released from cells and promote inflammatory responses or cause tissue damage.

Reagin IgE antibody that mediates an immediate hypersensitivity reaction.

Receptor editing A process by which some immature B cells that recognize self antigens in the bone marrow may be induced to change their Ig specificities. Receptor editing involves reactivation of the *RAG* genes, additional light chain *VJ* recombinations, and new Ig light chain production, which allows the cell to express a different Ig receptor that is not self-reactive.

Recombination-activating genes 1 and 2 (*RAG1* and *RAG2*) The genes encoding RAG-1 and RAG-2 proteins, which make up the V(D)J recombinase and are expressed in developing B and T cells. RAG proteins bind to recombination signal sequences and are critical for DNA recombination events that form functional *Ig* and *TCR* genes. Therefore, RAG proteins are required for expression of antigen receptors and for the maturation of B and T lymphocytes.

Recombination signal sequences Specific DNA sequences found adjacent to the *V, D,* and *J* segments in antigen receptor loci that are recognized by the RAG-1/RAG-2 complex during *V(D)J* recombination. The recognition sequences consist of a conserved stretch of 7 nucleotides, called the heptamer, located adjacent to the *V, D,* or *J* coding sequence, followed by a spacer of 12 or 23 nonconserved nucleotides and a conserved stretch of 9 nucleotides, called the nonamer.

Red pulp An anatomic and functional compartment of the spleen composed of vascular sinusoids, scattered among which are large numbers of erythrocytes, macrophages, dendritic cells, sparse lymphocytes, and plasma cells. Red pulp macrophages clear the blood of microbes, other foreign particles, and damaged red blood cells.

Regulatory T cells A population of T cells that inhibits the activation of other T cells and is necessary to maintain peripheral tolerance to self-antigens. Most regulatory T cells are $CD4^+$ and express the transcription factor FOXP3, the α chain of the IL-2 receptor (CD25), CTLA-4, and the transcription factor FOXP3.

Respiratory burst The process by which reactive oxygen species such as superoxide anion, hydroxyl radical, and hydrogen peroxide are produced in neutrophils and macrophages. The respiratory burst is mediated by the enzyme phagocyte oxidase and is usually triggered by bacterial products such as LPS.

Reverse transcriptase An enzyme encoded by retroviruses, such as HIV, that synthesizes a DNA copy of the viral genome from an RNA template. The process is essential for replication of RNA viruses, and reverse transcriptase inhibitors are used as drugs to treat HIV-1 infection. Purified reverse transcriptase was used widely in molecular biology research for purposes of cloning complementary DNAs encoding a gene of interest from messenger RNA.

Rh blood group antigens A system of protein alloantigens expressed on red blood cell membranes that are the cause of transfusion reactions and hemolytic disease of the fetus and newborn. The most clinically important Rh antigen is designated RhD.

Rheumatoid arthritis An autoimmune disease characterized primarily by inflammatory damage to joints and sometimes inflammation of blood vessels, lungs, and other tissues. $CD4^+$ T cells, activated B lymphocytes, and plasma cells are found in the inflamed joint lining (synovium), and numerous proinflammatory cytokines, including IL-1, IL-6 and TNF, are present in the synovial (joint) fluid.

RNA vaccine See **mRNA vaccine**

RIG-like receptors (RLRs) Cytosolic receptors of the innate immune system that recognize viral RNA and induce production of type I interferons. The two best characterized RLRs are RIG-I (retinoic acid–inducible gene I) and MDA5 (melanoma differentiation-associated gene 5).

RORγT (retinoid-related orphan receptor γ T) A transcription factor expressed in and required for differentiation of Th17 cells and group 3 innate lymphoid cells, encoded by the *RORC* gene.

S

SARS-CoV-2 *See* **COVID-19.**

Scavenger receptors A family of cell surface receptors expressed on macrophages, originally defined as receptors that mediate endocytosis of oxidized or acetylated low-density lipoprotein particles but that also bind and mediate the phagocytosis of a variety of microbes.

SCID See **Severe combined immunodeficiency.**

SCID mouse A mouse strain in which B and T cells are absent because of an early block in maturation from bone marrow precursors. SCID mice carry a mutation in a component of the enzyme DNA-dependent protein kinase, which is required for double-stranded DNA break repair. Deficiency of this enzyme results in abnormal joining of *Ig* and *TCR* gene segments during recombination and therefore failure to express antigen receptors.

Secondary immune response An adaptive immune response that occurs on second or subsequent exposure to an antigen. A secondary response is characterized by more rapid kinetics and greater magnitude relative to the primary immune response, which occurs on first exposure.

Secondary immunodeficiency See **Acquired immunodeficiency**.

Secondary lymphoid organ Organized collections of lymphocytes and accessory cells, including the spleen, lymph nodes, and mucosa-associated lymphoid tissues, in which adaptive immune responses are initiated. Synonymous with **Peripheral lymphoid organ**.

Secretory component The proteolytically cleaved portion of the extracellular domain of the poly-Ig receptor that remains bound to an IgA molecule in mucosal secretions.

Selectin Any one of three separate but closely related carbohydrate-binding proteins that mediate low-affinity adhesion of leukocytes to postcapillary venule endothelial cells, leading to rolling of the leukocytes on endothelial surfaces of the venules. Each of the selectin molecules is a single-chain transmembrane glycoprotein with a similar modular structure, including an extracellular calcium-dependent lectin domain. The selectins include L-selectin (CD62L), expressed on leukocytes; P-selectin (CD62P), expressed on platelets and activated endothelium; and E-selectin (CD62E), expressed on activated endothelium.

Selective immunoglobulin (Ig) deficiency Immunodeficiencies characterized by a lack of only one or a few Ig classes or subclasses. IgA deficiency is the most common selective Ig deficiency, followed by IgG3 and IgG2 deficiencies. Patients with these disorders may be at increased risk for bacterial infections, but many are healthy.

Self MHC restriction The limitation (or restriction) of T cells to recognize antigens displayed by MHC molecules that the T cell encountered during maturation in the thymus (and thus sees as self MHC).

Self-tolerance Unresponsiveness of the adaptive immune system to self antigens, largely as a result of inactivation or death of self-reactive lymphocytes induced by exposure to these antigens or suppression by regulatory T cells. Self-tolerance is a cardinal feature of the normal immune system, and failure of self-tolerance leads to autoimmune diseases.

Septic shock A severe complication of bacterial or fungal infections that spread to the bloodstream (sepsis), and is characterized by hypotension and shock, disseminated intravascular coagulation, and metabolic disturbances. In bacterial infections, the syndrome is most often due to the effects of bacterial cell wall components, such as LPS or peptidoglycan, that bind to TLRs on various cell types and induce expression of large amounts inflammatory cytokines, including TNF, IL-6 and IL-12.

Seroconversion The production of detectable antibodies in the serum specific for a microorganism during the course of an infection or in response to immunization.

Serology The study of blood (serum) antibodies and their reactions with antigens. The term *serology* is often used to refer to the diagnosis of infectious diseases by detection of microbe-specific antibodies in the serum.

Serotype An antigenically distinct subset of a species of an infectious organism that is distinguished from other subsets by serologic (i.e., serum antibody) tests. Humoral immune responses to one serotype of microbes (e.g., influenza virus) may not be protective against another serotype.

Serum The cell-free fluid that remains when blood or plasma forms a clot. Blood antibodies are found in the serum fraction.

Serum amyloid A (SAA) An acute-phase protein whose serum concentration rises significantly in the setting of infection and inflammation, mainly because of cytokine-induced synthesis by the liver. SAA activates leukocyte chemotaxis and phagocytosis.

Serum sickness A disease caused by the injection of large doses of a protein antigen into the blood and characterized by the deposition of antigen-antibody (immune) complexes in blood vessel walls, especially in the kidneys and joints. Immune complex deposition leads to complement fixation and leukocyte recruitment and subsequently to glomerulonephritis and arthritis. Serum sickness was originally described as a disorder that occurred in patients receiving injections of animal (horse or goat) serum containing antitoxin antibodies to prevent diphtheria.

Severe combined immunodeficiency (SCID) Immunodeficiency diseases in which both B and T lymphocytes do not develop or do not function properly, and therefore both humoral immunity and cell-mediated immunity are impaired. Children with SCID usually have infections during the first year of life and succumb to these infections unless the immunodeficiency is treated. SCID has several different genetic causes.

Signal transducer and activator of transcription (STAT) A member of a family of seven mammalian proteins that function as transcription factors in response to binding of cytokines to type I and type II cytokine receptors. STATs are present as inactive monomers in the cytosol of cells and are recruited to the cytoplasmic tails of cross-linked cytokine receptors, where they are tyrosine phosphorylated by JAKs. The phosphorylated STAT proteins dimerize and move to the nucleus, where they bind to specific sequences in the promoter regions of various genes and stimulate their transcription. Different STATs are activated by different cytokines.

Single-chain variable fragment (scFv) A genetically engineered polypeptide that includes only Ig heavy chain and light chain V domains that fold to form an antibody binding site of known specificity, used as a research reagent, or in tumor immunotherapy (e.g., to recognize tumor antigens in bispecific T cell engagers [BiTEs] and chimeric antigen receptors [CARs]).

Single-positive thymocyte A maturing T cell precursor in the thymus that expresses CD4 or CD8 molecules but not both. Single-positive thymocytes are found mainly in the medulla and have matured from the double-positive stage, during which thymocytes express both CD4 and CD8 molecules.

Smallpox A disease caused by variola virus. Smallpox was the first infectious disease shown to be preventable by vaccination and the first disease to be completely eradicated by a worldwide vaccination program.

Somatic hypermutation High-frequency point mutations in Ig heavy and light chains that occur in germinal center B cells in response to signals from T follicular helper (Tfh) cells. Mutations that result in increased affinity of antibodies for antigen impart a selective survival advantage to the B cells producing those antibodies and lead to affinity maturation of a humoral immune response.

Somatic recombination The process of DNA recombination by which the functional genes encoding the variable regions of antigen receptors are formed during lymphocyte development. A relatively limited set of inherited, or germline, DNA segments that are initially separated from one another are brought together by enzymatic deletion of intervening sequences and ligation of the segments. This process occurs only in developing B or T lymphocytes and is mediated by RAG-1 and RAG-2 proteins. This process is also called **V(D)J recombination**.

Specificity A cardinal feature of the adaptive immune system, namely, that immune responses are directed toward and able to distinguish between distinct antigens or small parts of macromolecular antigens. This fine specificity is attributed to lymphocyte antigen receptors that may bind to one molecule but not to another, even closely related, molecule.

Spleen A secondary lymphoid organ in the left upper quadrant of the abdomen. The spleen is the major site of adaptive immune responses to blood-borne antigens. The red pulp of the spleen is composed of blood-filled vascular sinusoids lined by phagocytes that ingest opsonized antigens and damaged red blood cells. The white pulp of the spleen contains lymphocytes and lymphoid follicles where B cells are activated.

SRC family kinases A family of protein tyrosine kinases, homologous to the SRC tyrosine kinase, which initiate signaling downstream of immune receptors by phosphorylating tyrosine residues on ITAM motifs. LCK is a prominent SRC-family kinase in T cells and LYN in B cells.

SRC homology 2 (SH2) domain A three-dimensional domain structure of approximately 100 amino acid residues present in many signaling proteins that permits specific noncovalent interactions with other proteins by binding to phosphotyrosines. Each SH2 domain has a unique binding specificity that is determined by the amino acid residues adjacent to the phosphotyrosine on the target protein. Several proteins involved in early signaling events in T and B lymphocytes interact with one another through SH2 domains.

SRC homology 3 (SH3) domain A three-dimensional domain structure of approximately 60 amino acid residues present in many signaling proteins that mediates protein-protein binding. SH3 domains bind to proline residues and function cooperatively with the SH2 domains of the same protein. For instance, SOS, the guanine nucleotide exchange factor for RAS, contains both SH2 and SH3 domains, and both are involved in SOS binding to the adaptor protein GRB-2.

Stem cell An undifferentiated cell that divides continuously and gives rise to additional stem cells and to cells of multiple different lineages. For example, all blood cells arise from a common hematopoietic stem cell.

STING (Stimulator of IFN Genes) An adaptor protein located in the endoplasmic reticulum membrane, which is used by several cytoplasmic DNA sensor molecules to transduce signals that activate the IRF3 transcription factor, leading to type I IFN gene expression.

Subunit vaccine A vaccine composed of purified antigens or portions (subunits) of microbes. Examples of this type of vaccine include diphtheria and tetanus toxoids, pneumococcus and *Haemophilus influenzae* polysaccharide vaccines, and purified polypeptide vaccines against hepatitis B and influenza virus. Subunit vaccines may stimulate antibody and helper T cell responses, but they often do not generate strong CTL responses, and boosters may be needed for long-lasting immunity.

Superantigens Microbial proteins that bind to and activate all of the T cells in an individual that express a particular set or family of Vβ *TCR* genes. Superantigens are presented to T cells by binding to nonpolymorphic regions of class II MHC molecules on APCs, and they interact with conserved regions of TCR Vβ domains. Several staphylococcal enterotoxins are superantigens. Their importance lies in their ability to activate many T cells, which results in production of large amounts of cytokine and a clinical syndrome that is similar to septic shock.

Suppressor T cells T cells that block the activation and function of other T lymphocytes. These cells were described in the 1970s, but because it has been difficult to clearly define suppressor T cells or their mode of action, the term is no longer used.

Surrogate light chains Two nonvariable proteins that associate with Ig μ heavy chains in pre-B cells to form the pre-B cell receptor. The two surrogate light chain proteins include the V pre-B protein, which is homologous to a light-chain V domain, and λ5, which is covalently attached to the μ heavy chain by a disulfide bond.

Switch recombination The molecular mechanism underlying Ig isotype switching in which a rearranged VDJ gene segment in an antibody-producing B cell recombines with a downstream C gene and the intervening C gene or genes are deleted. DNA recombination events in switch recombination are triggered by CD40 and cytokines, which activate an enzyme called activation-induced cytidine deaminase (AID), and involve nucleotide sequences called switch regions located in the introns at the 5′ end of each C_H locus.

SYK A cytoplasmic protein tyrosine kinase in B cells, similar to ZAP-70 in T cells, that is critical for early signaling steps in antigen-induced B cell activation. SYK binds to phosphorylated tyrosines in the cytoplasmic tails of the Igα and Igβ chains of the BCR complex and in turn phosphorylates adaptor proteins that recruit other components of the signaling cascade.

Syngeneic Genetically identical. All animals of an inbred strain and monozygotic twins are syngeneic.

Syngeneic graft A graft from a donor who is genetically identical to the recipient. Syngeneic grafts are not rejected.

Synthetic vaccine Vaccines composed of recombinant DNA-derived protein antigens. Synthetic vaccines for hepatitis B virus and herpes simplex virus are now in use.

Systemic inflammatory response syndrome (SIRS) The systemic changes observed in patients who have disseminated bacterial infections and other conditions that induce widespread inflammation, such as burns. In its mild form, SIRS consists of neutrophilia, fever, and a rise in acute-phase reactants in the plasma. These changes are stimulated by bacterial products such as LPS and are mediated by cytokines of the innate immune system. In severe cases, SIRS may include disseminated intravascular coagulation, acute respiratory distress syndrome, and shock.

Systemic lupus erythematosus (SLE) A chronic systemic autoimmune disease that affects predominantly women and is characterized by rashes, arthritis, glomerulonephritis, hemolytic anemia, thrombocytopenia, and central nervous system involvement. Many different autoantibodies are found in patients with SLE, particularly anti-DNA antibodies. Many of the manifestations of SLE are due to the formation of immune complexes composed of autoantibodies and their specific antigens, with deposition of these complexes in small blood vessels in various tissues. The underlying mechanism for the breakdown of self-tolerance in SLE is not understood.

T

T cell receptor (TCR) The clonally distributed antigen receptor on T lymphocytes. The most common form of TCR is composed of a heterodimer of two disulfide-linked transmembrane polypeptide chains, designated α and β, each containing one N-terminal Ig-like variable (V) domain, one Ig-like constant (C) domain, a hydrophobic transmembrane region, and a short cytoplasmic region. The $\alpha\beta$ TCR is expressed on $CD4^+$ and $CD8^+$ T cells and recognizes complexes of foreign peptides bound to self MHC molecules on the surface of APCs. (Another less common type of TCR, composed of γ and δ chains, is found on a small subset of T cells and recognizes different types of antigens.)

T cell receptor (TCR) transgenic mouse A genetically engineered strain of mouse that expresses transgenically encoded TCR α and β genes encoding a TCR of a single defined specificity. Because of allelic exclusion of endogenous TCR genes, most or all of the T cells in a TCR transgenic mouse have the same antigen specificity, which is a useful property for various research purposes.

T follicular helper (Tfh) cells A subset of $CD4^+$ helper T cells present within lymphoid follicles that are critical in providing signals to B cells in the germinal center reaction that stimulate somatic hypermutation, isotype switching, and the generation of memory B cells and long-lived plasma cells. Tfh cells express CXCR5, ICOS, PD-1, IL-21, and BCL-6.

T lymphocyte The key component of cell-mediated immune responses in the adaptive immune system. T lymphocytes mature in the thymus, circulate in the blood, populate secondary lymphoid tissues, and are recruited to peripheral sites of antigen exposure. They express antigen receptors (TCRs) that recognize peptide fragments of foreign proteins bound to self MHC molecules. Functional subsets of T lymphocytes include $CD4^+$ helper T cells and $CD8^+$ CTLs.

T-BET A T-box family transcription factor that is required for the differentiation of Th1 cells and ILC1s.

T-dependent antigen An antigen that requires both B cells and helper T cells to stimulate an antibody response. T-dependent antigens are protein antigens that contain some epitopes recognized by T cells and other epitopes recognized by B cells. Helper T cells produce cytokines and cell surface molecules that stimulate B cell proliferation and differentiation into antibody-secreting cells. Humoral immune responses to T-dependent antigens are characterized by isotype switching, affinity maturation, and memory.

Tacrolimus An immunosuppressive drug (also known as FK506) of the calcineurin inhibitor class, used to treat allograft rejection, that blocks T cell cytokine gene transcription, similar to cyclosporine. Tacrolimus binds to a cytosolic protein called FK506-binding protein, and the resulting complex inhibits the phosphatase calcineurin, thereby inhibiting activation and nuclear translocation of the transcription factor NFAT.

Tertiary lymphoid organ A collection of lymphocytes and antigen-presenting cells organized into B cell follicles and T cell zones that develop in sites of chronic immune-mediated inflammation, such as the joint synovium of rheumatoid arthritis patients.

Th1 cells A subset of $CD4^+$ helper T cells that secrete a particular set of cytokines, including IFN-γ, and whose principal function is to stimulate phagocyte-mediated defense against infections, especially with intracellular microbes.

Th2 cells A subset of $CD4^+$ helper T cells that secrete a particular set of cytokines, including IL-4, IL-5, and IL-13, and whose principal function is to stimulate IgE and eosinophil/mast cell–mediated immune reactions.

Th17 cells A subset of $CD4^+$ helper T cells that secrete a particular set of inflammatory cytokines, including IL-17 and IL-22, that are protective against bacterial and fungal infections and also mediate inflammatory reactions in autoimmune and other inflammatory diseases.

Thymic epithelial cells Epithelial cells abundant in the cortical and medullary stroma of the thymus that play a critical role in T cell development. In the process of positive selection, maturing T cells that weakly recognize self peptides bound to MHC molecules on the surface of thymic epithelial cells are rescued from programmed cell death.

Thymocyte A precursor of a mature T lymphocyte present in the thymus.

Thymus A bilobed organ situated in the anterior mediastinum that is the site of maturation of T lymphocytes from bone marrow–derived precursors. The thymus is divided into an outer cortex and an inner medulla and contains stromal thymic epithelial cells, macrophages, dendritic cells, and numerous T cell precursors (thymocytes) at various stages of maturation.

T-independent antigen Nonprotein antigens, such as polysaccharides and lipids, which can stimulate antibody responses without a requirement for antigen-specific helper T lymphocytes. T-independent antigens usually contain multiple identical epitopes that can cross-link membrane Ig on B cells and thereby activate the cells. Humoral immune responses to T-independent antigens show relatively little heavy-chain isotype switching or affinity maturation, two processes that require signals from helper T cells.

Tissue typing The determination of the particular MHC alleles expressed by an individual for the purpose of matching allograft donors and recipients. Tissue typing, also called HLA typing, is usually done by molecular (PCR-based) sequencing of HLA alleles or by serologic methods (lysis of an individual's cells by panels of anti-HLA antibodies).

TNF receptor-associated factors (TRAFs) A family of adaptor molecules that interact with the cytoplasmic domains of various receptors in the TNF receptor family, including TNF-RII, lymphotoxin (LT)-β receptor, and CD40. Each of these receptors contains a cytoplasmic motif that binds different TRAFs, which in turn engage other signaling molecules, leading to activation of the transcription factors AP-1 and NF-κB.

Tolerance Unresponsiveness of the adaptive immune system to antigens, as a result of inactivation or death of antigen-specific lymphocytes, induced by exposure to the antigens. Tolerance to self antigens is a normal feature of the adaptive immune system, and tolerance to foreign antigens may be induced under certain conditions of antigen exposure.

Tolerogen An antigen that induces immunologic tolerance, in contrast to an immunogen, which induces an immune response. Many antigens can be either tolerogens or immunogens, depending on how they are administered. All self antigens are tolerogenic. Tolerogenic forms of foreign antigens include large doses of proteins administered without adjuvants and orally administered antigens.

Toll-like receptors (TLRs) A family of pattern recognition receptors of the innate immune system that are expressed by many cell types and recognize microbial structures, such as flagellin, lipopolysaccharide, peptidoglycan, dsRNA, and CpG DNA. TLRs transduce signals that lead to the expression of inflammatory and antiviral genes. There are 10 human TLRs, 7 of which are expressed on the plasma membrane of cells and 3 are located in endosomal membranes.

Tonsils Partially encapsulated secondary lymphoid tissues located beneath barrier epithelium in the nasopharynx and oropharynx, including adenoids (pharyngeal tonsils), palatine tonsils, and lingual tonsils. Tonsils are sites of initiation of adaptive immune responses to microbes in the upper respiratory tract.

Toxic shock syndrome An acute illness characterized by shock, skin exfoliation, conjunctivitis, and diarrhea that is associated with tampon use and caused by a *Staphylococcus aureus* superantigen, which is a polyclonal activator of all T cells expressing TCRs using a particular subgroup of Vβ genes.

Transfusion Transplantation of circulating blood cells, platelets, or plasma from one individual to another. Transfusions are performed to treat blood loss from hemorrhage or to treat a deficiency in one or more blood cell types (such as red blood cells and platelets) resulting from inadequate production or excess destruction.

Transfusion reactions An immunologic reaction against transfused blood products, usually mediated by preformed antibodies in the recipient that bind to donor blood cell antigens, such as ABO blood group antigens or histocompatibility antigens. Transfusion reactions can lead to intravascular lysis of red blood cells and, in severe cases, kidney damage, fever, shock, and disseminated intravascular coagulation.

Transgenic mouse A mouse that expresses an exogenous gene that has been introduced into the genome by injection of a specific DNA sequence into the pronuclei of fertilized mouse eggs. Transgenes insert randomly at chromosomal break points and are subsequently inherited as simple Mendelian traits. By the design of transgenes with tissue-specific regulatory sequences, mice can be produced that express a particular gene only in certain tissues. Transgenic mice are used extensively in immunology research to study the functions of various cytokines, cell surface molecules, and intracellular signaling molecules.

Transplantation The process of transferring cells, tissues, or organs (i.e., grafts) from one individual to another or from one site to another in the same individual. Transplantation is used to treat a variety of diseases in which there is a functional disorder of a tissue or organ. The major barrier to successful transplantation between individuals is immunologic reaction to the transplanted graft (rejection).

Transporter associated with antigen processing (TAP) An ATP-dependent peptide transporter that mediates the active transport of peptides from the cytosol to the site of assembly of class I MHC molecules inside the endoplasmic reticulum. TAP is a heterodimeric molecule composed of TAP-1 and TAP-2 polypeptides, both encoded by genes in the MHC. Because antigenic peptides are required for stable assembly of class I MHC molecules, TAP-deficient animals express few cell surface class I MHC molecules, which results in diminished development and activation of $CD8^+$ T cells.

Tuft cell A specialized intestinal epithelial cell type that is involved in enhancing mucus secretion in response to helminth infection. Tuft cells are activated by helminths to secrete IL-25, which stimulates ILC2s to secrete IL-13, which in turn promotes formation of mucus secreting goblet cells.

Tumor immunity Protection against the development or progression of tumors by the immune system. Although immune responses to naturally occurring tumors can frequently be demonstrated, tumors often escape these responses. New therapies that target T cell inhibitory molecules, such as PD-1, are proving effective in enhancing T cell–mediated antitumor immunity.

Tumor-infiltrating lymphocytes (TILs) Lymphocytes isolated from the inflammatory infiltrates present in and around surgical resection samples of solid tumors that are enriched with tumor-specific CTLs and NK cells. In an experimental mode of cancer treatment, TILs are grown in vitro in the presence of high doses of IL-2 and are then adoptively transferred back into patients with the tumor.

Tumor necrosis factor receptor superfamily (TNFRSF) A large family of structurally homologous transmembrane proteins that bind TNF superfamily proteins and generate signals that regulate proliferation, differentiation, apoptosis, and inflammatory gene expression (see Appendix II).

Tumor necrosis factor superfamily (TNFSF) A large family of structurally homologous transmembrane proteins that regulate diverse functions in responding cells, including proliferation, differentiation, apoptosis, and inflammatory gene expression. TNFSF members typically form homotrimers, either within the plasma membrane or after proteolytic release from the membrane, and bind to homotrimeric TNF receptor superfamily (TNFRSF) molecules, which then initiate a variety of signaling pathways (see Appendix II).

Tumor-specific antigen An antigen whose expression is restricted to a particular tumor and is not expressed by normal cells. Tumor-specific antigens may serve as target antigens for antitumor immune responses.

Tumor-specific transplantation antigen (TSTA) An antigen expressed on experimental animal tumor cells that can be detected by induction of immunologic rejection of tumor transplants. TSTAs were originally defined on chemically induced rodent sarcomas and shown to stimulate CTL-mediated rejection of transplanted tumors.

Two-signal hypothesis A now-proven hypothesis that states that the activation of lymphocytes requires two distinct signals, the first being antigen and the second either microbial products or components of innate immune responses to microbes. The requirement for antigen (so-called signal 1) ensures that the ensuing immune response is specific. The requirement for additional stimuli triggered by microbes or innate immune reactions (signal 2) ensures that immune responses are induced when they are needed, that is, against microbes and other noxious substances and not against harmless substances, including self antigens. In T cells, signal 2 is referred to as costimulation and is often mediated by membrane molecules on APCs, such as B7 proteins.

Type 1 diabetes A disease characterized by a lack of insulin that leads to various metabolic and vascular abnormalities. The insulin deficiency results from autoimmune destruction of the insulin-producing β cells of the islets of Langerhans in the pancreas, usually during childhood. $CD4^+$ and $CD8^+$ T cells, antibodies, and cytokines have been implicated in the islet cell damage.

U

Ubiquitination Covalent linkage of one or several copies of a small polypeptide called ubiquitin to a protein. Ubiquitination frequently serves to target proteins for proteolytic degradation mainly by proteasomes, a critical step in the class I MHC pathway of antigen processing and presentation.

Uracil N-glycosylase (UNG) An enzyme that removes uracil residues from DNA, leaving an abasic site. UNG is a key participant in isotype switching, and homozygous UNG mutations result in a hyper-IgM syndrome.

Urticaria Localized transient itchy swelling and redness of the skin caused by leakage of fluid and plasma proteins from small vessels into the upper dermis during an immediate hypersensitivity reaction.

V

V gene segments A DNA sequence that encodes most of the variable domain of an Ig heavy chain or light chain or a TCR α, β, γ, or δ chain. Each antigen receptor locus contains many different *V* gene segments, any one of which may recombine with downstream *D* or *J* segments during lymphocyte maturation to form functional antigen receptor V genes.

V(D)J recombinase The complex of RAG1 and RAG2 proteins that catalyzes lymphocyte antigen receptor gene recombination.

Vaccine A preparation of microbial antigen, often combined with adjuvants, which is administered to individuals to induce protective immunity against microbial infections. Vaccines may be in the form of live but avirulent microorganisms, killed microorganisms, purified macromolecular components of a microorganism, a nonpathogenic virus carrying genes encoding microbial pathogen antigens, or a lipid nanoparticle carrying RNA encoding viral antigens. Vaccines containing tumor antigens (cancer vaccines) are being developed to treat cancers.

Variable region The extracellular, N-terminal region of an Ig heavy or light chain or a TCR α, β, γ, or δ chain that contains variable amino acid sequences that differ between every clone of lymphocytes and that are responsible for the specificity for antigen. The antigen-binding variable sequences are localized to extended loop structures or hypervariable segments.

Vasoactive amines Low-molecular-weight nonlipid compounds, such as histamine, that all have an amine group, are stored in and released from the cytoplasmic granules of mast cells, and mediate many of the biologic effects of immediate hypersensitivity (allergic) reactions. (Also called biogenic amines.)

Viral vector vaccine A vaccine that consists of a live nonpathogenic virus that has been engineered to be nonreplicating and incorporates genes encoding antigens of a pathogenic virus. Vaccination results in short-lived infection by the hybrid virus and expression of the pathogen antigens, which induces a protective immune response against the pathogen. The most widely used viral vector vaccines are used to protect against COVID-19, and are composed of adenovirus vectors carrying SARS-CoV-2 spike protein genes. Anti-SARS-CoV-2 virial vector vaccines have been used to immunize millions of people and have proven to offer robust protection against serious disease after infection by several variants of the virus.

Virus An obligate intracellular microorganism or infectious particle that consists of a simple nucleic acid genome packaged in a protein capsid, sometimes surrounded by a membrane envelope. Many pathogenic animal viruses cause a wide range of diseases. Humoral immune responses to viruses can be effective in blocking infection of cells, and NK cells and CTLs are necessary to kill cells already infected.

W

Western blot An immunologic technique to determine the presence of a protein in a biologic sample. The method involves separation of proteins in the sample by electrophoresis, transfer of the protein array from the electrophoresis gel to a support membrane by capillary action (blotting), and finally detection of the protein by binding of an enzymatically or radioactively labeled antibody specific for that protein.

Wheal-and-flare reaction Local swelling and redness in the skin at a site of an immediate hypersensitivity reaction. The wheal reflects increased vascular permeability, and the flare results from increased local blood flow, both changes resulting from mediators such as histamine released from activated dermal mast cells.

White pulp The part of the spleen that is composed predominantly of lymphocytes, arranged in periarteriolar lymphoid sheaths and follicles, and other leukocytes. The remainder of the spleen contains sinusoids lined with phagocytic cells and filled with blood, called the **red pulp**.

Wiskott-Aldrich syndrome An X-linked disease characterized by eczema, thrombocytopenia (reduced blood platelets), and immunodeficiency manifested as susceptibility to bacterial infections. The defective gene encodes a cytosolic protein involved in signaling cascades and regulation of the actin cytoskeleton.

X

Xenoantigen An antigen in a graft from another species.

Xenograft (xenogeneic graft) An organ or tissue graft derived from a species different from the recipient. Transplantation of xenogeneic grafts (e.g., from a pig) to humans is not yet practical because of special problems related to immunologic rejection.

Xenoreactive Describing a T cell or antibody that recognizes and responds to an antigen on a graft from another species (a xenoantigen). The T cell may recognize an intact xenogeneic MHC molecule or a peptide derived from a xenogeneic protein bound to a self MHC molecule.

X-linked agammaglobulinemia An immunodeficiency disease, also called Bruton agammaglobulinemia, characterized by a block in early B cell maturation and an absence of serum Ig. Patients suffer from pyogenic bacterial infections. The disease is caused by mutations or deletions in the gene encoding BTK, an enzyme involved in signal transduction in developing B cells.

X-linked hyper-IgM syndrome A rare immunodeficiency disease caused by mutations in the CD40 ligand gene and characterized by failure of B cell heavy-chain isotype switching and defective cell-mediated immunity. Patients suffer from pyogenic bacterial and intracellular infections.

Z

ζ Chain A transmembrane protein expressed in T cells as part of the TCR complex that contains ITAMs in its cytoplasmic tail and binds the ZAP-70 protein tyrosine kinase during T cell activation.

Zeta-associated protein of 70 kD (ZAP-70) A cytoplasmic protein tyrosine kinase, similar to SYK in B cells, that is critical for early signaling steps in antigen-induced T cell activation. ZAP-70 binds to phosphorylated tyrosines in the cytoplasmic tails of the ζ chain and CD3 chains of the TCR complex and in turn phosphorylates adaptor proteins that recruit other components of the signaling cascade.

APPENDIX I

Principal Features of Selected CD Molecules

The following list includes selected CD molecules that are referred to in the text. Many cytokines and cytokine receptors have been assigned CD numbers, but we refer to these by the more descriptive cytokine designation, and these are listed in Appendix II. A complete and up-to-date listing of CD molecules may be found at http://www.hcdm.org.

CD Number (Other Names)	Molecular Structure, Family	Main Cellular Expression	Known or Proposed Function(s)
CD1a–e	35-44 kD; class I MHC-like Ig superfamily; β_2-microglobulin associated	Thymocytes, dendritic cells (including Langerhans cells)	Presentation of nonpeptide (lipid and glycolipid) antigens to some iNKT cells
CD2 (LFA-2)	50 kD; Ig superfamily	T cells, NK cells	Adhesion molecule (binds CD58); T cell activation; cytotoxic T lymphocyte (CTL) and natural killer (NK) cell–mediated lysis
CD3γ (CD3g)	25–28 kD; associated with CD3 δ and CD3ε in TCR complex; Ig superfamily; ITAM in cytoplasmic tail	T cells	Cell surface expression of and signal transduction by the T cell antigen receptor
CD3δ (CD3d)	20 kD; associated with CD3γ and CD3ε in TCR complex; Ig superfamily; ITAM in cytoplasmic tail	T cells	Cell surface expression of and signal transduction by the T cell antigen receptor
CD3ε (CD3e)	23 kD; associated with CD3δ and CD3γ in TCR complex; Ig superfamily; ITAM in cytoplasmic tail	T cells	Cell surface expression of and signal transduction by the T cell antigen receptor
CD4	55 kD; Ig superfamily	Class II MHC-restricted T cells; some macrophages	Coreceptor in class II MHC-restricted antigen-induced T cell activation (binds to class II MHC molecules); thymocyte development; receptor for HIV

Continued

CD Number (Other Names)	Molecular Structure, Family	Main Cellular Expression	Known or Proposed Function(s)
CD5	67 kD; scavenger receptor family	T cells; B-1 B cell subset	Signaling molecule; binds CD72
CD8α (CD8a)	34 kD; expressed as a homodimer or heterodimer with CD8β chain	Class I MHC-restricted T cells; subset of dendritic cells	Coreceptor in class I MHC-restricted antigen-induced T cell activation (binds to class I MHC molecules); thymocyte development
CD8β (CD8b)	34 kD; expressed as a heterodimer with CD8α chain Ig superfamily	Class I MHC-restricted T cells	Same as CD8α
CD10	100 kD; type II membrane protein	Immature and some mature B cells; lymphoid progenitors, granulocytes	Metalloproteinase; unknown function in the immune system
CD11α (LFA-1α chain)	180 kD; noncovalently linked to CD18 to form LFA-1 integrin	Leukocytes	Cell-cell adhesion; binds to ICAM-1 (CD54), ICAM-2 (CD102), and ICAM-3 (CD50)
CD11β (MAC-1; CR3)	165 kD; noncovalently linked to CD18 to form MAC-1 integrin	Granulocytes, monocytes, macrophages, dendritic cells, NK cells	Phagocytosis of iC3b-coated particles; neutrophil and monocyte adhesion to endothelium (binds CD54) and extracellular matrix proteins
CD11c (p150,95; CR4α chain; integrin αX)	145 kD; noncovalently linked to CD18 to form p150,95 integrin	Dendritic cells, monocytes, macrophages, granulocytes, NK cells	Similar functions as CD11b
CD14	53 kD; GPI linked	Dendritic cells, monocytes, macrophages, granulocytes	Binds complex of LPS and LPS-binding protein and displays LPS to TLR4; required for LPS-induced macrophage activation
CD16a (FcγRIIIA)	50–70 kD; transmembrane protein; Ig superfamily	NK cells, macrophages	Binds Fc region of IgG; phagocytosis and antibody-dependent cellular cytotoxicity
CD16b (FcγRIIIB)	50–70 kD; GPI linked; Ig superfamily	Neutrophils	Binds Fc region of IgG; synergizes with Fcγ RII in immune complex–mediated neutrophil activation
CD18	95 kD; noncovalently linked to CD11a, CD11b, or CD11c to form β_2 integrins	Leukocytes	See CD11a, CD11b, CD11c

CD Number (Other Names)	Molecular Structure, Family	Main Cellular Expression	Known or Proposed Function(s)
CD19	95 kD; Ig superfamily	Most B cells	B cell activation; forms a coreceptor complex with CD21 and CD81 that delivers signals activating signals in B cells
CD20	35–37 kD; membrane-spanning 4A (MS4A) family	B cells	Possible role in B cell activation or regulation; calcium ion channel
CD21 (CR2; C3d receptor)	145 kD; regulators of complement activation (RCA) family	Mature B cells, follicular dendritic cells	Receptor for complement fragment C3d; forms a coreceptor complex with CD19 and CD81 that delivers activating signals in B cells; receptor for Epstein-Barr virus
CD22 (Siglec-2)	130–140 kD; Ig superfamily; Siglec family; ITIM in cytoplasmic tail	B cells	Inhibits B cell activation
CD23 (FcεRIIB)	45 kD; C-type lectin	Activated B cells, monocytes, macrophages	Low-affinity Fcε receptor, induced by IL-4; unknown function
CD25 (IL-2 receptor α chain)	55 kD; noncovalently associated with IL-2Rβ (CD122) and IL-2Rγ (CD132) chains to form the high-affinity IL-2 receptor	Activated T and B cells, regulatory T cells (Treg)	Together with CD122 and CD132, binds IL-2 and promotes responses to low concentrations of IL-2; delivers signals required for Treg differentiation and survival
CD28	Homodimer of 44-kD chains; Ig superfamily	T cells (all $CD4^+$ and >50% of $CD8^+$ cells in humans; all mature T cells in mice)	Receptor on T cells for costimulatory molecules CD80 (B7-1) and CD86 (B7-2)
CD29	130 kD; noncovalently linked to CD49a–d chains to form VLA (β_1) integrins	T cells, B cells, monocytes, granulocytes	Leukocyte adhesion to extracellular matrix proteins and endothelium (see CD49)
CD30 (TNFRSF8)	120 kD; TNFR superfamily	Activated T and B cells; NK cells, monocytes, Reed-Sternberg cells in Hodgkin disease	Not established
CD31 (platelet/endothelial cell adhesion molecule 1 [PECAM-1])	130–140 kD; Ig superfamily	Platelets, monocytes, granulocytes, B cells, endothelial cells	Adhesion molecule involved in leukocyte transmigration through endothelium

Continued

CD Number (Other Names)	Molecular Structure, Family	Main Cellular Expression	Known or Proposed Function(s)
CD32 (FcγRII)	40 kD; Ig superfamily; A, B, and C forms are products of different but homologous genes; ITAM in cytoplasmic tail of A form; ITIM in cytoplasmic tail of B form.	B cells, macrophages, dendritic cells, granulocytes	Fc receptor for antigen-complexed IgG; B form acts as inhibitory receptor that blocks activation signals in B cells and other cells
CD34	105–120 kD; sialomucin	Hematopoietic stem and progenitor cells; endothelial cells in high endothelial venules	Marker for hematopoietic stem cells; function not established
CD35 (type 1 complement receptor, CR1)	190–285 kD (four products of polymorphic alleles); regulator of complement activation (RCA) family	Granulocytes, monocytes, erythrocytes, B cells, follicular dendritic cells, some T cells	Binds C3b and C4b; promotes phagocytosis of C3b- or C4b-coated particles and immune complexes; regulates complement activation
CD36	85–90 kD	Platelets, monocytes, macrophages, endothelial cells	Scavenger receptor for oxidized low-density lipoprotein; platelet adhesion; phagocytosis of apoptotic cells
CD40	Homodimer of 44- to 48-kD chains; TNFR superfamily	B cells, macrophages, dendritic cells, endothelial cells	Binds CD154 (CD40L); role in T cell–mediated activation of B cells, macrophages, and dendritic cells
CD44	80–>100 kD, highly glycosylated	Leukocytes, erythrocytes	Binds hyaluronan; involved in leukocyte adhesion to endothelial cells and extracellular matrix
CD45 (leukocyte common antigen [LCA], B220)	Multiple isoforms, 180–220 kD (see CD45R); protein tyrosine phosphatase receptor family; fibronectin type III family	Hematopoietic cells	Tyrosine phosphatase that regulates T and B cell activation
CD45R	CD45RO: 180 kD CD45RA: 220 kD CD45RB: 190, 205, and 220 kD isoforms	CD45RO: memory T cells; subset of B cells, monocytes, macrophages CD45RA: naive T cells, B cells, monocytes CD45RB: B cells, subset of T cells	See CD45

CD Number (Other Names)	Molecular Structure, Family	Main Cellular Expression	Known or Proposed Function(s)
CD46 (membrane cofactor protein [MCP])	52–58 kD; regulators of complement activation (RCA) family	Leukocytes, epithelial cells, fibroblasts	Regulation of complement activation
CD47	47–52 kD; Ig superfamily	All hematopoietic cells, epithelial cells, endothelial cells, fibroblasts	Leukocyte adhesion, migration, activation; ligand for signal regulatory protein α (SIRPα); "don't eat me" signal to phagocytes
CD49d	150 kD; noncovalently linked to CD29 to form VLA-4 ($\alpha_4\beta_1$ integrin)	T cells, monocytes, B cells, NK cells, eosinophils, dendritic cells, thymocytes	Leukocyte adhesion to endothelium and extracellular matrix; binds to VCAM-1 and MadCAM-1; binds fibronectin and collagens
CD54 (ICAM-1)	75–114 kD; Ig superfamily	T cells, B cells, monocytes, endothelial cells (cytokine inducible)	Cell-cell adhesion; ligand for CD11aCD18 (LFA-1) and CD11bCD18 (Mac-1); receptor for rhinovirus
CD55 (decay-accelerating factor [DAF])	55–70 kD; GPI linked; regulators of complement activation (RCA) family	Broad	Regulation of complement activation
CD58 (Leukocyte function –associated antigen 3 [LFA-3])	55–70 kD; GPI-linked or integral membrane protein	Broad	Leukocyte adhesion; binds CD2
CD59	18–20 kD; GPI linked	Broad	Binds C9; inhibits formation of complement membrane attack complex
CD62E (E-selectin)	115 kD; selectin family	Endothelial cells	Leukocyte-endothelial adhesion
CD62L (L-selectin)	74–95 kD; selectin family	B cells, T cells, monocytes, granulocytes, some NK cells	Leukocyte-endothelial adhesion; homing of naive T and B cells to lymph nodes
CD62P (P-selectin)	140 kD; selectin family	Platelets, endothelial cells (present in granules, translocated to cell surface on activation)	Leukocyte adhesion to endothelium, platelets; binds CD162 (PSGL-1)
CD64 (FcγRI)	72 kD; Ig superfamily; noncovalently associated with the FcR common γ chain	Monocytes, macrophages, activated neutrophils	High-affinity Fcγ receptor; role in phagocytosis, ADCC, macrophage activation
CD66e (carcinoembryonic antigen	180–220 kD; Ig superfamily; CEA	Colonic and other epithelial cells	? Adhesion; clinical marker of carcinoma burden

Continued

CD Number (Other Names)	Molecular Structure, Family	Main Cellular Expression	Known or Proposed Function(s)
CD69	23 kD; C-type lectin	Activated B cells, T cells, NK cells, neutrophils	Binds to and reduces surface expression of S1PR1, thereby promoting retention of recently activated lymphocytes in lymphoid organs
CD74 (Class II MHC invariant chain [I_i])	33-, 35-, and 41-kD isoforms	B cells, dendritic cells, monocytes, macrophages; other class II MHC-expressing cells	Binds to and directs intracellular sorting of newly synthesized class II MHC molecules
CD79a (Igα)	33, 45 kD; forms dimer with CD79b; Ig superfamily; ITAM in cytoplasmic tail	Mature B cells	Required for cell surface expression of and signal transduction by the B cell antigen receptor complex
CD79b (Igβ)	37–39 kD; forms dimer with CD79α; Ig superfamily; ITAM in cytoplasmic tail	Mature B cells	Required for cell surface expression of and signal transduction by the B cell antigen receptor complex
CD80 (B7-1)	60 kD; Ig superfamily	Dendritic cells, activated B cells and macrophages	Costimulator for T lymphocyte activation; ligand for CD28 and CD152 (CTLA-4)
CD81 (target for antiproliferative antigen 1 [TAPA-1], TSPAN-28)	26 kD; tetraspanin family (TM4SF)	T cells, B cells, NK cells, dendritic cells, thymocytes, endothelial cells	B cell activation; forms a coreceptor complex with CD19 and CD21 that delivers activating signals in B cells
CD86 (B7-2)	80 kD; Ig superfamily	B cells, monocytes; dendritic cells; some T cells	Costimulator for T lymphocyte activation; ligand for CD28 and CD152 (CTLA-4)
CD88 (C5a receptor)	43 kD; G protein–coupled, seven membrane-spanning receptor family	Granulocytes, monocytes, dendritic cells, mast cells	Receptor for C5a complement fragment; role in complement-induced inflammation
CD89 (Fcα receptor)	55–75 kD; Ig superfamily; noncovalently associated with the common FcRγ chain	Granulocytes, monocytes, macrophages, T cell subset, B cell subset	Binds IgA; mediates IgA-dependent cellular cytotoxicity
CD90 (Thy-1)	25–35 kD; GPI linked; Ig superfamily	Thymocytes, mature T cells (mice), $CD34^+$ hematopoietic progenitor cells, neurons	Marker for T cells; unknown function
CD94	43 kD; C-type lectin; on NK cells, covalently assembles with other C-type lectin molecules (NKG2)	NK cells; subset of $CD8^+$ T cells	CD94/NKG2 complex functions as an NK cell inhibitory receptor; binds human leukocyte antigen E (HLA-E) class I MHC molecules

CD Number (Other Names)	Molecular Structure, Family	Main Cellular Expression	Known or Proposed Function(s)
CD95 (FAS)	Homotrimer of 45-kD chains; TNFR superfamily	Broad	Binds FAS ligand; delivers signals leading to apoptotic death
CD102 (ICAM-2)	55–65 kD; Ig superfamily	Endothelial cells, lymphocytes, monocytes, platelets	Ligand for CD11aCD18 (LFA-1); cell–cell adhesion
CD103 (α_E integrin subunit)	Dimer of 150- and 25-kD subunits; noncovalently linked to β_7 integrin subunit to form $\alpha_E\beta_7$ integrin	Intraepithelial lymphocytes, other cell types	Role in T cell homing to and retention in mucosa; binds E-cadherin
CD106 (VCAM-1)	100–110 kD; Ig superfamily	Endothelial cells, macrophages, follicular dendritic cells, marrow stromal cells	Adhesion of cells to endothelium; receptor for CD49dCD29 (VLA-4) integrin; role in lymphocyte trafficking
CD134 (OX40, TNFRSF4)	29 kD; TNFR superfamily	Activated T cells	Receptor for T cell CD252; T cell costimulation
CD141 (BDCA-3, thrombomodulin)	60 kD; EGF-like domains	Cross-presenting dendritic cells, monocytes, endothelial cells	Binds thrombin and prevents blood coagulation
CD150 (signaling lymphocyte activation molecule [SLAM])	37 kD; Ig superfamily	Thymocytes, activated lymphocytes, dendritic cells, endothelial cells	Regulation of B cell–T cell interactions and lymphocyte activation
CD152 (cytotoxic T lymphocyte–associated protein 4 [CTLA-4])	33, 50 kD; Ig superfamily	Activated T lymphocytes, regulatory T cells	Mediates suppressive function of regulatory T cells; inhibits T cell responses; binds and internalizes CD80 (B7-1) and CD86 (B7-2) on antigen-presenting cells
CD154 (CD40 ligand [CD40L])	Homotrimer of 32- to 39-kD chains; TNF superfamily	Activated CD4$^+$ T cells	Activation of B cells, macrophages, and endothelial cells; binds to CD40
CD158 (killer Ig-like receptor [KIR])	50, 58 kD; Ig superfamily; KIR family; ITIMs or ITAMs in cytoplasmic tail	NK cells, T cell subset	Inhibition or activation of NK cells on interaction with appropriate class I HLA molecules
CD159a (NKG2A)	43 kD; C-type lectin; ITIM in cytoplasmic tail; forms heterodimer with CD94	NK cells, T cell subset	Inhibition or activation of NK cells on interaction with class I HLA molecules
CD159c (NKG2C)	40 kD; C-type lectin; forms heterodimer with CD94	NK cells	Activation of NK cells on interaction with the appropriate class I HLA molecules

Continued

CD Number (Other Names)	Molecular Structure, Family	Main Cellular Expression	Known or Proposed Function(s)
CD162 (P-selectin glycoprotein ligand 1 [PSGL-1])	Homodimer of 120-kD chains; sialomucin	T cells, monocytes, granulocytes, some B cells	Ligand for selectins (CD62P, CD62L); adhesion of leukocytes to endothelium
CD178 (FAS ligand [FASL])	Homotrimer of 31-kD subunits; TNF superfamily	Activated T cells	Ligand for CD95 (FAS); triggers apoptotic death
CD206 (mannose receptor)	166 kD; C-type lectin	Macrophages	Binds high-mannose-containing glycoproteins on pathogens; mediates macrophage endocytosis of glycoproteins and phagocytosis of bacteria, fungi, and other pathogens
CD223 (lymphocyte-activation gene 3 [LAG3])	57.4 kD; Ig superfamily	T cells, NK cells, B cells, plasmacytoid DCs	Binds class II MHC; inhibits T cell activation
CD244 (2B4)	41 kD; Ig superfamily; CD2/CD48/CD58 family; SLAM family	NK cells, CD8 T cells, $\gamma\delta$ T cells	Receptor for CD148; modulates NK cell cytotoxic activity
CD247 (TCR ζ chain)	18 kD; ITAMs in cytoplasmic tail	T cells; NK cells	Signaling chain of TCR complex— and NK cell—activating receptors
CD252 (OX40 ligand)	21 kD; TNF superfamily	Dendritic cells, macrophages, B cells	Ligand for CD134 (OX40, TNFRSF4); costimulates T cells
CD267 (TACI)	31 kD; TNFR superfamily	B cells	Receptor for cytokines BAFF and APRIL; promotes T-independent B cell responses and B cell survival
CD268 (BAFF receptor)	19 kD; TNFR superfamily	B cells	Receptor for BAFF; promotes B cell survival
CD269 (B cell maturation antigen [BCMA])	20 kD; TNFR superfamily	B cells	Receptor for BAFF and APRIL; promotes plasma cell survival
CD273 (PD-L2)	25 kD; Ig superfamily; structurally homologous to B7	Dendritic cells, monocytes, macrophages	Ligand for PD-1; inhibits T cell activation
CD274 (PD-L1)	33 kD; Ig superfamily; structurally homologous to B7	Leukocytes, other cells	Ligand for PD-1; inhibits T cell activation
CD275 (ICOS ligand)	60 kD; Ig superfamily; structurally homologous to B7	B cells, dendritic cells, monocytes	Binds ICOS (CD278); T cell costimulation

CD Number (Other Names)	Molecular Structure, Family	Main Cellular Expression	Known or Proposed Function(s)
CD278 (inducible costimulator [ICOS])	55–60 kD; Ig superfamily; structurally homologous to CD28	Activated T cells	Binds ICOS-L (CD275); T cell costimulation and follicular T helper cell differentiation
CD279 (PD-1)	55 kD; Ig superfamily; structurally homologous to CD28; ITIM and ITSM in cytoplasmic tail	Activated T and B cells	Binds PD-L1 and PD-L2; inhibits T cell activation
CD303 (BDCA2, C-type lectin domain family 4 member C [CLEC4C])	25 kD; C-type lectin superfamily	Plasmacytoid dendritic cells	Binds to microbial carbohydrates; inhibits dendritic cell activation
CD304 (BDCA4, Neuropilin)	103 kD; complement-binding, coagulation factor V/VIII, and meprin domains	Plasmacytoid dendritic cells, many other cell types	Vascular endothelial growth factor A receptor
CD314 (NKG2D)	42 kD; C-type lectin	NK cells, activated $CD8^+$ T cells, NK-T cells, some myeloid cells	Binds MHC class I, and the class I–like molecules MIC-A, MIC-B, Rae1, and ULBP4; role in NK cell and CTL activation
CD357 (GITR, TNFRSF18)	26 kD; TNFR superfamily	$CD4^+$ and $CD8^+$ T cells, Treg	? Role in T cell/Treg function
CD363 (type 1 sphingosine-1-phosphate receptor 1 [S1PR1])	42.8 kD; G protein–coupled, seven membrane-spanning receptor family	Lymphocytes, endothelial cells	Binds sphingosine 1-phosphate and mediates chemotaxis of lymphocytes out of lymphoid organs
CD365 (hepatitis A virus cellular receptor 1 [HAVCR1], TIM-1)	38.7 kD; Ig superfamily, T cell transmembrane, Ig, and mucin family	T cells, kidney and testis	Receptor for several viruses
CD366 (hepatitis A virus cellular receptor 2 [HAVCR2], TIM-3)	33.4 kD; Ig superfamily, Ig superfamily, T cell transmembrane, Ig, and mucin family	T cells, macrophages, dendritic cells, NK cells	Receptor for several viruses; binds phosphatidylserine on apoptotic cells; inhibits T cell responses
CD369 (CLEC7A, DECTIN 1)	27.6 kD; C-type lectin family	Dendritic cells, monocytes, macrophages, B cells	Pattern recognition receptor specific for fungal and bacterial cell wall glucans

The lowercase letters affixed to some CD numbers refer to CD molecules that are encoded by multiple genes or that belong to families of structurally related proteins.

ADCC, Antibody-dependent cell-mediated cytotoxicity; *APRIL,* a proliferation-inducing ligand; *BAFF,* B cell–activating factor belonging to the TNF family; *CTL,* cytotoxic T lymphocyte; *gp,* glycoprotein; *GITR,* glucocorticoid-induced TNFR-related; *GPI,* glycophosphatidylinositol; *ICAM,* intercellular adhesion molecule; *Ig,* immunoglobulin; *IL,* interleukin; *ITAM,* immunoreceptor tyrosine-based activation motif; *ITIM,* immunoreceptor tyrosine-based inhibition motif; *ITSM,* immunoreceptor tyrosine-based switch motif; *LFA,* lymphocyte function–associated antigen; *LPS,* lipopolysaccharide; *MadCAM,* mucosal addressin cell adhesion molecule; *MHC,* major histocompatibility complex; *NK* cells, natural killer cells; *TACI,* transmembrane activator and CAML interactor; *TCR,* T cell receptor; *TLR,* Toll-like receptor; *TNF,* tumor necrosis factor; *TNFR,* TNF receptor; *VCAM,* vascular cell adhesion molecule; *VLA,* very late activation.

II APPENDIX

Cytokines

Cytokine and Subunits	Principal Cell Source	Cytokine Receptor and Subunits*	Principal Cellular Targets and Biologic Effects
Type I Cytokine Family Members			
Interleukin-2 (IL-2)	T cells	CD25 (IL-2Rα) CD122 (IL-2Rβ) CD132 (γc)	*T cells:* proliferation and differentiation into effector and memory cells; regulatory T cell development, survival, and function *NK cells:* proliferation, activation
Interleukin-3 (IL-3)	T cells	CD123 (IL-3Rα) CD131 (βc)	*Immature hematopoietic progenitors:* maturation of all hematopoietic lineages
Interleukin-4 (IL-4)	CD4$^+$ T cells (Th2, Tfh), mast cells	CD124 (IL-4Rα) CD132 (γc)	*B cells:* isotype switching to IgE, IgG4 (in humans; IgG1 in mice) *T cells:* Th2 differentiation, proliferation *Macrophages:* alternative activation and inhibition of IFN-γ–mediated classical activation
Interleukin-5 (IL-5)	CD4$^+$ T cells (Th2), group 2 ILCs	CD125 (IL-5Rα) CD131 (βc)	*Eosinophils:* activation, increased generation
Interleukin-6 (IL-6)	Macrophages, endothelial cells, T cells, fibroblasts	CD126 (IL-6Rα) CD130 (gp130)	*Liver:* synthesis of acute phase proteins *B cells:* proliferation of antibody-producing cells *T cells:* Th17 differentiation

Continued

Cytokine and Subunits	Principal Cell Source	Cytokine Receptor and Subunits*	Principal Cellular Targets and Biologic Effects
Interleukin-7 (IL-7)	Fibroblasts, bone marrow stromal cells	CD127 (IL-7R) CD132 (γc)	*Immature lymphoid progenitors:* proliferation of early T and B cell progenitors *T lymphocytes:* survival of naive and memory cells
Interleukin-9 (IL-9)	$CD4^+$ T cells	CD129 (IL-9R) CD132 (γc)	*Mast cells, B cells, T cells, and epithelial cells:* survival and activation
Interleukin-11 (IL-11)	Bone marrow stromal cells	IL-11Rα CD130 (gp130)	Production of platelets
Interleukin-12 (IL-12): IL-12A (p35) IL-12B (p40)	Macrophages, dendritic cells	CD212 (IL-12Rβ1) IL-12Rβ2	*T cells:* Th1 differentiation *NK cells and T cells:* IFN-γ synthesis, increased cytotoxic activity
Interleukin-13 (IL-13)	$CD4^+$ T cells (Th2), NKT cells, group 2 ILCs, mast cells	CD213a1 (IL-13Rα1) CD213a2 (IL-13Rα2) CD132 (γc)	*B cells:* isotype switching to IgE *Epithelial cells:* increased mucus production *Macrophages:* alternative activation
Interleukin-15 (IL-15)	Macrophages, other cell types	IL-15Rα CD122 (IL-2Rβ) CD132 (γc)	*NK cells:* proliferation *T cells:* survival and proliferation of memory $CD8^+$ cells
Interleukin-17A (IL-17A) Interleukin-17F (IL-17F)	$CD4^+$ T cells (Th17), group 3 ILCs	CD217 (IL-17RA) IL-17RC	*Epithelial cells, macrophages and other cell types:* increased chemokine and cytokine production; GM-CSF and G-CSF production
Interleukin-21 (IL-21)	Tfh cells	CD360 (IL-21R) CD132 (γc)	*B cells:* activation, proliferation, differentiation
Interleukin-23 (IL-23): IL-23A (p19) IL-12B (p40)	Macrophages, dendritic cells	IL-23R CD212 (IL-12Rβ1)	*T cells:* differentiation and proliferation of Th17 cells
Interleukin-25 (IL-25; IL-17E)	T cells, dendritic cells, macrophages, epithelial cells, mast cells, eosinophils	IL-17RB	*T cells:* Th2 differentiation *ILCs:* ILC2 activation

Continued

Cytokine and Subunits	Principal Cell Source	Cytokine Receptor and Subunits*	Principal Cellular Targets and Biologic Effects
Interleukin-27 (IL-27): IL-27 (p28), EBI-3	Macrophages, dendritic cells	IL-27Rα CD130 (gp130)	*T cells:* enhancement of Th1 differentiation; inhibition of Th17 differentiation *NK cells:* IFN-γ synthesis?
Stem cell factor (c-Kit ligand)	Bone marrow stromal cells	CD117 (KIT)	*Pluripotent hematopoietic stem cells:* maturation of all hematopoietic lineages
Granulocyte-monocyte CSF (GM-CSF)	T cells, macrophages, endothelial cells, fibroblasts	CD116 (GM-CSFRα) CD131 (βc)	*Immature and committed progenitors, mature macrophages:* maturation of granulocytes and monocytes, macrophage activation
Monocyte CSF (M-CSF, CSF1)	Macrophages, endothelial cells, bone marrow endothelial cells, fibroblasts	CD115 (CSF1R)	*Committed hematopoietic progenitors:* maturation of monocytes
Granulocyte CSF (G-CSF, CSF3)	Macrophages, fibroblasts, endothelial cells	CD114 (CSF3R)	*Committed hematopoietic progenitors:* maturation of granulocytes
Thymic stromal lymphopoietin (TSLP)	Keratinocytes, bronchial epithelial cells, fibroblasts, smooth muscle cells, endothelial cells, mast cells, macrophages, granulocytes and dendritic cells	TSLP-receptor CD127 (IL-7R)	*T cells:* Th2 differentiation *ILCs:* ILC2 activation *Dendritic cells:* activation *Eosinophils:* activation *Mast cells:* cytokine production
Type II Cytokine Family Members			
Interferon-α (IFN-α, type I IFN) (multiple proteins)	Plasmacytoid dendritic cells, macrophages	IFNAR1 CD118 (IFNAR2)	*All cells:* antiviral state, increased class I MHC expression *NK cells:* activation
Interferon-β (IFN-β, type I IFN)	Fibroblasts, plasmacytoid dendritic cells	IFNAR1 CD118 (IFNAR2)	*All cells:* antiviral state, increased class I MHC expression *NK cells:* activation
Interferon-γ (IFN-γ, type II IFN)	T cells (Th1, $CD8^+$ T cells), NK cells, group 1 ILCs	CD119 (IFNGR1) IFNGR2	*Macrophages:* classical activation (increased microbicidal functions)

Continued

Cytokine and Subunits	Principal Cell Source	Cytokine Receptor and Subunits*	Principal Cellular Targets and Biologic Effects
			B cells: isotype switching to opsonizing and complement-fixing IgG subclasses (established in mice, not humans) *T cells:* Th1 differentiation *Various cells:* increased expression of class I and class II MHC molecules, increased antigen processing and presentation to T cells
Interleukin-10 (IL-10)	Macrophages, T cells (mainly regulatory T cells)	CD210 (IL-10Rα) IL-10Rβ	*Macrophages, dendritic cells:* inhibition of expression of IL-12, costimulators, and class II MHC
Interleukin-22 (IL-22)	Th17 cells, group 3 ILCs	IL-22Rα1 *or* IL-22Rα2 IL-10Rα2	*Epithelial cells:* production of defensins, increased barrier function *Hepatocytes:* survival
Interferon-λs (IFN-λa; type III IFNs)	Dendritic cells	IFNLR1 (IL-28Rα) CD210B (IL-10Rβ2)	*Epithelial cells:* antiviral state
Leukemia inhibitory factor (LIF)	Embryonic trophectoderm, bone marrow stromal cells	CD118 (LIFR) CD130 (gp130)	*Stem cells:* block in differentiation
Oncostatin M	Bone marrow stromal cells	OSMR CD130 (gp130)	*Endothelial cells:* upregulation of cytokine and adhesion molecule expression *Intestinal stromal cells:* production of inflammatory cytokines, chemokines
TNF Superfamily Cytokines[†]			
APRIL (CD256, TNFSF13)	T cells, dendritic cells, monocytes, follicular dendritic cells	TACI (TNFRSF13B) *or* BCMA (TNFRSF17)	*B cells:* survival, proliferation
BAFF (CD257, TNFSF13B)	Dendritic cells, monocytes, follicular dendritic cells, B cells	BAFF-R (TNFRSF13C) *or* TACI (TNFRSF13B) *or* BCMA (TNFRSF17)	*B cells:* survival, proliferation

Continued

Cytokine and Subunits	Principal Cell Source	Cytokine Receptor and Subunits*	Principal Cellular Targets and Biologic Effects
Lymphotoxin-α (LTα, TNFβ, TNFSF1)	T cells, B cells	CD120a (TNFRSF1) or CD120b (TNFRSF2)	Same as TNF
Lymphotoxin-αβ (LTαβ)	T cells, NK cells, follicular B cells, lymphoid inducer cells	LTβR	*Lymphoid tissue stromal cells and follicular dendritic cells:* chemokine expression and lymphoid organogenesis
Tumor necrosis factor (TNF, TNFα TNFSF2)	Macrophages, NK cells, T cells	CD120a (TNFRSF1) *or* CD120b (TNFRSF2)	*Vascular endothelium:* activation (inflammation, coagulation), increased permeability *Neutrophils:* activation *Hypothalamus:* fever *Muscle, fat:* catabolism (cachexia) *Heart:* reduced cardiac output
Osteoprotegerin (OPG, TNFRSF11B)	Osteoblasts	RANKL	*Osteoclast precursor cells:* inhibits osteoclast differentiation
IL-1 Family Cytokines			
Interleukin-1α (IL-1α)	Macrophages, dendritic cells, fibroblasts, endothelial cells, keratinocytes, hepatocytes	CD121a (IL-1R1) IL-1RAP *or* CD121b (IL-1R2)	*Endothelial cells:* activation (inflammation, coagulation) *Hypothalamus:* fever
Interleukin-1β (IL-1β)	Macrophages, dendritic cells, fibroblasts, endothelial cells, keratinocytes; major type of biologically active IL-1	CD121a (IL-1R1) IL-1RAP *or* CD121b (IL-1R2)	*Endothelial cells:* activation (inflammation, coagulation) *Hypothalamus:* fever *Liver:* synthesis of acute-phase proteins *T cells:* Th17 differentiation
Interleukin-1 receptor antagonist (IL-1RA)	Macrophages	CD121a (IL-1R1) IL-1RAP	*Various cells:* competitive antagonist of IL-1
Interleukin-18 (IL-18)	Monocytes, macrophages, dendritic cells, Kupffer cells, keratinocytes, chondrocytes, synovial fibroblasts, osteoblasts	CD218a (IL-18Rα) CD218b (IL-18Rβ)	*NK cells and T cells:* IFN-γ synthesis *Monocytes:* expression of GM-CSF, TNF, IL-1β *Neutrophils:* activation, cytokine release
Interleukin-33 (IL-33)	Epithelial cells, endothelial cells, dendritic cells, smooth muscle cells, fibroblasts	ST2 (IL1RL1) IL-1 receptor accessory protein (IL1RAP)	*T cells:* Th2 differentiation *ILCs:* ILC2 activation

Continued

Cytokine and Subunits	Principal Cell Source	Cytokine Receptor and Subunits*	Principal Cellular Targets and Biologic Effects
Other Cytokines			
Transforming growth factor-β (TGF-β)	T cells (mainly Tregs), macrophages, other cell types	TGF-β R1 TGF-β R2 TGF-β R3	*T cells:* inhibition of proliferation and effector functions; differentiation of Th17 and Treg *B cells:* inhibition of proliferation; IgA production *Macrophages:* inhibition of activation; stimulation of angiogenic factors *Fibroblasts:* increased collagen synthesis

*Most cytokine receptors are dimers or trimers composed of different polypeptide chains, some of which are shared between receptors for different cytokines. The set of polypeptides that compose a functional receptor (cytokine binding plus signaling) for each cytokine is listed. The functions of each subunit polypeptide are not listed.

†All TNF superfamily (TNFSF) members are expressed as cell surface transmembrane proteins, but only the subsets that are predominantly active as proteolytically released soluble cytokines are listed in the table. Other TNFSF members that function predominantly in the membrane-bound form and are not, strictly speaking, cytokines are not listed in the table. These membrane-bound proteins and the TNFRSF receptors they bind to include OX40L (CD252, TNFSF4):OX40 (CD134, TNFRSF4); CD40L (CD154, TNFSF5):CD40 (TNFRSF5); FasL (CD178, TNFSF6):Fas (CD95, TNFRSF6); CD70 (TNFSF7):CD27 (TNFRSF27); CD153 (TNFSF8):CD30 (TNFRSF8); TRAIL (CD253, TNFSF10):TRAIL-R (TNFRSF10A-D); RANKL (TNFSF11):RANK (TNFRSF11); TWEAK (CD257, TNFSF12):TWEAKR (CD266, TNFRSF12); LIGHT (CD258, TNFSF14):HVEM (TNFRSF14); GITRL (TNFSF18):GITR (CD357 TNFRSF18); and 4-IBBL:4-IBB (CD137).

APRIL, A proliferation-inducing ligand; *BAFF,* B cell–activating factor belonging to the TNF family; *BCMA,* B cell maturation protein; *CSF,* colony-stimulating factor; *IFN,* interferon; *ILCs,* innate lymphoid cells; *MHC,* major histocompatibility complex; *NK cell,* natural killer cell; *NKT cell,* natural killer T cell; *OSMR,* oncostatin M receptor; *RANK,* receptor activator of nuclear factor κB; *RANKL,* RANK ligand; *TACI,* transmembrane activator and calcium modulator and cyclophilin ligand interactor; *Th,* T helper; *Tfh,* T follicular helper; *TNF,* tumor necrosis factor; *TNFSF,* TNF superfamily; *TNFRSF,* TNF receptor superfamily; *Treg,* regulatory T cell.

APPENDIX

Clinical Cases

This appendix presents six clinical cases illustrating various diseases involving the immune system. These cases are not meant to teach clinical skills but rather to show how the basic science of immunology contributes to our understanding of human diseases. Each case illustrates typical ways in which a disease manifests, what tests are used in diagnosis, and common modes of treatment. The appendix was compiled with the assistance of Dr. Richard Mitchell and Dr. Jon Aster, Department of Pathology, Brigham and Women's Hospital, Boston; Dr. Robin Colgrove, Harvard Medical School, Boston; Dr. George Tsokos, Department of Medicine, Beth Israel-Deaconess Hospital, Boston; Dr. David Erle and Dr. Laurence Cheng, Department of Medicine, University of California San Francisco; Dr. Caroline Sokol, Dr. Zachary Wallace, Dr. Seth Bloom, and Dr Jonathan Hermann, Massachusetts General Hospital, Boston

CASE 1: LYMPHOMA

E.B. was a 58-year-old chemical engineer who had been well all his life. One morning, he noticed a lump in his left groin while showering. It was not tender, and the overlying skin appeared healthy. After a few weeks, he began to worry about it because it did not go away, and he finally made an appointment with a physician after 2 months. On physical examination, the physician noted a subcutaneous firm, movable nodule, approximately 3 cm in diameter, in the left inguinal region. The physician asked E.B. if he had recently noticed any infections of his left foot or leg; E.B. had not. E.B. did complain that he had been waking up frequently at night drenched in perspiration. The physician also found some slightly enlarged lymph nodes in E.B.'s right neck. Otherwise, the physical examination findings were normal. The physician explained that the inguinal mass probably was a lymph node that was enlarged as a result of a reaction to some infection. However, he drew blood for tests and referred E.B. to a clinic, where a fine-needle aspiration of cells from the lymph node was performed. Examination of smears prepared from aspirated cells revealed mainly small, lymphocytes. Flow cytometric evaluation of these cells showed a 10-fold excess of B cells expressing λ immunoglobulin (Ig) light chain compared with B cells expressing κ Ig light chain.

Because of the suspicion of B cell lymphoma, a malignant tumor of cells of the B lymphocyte lineage, the surgeon elected to remove the entire lymph node. Histologic examination revealed an expansion of the node by follicular structures composed of mainly small- to intermediate-sized lymphocytes with irregular or "cleaved" nuclear contours mixed with smaller numbers of large lymphocytes with prominent nucleoli (Fig. A.1). Flow cytometric analysis of these cells showed a predominant population of B cells expressing IgM, λ light chain, CD10, and CD20, and immunohistochemical stains performed on slides showed strong cytoplasmic staining for BCL-2. On this basis, the diagnosis of follicular lymphoma of low histologic grade was made.

1. Why does the presence of a B cell population in which a large majority of the cells express λ light chain indicate a neoplasm rather than a response to an infection?
2. If the lymph node cells were analyzed by polymerase chain reaction (PCR) to assess Ig heavy-chain gene rearrangements, what abnormal finding would you expect?
3. Normal follicular B cells fail to express the BCL-2 protein. Why might the tumor cells express BCL-2?

E.B.'s blood tests indicated that he was anemic (low red blood cell count). He underwent staging tests to determine the extent of his lymphoma. Positron emission tomography (PET) and computed tomography

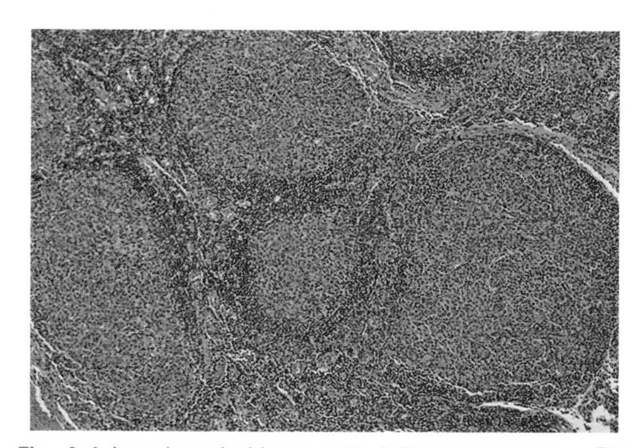

Fig. A.1 Lymph node biopsy with follicular lymphoma. The microscopic appearance of the patient's inguinal lymph node is shown. The follicular structures are abnormal, composed of a monotonous collection of neoplastic cells. By contrast, a lymph node with reactive hyperplasia would have follicles with germinal center formation, containing a heterogeneous mixture of cells.

(CT) scanning showed enlarged hilar and mediastinal lymph nodes, an enlarged spleen, and lesions in the liver. A bone marrow biopsy also showed presence of lymphoma. E.B. was treated with injections of a mouse/human chimeric monoclonal IgG antibody called rituximab, which is specific for human CD20. Imaging studies performed 6 months after the rituximab treatment was begun showed regression in the size of lesions, and E.B. felt well enough to continue working.

4. By what mechanisms would the anti-CD20 antibody help this patient?
5. What are the advantages of using a "humanized" antibody, such as rituximab, as a drug instead of a mouse antibody?

Answers to Questions for Case 1

1. During the maturation of B cells, the cells first express a rearranged μ heavy chain gene, which associates with the surrogate light chain to produce the pre–B cell receptor (see Chapter 4). The cells then rearrange a light chain gene: first κ, then λ. If the κ protein is produced in any B cell, the λ gene does not rearrange; λ rearrangement occurs only if the κ rearrangement is unsuccessful or if the assembled Ig molecule is strongly self-reactive. So, any B cell can produce only one of the two light chains. In humans, about 50% to 60% of the mature B cells express κ and 40% to 50% express λ. In a polyclonal response to an infection or other stimulus, many B cells respond and this ratio is maintained. However, if there is a marked overrepresentation of one light chain (in this case, λ), it usually indicates that a λ-producing B cell clone has proliferated. This is characteristic of a B cell tumor (lymphoma), which arises from a single B cell.
2. Each clone of B cells has a unique rearrangement of V, (D) and J gene segments, forming the genes that encode V regions of heavy and light chains. B cell lymphomas are monoclonal, being composed of cells that all contain the same Ig heavy-chain and light-chain gene rearrangements. Such tumors can be reliably distinguished by the use of PCR amplification of rearranged Ig heavy-chain (IgH) gene segments. This method uses consensus PCR primers that hybridize with virtually all IgH variable (V) gene segments and joining (J) gene segments, allowing these primers to amplify essentially all heavy-chain gene rearrangements in a sample (e.g., DNA prepared from enlarged lymph node). The size of the amplified products is then analyzed by capillary electrophoresis, which can separate PCR products that differ in size by as little as a single nucleotide. When the V, D, and J segments of IgH genes (as well as other antigen receptor genes) are joined during antigen receptor rearrangement in pre–B cells, the rearranged segments are of differing length in each cell due to the action of enzymes that remove nucleotides (nucleases) and add bases (a specialized DNA polymerase called terminal deoxyribonucleotide transferase [TdT]). Within a normal population of B cells, fragments of differing size are generated by PCR with consensus IgH primers. By contrast, in the case of a B cell lymphoma, all the B cells have the same VDJ rearrangement, and one or two PCR products (if both IgH alleles are rearranged) are preferentially amplified, each appearing as a sharp peak of a particular size.
3. Many lymphomas have characteristic underlying acquired chromosomal translocations or mutations that dysregulate specific oncogenes. More than 90% of follicular lymphomas have an acquired 14;18 chromosomal translocation that brings the coding sequence of *BCL2*, a gene on chromosome 18 encoding a protein that inhibits programmed cell death (apoptosis), adjacent to a transcriptional enhancer within the Ig heavy chain gene locus

located on chromosome 14. As a result, BCL-2 is overexpressed in follicular lymphoma cells. In most instances the chromosomal breakpoint in the IgH gene involved in the translocation is located precisely at the point where RAG proteins normally cut the DNA of pre-B cells that are undergoing Ig gene rearrangement, suggesting that the translocation stems from a mistake that occurs during normal antigen receptor gene rearrangement. Clinically, the presence of a *BCL-2/IgH* fusion gene, the consequence of the t(14;18), may be determined by fluorescent in situ hybridization using probes of different colors that are specific for *IgH* and *BCL-2*. These probes are hybridized to prepared sections of tissues involved by follicular lymphoma, and spatial superimposition of the probes within the nuclei of tumor cells indicates the existence of an *IgH/BCL-2* fusion gene. Alternatively, it is possible to perform PCR on DNA isolated from the tumor with primer pairs in which one primer is specific for *IgH* and the other specific for *BCL-2*. These primers will generate a product only when the *IgH* and *BCL-2* genes are joined to one another, which is taken as indirect evidence of a t(14;18).

4. CD20 is expressed on most mature B cells and is also uniformly expressed by all the tumor cells in follicular lymphomas. Injected rituximab (Rituxan) will therefore bind to the lymphoma cells and facilitate their destruction, likely through similar mechanisms by which antibodies normally destroy microbes. These mechanisms involve binding of the Fc portion of rituximab to different proteins in the patient, including Fc receptors on natural killer cells, leading to cytotoxic killing of the lymphoma cells, and to complement proteins leading to complement-mediated killing of the lymphoma cells (see Chapter 8). Most normal B cells will also be depleted by rituximab, although antibody-secreting plasma cells, which do not express CD20, are not affected. If necessary, the immune deficiency caused by loss of normal B cells can be corrected by administration of pooled IgG from healthy donors, a form of passive immunity.
5. Monoclonal antibodies (mAbs) derived from nonhuman B cells (e.g., mouse) will appear foreign to the human immune system. When injected multiple times with these mAbs, humans will mount humoral immune responses and produce antibodies specific for the injected foreign mAb. These anti-antibody responses will promote clearance of the mAb from the circulation and therefore reduce the therapeutic benefits of the mAb. Furthermore, the Fc regions of human IgG bind better than mouse IgG to human Fc receptors and complement proteins, both of which are important for the effectiveness of mAb drugs (see Answer 3). For these reasons, most recently developed mAbs used as drugs have been genetically engineered to contain mainly or all human Ig amino acid sequences. Patients will generally not react against these drugs, just as they do not respond to their own antibodies. Rituximab is a chimeric mAb, with the CD20-binding variable regions originating from mouse IgG, and the remainder of the antibody including the Fc region from human IgG. The small amount of mouse sequences in rituximab do not appear to induce anti-antibody responses in patients, perhaps because potentially responding B cells are destroyed by the drug.

CASE 2: HEART TRANSPLANTATION COMPLICATED BY ALLOGRAFT REJECTION

C.M., a computer software salesman, was 48 years old when he came to his primary care physician because of fatigue and shortness of breath. He had not seen a doctor on a regular basis before this visit and felt well until 1 year ago, when he began experiencing difficulty climbing stairs or playing basketball with his children. Over the past 6 months he had trouble breathing when he was recumbent. He did not remember ever experiencing significant chest pain and had no family history of heart disease. He did recall that approximately 18 months ago he had to take 2 days off from work because of a severe flulike illness.

On examination, he had a pulse of 105 beats per minute, a respiratory rate of 32 breaths per minute, and a blood pressure of 100/60 mm Hg and was afebrile. His physician heard crackles (evidence of abnormal fluid accumulation) in the bases of both lungs. His feet and ankles were swollen. A chest x-ray showed pulmonary edema and pleural effusions and a significantly enlarged left ventricle. These findings were consistent with right and left ventricular congestive heart failure, which is a reduced capacity of the heart to pump normal volumes of blood, resulting in fluid accumulation in various tissues. C.M. was admitted to the cardiology service of the University Hospital. On the basis of further tests, including coronary angiography and echocardiography, C.M. was given the diagnosis of dilated cardiomyopathy, a progressive and fatal form of heart failure in which the heart chambers become dilated and inefficient at pumping blood (reduced ejection fraction). His physicians told him he may benefit from aggressive medical management, including drugs that enhance heart muscle contraction, reduce the workload of the heart, and enhance excretion of accumulated fluid, but if his underlying heart disease continued to progress, the best long-term option would be to receive a heart transplant. Unfortunately, despite optimal medical management, his symptoms of congestive heart failure continued to worsen until he was no longer able to manage even routine activities of daily living, and he was listed for heart transplantation.

A panel-reactive antibody (PRA) test was performed on C.M.'s serum to determine whether he had been previously sensitized to alloantigens. This test (performed monthly) showed C.M. had no circulating antibodies against human leukocyte antigens (HLAs), and there was no further immunologic testing done at that time. Two weeks later in a nearby city, a donor heart was removed from a victim of a construction site accident. The donor had the same ABO blood group type as C.M. The transplant surgery, performed 4 hours after the donor heart was removed, went well, and the allograft was functioning properly postoperatively.

1. What problems might arise if the transplant recipient and the donor have different blood types or if the recipient has high levels of anti-HLA antibodies?

C.M. was placed on intensive immunosuppressive therapy beginning even as he was being transported to the operating room for the transplant; this included daily doses of tacrolimus, mycophenolic acid, and prednisone. Endomyocardial biopsy was performed 1 week after surgery and showed no evidence of myocardial injury or inflammatory cells. He was sent home 10 days after surgery, and within a month he was able to do light exercise without problems. Routinely scheduled endomyocardial biopsies were initially performed weekly, and then every other week within the first 3 months after transplantation—all of which showed no evidence of rejection. However, a biopsy performed 14 weeks after surgery showed the presence of numerous lymphocytes and macrophages within the myocardium with associated cardiomyocyte injury (Fig. A.2). The findings were interpreted as evidence of acute cellular allograft rejection.

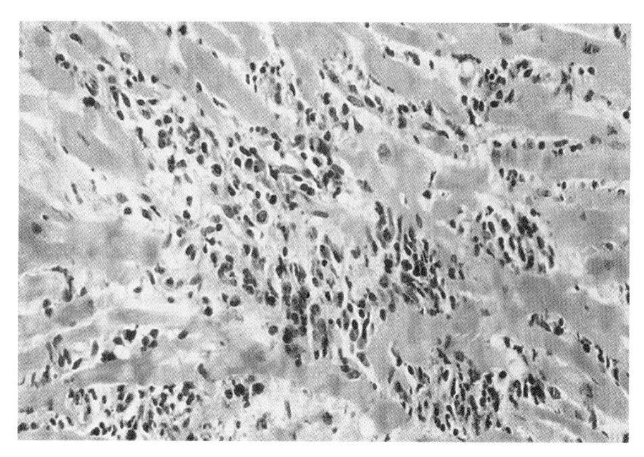

Fig. A.2 Endomyocardial biopsy showing acute cellular rejection. The heart muscle is infiltrated by lymphocytes, and necrotic muscle fibers are present. (Courtesy Dr. Richard Mitchell, Department of Pathology, Brigham and Women's Hospital, Boston, Massachusetts.)

2. What was the patient's immune system responding to, and what were the effector mechanisms in the acute cellular rejection episode?

C.M.'s serum creatinine level, an indicator of renal function, was high (2.2 mg/dL; normal, <1.5 mg/dL). His physicians therefore did not want to increase his tacrolimus dose because this drug can be toxic to the kidneys. He was given three additional doses of methyl prednisolone (a steroid drug) over 18 hours, and a repeat endomyocardial biopsy 1 week later showed only a few scattered macrophages and a small focus of healing tissue. C.M. went home feeling well, and he was able to live a relatively normal life, taking tacrolimus, mycophenolic acid, and prednisone daily.

3. What is the goal of the immunosuppressive drug therapy?

Coronary angiograms performed yearly since the transplant showed a gradual diffuse narrowing of the lumens of the coronary arteries. In the sixth year after transplantation, C.M. began experiencing shortness of breath after mild exercise and showed left ventricular dilation on radiographic examination. An intravascular ultrasound examination demonstrated significant diffuse thickening of the coronary arterial walls with luminal narrowing (Fig. A.3). An endomyocardial biopsy showed areas of microscopic subendocardial

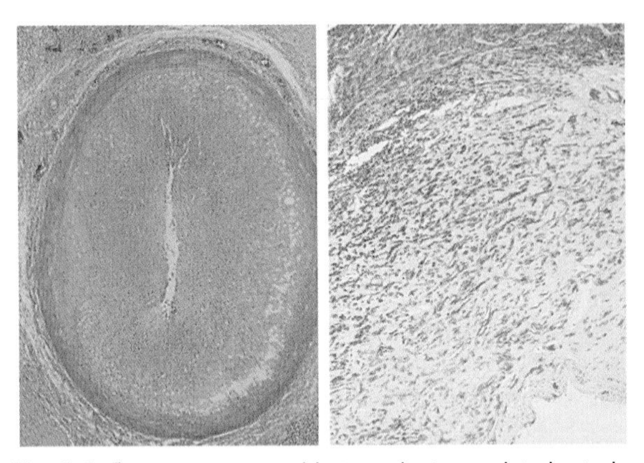

Fig. A.3 Coronary artery with transplant-associated arteriosclerosis. This histologic section was taken from a coronary artery of a cardiac allograft that was removed from a patient 5 years after transplantation because of graft failure. The lumen is greatly narrowed by the presence of intimal smooth muscle cells. (Courtesy Dr. Richard Mitchell, Department of Pathology, Brigham and Women's Hospital, Boston, Massachusetts.)

infarction, as well as evidence of sublethal ischemia (myocyte vacuolization). C.M. and his physicians are now considering the possibility of a second cardiac transplant.

4. What process has led to failure of the graft after 6 years?

Answers to Questions for Case 2

1. If the recipient and the heart donor had different blood types, or if the recipient had high levels of anti-HLA antibodies, hyperacute rejection might occur after transplantation (see Chapter 10). People with type A, B, or O blood groups have pre-formed circulating IgM antibodies against the antigens they do not possess (B, A, or both, respectively). People who have received previous blood transfusions or transplants or were previously pregnant may have circulating anti-HLA antibodies. Blood group and HLA antigens are present on endothelial cells. If the antibodies are already present in the recipient at the time of transplantation, they can bind to the antigens on graft endothelial cells, causing complement activation, leukocyte recruitment, and thrombosis. As a result, the graft blood supply becomes impaired, and the organ can rapidly undergo ischemic necrosis. The PRA test is typically performed to determine whether a patient needing a transplant has preexisting antibodies specific for a broad panel of HLA antigens. The test is done by mixing the patient's serum with a collection of HLA-coated microbeads; antibody binding is detected by flow cytometry of the beads, after addition of fluorescently labeled antibodies directed against human Ig. The results are expressed as a percentage (0% to 100%) of the various HLA-coated beads that have bound to the patient's serum antibodies. The higher the PRA value obtained, the greater the chance that the recipient will have an antibody that can potentially react with a graft and cause hyperacute rejection. The test is typically performed on a monthly basis as the patient is awaiting a heart. This is because many events can induce new anti-HLA antibodies, including a blood transfusion, or new exposures to microbes or drugs, which can potentially elicit antibodies that by chance cross-react with donor HLA.
2. In the acute cellular rejection episode, the patient's immune system is responding to alloantigens in the graft. The main antigens are donor major histocompatibility complex (MHC) molecules encoded by alleles not shared by the recipient; milder reactions may also occur against unshared allelic variants of other proteins (minor histocompatibility antigens). These alloantigens may be expressed on the donor endothelial cells, leukocytes, and parenchymal cells within the donor heart. The effector mechanisms of the acute rejection episode include both cell-mediated and antibody-mediated reactions. With acute cellular rejection, activated alloreactive recipient $CD4^+$ T cells secrete cytokines that promote macrophage activation and inflammation and can cause myocyte or endothelial cell injury and dysfunction, and alloreactive $CD8^+$ cytotoxic T lymphocytes can directly kill graft cells. Antibody-mediated rejection can occur when the recipient develops new circulating donor-specific antibodies, predominantly directed against donor MHC molecules. Such recipient antibodies bind to graft cells (particularly endothelium), leading to complement activation and leukocyte recruitment.
3. The goal of immunosuppressive drug therapy is to suppress the recipient's immune response to alloantigens present in the graft, thereby preventing or treating rejection. The drugs work by depleting T cells (antithymocyte globulin) or by blocking

T cell activation or proliferation (tacrolimus, cyclosporine, rapamycin, mycophenolic acid), and/or inflammatory cytokine production (prednisone). A combination of drugs is given since each can be administered at lower doses at which the risks of adverse side-effects are reduced relative to what is required in a single-drug regimen. An attempt is made to preserve some immune function to combat infections.

4. The graft has failed because of thickening of the walls and narrowing of the lumens of the graft arteries (see Chapter 10). This vascular change, called graft arteriosclerosis or transplant-associated arteriosclerosis, diffusely involves the coronary vasculature and leads to downstream ischemic damage to the heart; it is the most frequent reason for long-term graft failure. It may be caused by a T cell–mediated inflammatory reaction directed against vessel wall alloantigens, which subsequently smolders as a chronic macrophage-mediated injury that results in cytokine-stimulated smooth muscle cell migration into the intima, with smooth muscle cell proliferation and increased matrix synthesis.

CASE 3: ALLERGIC ASTHMA

Ten-year-old I.E. was brought to her pediatrician's office in November because of frequent coughing for the past 2 days, audible wheezing, and a feeling of tightness in her chest. Her symptoms had been especially severe at night. In addition to her routine checkups, she had visited the physician in the past for occasional ear and upper respiratory tract infections but had not previously experienced wheezing or chest tightness. She had eczema, but otherwise she was in good health and was developmentally normal. Her immunizations were up to date. She lived at home with her mother, father, two sisters aged 12 and 4, and a pet cat. Both of her parents smoked cigarettes, and her father suffered from allergic rhinitis.

At the time of her physical examination, I.E. had a temperature of 37° C (98.6° F), blood pressure of 105/65 mm Hg, and a respiratory rate of 30 breaths per minute. She did not appear short of breath but had mild subcostal retraction. There were no signs of ear infection or pharyngitis. Auscultation of the chest revealed diffuse wheezing in both lungs. There was no evidence of pneumonia. The physician made a presumptive diagnosis of bronchospasm and referred I.E. to a pediatric allergist-immunologist. In the meantime, she was given a prescription for a short-acting β2-adrenergic agonist bronchodilator inhaler and was instructed to administer the drug every 4 hours to relieve symptoms. This drug binds to β2-adrenergic receptors on bronchial smooth muscle cells and causes them to relax, resulting in dilation of the bronchioles. The family was also prescribed a spacer, a device to optimize delivery of the medication, and taught to administer the inhaler using the spacer.

1. Asthma is often an atopic disease, particularly in patients older than 6 to 8 years of age. What are the different ways in which atopy may manifest clinically?

One week later, I.E. was seen again by the allergist. He auscultated her lungs and confirmed the presence of wheezing. I.E. was instructed to blow into a spirometer, and the physician determined that her forced expiratory volume in 1 second (FEV1) was 65% of normal, indicating airway obstruction. The physician then administered a nebulized bronchodilator and 10 minutes later performed the test again. The repeat FEV1 was 85% of normal, indicating reversibility of the airway obstruction. Blood was drawn and sent for total and differential blood cell count and determination of IgE levels. In addition, a skin test was performed to determine hypersensitivity to various antigens and showed a positive result for cat dander, house dust mites, and tree pollens (Fig. A.4). The patient was instructed to begin using an inhaled corticosteroid and to use her bronchodilator as needed for respiratory symptoms. Her parents were instructed to make a return appointment 2 weeks later for reevaluation of I.E. and discussion of blood test results.

2. What is the immunologic basis for a positive skin test?

At I.E.'s return appointment 2 weeks later, laboratory tests revealed that she had a serum IgE level of 1200 kU/mL (normal range, 0 to 180) and a total white blood cell count of 7000/mm^3 (normal, 4300 to 10,800/mm^3), with an absolute eosinophil count of 700/mm^3 (normal, <500). When she returned to the allergist's office 1 week later, her respiratory status on physical examination was significantly improved, with no audible wheezing. I.E.'s FEV1 had improved to 85% of normal. The family was told that I.E. had reversible airway obstruction, possibly triggered by a viral illness and possibly related to cat and dust allergies. The

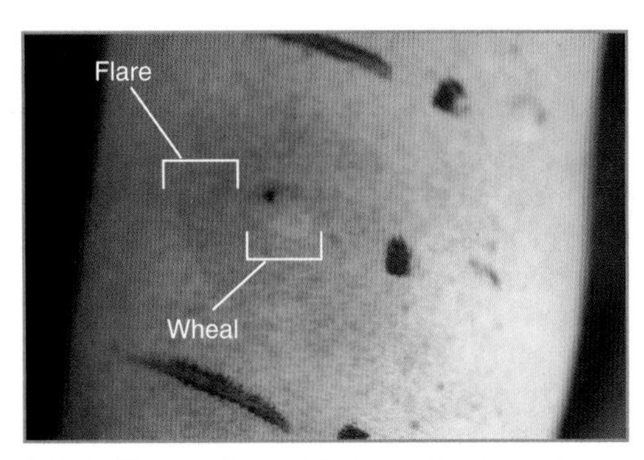

Fig. A.4 **Positive result on prick skin testing for environmental antigens.** Small amounts of the antigens are applied into the superficial layers of the skin using a short needle to prick the skin. If mast cells are present with bound immunoglobulin E specific for the test antigen, the antigen will cross-link the Fc receptors to which the IgE is bound. This induces degranulation of the mast cells and the release of mediators that cause the wheal and flare reaction.

physician advised that, although rehoming the cat is ideal, at the very least the cat should be kept out of I.E.'s bedroom. The mother was told that smoking in the house probably was contributing to I.E.'s symptoms. The physician recommended that I.E. continue to use the short-acting inhaler for acute episodes of wheezing or shortness of breath. She was asked to return in 3 months, or sooner if she used the inhaler more than 2 days per week or if she awakened at night with symptoms more than once a month.

3. What is the mechanism for the increased IgE levels seen in patients who have allergic symptoms?

The family cat was given to a neighbor, and I.E. did well on the therapy for approximately 6 months, experiencing only mild wheezing a few times. The next spring, she began to have more frequent episodes of coughing and wheezing. During a soccer game one Saturday, she became very short of breath, and her parents brought her to the emergency department (ED) of the local hospital. After confirming that she was wheezing and showed signs of accessory respiratory muscle use, the ED physician treated her with a nebulized β2-agonist bronchodilator and an oral corticosteroid. After 6 hours, her symptoms resolved, and she was sent home. The following week, I.E. was brought to her allergist, who changed her inhaler prescription to a combined corticosteroid/long-acting β agonist. She has subsequently been well, with occasional mild attacks that are cleared by the inhaler.

4. What are the therapeutic approaches to allergic asthma?

Answers to Questions for Case 3

1. Atopic reactions to environmental antigens (allergens) are mediated by IgE and mast cells and may manifest in a variety of ways (see Chapter 11). The signs and symptoms usually reflect the site of entry of the allergen. Hay fever (allergic rhinitis) and asthma usually are responses to inhaled allergens, whereas urticaria and eczema more often occur with skin exposure or ingestion. Food allergies may also cause gastrointestinal or respiratory symptoms. The most dramatic presentation of allergies to insect venom, foods, or drugs is anaphylaxis, a reaction characterized by systemic vasodilation, increased vascular permeability, and airway obstruction (laryngeal edema or bronchoconstriction). Without intervention, patients with anaphylaxis may progress to asphyxia and cardiovascular collapse.
2. If an individual with an allergy is challenged with a small dose of the allergen injected into the skin, there is immediate release of histamine from triggered mast cells, which produces a central wheal of edema (from leakage of plasma) and the surrounding flare of vascular congestion (from vessel dilation). The injected allergen binds to previously produced IgE antibodies, which coat mast cells by attaching to Fcε receptors. The allergy skin test should not be confused with the skin test used to assess prior sensitization to certain infectious agents, such as *Mycobacterium tuberculosis.* A positive tuberculosis skin test is an example of a delayed-type hypersensitivity (DTH) reaction, mediated by antigen-stimulated type 1 helper T (Th1) cells, which release cytokines such as interferon-γ (IFN-γ), leading to macrophage activation and inflammation (see Chapter 6). Serum allergen-specific IgE tests are also routinely performed and give complementary information to traditional allergy skin testing.
3. For unknown reasons, patients with atopy mount type 2 helper T (Th2) cell responses to a variety of essentially harmless protein antigens, in which Th2 cells produce interleukin-4 (IL-4), IL-5, and IL-13 and follicular T helper (Tfh) cells produce IL-4 and IL-13. IL-4 and IL-13 induce IgE class switching in B cells, IL-5 activates eosinophils, and IL-13 stimulates mucus production (see Chapters 6 and 11). Atopy tends to

run in families, and genetic susceptibility is clearly involved. Attention has been focused especially on genes on the long arm of chromosome 5 (5q) that encode several Th2 cytokines; on 11q, where the gene for the α chain of the IgE receptor is located; and on genes on chromosomes 2 and 9, which encode the IL-33 receptor (ST2) and IL-33, respectively. IL-33 is a cytokine secreted by epithelial cells that activates group 2 innate lymphoid cells (ILC2), which may play a role in inducing strong Th2 responses.

4. A major therapeutic approach for allergies is prevention by avoiding exposure to precipitating allergens, identified through either allergy skin testing or serum IgE measurement. Although pharmacologic therapy previously has been focused on treating the symptoms of bronchoconstriction by elevating intracellular cyclic adenosine monophosphate (cAMP) levels (using β2-adrenergic agents and inhibitors of cAMP degradation), the balance of therapy has shifted to the use of antiinflammatory agents. These include corticosteroids (which block cytokine release) and receptor antagonists for lipid mediators (e.g., leukotrienes). Newer treatments that have been developed for treatment of asthma and other allergies include mAbs targeting IgE, IL-4/IL-13 receptors, IL-5, IL-33, and thymic stromal lymphopoietin (TSLP). The anticytokine therapies are often effective in treating patients with severe asthma characterized by evidence of strong type 2 immune responses, such as high eosinophil counts and eosinophilic airway inflammation (called T2-high asthma). The most effective treatment for anaphylaxis is the administration of epinephrine through intramuscular injection. Epinephrine causes blood vessel constriction, dilation of bronchioles, and increased cardiac output, thereby reversing the fall in blood pressure and airway obstruction.

CASE 4: SYSTEMIC LUPUS ERYTHEMATOSUS

N.Z., a 25-year-old woman, presented to her primary care physician with complaints of joint pain involving her wrists, fingers, and ankles. When seen in the physician's office, N.Z. had normal body temperature, heart rate, blood pressure, and respiratory rate. There was a noticeable red rash on her cheeks, most marked around her nose, sparing the nasolabial folds, and on questioning she said the redness worsened after being in the sun for 1 or 2 hours. The joints of her hands and wrists were swollen and tender. The other physical examination findings were unremarkable.

Her physician took a blood sample for various tests. Her hematocrit was 35% (normal, 37% to 48%). The total white blood cell count was 9800/mm^3 (within normal range), with a normal differential count. The erythrocyte sedimentation rate (ESR) was 40 mm/hr (normal, 1 to 20), reflecting systemic inflammation with cytokine-induced production of acute phase proteins. Her serum antinuclear antibody (ANA) test was positive at 1:2560 dilution (normally, negative at 1:40 dilution) with a homogeneous pattern. Other laboratory findings were unremarkable. On the basis of these findings, a diagnosis of systemic lupus erythematosus (SLE) was made. N.Z.'s physician prescribed oral prednisone (a corticosteroid) and hydroxychloroquine; with this treatment, her joint pain subsided.

1. What is the significance of the positive result for the ANA test?

Three months later, N.Z. began feeling unusually tired and thought she had the flu. For approximately a week she had noticed that her ankles were swollen, and she had difficulty putting on her shoes. She returned to her primary care physician. Her ankles and feet showed severe edema (swollen as a result of extra fluid in the tissue). Her abdomen appeared slightly distended, with a mild shifting dullness to percussion (a sign of an abnormally large amount of fluid in the peritoneal cavity). Her physician ordered several laboratory tests. Serum anti-dsDNA was 25 IU/ml (normal $<$10 U/ml), and her ESR was 120 mm/hr. Serum albumin was 0.8 g/dl (normal, 3.5 to 5.0). Measurement of serum complement proteins revealed a C3 of 42 mg/dl (normal, 80 to 180) and a C4 of 5 mg/dl (normal, 15 to 45). Urinalysis showed 4+ proteinuria, both red and white blood cells, and numerous hyaline and granular casts. A 24-hour urine sample contained 4 g of protein.

2. What is the likely reason for the decreased complement levels and the abnormalities in blood and urinary proteins?

Because of the abnormal urinalysis findings, the physician recommended a renal biopsy, which was performed 1 week later. The biopsy specimen was examined by routine histologic methods, immunofluorescence, and electron microscopy (Fig. A.5).

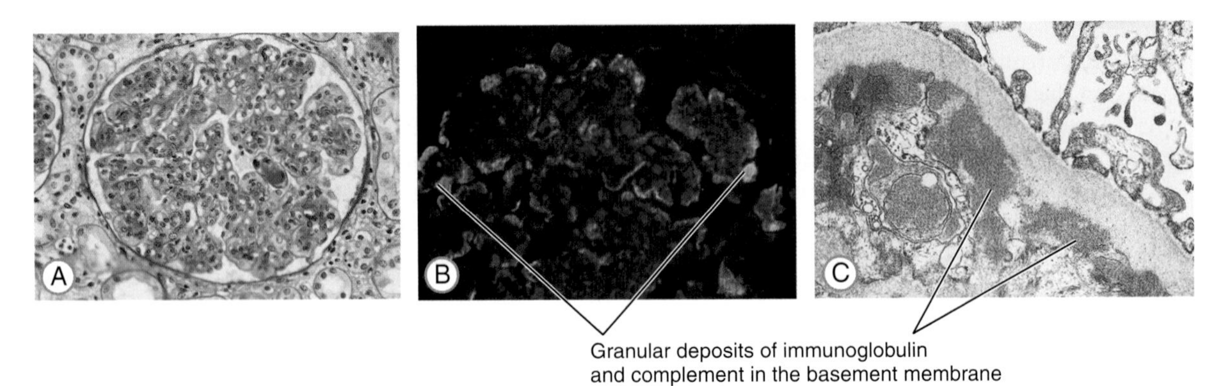

Fig. A.5 Glomerulonephritis with immune complex deposition in systemic lupus erythematosus. **A,** Light micrograph of a renal biopsy specimen in which neutrophilic infiltration in a glomerulus can be seen. **B,** Immunofluorescence micrograph showing granular deposits of immunoglobulin G (IgG) along the basement membrane. (In this technique, called immunofluorescence microscopy, a frozen section of the kidney is incubated with a fluorescein-conjugated antibody against IgG, and the site of deposition of the IgG is defined by determining where the fluorescence is located.) **C,** Electron micrograph of the same tissue revealing immune complex deposition. (Courtesy Dr. Helmut Rennke, Department of Pathology, Brigham and Women's Hospital, Boston, Massachusetts.)

3. What is the explanation for the pathologic changes seen in the kidney?

The physician made the diagnosis of proliferative lupus glomerulonephritis, prescribed a higher dose of prednisone, and recommended treatment with a cytotoxic drug (mycophenolate). N.Z.'s proteinuria and edema subsided over a 2-week period, and serum C3 levels returned to normal. Her corticosteroid dose was tapered to a lower amount. Over the next few years, she had intermittent flare-ups of her disease, with joint pain and edema and laboratory tests indicating depressed C3 levels and proteinuria. These were effectively managed with corticosteroids. Over the past 2 years she has received a calcineurin inhibitor called voclosporin twice a day orally, her renal function has stabilized, and N.Z. has been able to lead an active life.

Answers to Questions for Case 4

1. A positive ANA test reveals the presence of serum antibodies that bind to components of cellular nuclei. The test is performed by placing different dilutions of the patient's serum on top of a monolayer of human cells on a glass slide. A second fluorescently labeled anti-Ig antibody is then added, and the cells are examined with a fluorescent microscope to detect if any serum antibodies bound to the nuclei. The ANA titer is the maximum dilution of the serum that still produces detectable nuclear staining. Almost all patients with SLE have ANAs, which may be specific for histones, other nuclear proteins, or double-stranded DNA. These are autoantibodies, and their production is evidence of autoimmunity. ANAs are not specific for SLE, and this test is typically supplemented with a more specific test for antibodies against double-stranded DNA and Smith, ribonucleoprotein, SS-A, and SS-B antigens. Autoantibodies also may be produced against various cell membrane protein antigens. The development of autoantibodies generally precedes the clinical onset of SLE by as much as 9 to 10 years, and anti-dsDNA titers can be used to assess disease activity.
2. Some of the autoantibodies form circulating immune complexes by binding to antigens in the blood. Nuclear antigens may be increased in the circulation of patients with SLE because of increased apoptosis of several cell types (e.g., white blood cells, keratinocytes) and defective clearance of apoptotic cells. When these immune complexes deposit in the basement membranes of vessel walls, they may activate the classical pathway of complement, leading to inflammation, and depletion of complement proteins through consumption. Inflammation caused by the immune complexes in the kidney leads to leakage of protein and red blood cells into

the urine. The loss of protein in the urine results in reduced plasma albumin, reduction of osmotic pressure of the plasma, and fluid loss into the tissues. This explains the edema of the feet and abdominal distention.

3. The pathologic changes in the kidney result from the deposition of circulating immune complexes in the basement membranes of renal glomeruli. In addition, autoantibodies may bind directly to tissue antigens and form in situ immune complexes. These deposits can be seen by immunofluorescence (indicating type of antibody deposited) and electron microscopy (showing exact localization). The immune complexes activate complement, and leukocytes are recruited by complement by-products (C3a, C5a) and by binding of leukocyte Fc receptors to the IgG molecules in the complexes. These leukocytes become activated, and they produce reactive oxygen species and lysosomal enzymes that damage the glomerular basement membrane. These findings are characteristic of immune complex–mediated tissue injury, and complexes may deposit in joints and small blood vessels anywhere in the body, as well as in the kidney. SLE is a prototype of an immune complex disease (see Chapter 11).

CASE 5: HUMAN IMMUNODEFICIENCY VIRUS INFECTION: ACQUIRED IMMUNODEFICIENCY SYNDROME

J.C. was a 28-year-old assistant carpenter who presented to a clinic physician with 3 weeks of low-grade fevers, sore throat, and lymphadenopathy. Physical examination revealed "track marks," and when asked, the patient stated that 2 months earlier, he had begun using heroin with shared needles because he could no longer afford the cost of escalating doses of street oxycodone. Other findings on physical examination included lymphadenopathy and a faint, diffuse rash. Point-of-care tests for Epstein-Barr virus infection (monospot) and oropharyngeal streptococcal infection (rapid strep) were negative, as were blood cultures for bacteria and fungi. He was discharged with a diagnosis of a presumed viral syndrome.

1. What is the significance of 3 weeks of low-grade fevers and lymphadenopathy?

J.C. was seen the next week in the infectious diseases clinic, where a fourth-generation enzyme-linked immunosorbent assay (ELISA) performed on his serum was found to be negative for anti–human immunodeficiency virus (HIV) antibody but positive for HIV nucleocapsid p24. A follow-up HIV-1/HIV-2 differentiation assay did not detect antibodies to either HIV-1 or HIV-2, suggesting viral antigenemia without seropositivity. The concentration of HIV viral genomes in his blood (viral load) was determined to be 700,000/ml, and his blood $CD4^+$ T cell count was $300/mm^3$, with a reversal of the normal CD4/CD8 ratio (Fig. A.6). Hepatitis B virus (HBV) ELISAs were negative for antibodies specific for HBV surface and core antigens. HIV genotyping showed a lysine-to-asparagine mutation at codon 103 (K103N) of the HIV reverse transcriptase gene. Antiretroviral therapy (ART) was recommended, however clinical follow-up and care engagement was not established and treatment was never initiated.

2. What was this patient's major risk factor for acquiring HIV infection? What are other risk factors for HIV infection?
3. Why do the HIV tests include testing for the presence of both HIV antibodies and p24 protein?

Six months later, J.C. was seen at a community hospital for an abscess at an injection site. After incision and drainage, he declined the advice of his treating physicians to pursue additional medical evaluation. A $CD4^+$ T cell count obtained at that time was $500/mm^3$, and viral load was 15,000. He again declined to initiate ART. Six years later, J.C. was admitted to the hospital after a week of fevers and shortness of breath. A chest x-ray showed faint, diffuse infiltrates, and oxygen saturation was 90%. Initial microscopic examination of sputum stained for fungi (silver stain) was unrevealing, but he was started on antibiotics (trimethoprim-sulfamethoxazole) plus prednisone. PCR testing of sputum was positive for *Pneumocystis jirovecii.* J.C.'s condition initially worsened, but eventually he recovered fully. A repeat $CD4^+$ T cell count was now 150, with a viral load of 50,000 copies/ml. At this point, J.C. expressed desire to initiate ART and was started on dolutegravir (an HIV integrase inhibitor), plus two nucleoside/nucleotide analog inhibitors of the HIV

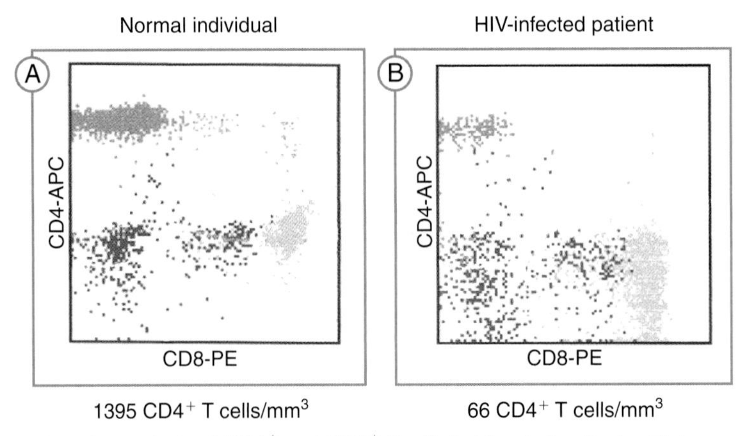

Fig. A.6 Flow cytometry analysis of CD4$^+$ and CD8$^+$ T cells in blood of patient with human immunodeficiency virus (HIV) infection. A suspension of the patient's white blood cells was incubated with monoclonal antibodies specific for CD4 and CD8. The anti-CD4 antibody was labeled with the fluorochrome allophycocyanin (APC), and the anti-CD8 antibody was labeled with the fluorochrome phycoerythrin (PE). These two fluorochromes emit light of different colors when excited by the appropriate wavelengths. The cell suspensions were analyzed in a flow cytometer, which can enumerate the number of cells stained by each of the differently labeled antibodies. In this way, the number of CD4$^+$ and CD8$^+$ T cells can be determined. Shown here are two-color plots of a control blood sample **(A)** and that of the patient **(B).** The CD4$^+$ T cells are shown in *orange (upper left quadrant)*, and the CD8$^+$ T cells are shown in *green (lower right quadrant)*. Note that these are not the colors of light emitted by the APC and PE fluorochromes.

reverse transcriptase (nucleoside/nucleotide analog reverse transcriptase inhibitors [NRTIs]), tenofovir/emtricitabine. He was also continued on trimethoprim-sulfamethoxazole at lower doses for prophylaxis. He was advised to stop smoking.

4. Why does ART therapy for HIV typically include two or three different antiviral drugs?
5. What caused the gradual decline in J.C.'s CD4$^+$ T cell count?
6. Why were antibiotics and prednisone started in the patient before a diagnosis of *P. jirovecii* infection was established by PCR?

One year later, his CD4 count was 800 and his viral load was undetectable, but he developed methicillin-resistant *Staphylococcus aureus* (MRSA) infection of his mitral valve (staphylococcal infective endocarditis), requiring surgical replacement with a bioprosthetic valve. Preoperative cardiac catheterization showed significant coronary artery disease. Postoperatively, he was able to discontinue heroin use with methadone maintenance. His antiretroviral drugs were continued, but the trimethoprim-sulfamethoxazole was stopped. He has remained in good health since. His long-term partner remains HIV negative.

7. What are the main risks to J.C.'s life at this point?

Answers to Questions for Case 5

1. This pattern is referred to as acute HIV syndrome. Although a very large number of infectious agents can cause an acute viral syndrome for a few days, the persistence in this case suggests one of a relatively small number of causes in a young, previously healthy person, including HIV infection. However, acute HIV infection also can be asymptomatic or substantially less symptomatic than what this patient experienced.
2. Intravenous drug use with sharing of needles and syringes is the major risk factor for HIV infection in this patient. Shared needles among people who inject drugs transmit blood-borne viral particles from one infected person to others. Other major risk factors for HIV infection include sexual intercourse with an infected person, transfusion of contaminated blood products, and birth from an infected mother (see Chapter 12). Intravenous drug use accounts for less than 10% of HIV cases in the United States. Most infections (70%) are in men who have sex with other men, and the

remainder are typically acquired via penile-vaginal intercourse (approximately 25%). Globally, more than 90% of new infections occur in heterosexuals. The demographics of the epidemic have changed over the past few decades.

3. In some patients presenting with acute infection, there often has been insufficient time to develop an antibody response, but levels of virus are high, so viral proteins can be readily detected. So-called fourth-generation tests detect both viral p24 protein antigens and anti-HIV antibodies and thus can be used to diagnose HIV infection in the window period before antibodies develop. These combined tests were approved in the United States in 2010, several years later than in other countries. If the screening test is positive, it would be followed up with more specific assays to distinguish HIV-1 from HIV-2 infection, and PCR tests to determine numbers of circulating viral particles and genotype of viral nucleic acid.
4. HIV has a very high mutation rate. Mutations in the reverse-transcriptase gene that render the enzyme resistant to nucleoside inhibitors occur frequently in patients receiving these drugs. Resistance to protease inhibitors may come about by similar mechanisms. Triple-drug therapy including drugs from at least two different classes of antiretroviral agents (or more recently, dual-drug therapy with an integrase inhibitor and a drug from another class in select patient groups) greatly reduces the chances of the virus developing drug resistance, but poor adherence can contribute to emergence of resistant strains. Nonnucleoside analog reverse transcriptase inhibitors (NNRTIs) are also effective anti-HIV drugs, but the lysine-to-asparagine mutation at codon 103 (K103N) of the HIV reverse transcriptase gene, discovered at the time of diagnosis, would make this patient's virus resistant to many first-generation NNRTIs. (Notably, unlike many other resistance mutations, viruses with the K103N mutation are often transmitted between individuals because this mutation does not impose significant fitness costs on the virus and therefore it can be maintained even in the absence of selective pressure from ART.) Integrase inhibitors (specifically integrase strand transfer inhibitors [INSTIs]) are another major class of anti-HIV drugs used in combination therapy that have more recently become a key component of first-line ART regimens. Other drug classes include viral protease inhibitors and inhibitors of HIV entry into and fusion with cells.
5. After initial infection, which often starts in mucosal tissues, HIV rapidly enters various types of cells in the body, mainly $CD4^+$ T lymphocytes, but also dendritic cells and mononuclear phagocytes. The gradual decline in $CD4^+$ T cells in this patient was caused by repetitive cycles of HIV infection of $CD4^+$ T cells in lymphoid organs, leading to death of the cells. The symptoms of acquired immunodeficiency syndrome (AIDS), including development of most AIDS-defining infections and cancers, more commonly occur after the blood count of $CD4^+$ T cells falls below 200 cells/mm^3, reflecting a severe depletion of T cells in the lymphoid organs.
6. This presentation in a person with known HIV infection is so highly suggestive of *P. jirovecii* pneumonia (PJP) that initiation of presumptive PJP treatment is appropriate pending additional diagnostic evaluation. The deficiencies in T cell–mediated immunity in patients with AIDS lead to impaired immunity to viruses, fungi, and protozoa that otherwise are easily controlled by an intact immune system, allowing for the development of opportunistic infections. *P. jirovecii* is a fungal organism that can live within phagocytes, but usually it is eradicated by the action of activated $CD4^+$ T cells. In the first days of PJP treatment, a potent inflammatory response to the dying microorganisms can cause dangerous clinical worsening, so steroid antiinflammatories are started immediately for severe cases.
7. With well-controlled HIV infection, patients can have a near-normal life expectancy, and most deaths are from causes not directly related to HIV infection. Both HIV infection itself and some of the antiretroviral drugs accelerate coronary artery disease; hence, infected persons who are effectively treated with antiretrovirals tend to die more frequently of disorders not directly related to the viral infection. The highest non–HIV-related risk to this patient's health was active intravenous drug use, now discontinued. In addition, people with well-suppressed HIV infection very rarely transmit the virus to others, so treatment can both control infection and prevent ongoing chains of transmission ("treatment as prevention").

CASE 6: SARS CoV-2 INFECTION: COVID-19

C.V. was a 63-year-old convenience store manager with a history of type 2 diabetes and hypertension who developed new-onset muscle aches, cough, and headache, which persisted over 4 days. He still had an appetite and could smell and taste his food, and therefore he thought it was unlikely that he had COVID-19. He received one dose of an mRNA SARS-CoV-2 vaccine 8 months ago, but because the wave of infections in the community had started to decline over the past several months, he decided not to follow up for additional doses, and he rarely wore a mask in the store where he worked. On the fifth day after symptom onset, he felt much worse, and fearing he may have COVID-19, he called his primary care physician's office. They advised he get tested for SARS-CoV-2 infection at a drive-through testing center in his neighborhood and then isolate himself from his partner and others until the test results were available. After returning from the testing center, he felt fatigued, short of breath while climbing the stairs in his home to his bedroom, and experienced difficulty breathing that night. The next morning, his partner drove him to the ER of the community hospital, where initial evaluation revealed a fever of 37.5°C, a rapid respiratory rate, a blood O_2 saturation on room air of 88% (normal >95). A chest CT scan showed bilateral ground-glass opacifications (Fig. A.7). The results of the SARS-CoV-2 PCR test from the day before came back positive. He was admitted to the intensive care unit, intubated, and treated with dexamethasone (a corticosteroid) and remdesivir (a viral RNA-dependent RNA polymerase inhibitor).

1. SARS-CoV-2, like many viruses, stimulates innate immune responses in infected hosts. How does this occur and how does the virus evade innate immunity?

C.V.'s partner, W.V., a 65-year-old woman with no significant chronic health problems, had received two doses of an mRNA COVID vaccine the previous year, and a booster shot 3 months ago. She developed a mild sore throat and runny nose on the day C.V. was admitted, and she was tested for SARS-CoV-2. Her results were also positive, but her upper respiratory symptoms abated over 2 days.

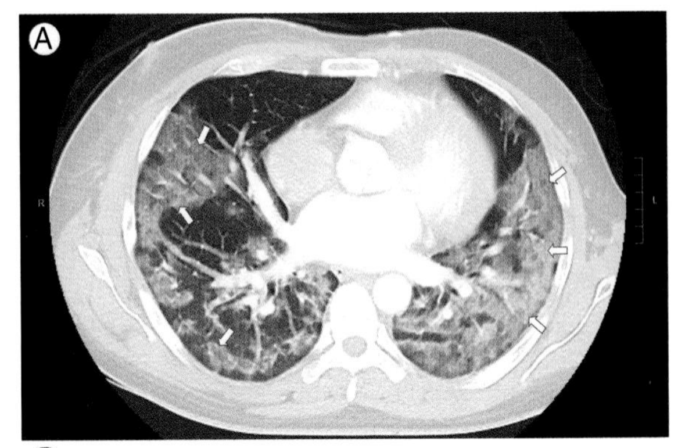

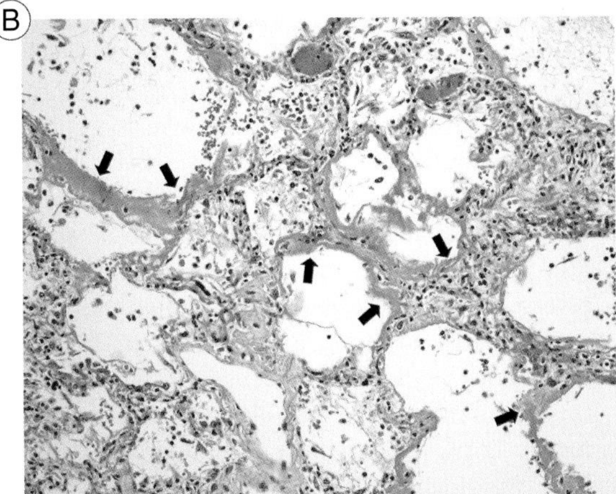

Fig. A.7 Pulmonary disease in severe COVID-19. **A,** A chest CT scan showing the appearance of lungs damaged in a lethal case of severe disease caused by SARS-CoV-2 infection. *White-appearing areas,* so called "ground-glass opacities" *(arrows)* represent airspaces filled with inflammatory infiltrates and fluid, which are incapable of gas exchange. These airspaces would be dark (radiolucent) in normal lungs. **B,** An H&E stained histologic section of lung tissue obtained at autopsy showing hyaline membranes composed of plasma proteins that have leaked out of damaged blood vessels and line the alveolar walls *(arrows),* typical of diffuse alveolar damage that corresponds to the ground-glass opacities seen in panel **B.** This histopathology corresponds to a clinical diagnosis of acute respiratory distress syndrome (ARDS), which is seen in most cases of lethal COVID-19. (Courtesy Dr. Robert Padera, Department of Pathology, Brigham and Women's Hospital, Boston, Massachusetts.)

2. What is the major way the adaptive immune system protects against SARS-CoV-2; how does vaccination work to enhance this protection?

C.V.'s clinical course in the ICU was complicated by a persistent need for mechanical ventilation, pulmonary embolism requiring anticoagulation, and heart failure. He gradually improved, was extubated after 4 weeks, and was discharged from the hospital after 6 weeks. He was vaccinated 2 weeks later. Since then, he has suffered from persistent fatigue, chest pain, and brain fog.

3. What are the major challenges for vaccine development and implementation that have been elucidated by the COVID COVD-19 pandemic?

Answers to Questions for Case 6

1. The major way viruses such as SARS-CoV-2 activate innate immune responses is through viral nucleic acids binding to intracellular pattern recognition receptors, including endosomal Toll-like receptor 7, which recognizes ssRNA, and cytosolic RIG-I and MDA5, which recognize features typical of viral but not mammalian RNAs. When they bind their ligands, these receptors activate signaling pathways that result in host cell production of type I IFNs, which in turn bind to receptors on host cells to induce an antiviral state. SARS-CoV-2 evades innate immunity by structural features of its genomic RNA that shield it from innate immune recognition and by blocking signaling that induces type I IFNs and signaling downstream of type I IFN receptors.
2. A major way adaptive immunity protects against SARS-CoV-2 infection is by high-affinity neutralizing antibodies that recognize parts of the viral spike proteins that bind to host angiotensin-converting enzyme-2 (ACE2). SARS-CoV-2 enters host cells by binding to ACE2, and thus anti–spike protein antibodies block viral infection of the cells. Non-neutralizing antibodies that recognize viral spike proteins can activate complement, help phagocytose free virus, or trigger other Fc receptor–directed activities. The production of effective antibodies depends on stimulation of $CD4^+$ Tfh cells, which recognize peptides derived from viral proteins, which are processed and bound to host class II MHC molecules. The Tfh cells collaborate with spike protein–specific B cells to induce germinal center reactions yielding long-lived plasma cells that produce high-affinity anti–spike protein antibodies and memory B cells. $CD8^+$ cytotoxic T lymphocytes (CTL) specific for peptides derived from other viral proteins also play a role in combatting SARS-CoV-2 infection by killing infected cells.
3. Like other RNA viruses, replication of the SARS-CoV-2 RNA genome is highly error prone, thereby generating many mutations. During the course of the COVID-19 pandemic, several successive waves of infection have rapidly spread worldwide, each with a major variant of SARS-CoV-2 carrying mutations that alter the spike protein structure, and are thus capable, at least in part, of evading neutralization by antibodies generated in responses to infection by previous forms of the virus. (The mutations likely increase the ability of the virus to infect and spread.) The dominance of these variants has evolved under the selective pressures of adaptive immune responses to viral infection as well as to SARS-CoV-2 vaccines. This is the major way SARS-CoV-2 has evaded adaptive immunity and is the major challenge for SARS-CoV-2 vaccine development going forward. Nonetheless, the vaccines used in the mass vaccination program, which are based on the spike protein sequences found in the first wave of infections in 2019 and 2020, remain highly effective in preventing serious illness and death from later SARS-CoV-2 variants. The second major challenge for vaccination programs is achieving a high rate of vaccination worldwide, which is hampered logistically by poverty, reluctance, and disinformation.

ANSWERS TO REVIEW QUESTIONS

CHAPTER 1

1. Innate immunity responds immediately to infections and injury with effector cells and molecules that are always present and functional, whereas adaptive immune responses require activation of clones of lymphocytes to expand and differentiate into effector cells that can fight infection, a process that takes several days. The innate immune system uses a limited number of receptors, which recognize different molecular patterns common to many species of microbes, whereas the adaptive immune system uses two types of highly specific antigens receptors, each with millions of variations and each expressed by a different clone of lymphocytes, which recognize distinct molecular features of antigens. Innate immune responses are mostly identical in quality and magnitude upon repeated exposures to the same type of microbe, whereas the adaptive immune system generates long-lived memory that protects against subsequent infections by the same species and responds more rapidly in a specialized manner to eliminate repeat infections.
2. The two types of adaptive immunity are cell-mediated immunity and humoral immunity. Cell-mediated immunity, mediated by T cells, is essential for protection against pathogens that infect tissue cells or phagocytes. Humoral immunity, which is mediated by antibodies, provides protection primarily against pathogens that are located outside of cells. Many pathogens have extracellular and intracellular phases of their life cycle, such as viruses, and are defended against by both cell-mediated and humoral immunity.
3. B lymphocytes express surface immunoglobulin (Ig), which functions as their antigen receptor, and mediate humoral immunity. Following activation, B lymphocytes differentiate into antibody-secreting plasma cells. T lymphocytes express the T cell antigen receptor (TCR) and either CD4 or CD8 and mediate cell-mediated immune responses. After activation by peptide antigens displayed by cell surface major histocompatibility complex (MHC) molecules, $CD4^+$ T cells secrete cytokines and express membrane-bound activating ligands, which induce inflammation, enhance the functions of phagocytes, and promote B cell antibody responses. After activation by peptide antigens displayed by MHC molecules, $CD8^+$ T cells release cytotoxic proteins that kill infected cells and tumor cells.
4. Naive lymphocytes are mature B or T cells that have not yet encountered a foreign antigen. Following activation by antigen, naive lymphocytes proliferate and differentiate into cells that acquire the ability to protect against or eliminate pathogens. These lymphocytes are known as effector cells. Most effector cells die after the antigen is eliminated, but a subset of previously activated lymphocytes known as memory cells live for extended periods. Memory lymphocytes not only survive for long times but also respond more rapidly and vigorously than do naive lymphocytes when challenged by antigen.
5. B lymphocytes reside in follicles in secondary (peripheral) lymphoid organs. T cells reside in the parafollicular cortex of lymph nodes and the periarteriolar lymphoid sheaths of the spleen. B and T cells are maintained in these locations by the action of specific cytokines called chemokines, which are secreted by stromal cells in the different regions of the lymphoid organ and bind to different chemokine receptors expressed on B and T cells.
6. Naive lymphocytes home from the blood into secondary lymphoid organs, and then, via lymphatics exiting lymph nodes or through blood vessels in the spleen, they migrate back into the blood and recirculate through other secondary lymphoid organs. Effector lymphocytes are generated in secondary lymphoid organs, and most migrate into blood and then home to the tissue site where the activating antigen may be located.

CHAPTER 2

1. Innate immunity is directed against common molecular patterns shared by different microbes and the products of damaged cells and is mediated by cellular receptors located in the plasma membrane

(where they recognize extracellular microbes), endosomal vesicles (ingested microbes), and cytosol. Some secreted proteins of limited diversity also recognize microbes. Adaptive immunity uses an extremely diverse set of antigen receptors (cell surface and secreted antibodies and cell surface TCRs) to recognize a wide range of microbial and nonmicrobial antigens.

2. Examples of microbial substances recognized by the innate immune system include lipopolysaccharide recognized by Toll-like receptor 4 (TLR-4); peptidoglycan recognized by TLR-2; flagellin recognized by TLR-5; microbial DNA recognized by TLR-9 and by cytoplasmic DNA sensors; viral RNAs recognized by endosomal TLR-3, -7, and -8 and by cytosolic RIG-like receptors; bacterial peptidoglycans recognized by cytoplasmic NOD-like receptors; and mannans recognized by the cell surface mannose receptor and by mannose-binding protein in the blood.
3. Inflammasomes are multiprotein complexes found in the cytoplasm of phagocytes, dendritic cells (DCs), and other cell types that respond to pathogens or cell stress by inflammatory cytokine secretion or cell death. Several types of inflammasomes generate an enzyme that proteolytically cleaves a precursor of the cytokine interleukin-1β (IL-1β), producing an active proinflammatory form of IL-1β that is released from the cell. One example is the inflammasome that contains a NOD family molecule called NLRP3 and the proteolytic enzyme caspase-1. NLRP3 responds to many different stimuli that indicate cell infection or injury, leading to activation of caspase-1, which then cleaves the IL-1β precursor. Stimuli that activate the NLRP3 inflammasome include various microbial products, crystals such as sodium urate and cholesterol, reduced potassium concentration, and reactive oxygen species.
4. The skin provides a relatively impermeable multilayered physical epithelial barrier by virtue of a surface layer of keratin produced by the skin epithelial cells called keratinocytes and by tight junctions between the keratinocytes. The intestinal tract is lined by a single layer of epithelial cells, also held together by tight junctions. Some of the intestinal epithelial cells secrete a layer of mucus that serves as a microbial barrier. Both skin and intestinal epithelial cells secrete antimicrobial peptide antibiotics, and both of these epithelial barriers also contain intraepithelial lymphocytes that may aid in antimicrobial defense.
5. Phagocytes express a variety of receptors that recognize microbial carbohydrates, Fc receptors that bind microbes coated (opsonized) by antibodies, and complement receptors that bind microbes opsonized by complement proteins. Microbes that bind to these receptors are internalized into phagosomes, which fuse with lysosomes, where the microbes are destroyed by reactive oxygen and nitrogen species and lysosomal enzymes.
6. Natural killer (NK) cells express inhibitory receptors that recognize MHC class I molecules on healthy host cells and can then inhibit NK cell activation. In virally infected cells, MHC class I molecules may be downregulated and therefore fail to engage inhibitory receptors, and at the same time, ligands for activating NK cell receptors are expressed. As a result, NK cells are activated to kill these infected cells.
7. Tumor necrosis factor (TNF) and IL-1 stimulate inflammation in part by activating endothelial cells that line blood vessels to express molecules that recruit neutrophils and monocytes out of the blood vessels and into sites of infection. IL-12 made by macrophages and dendritic cells contributes to NK cell and T cell activation. Type I interferons inhibit viral replication, thus inducing an antiviral state in infected and adjacent cells, and also may enhance viral antigen display by infected cells for recognition by T cells.
8. Innate immune responses induce the expression of costimulators on dendritic cells that can provide signals for T cell activation that work together with signals produced by antigen recognition. Innate immune cells also make cytokines that promote the adaptive immune responses. Complement activation as part of the innate immune response can lead to the generation of complement fragments that enhance B lymphocyte activation.

CHAPTER 3

1. Antigens that enter through epithelial barriers, such as skin or intestines, are captured by dendritic cells that reside in or below the epithelium, and the dendritic cells transport the antigens to the draining lymph nodes, where the antigens are displayed to

lymphocytes. Cell-free antigens also may enter secondary lymphoid organs and be captured by resident dendritic cells.

2. Major histocompatibility complex (MHC) molecules are cell surface proteins that bind peptides derived from protein antigens and display them for recognition by T cells. Human MHC proteins are called human leukocyte antigen (HLA) molecules. Their physiologic function is peptide antigen presentation to T cells. They were initially discovered as products of polymorphic genes that mediate transplant rejection (hence the name MHC) or induce antileukocyte antibody responses in multiparous women (hence the name HLA).
3. Proteins that are produced in the cytosol or are internalized from outside the cell into endosomes and are then transported to the cytosol are digested by cytosolic organelles called proteasomes, and the peptides generated by the proteasomes are presented by class I MHC molecules. Proteins from outside the cell that are internalized into endocytic vesicles may be processed by lysosomal proteases, and the peptides generated in this way are presented by class II MHC molecules.
4. Protein antigens in the cytosol are cleaved into peptides by proteasomes, and the peptides are transported into the endoplasmic reticulum (ER) by the TAP molecule. Once inside the ER, these peptides bind to newly produced class I MHC molecules. The peptide–class I MHC complex is then transported to and displayed on the cell surface. MHC class II α and β chains are produced in the ER, where they assemble with each other and with an invariant chain that occludes the antigen binding cleft. The MHC class II–invariant chain complex is transported to a late endosomal/lysosomal compartment, where the invariant chain is degraded, leaving a peptide called CLIP in the cleft. Proteins internalized by the endocytic pathway may be degraded into peptides by proteases in late endosomes and lysosomes. Many peptides generated in this way displace CLIP and bind tightly to the cleft of the class II MHC molecules, which are then transported to and displayed on the cell surface.
5. $CD4^+$ T cells (both naive cells and helper T (Th) cells derived from the naive cells) recognize peptide antigens bound to class II MHC molecules, and $CD8^+$ T lymphocytes (both naive cells and the cytotoxic T lymphocytes derived from the naive cells) recognize MHC class I–peptide complexes. The CD4 coreceptor on the T cells binds to class II MHC molecules on antigen-presenting cells, and the CD8 coreceptor on T cells binds to class I MHC molecules on antigen-presenting cells and infected target cells.

CHAPTER 4

1. Antibody and TCR proteins contain variable domains that are involved in antigen recognition and constant domains that, in the case of antibodies, mediate effector functions. Variable domains contain hypervariable regions (sequences that differ among different antibodies or TCRs) that form the binding sites for antigens.
2. Antibodies can recognize many types of molecules, including small chemicals, proteins, carbohydrates, lipids, and nucleic acids. In proteins, antibodies can recognize conformational or linear features, called epitopes. TCRs can recognize only linear peptides ranging from 8 to 20 amino acid residues, that are proteolytically generated from proteins and bound to the clefts of MHC molecules.
3. Diversity of antibodies and TCRs is generated by V-D-J recombination, which is the joining of individual V, D and J DNA segments in developing lymphocytes from a choice of many such segments that are spatially separated in the inherited DNA of antibody and TCR gene loci. Variations in nucleotide sequences introduced by the use of different V, D, and J segment combinations (combinatorial diversity), and loss or enzymatic introduction of uninherited sequence variations between the segments during V-D-J joining (junctional diversity) contribute to diversity, with junctional alterations making the largest contribution.
4. Checkpoints in lymphocyte development are stages that must be successfully completed to permit survival and further maturation of the cells. The first checkpoint in B and T cell maturation involves the selection of pre-B and pre-T cells that have productively rearranged the μ heavy-chain gene in the case of B lineage cells and the TCR β chain gene in the case of developing T cells. The second checkpoint is after production of complete antigen receptors

and ensures that only cells with the proper V-D-J recombination can mature. Positive selection is a process in which T cells that can recognize self MHC molecules weakly are allowed to survive and express the type of coreceptor (CD4 or CD8) that matches the type of MHC molecule recognized.

5. Negative selection results in the deletion or editing of strongly self-reactive lymphocytes, in the thymus for T cells and in the bone marrow for B cells. This process eliminates many self antigen–reactive lymphocytes.

CHAPTER 5

1. The TCR complex is made up of the TCR α and β chains, which are responsible for antigen recognition, and the CD3 and ζ proteins, which are required for signal transduction.
2. Molecules other than the TCR used by T cells to respond to antigens include the CD4 and CD8 coreceptors, which bind to class II and class I MHC molecules, respectively; costimulatory receptors such as CD28, which bind to costimulators expressed on activated antigen-presenting cells (APCs); and adhesion molecules such as the integrin LFA-1, which mediate T cell adhesion to APCs (and also control the migration of the T cells).
3. Costimulation refers to signals delivered to a lymphocyte that are required for activation in addition to but independent of antigen receptor signaling. Costimulatory signals are commonly referred to as a "second signal" (antigen being "signal 1") and provide lymphocytes with the information that the antigen they are recognizing may be of microbial origin. B7-1 (CD80) and B7-2 (CD86) are the major costimulators on APCs, which bind to CD28 on T cells.
4. Antigen recognition results in the CD4 or CD8 coreceptors in T cells bringing the LCK tyrosine kinase bound to their cytosolic tails in proximity to CD3 and ζ chain ITAMs. Phosphorylation of the ITAMs by LCK results in the recruitment and activation of the ZAP-70 tyrosine kinase, which in turn phosphorylates several other adaptor protein and enzymes, thereby initiating many different signaling pathways by activating different downstream enzymes. Some of these pathways include activation of phospholipase Cγ, resulting in calcium signaling and the subsequent activation of the NFAT transcription factor; activation of PKCΦ, resulting in the activation of the nuclear factor (NF)–κB transcription factor; and activation of MAP kinases, leading to the production of the AP-1 transcription factor. These transcription factors enter the nucleus and promote expression of many genes that encode proteins required for T cell clonal expansion, differentiation, and effector functions.
5. The major growth factor for T cells is interleukin-2 (IL-2). It is produced by T cells in response to antigen receptor signals and costimulation. T cells that have recognized antigens express increased levels of receptors for IL-2 and thus preferentially respond to the growth factor during immune responses to the antigens. Regulatory T cells (Tregs) also need IL-2 for their survival and function.
6. CD4^{+} helper T cells activate other cells (B lymphocytes, macrophages) by the surface molecule CD40-ligand engaging CD40 on the other cells and by the actions of secreted cytokines.
7. Memory cells survive after the antigen is cleared, slowly proliferate to maintain their numbers for months to years, and respond more rapidly and strongly to antigen exposure than do naive cells.
8. Two proteins in the CD28 family that are expressed on T cells and act to inhibit T cell responses are CTLA-4 and PD-1. CTLA-4 is expressed on activated conventional T cells and is always expressed on regulatory T cells. It binds to B7-1 and B7-2 with higher affinity than CD28 binds these molecules, and therefore CTLA-4 prevents the B7 proteins from costimulating T cells. PD-1 is expressed on activated T cells, and upon binding PD-Ll or PD-L2 on antigen-presenting cell, delivers inhibitory signals that block the activating signals generated by the TCR and CD28.
9. Naive T cells express the adhesion molecule L-selectin and the chemokine receptor CCR7, which mediate homing to secondary lymphoid organs, such as lymph nodes. Differentiated effector cells lose expression of these molecules and instead express adhesion molecules that bind to molecules on endothelium exposed to inflammatory cytokines. The effector cells also express receptors for chemokines produced at sites of inflammation, thus preferentially migrating to these sites.

CHAPTER 6

1. Intracellular microbes that reside in phagosomes of macrophages, including several bacterial and fungal species, are eliminated by Th cells, especially those of the Th1 subset that activate the phagocytes to destroy ingested microbes. Microbes whose life cycle includes presence in the cytosol, such as viruses, may be eliminated by $CD8^+$ T cell–mediated killing of the infected cells, thus eliminating the reservoir of infection.
2. Th1 cells secrete the cytokine interferon-γ (IFN-γ), which activates macrophages to kill phagocytosed microbes. Th2 cells secrete IL-4 and IL-13, which stimulate intestinal mucus production and gut peristalsis, and IL-5, which activates eosinophils. IL-4 and IL-13 secreted by follicular helper T (Tfh) cells induce B cell IgE production, which contributes to immunity against helminths. Th2 cells are involved in defense against helminths. Th17 cells secrete IL-17, which enhances neutrophil responses, that ingest and destroy extracellular fungi and bacteria, and IL-22, which promotes repair of epithelial barriers injured by microbes.
3. In additon to activating macrophges by secreting interferon-γ, Th1 cells also express CD40 ligand, which activates macrophages by engaging CD40. Macrophages activated by Th1 cells make increased nitric oxide and reactive oxygen species. These free radicals can destroy ingested microbes. Activated macrophages also produce increased amounts of lysosomal enzymes, which help to destroy microbes, and cytokines such as IL-1, TNF, IL-6 and chemokines, all of which promote inflammation and call more leukocytes into the reaction.
4. $CD8^+$ cytotoxic T lymphocytes (CTL) that recognize micobial peptide antigen displayed by class I MHC on an infected tissue cell release granules that contain perforin and granzymes, which enter the infected cells and induce their death by apoptosis.
5. Some intracellular microbes evade immunity by preventing phagolysosomal fusion. Many viruses inhibit antigen presentation and some may inactivate effector T cells.

CHAPTER 7

1. The signals that induce B cell responses to protein antigens include binding of the protein to membrane immunoglobulin (Ig) on the B cell, and subsequent signals delivered by Tfh cells, including secreted cytokines that bind to cytokine receptors on the B cell, and CD40 ligand on activated Th cells, which bind to CD40 on the B cell. The signals that induce B cell responses to a polysaccharide antigen are generated by the binding of the polysaccharide, which is polyvalent, to multiple membrane Ig molecules on the B cell, thereby cross-linking B cell receptors and activating signal transduction pathways. Complement fragments bound to antigens engage the complement receptor CR2 (CD21) on B cells, which generates signals that increase B cell activation. This is especially important for polysaccharide and other nonprotein antigens, which cannot activate helper T cells. Activation of Toll-like receptors on B cells by microbial molecules at the same time as B cell receptor (BCR) antigen recognition may also contribute to B cell activation.
2. Secondary antibody responses develop more quickly and are of greater magnitude than primary immune responses. Secondary responses to protein antigens also differ from primary responses in that the antibodies produced are higher-affinity IgG, IgA, or IgE antibodies, whereas low-affinity IgM antibodies are mainly produced in the primary response.
3. B cells express membrane Ig molecules that bind intact proteins and facilitate their endocytosis. The internalized proteins are processed into peptides, and the peptides are bound to class II MHC molecules and displayed on the B cell surface. Th cells specific for peptide-MHC complexes presented by a B cell lead to activation of the T cell. Thus, a B cell and a T cell recognize different parts of the same protein in sequence. The B cell recognition occurs first and is independent of the T cell, and the T cell recognition is second and requires B cell presentation of a peptide fragment of the antigen. The initial B–T interactions occur at the interface of the B and T cell zones of lymph nodes or spleen, just outside the follicles. These interactions drive

differentiation of helper T cells into Tfh cells, and then both the activated B cells and Tfh cells migrate into the follicle, where a germinal center reaction ensues. B cell presentation of peptide-MHC antigen to Tfh cells and Tfh cell activation of the B cells via cytokines and CD40L continue in the germinal center.

4. Signals delivered by Th cells induce heavy-chain isotype (class) switching in B cells. These signals include CD40 ligand, which binds to CD40 on B cells, and cytokines secreted by the Th cells, which bind to cytokine receptors on the B cell. The cytokine signals determine which heavy-chain gene locus will become accessible for switch recombination, and the CD40 signal induces expression of the AID enzyme, which is responsible for initiating the DNA breaks that are required for switch recombination. Heavy-chain isotype switching is important because it allows the antibody response to be specialized to particular locations and type of microbes. For example, IgE is important for eradicating worm infections; IgA secreted into the gut is important to combat intestinal pathogens; and IgG is transported through the placenta and is important for protecting newborns from infections. Switching to some subtypes of IgG also enhances complement- and phagocyte-mediated defense against microbes because these subtypes bind most avidly to phagocyte Fc receptors or to complement proteins.
5. Affinity maturation is the increase in the average affinity of antibodies for a protein antigen that occurs as an immune response develops over time. The process occurs in the germinal center and requires signals from Th cells, which induces expression of the AID enzyme in the B cells, which causes DNA breaks and error-prone repair. Thus, rapidly dividing B cells undergo point mutations in the variable-region genes of the heavy-chain and light-chain loci. B cells in which these mutations result in increased affinity of the antibodies they produce have a selective advantage for binding to the antigen displayed by follicular dendritic cells and for presenting the antigen to Tfh cells. These B cells receive signals that prevent apoptotic death, and thus the highest-affinity B cells are selected to survive and develop into antibody-secreting plasma cells.
6. Antibodies produced in response to T-independent polysaccharide and lipid antigens are predominantly IgM antibodies of relatively low affinity. These antigens are inefficient at generating long-lived plasma cells and memory B cells because of the absence of Th cell signals, so the IgM response to TI antigens wanes relatively quickly.

CHAPTER 8

1. The N-terminal variable regions of antibodies are involved in antigen binding and neutralization of microbes and toxins. The Fc portion of the heavy-chain constant region is involved in binding and activating complement and binding to Fc receptors in various cells, which is important for phagocytosis, antibody-mediated cellular cytotoxicity by NK cells, transport across mucosal epithelia and placenta, and maintenance of prolonged half-life in the blood.
2. Class switching allows antibodies to perform different effector functions that are particularly suited to certain infections, and it allows delivery of the antibody to certain sites of infection. For example, some IgG subclasses bind well to Fc receptors on phagocytes, permitting internalization and killing of extracellular microbes. IgG antibodies are also transported through the placenta into the fetus, and protect the newborn from infections. IgA antibodies are secreted into the lumen of the gut, where they can bind to pathogenic microbes and prevent them from invading through the intestinal epithelial barrier. Affinity maturation improves the ability of antibodies to bind tightly to pathogens and therefore more effectively neutralize the microbes and target them for destruction by complement or phagocytes.
3. Neutralization prevents microbes located in mucosal secretions, blood, or extracellular tissue fluid from binding to cellular receptors, which is a first step in infecting cells. For example, entry of viruses into cells requires binding to specific cell surface receptors. Antibody binding to viral envelope antigens blocks the virus from binding to their receptors. Neutralization also inhibits the spread of microbes from an infected cell to another cell.
4. The variable domains of IgG antibodies specifically bind to antigens on microbial surfaces, a process called opsonization, and then constant domains in the Fc region of the IgG antibodies bind to Fc receptors on macrophages or neutrophils. Binding of the

antibody to the Fc receptor stimulates internalization of the microbe by phagocytosis and activates the phagocyte, and the microbe is killed by various mechanisms inside the cell.

5. The classical pathway of complement is activated when the complement protein C1 binds to the Fc regions of IgG or IgM molecules in antibody-antigen complexes. In the alternative pathway, the complement protein C3 is spontaneously hydrolyzed to form C3b, which then binds covalently to microbial cell surfaces. In the lectin pathway, the first step is binding of the protein mannose-binding lectin (MBL) to mannose residues on microbial surfaces. In all three pathways, the first step is followed by activation of a cascade of proteases, generating an enzymatic complex called C3 convertase, which is covalently attached to the microbial surface. This enzyme cleaves C3 to produce a number of active proteins and initiate the late steps of complement activation.
6. Host cells have regulatory proteins on their cell surfaces, including decay-accelerating factor (DAF), complement receptor 1 (CR1), and C4-binding protein (C4bp) that prevent the formation of the C3 convertase on healthy host cells. These regulatory proteins are not expressed by microbes. Alternative pathway complement proteins also tend to not bind to normal host cells. The regulatory proteins may be overwhelmed if large amounts of antibodies attach to host cells, leading to complement activation on these cells, as occurs in some autoimmune diseases.
7. The main functions of the complement system are to promote inflammation, opsonize microbes for phagocyte clearance, and directly lyse microbes. Inflammation is promoted by the complement protein fragments C5a and C3a. Opsonization is mediated mainly by C3b. Lysis is mediated by the membrane attack complex, which is composed of C5b, C6, C7, C8, and polymerized C9.
8. IgA and some IgM antibodies are transported by the poly-Ig receptor from the lamina propria (where they are produced), through mucosal epithelial cells, into the lumen of the gut or the airways, where they neutralize pathogens.
9. Maternal IgG is transported by the neonatal Fc receptor into the fetal circulation through the placenta, so the baby is born with a full range of antibodies against microbes that the mother has been exposed to in the past. Maternal IgA and IgG in breast milk are ingested by the nursing baby and protect against intestinal pathogens.

CHAPTER 9

1. The adaptive immune system does not normally mount effective immune responses to self molecules. This state of immune unresponsiveness is called tolerance and is important because T and B cells expressing antigen receptors that may recognize self-antigens arise during lymphocyte development, and these lymphocytes must be controlled or eliminated to prevent autoimmune disease. In addition, the immune system has to be tolerant of foreign (paternal) antigens in the fetus and commensal microbes. The mechanisms of tolerance induction may be therapeutically exploited to inhibit harmful immune responses to allergens, self antigens, and transplants.
2. Central tolerance is the elimination or inactivation of self-reactive T and B cells during their development in the thymus or bone marrow, respectively. Central tolerance is induced in immature T cells in the thymus after they express TCRs. If a developing T cell recognizes, with high avidity, peptides derived from self proteins bound to self MHC presented by thymic antigen-presenting cells (APCs), signals will be generated that lead to apoptosis of the T cell (called clonal deletion or negative selection). Surviving $CD4^+$ T cells may develop into protective Tregs. Furthermore, some proteins mainly expressed by cells in a particular peripheral tissue type or organ may be also expressed by medullary thymic epithelial cells (MTECs) under the control of the AIRE protein. The developing T cells that recognize peptides from these self proteins in complex with self MHC are deleted. Central tolerance develops in immature B cells after they express a functional membrane B cell receptor complex. Recognition of self antigens by immature B cells leads to apoptosis or to receptor editing, whereby a new round of V-J recombination in the light-chain genes generates new specificities that are not self-reactive.
3. Most Tregs are $CD4^+$ T cells that express the IL-2 receptor α chain CD25 and the transcription factor FOXP3. Tregs develop in the thymus from

immature thymocytes as a consequence of self antigen recognition. Tregs can also differentiate from mature naive T cells in peripheral lymphoid tissues as a result of antigen recognition together with signals from cytokines such as transforming growth factor β (TGF-β). Tregs protect against autoimmunity by suppressing activation of self-reactive T cells by antigen-presenting cells (APCs) or by directly inhibiting the T cells. The principal mechanisms by which Tregs suppress immune responses include blocking and removal of B7 costimulators on APCs by CTLA-4 (which is expressed constitutively on Tregs), secretion of immunosuppressive cytokines (e.g., TGF-β, IL-10), and consumption of the growth factor IL-2.

4. Peripheral tolerance may be induced by several mechanisms that lead to dysfunctional states of self-reactive T cells. Anergy is a long-lasting condition in which a T cell will not respond to antigen stimulation. Anergy is induced in naive T cells when they recognize peptide–MHC antigen without costimulation. This may also occur when "immature" dendritic cells, which have not been exposed to microbial stimuli, process and present self peptide–MHC to T cells. Such DCs will not express sufficient levels of B7-1, B7-2, or other molecules to provide costimulation, and therefore the self-reactive T cell will become anergic. Alternatively, anergy may be induced when Tregs block costimulation via CTLA-4. Anergy may fail as a peripheral tolerance mechanism during an infection when a T cell recognizes self peptide–MHC on a DC that has been activated by innate responses to the microbe. A related phenomenon, called "exhaustion," occurs when T cells are repeatedly stimulated (e.g., by tumors and chronic infections). These T cells respond initially but begin to express inhibitory receptors such as PD-1 and can no longer effectively respond to antigen. Exhaustion may protect against self tissue damage by cross-reactive T cells during exuberant or prolonged responses to infection.
5. The mechanisms that prevent immune responses to commensal organisms include abundant IL-10–producing Tregs, inhibitory signaling of Toll-like receptors in gut dendritic cells, and mucus and epithelial barriers keeping microbes away from the intestinal immune system. Tolerance to the allogeneic fetus is maintained by Tregs, exclusion of inflammatory cells from the pregnant uterus, impaired antigen presentation, and inhibition of Th1 responses in the placenta.
6. Multiple genes likely contribute to the development of common autoimmune diseases. Particular MHC alleles are frequently associated with autoimmunity. MHC genes may be important in the development of autoimmunity because they present native or chemically modified self peptides to T cells. Many non-MHC genes have been implicated in various autoimmune diseases, but their roles are largely undefined. Several rare autoimmune diseases are caused by single-gene mutations that interfere with mechanisms of tolerance. These include mutations in the genes encoding AIRE, CTLA-4, FOXP3, FAS, and complement C2.
7. Infections may promote the development of autoimmunity by (a) inducing costimulatory molecule expression by APCs that present self antigens to lymphocytes; (b) causing inflammation and tissue damage, which exposes normally sequestered self antigens to the immune system; and (c) molecular mimicry, if the microbe expresses an antigen molecularly similar to a self antigen and thereby simulates an immune response (antibodies or T cells) that cross-reacts with self antigens.

CHAPTER 10

1. Because genomic instability is a hallmark of cancer cells, tumors often contain many mutated genes that produce proteins (neoantigens) that are not normally present and thus may appear foreign to the immune system and induce immune responses. Tumors may overexpress or inappropriately express antigens that are normally expressed only at low levels in normal tissues or only during development and therefore do not induce tolerance. Some tumors caused by oncogenic viruses may express viral antigens that induce immune responses.
2. Some tumors occur more frequently in immunocompromised hosts than in people with normal immune systems. The presence of abundant $CD8^+$, effector Th1, and memory T cells in or around some tumors is predictive of a better prognosis. In experimental animals, immunologic rejection of tumors can be demonstrated by transplanting tumor cells in animals that have been previously

immunized with the tumor cells or by transfer of T cells from tumor-bearing animals. Drugs that block T cell inhibitory molecules enhance patients' T cell responses to their tumors and prevent progression of tumor growth.

3. Naive $CD8^+$ T cells recognize tumor antigens in the same way they recognize microbial antigens: by TCR binding to tumor-derived peptides displayed on class I MHC molecules on dendritic cells. This means the DC must internalize tumor cells (or their products) and process the internalized tumor proteins by the class I MHC pathway, which involves proteasomal degradation of the proteins into peptides. The presentation of peptides derived from internalized proteins on class I MHC molecules is called cross-presentation. The DC displays not only tumor-derived peptide antigens to naive $CD8^+$ T cells but also costimulators. The combination of antigen and costimulators activates clonal expansion and differentiation of the naive $CD8^+$ T cells into effector cytotoxic T lymphocytes.
4. Tumor immune evasion mechanisms include down-regulation of MHC molecules to avoid T cell recognition of tumor antigens; loss of expression of tumor antigens; secretion of immunosuppressive cytokines (e.g., TGF-β); engagement of inhibitory receptors on T cells (e.g., CTLA-4, PD-1). Tumors may also promote other cellular suppressors of immune responses, including regulatory T cells (Tregs) and myeloid-derived suppressor cells.
5. Host immune responses to tumor antigens can be enhanced by treating the tumor-bearing patient with antibodies such as anti–CTLA-4 or anti–PD-1 that block T cell inhibitory receptors, so called "immune checkpoint blockade." These therapies are often complicated by autoimmune reactions against various tissues. Some tumors are treated by adoptive transfer of a patient's T cells that have been genetically engineered ex vivo to express chimeric antigen receptors (CARs) specific for a tumor antigen. CARs use an antibody-like domain to bind the tumor antigens linked to signaling domains that are active in T cells (signaling domains from the TCR complex and costimulatory receptors). CAR–T cell therapy is often complicated by severe systemic inflammation caused by release of cytokines from the activated T cells, as well as poorly explained neurotoxicity. Passive immunity to tumors can be induced by administering antitumor antibodies or T cells expressing tumor-specific antigen receptors.
6. Allogeneic MHC molecules with any bound peptide are likely to resemble self MHC plus a foreign peptide, and therefore normal T cells may cross-react with the allogeneic molecules. There are many thousands of allogeneic MHC molecules on every graft cell displaying many different donor peptides. Many of these peptide-MHC complexes may be recognized by a graft recipient's T cells. Individuals develop tolerance to self antigens (self peptide–self MHC complexes) but are not tolerant to the foreign antigens of a graft (foreign MHC with self or foreign peptides).
7. Allografts may be attacked by alloreactive T cells that are activated after transplantation. $CD8^+$ cytotoxic T lymphocytes recognize allogeneic class I MHC molecules on graft cells and directly kill those cells. $CD4^+$ T cells recognize allogeneic class II MHC molecules and initiate inflammatory responses that damage the graft cells. These T cell responses contribute to acute cellular rejection. In chronic rejection, alloreactive T cells may induce inflammation that promotes graft vascular disease, ultimately leading to graft failure from inadequate blood supply. Allografts may be rejected by antibodies against allogeneic MHC or other minor histocompatibility antigens. If the antibodies are preformed in the recipient as a result of prior pregnancy, transfusions, or transplantation, they may bind to graft endothelial cells and cause hyperacute rejection. If the antibodies form as a result of exposure to the allograft after transplantation, they may cause acute humoral rejection.
8. Patients in need of a transplant may be screened to test for the presence of serum antibodies that react with different MHC molecules. Donors whose MHC molecules are recognized by the patients' antibodies will not be used. Recipients may be typed for the HLA alleles they have, and organs can be chosen with the best-matched alleles. HLA matching is essential for hematopoietic stem cell transplantation but is not as important for solid organ transplants. Rejection of solid organ grafts is prevented mainly by treating the recipient with immunosuppressive drugs, such as calcineurin inhibitors (e.g., cyclosporine, tacrolimus), mTOR inhibitors (e.g., rapamycin), anti–T cell antibodies, corticosteroids, and antimetabolites (mycophenolate mofetil).

9. T cells transplanted with the hematopoietic stem cells can respond to minor histocompatibility molecules in the recipient, causing graft-versus-host disease. Recipients are also often immunodeficient as their immune systems are reconstituted.

CHAPTER 11

1. Hypersensitivity refers to tissue injury and disease caused by immune responses. Immediate hypersensitivity (type I hypersensitivity) is caused by the release of mediators from mast cells triggered by antigen cross-linking of immunoglobulin E (IgE) bound to IgE-specific Fc receptors. Antibodies specific for cell or tissue antigens can cause damage by activating complement and engaging phagocytes (type II hypersensitivity); antibodies can also destroy circulating cells and block essential molecules or their receptors. Antigen-antibody complexes (immune complexes) deposit in blood vessels, causing inflammation and thrombosis, leading to tissue injury (type III hypersensitivity). Reactions of T lymphocytes, often against self antigens in tissues, can cause inflammation and tissue damage (type IV hypersensitivity).
2. Damaging immune responses may be elicited by self antigens (autoimmunity), environmental antigens and chemicals (allergies and other types of hypersensitivity), and microbial infections.
3. Exposure to an environmental antigen induces differentiation of IL-4–producing T follicular helper T cells, which in turn induce IgE antibody responses to the antigen. The IgE binds to high-affinity IgE-specific Fcε receptors on mast cells in tissues throughout the body. On subsequent exposure to the same antigen, the mast cell–bound IgE molecules bind the antigen and become cross-linked, generating signals from the associated Fcε receptors that lead to mast cell granule release, production of leukotrienes and prostaglandins, and synthesis and secretion of cytokines. Vasoactive amines such as histamine, released from the granules, and prostaglandins cause acute vascular changes leading to increased blood vessel permeability and edema, usually within minutes of exposure to the antigen. The late-phase reaction is an inflammatory response that develops over hours in which blood leukocytes are recruited to the site of mast cell degranulation, caused by TNF and other cytokines secreted by the mast cells.
4. Allergic rhinitis and sinusitis are immediate hypersensitivity reactions to inhaled allergens, such as pollen proteins, leading to upper airway mucosal mast cell secretion of histamine, IL-13 production by Th2 cells, and long-lasting inflammation due to various cytokines. Food allergies are caused by ingested allergens leading to intestinal mucosal mast cell histamine release, causing increased peristalsis. Allergic bronchial asthma is caused by inhaled allergens inducing bronchial mast cell release of mediators, including leukotrienes, which cause bronchial constriction and airway obstruction, excessive secretion of mucus in the airways, and bronchial smooth muscle hypertrophy. Atopic dermatitis (eczema) develops in the setting of defective skin barrier function leading to local bacterial infections and the activation of keratinocytes to secrete cytokines that promote type 2 immune responses. Patients with eczema may develop chronic late-phase reactions in the skin, and children with eczema are at increased risk of developing food allergies and asthma. Anaphylaxis is a severe systemic immediate hypersensitivity reaction characterized by shock and airway obstruction resulting from mast cell degranulation in many tissue sites, usually after exposure to an antigen that is injected or ingested. Immediate hypersensitivity diseases are treated by inhibiting mast cell degranulation, antagonizing the effects of mast cell mediators, and reducing inflammation. Drugs include antihistamines for hay fever, inhaled β-adrenergic agonists and corticosteroids for asthma, and epinephrine in anaphylaxis. Some patients benefit from repeated administration of small doses of allergens, called desensitization. Antibodies that block IgE binding to their receptors on mast cells or block various cytokines or their receptors, including IL-4, IL-5, IL-33, and thymic stromal lymphopoietin (TSLP), are now approved for the treatment of some forms of asthma and atopic dermatitis.
5. Antibodies cause tissue injury and disease by activating cytotoxic and inflammatory effector functions,

mainly complement activation and opsonization and phagocytosis via Fc receptors. Some antibodies may cause disease by binding to and interfering with the normal function of a particular protein.

6. Examples of diseases caused by antibodies specific for cell surface or tissue antigens include autoimmune thrombocytopenia or anemia caused by antibodies specific for platelet or red cell membrane proteins and blistering diseases such as pemphigus vulgaris, caused by antibodies against cell adhesion proteins on skin keratinocytes. In pernicious anemia, autoantibodies specific for intrinsic factor, a protein required for absorption of vitamin B_{12}, decrease the intestinal uptake of this vitamin, leading to anemia and neurologic problems. In some forms of glomerulonephritis, antibodies bind to matrix proteins in the glomerular basement membrane and induce inflammatory damage. In rheumatic fever, cardiac inflammation is caused by an antibody specific for a streptococcal bacterial antigen that cross-reacts with a myocardial antigen.
7. Immune complexes deposit in the walls of blood vessels and cause inflammation of the vessel (vasculitis), which leads to blood clotting in the vessel lumen (thrombosis) and loss of blood supply to tissues supplied by the vessels. The site of immune complex deposition is not related to the specificity of the antibodies. Therefore, immune complex disease may simultaneously affect many different tissue sites, as occurs in systemic lupus erythematosus, arteritis syndromes associated with chronic infections, and serum sickness after therapeutic injection of antibodies from another species. Diseases caused by antibodies against cell surface or extracellular matrix proteins are usually characterized by injury and loss of function restricted to the particular organ or tissue that expresses the protein.
8. Type 1 diabetes is caused by $CD4^+$ and $CD8^+$ T cells that are specific for pancreatic islet cell proteins and that destroy insulin-producing cells, leading to impaired glucose metabolism and cardiovascular disease. Multiple sclerosis is caused by $CD4^+$ T cells specific for central nervous system (CNS) myelin sheath proteins that cause inflammation, demyelination, and CNS motor and sensory symptoms. Contact hypersensitivity (e.g., poison ivy, nickel hypersensitivity) is caused by T cells specific for skin proteins that are modified by plant toxins, metals, and other chemicals, leading to inflammation and blistering.

CHAPTER 12

1. Infections are the most common manifestation of immunodeficiency diseases. The type of infection may vary with the type of deficiency. B cell/antibody deficiencies result in increased infections with bacteria and fungi that live and replicate outside cells, as well as viral infections, such as respiratory and gastrointestinal viruses, that are normally combatted by neutralizing antibodies. In contrast, T cell deficiencies result in increased infections by microbes that live and reproduce inside cells, such as mycobacteria and certain fungi and viruses. Malignant tumors are also increased in patients with immunodeficiency disease.
2. Mutations of the common γ chain (γ_c) of cytokine receptors, adenosine deaminase, RAG1, and RAG2 all block both T cell and B cell maturation, leading to severe combined immunodeficiency disease (SCID). Mutations in Bruton tyrosine kinase block B cell maturation, causing X-linked agammaglobulinemia. Mutations in transcription factors needed to induce class II MHC expression cause the bare lymphocyte syndrome, leading to impaired $CD4^+$ T cell development and failure to activate the few $CD4^+$ T cells that do mature. In DiGeorge syndrome, a deletion of a part of chromosome 22 causes a defect in thymic development, leading to a failure of T cell maturation.
3. Class II MHC deficiency (see Question 2) results in poor cell-mediated immunity and poor T dependent B cell responses, leading to susceptibility to a variety of infections. Mutations in CD40 ligand (CD40L) result in X-linked hyper-IgM syndrome, characterized by an inability of helper T cells to activate both B cells and macrophages, resulting in defective T cell dependent B cell responses and defective activation of macrophages. Affected boys have low IgG and defective cell-mediated immunity and are susceptible to infections with extracellular bacterial the intracellular fungus *Pneumocystis jiroveci.* Many other mutations affecting various T and B cell signaling pathways and receptors have been described. Examples include mendelian susceptibility

to mycobacterial disease, which is characterized by impaired Th1-mediated immunity against intracellular infections as a result of mutations of genes required for IFN-γ–producing T cells or IFN-γ receptor signaling, and common variable immunodeficiency, which is due to mutations in genes involved in B cell maturation and activation, leading to poor antibody responses.

4. HIV enters T cells by binding to CD4 and the chemokine receptor CXCR4 or CCR5, leading to fusion of the virus with the host cell membrane. Once inside the cell, the virus is uncoated by viral protease, its RNA genome is copied into DNA by viral reverse transcriptase, and the DNA integrates into the host cell's DNA by the action of viral integrase. The integrated viral DNA is transcribed into viral mRNA by host enzymes, viral proteins are translated (also by host cell enzymes), and new viral particles are formed and released by cell surface budding.

5. The major clinical problems caused by HIV infections are opportunistic infections, certain tumors caused by oncogenic viruses (EBV, Kaposi sarcoma herpesvirus), neurocognitive defects, and wasting. The infections are caused by a profound loss of T cell–mediated and T-dependent antibody-mediated immunity, mainly due to the death of infected $CD4^+$ T cells. The increase in tumors reflects reduced T cell–mediated immune surveillance against oncogenic viruses and transformed cells. The neurocognitive abnormality reflects loss of microglial and perhaps neuronal function of unclear pathogenesis. The wasting syndrome is caused by altered metabolism and reduced caloric intake, possibly a result of cytokines produced during repeated and chronic infections. Many of these clinical manifestations have been greatly reduced by treatment with combinations of antiretroviral drugs.

INDEX

Page numbers followed by *f* and *t* indicate figures and tables, respectively.

J

L

U